AF847534

Nanomedicine for Bioactives

Mahfoozur Rahman • Sarwar Beg •
Vikas Kumar • Farhan Jalees Ahmad
Editors

Nanomedicine for Bioactives

Healthcare Applications

Springer

Editors
Mahfoozur Rahman
Department of Pharmaceutical Sciences
Shalom Institute of Health & Allied
Sciences, SHUATS
Allahabad, Uttar Pradesh, India

Sarwar Beg
Nanomedicine Research Lab
School of Pharmaceutical Education &
Research
Jamia Hamdard, New Delhi, India

Vikas Kumar
Department of Pharmaceutical Sciences
Shalom Institute of Health & Allied
Sciences, SHUATS
Allahabad, Uttar Pradesh, India

Farhan Jalees Ahmad
Nanomedicine Research Lab
School of Pharmaceutical Education &
Research
Jamia Hamdard, New Delhi, India

ISBN 978-981-15-1663-4 ISBN 978-981-15-1664-1 (eBook)
https://doi.org/10.1007/978-981-15-1664-1

© Springer Nature Singapore Pte Ltd. 2020
This work is subject to copyright. All rights are reserved by the Publisher, whether the whole or part of the material is concerned, specifically the rights of translation, reprinting, reuse of illustrations, recitation, broadcasting, reproduction on microfilms or in any other physical way, and transmission or information storage and retrieval, electronic adaptation, computer software, or by similar or dissimilar methodology now known or hereafter developed.
The use of general descriptive names, registered names, trademarks, service marks, etc. in this publication does not imply, even in the absence of a specific statement, that such names are exempt from the relevant protective laws and regulations and therefore free for general use.
The publisher, the authors, and the editors are safe to assume that the advice and information in this book are believed to be true and accurate at the date of publication. Neither the publisher nor the authors or the editors give a warranty, expressed or implied, with respect to the material contained herein or for any errors or omissions that may have been made. The publisher remains neutral with regard to jurisdictional claims in published maps and institutional affiliations.

This Springer imprint is published by the registered company Springer Nature Singapore Pte Ltd.
The registered company address is: 152 Beach Road, #21-01/04 Gateway East, Singapore 189721, Singapore

Preface

This book, "Nanomedicine for Bioactives: Healthcare Applications", is the product of our sincere effort, to provide scientific evidence for the extraordinary power of nature's wonder molecules—bioactives. Chapters by experts in different specific aspects of bioactive and its effective delivery by nanomedicines for better healthcare applications have been presented to our widespread readers and health professionals. Nanotechnology is an opening up for new perspectives in all scientific and technological fields. Among these applications, bioactive and nutraceuticals are the fast-growing fields in nano-research. A variety of nanoformulations such as polymeric nanoparticles, nanocapsules, nano emulsions, transferosomes and ethosomes, liposomes, lipospheres and lipid polymer hybrid nanoparticles have been reported for bioactive delivery and food materials. New herbal drugs and nutraceuticals-loaded nanomedicine are reported to have remarkable advantages over conventional formulations of plant actives and extracts which include enhancement of solubility, bioavailability, multiple drug delivery, expansion of stability, sustained delivery, improved tissue macrophages distribution, protection from toxicity, enhancement of pharmacological activity and protection from physical and chemical degradation.

A succinct account on key highlights of each of the chapter included in the book has been discussed in the below-mentioned text.

Chapter 1 on the topic entitled "*Ganoderic acid for anticancer applications: Scope of nanomedicine in its effective delivery*" extensively covers the history of *G. lucidum*, its traditional use and therapeutic application against cancer with molecular mechanism based on in vitro and in vivo experiments.

Chapter 2 of Dr Akhbari on "*Vitamin E-based nanomedicines for anticancer drug delivery*" covers the role of vitamin which is a promising candidate for achieving patients' dream: "Chemotherapy at Home". Furthermore, vitamin E-based nanomedicines are suitable candidates for the preparation of novel anticancer treatments in the future.

Chapter 3 of Dr Akhbari on the topic entitled as "*Rutin-based phytomedicines for cancer benefit*" emerging as therapeutic agent for fighting cancer because of its antioxidant and anti-inflammatory effects. It also has been shown to induce apoptosis, or cancer cell death, and display anti-tumour effects and Chap. 7 of Dr Ahmad on "*Nanomedicine Approaches for the delivery of herbal anticancer drugs*" explores the use of nanomedicine to revolutionize the way to discover and administer the phytoconstituents in the patients.

In Chap. 4, Dr B. Mukherjee describes the role of betulinic acid and its delivery by nanoformulation approach towards cancer therapy.

Dr S. Kumar in Chaps. 5 and 9 summarizes the role of quercetin-loaded nanomedicine in cancer management and as nutritional applications.

Dr Farhan J. Ahmad in Chap. 6 on "Polyunsaturated fatty acids-loaded nanomedicine for solid tumour" summarizes that the PUFA-loaded nanoformulations showing both anti-inflammatory and anti-neoplastic activities could make them more bioavailable. Whereas PUFAs helps in decreasing the adverse effects, lowering the toxicity, and targeting and affecting only the cancerous cell, without affecting the normal cells which are present in the human body.

Entesar Hanan in Chap. 8 discusses the prospects of "*Nutraceuticals-loaded chitosan nanoparticles for healthcare applications*" and also highlights the superior characteristics of the said nanoparticles and its gained versatility in the effective delivery of nutraceuticals.

In Chap. 10, Dr Deepika Singh discusses on "*Polyphenols-loaded nanomedicines against skin aging*" and broadly covers plant-derived polyphenols, its bioavailability, different types of nanocarriers for pharmaceutical and cosmetological purposes and product-oriented solutions used in skin aging.

Dr Kamalinder K. Singh in Chap. 11 describes the various skin infections and then comprehensively describes various lipid nanocarriers and their application in the treatment of utaneous infections. Dr Shammy Jindal in Chap. 12 discusses on the emergence of lipid nanocarriers in the effective dermal delivery of lutein.

Dr Biswajit Mukherjee in Chap. 13 describes the array of biological activities of apigenin and addresses the issues of delivering apigenin through nanocarriers to improve the efficacy of apigenin, more particularly in diabetes and various types of cancer.

In Chap. 14, Dr Dinesh K. Patel discusses the importance of *Aegle marmelos* and their phytochemical "aeglin" in various therapeutic healthcare applications.

Dr Nupur Garg in Chap. 15 summarizes the clinical applications and epigallocatechin-3-gallate which combined with the nanosystems hold a promising future for advancements in the therapeutic applications of natural products.

In Chap. 16, Mohamad Taleuzzaman describes the chemical and biological applications of eugenol and eugenol nanoformulations. It has abundant potential applications and are sure to be incorporated in future into commercially available products and new uses/processes are lying in wait to be explored.

Dr Madhu Khatri in Chap. 17 focuses on regulations for proper usage of nanomaterials and also highlights the currently existing regulations for safe delivery of drug molecules.

In Chap. 18, Dr Madhu Gupta discusses the intake of fruits and vegetables has enhanced the level of components having antioxidative effects such as vitamin C, vitamin E, selenium, lycopene and several phytochemicals in the treatment of cancer as nutritional value.

Dr Harshita, in Chap. 19, describes the role of thymoquinone, its molecular mechanism of action, clinical-pharmaceutical aspect and its potential targets for cancer prevention including breast cancer in preclinical models.

This book focuses on the advanced nanomaterials utilized for the encapsulation of nutrients/vitamins/phytoconstituents for nutrition and, beyond of it, for other healthcare benefits. This book, therefore, carries a lot of potential as a repertoire of knowledge and package of information for the herbal scientist, pharmaceutical scientists, nano-scientists and nanobiotechnologists to provide holistic information on the subject of interest.

Finally, volume editors would like to extend their appreciation to Springer and their staff for providing professional platform for communication with the experts in the field.

Allahabad, Uttar Pradesh, India — Mahfoozur Rahman
Jamia Hamdard, New Delhi, India — Sarwar Beg
Allahabad, Uttar Pradesh, India — Vikas Kumar
Jamia Hamdard, New Delhi, India — Farhan Jalees Ahmad

Acknowledgements

We thank Dr Uma Maheshwari, project coordinator, for regularly communicating with the editors and attending to address all the queries raised during communication with the authors and publisher.

We also thank Ms Camilya Anitta, project manager, for her help in manuscript proofreading process till the final publication of book.

We would like to express heartfelt thanks to our family members for their understanding and support.

Contents

About the Editors

Mahfoozur Rahman, PhD is an Assistant Professor at the Department of Pharmaceutical Sciences, Faculty of Health Science, Sam Higginbottom University of Agriculture, Technology & Sciences (SHUATS), Allahabad, India. His major areas of research interest include development and characterization of nanosized drug delivery systems for inflammatory disorders including psoriasis, arthritis, neurodegenerative disorders and cancer. Besides, he is also working on amalgamation of herbal medicinal plants with modern therapeutics in order to deliver a scientifically acceptable therapy for various diseases management. To date he has published over 100 publications in peer-reviewed journals.

Sarwar Beg is an Assistant Professor at the Department of Pharmaceutics, School of Pharmaceutical Education and Research, Jamia Hamdard, New Delhi. He has over 10 years of teaching and research experience in the field of pharmaceutics and biopharmaceutics, especially in the systematic development and characterization of novel and nanostructured drug delivery systems employing Quality by Design paradigms. Dr Beg has authored 140 publications, 45 book chapters, 10 books and 3 Indian patent applications.

Vikas Kumar Assistant Professor (Pharmacognosy) at the Department of Pharmaceutical Sciences, Shalom Institute of Health and Allied Sciences, Sam Higginbottom University of Agriculture, Technology & Sciences (SHUATS), Allahabad, India, has been working on traditional medicine in drug discovery leading to the development of therapeutic leads from natural resources. His research work is highlighted on screening, evaluation, formulation and standardization of herbal drugs with their validation to ensure quality, efficacy and safety.

Farhan Jalees Ahmad is currently a Professor at the School of Pharmaceutical Education & Research, Jamia Hamdard, New Delhi, India, and is an internationally known researcher in the area of Pharmaceutical Sciences. He has 26 years of rich experience in Research and Teaching. His domain of research therein included development, scale-up, technology transfer and launching of pharmaceutical products, both for domestic and international markets. He has published more than 300 research and review papers, 12 Book chapters and 9 books.

Ganoderic Acid for Anticancer Applications: Scope of Nanomedicine in Its Effective Delivery

Mahfoozur Rahman

1 Introduction

Cancer, also called malignancy, is an abnormal growth of group of cells, which leads to tumor. It can originate in any part of the body. The abnormal division of cells starts due to change in genetic makeup of cell. After reaching 2 mm^2, it moves to other places called metastasis [1]. There are more than 100 types of cancer including breast cancer, skin cancer, lung cancer, colon cancer, prostate cancer, and lymphoma. Cancers may be classified by the type of cells in which the cancer originates. Further, the types of cancer like carcinoma are derived from epithelial cells. Sarcoma is a cancer derived from the connective tissue like bone, cartilage, fat, and nerve [1]. Others are lymphoma and leukemia, which arise from hematopoietic (blood forming) cells that come out from the marrow and tend to mature in the lymph nodes and blood, respectively. Germ cell tumor, which is derived from pluripotent cells, is founds in the testicle or the ovary (seminoma, dysgerminoma). Blastoma is a type of cancer that is derived from the immature precursor cell or embryonic tissue [1]. There are many symptoms which depend on the types of cancer. Cancer therapy includes chemotherapy, radiation, and/or surgery. Chemotherapeutic drugs, also called as antineoplastic drugs, are used in the management of malignant or cancerous disease. Chemotherapeutic drugs include various classes of anticancer drugs such as alkylating agents, antimetabolites, natural products, and hormones. Apart from these, there are so many drugs which may not fall within those classes but utilize in the management of malignant disease [1].

Ganoderic acid (GA) is closely related to triterpenoids, which may derive from lanosterol. It is naturally obtained from *Ganoderma* mushrooms. Since several decades, the fruiting bodies of *Ganoderma* fungi have been widely used in the

M. Rahman (✉)
Department of Pharmaceutical Sciences, Shalom Institute of Health & Allied Sciences, SHUATS, Allahabad, Uttar Pradesh, India

© Springer Nature Singapore Pte Ltd. 2020

M. Rahman et al. (eds.), *Nanomedicine for Bioactives*,
https://doi.org/10.1007/978-981-15-1664-1_1

traditional medicines in East Asia. Continuous efforts have been made to identify the chemical constituents, which may be responsible for the pharmacological effects. There is large quantity of GA that has been isolated, of which ganoderic acid A and ganoderic acid B are the most characterized. The GA has been found to possess biological activities including hepatoprotective, antitumor effects, immunomodulating, anti-inflammatory, antidiabetic, antioxidative, free radical scavenging, and antiaging action [2]. In China, GA is known as *Lingzhi* that has been widely used through centuries for increasing health and longevity in Asian countries. Most mushrooms contain 90% water by weight. The remaining 10% weight contains 26–28% carbohydrate, 3–5% crude fat, 59% crude fiber, and 7–8% crude protein, and a wide variety of bioactives are comprised of terpenoids, steroids, phenolic, glycoproteins, etc. [2]. Many literatures revealed that triterpenes and polysaccharides are the major active pharmacological components of *G. lucidum*. Triterpenes are a class of chemical compound composed of three terpene units and are naturally occurring compounds with six isoprene unit, whereas terpenes are known for their anti-inflammatory, antitumorigenic, and hypolipidemic activity [3]. GA is a subtype of triterpenes which have four cyclic and two linear isoprene. To date, more than 140 species of triterpenes and triterpenoids are recognized in *G. lucidum* and have shown remarkable pharmacological and therapeutic activities on multiple human disease including cancer [3]. Researchers have found that various subtypes of triterpenes are extracted from *G. lucidum* which directly induce the apoptosis of human cancer cell lines. The primary active constituents of mushrooms are GA that is particularly known as GA-DM, obtained from *G. lucidum* [3]. Therefore, this chapter extensively reviews the potential therapeutic application of GA-DM in various cancer and its current and future perspectives as a therapeutic candidate in other diseases as well.

2 Source of Ganoderic Acid and Factors Affecting It

Mushrooms have long historical use predominately by Far East countries dating back more than 4000 years. A fair number of mushrooms have health benefits that are used in Asian countries, but the most popular mushroom is *Ganoderma lucidum*, which comes out from the shiny appearance of its fruiting body. It becomes more popular due to its health benefits rather than its physical appearance [3]. In China and Korea, *Ganoderma lucidum* is called herb of spiritual potency, whereas in Japan, it is called reishi or mannen. In the Far East countries, the consumption of *Ganoderma lucidum* for medicinal purpose is through different methods of ingestion [1–3]. Although mushroom is nontoxic, it is very difficult to ingest due to its very tough nature and its raw form that is incredibly bitter. Usually, its ingestion takes place by taking it in powder form. However, it is available as commercial product such as powders, supplements, and tea [1–3]. In the Far East countries, as herbal medicine practices, *G. lucidum* is prescribed in different ways. The approaches are used to inject a solution of powder mushroom and drink mushroom powder with soup, syrup, tea, capsule, tincture, or bolos. In recent years, more researches have been conducted and concluded with clinical benefits and developments of more pharmaceuticals to load it and treat against various ailments. From the recent research on *Ganoderma lucidum*, some interesting compounds are found that may have some clinical outputs [3]. This novel compound is triterpenoids, i.e., GA-DM. However, challenges like low yield of GA-DM upon artificial cultivation of the said fungus occur [4]. These challenges occur because it mostly takes several months to cultivate the fruiting body of the fungus and is very difficult to control the optimum amount of GA-DM produced by the said fungus during cultivation [4]. Researchers work on this particular aspect of enhancing GA-DM using two main branches [4]. The first is growing environment and examining the effect of GA production. Researchers had worked on optimizing the growing conditions such as medium, oxygen supply, pH, and method of cultivation by implementing submerged fermentation of the fungus. All these are used as an alternative method for the efficient production of GA-DM. The second method is regulating the gene expression levels of those gene which specifically involved the biosynthesis of GA-DM [4, 5]. Apart from these methods, the process of converting the analogue impurities to desired compound through chemical conversion is also used. Further, the extraction method is also important in getting the maximum concentration of GA-DM. Extraction of GA-DM is usually done by using methanol, ethanol, acetone, chloroform, ether, or a mixture of these solvents [4, 5]. Furthermore, the extracts can be further purified by various separation methods like normal- and reverse-phase HPLC.

3 Anticancer Activity of GA-DM

GA-DM has gained attention in research due to its several therapeutics. Recent research has been focused on the utilization of GA-DM in cancer therapies. Several literatures have been reported about GA-DM treating various cancers including prostate cancer, melanoma, breast cancer, and meningioma [6, 7]. In treating prostate cancer, conventional therapies are found effective in the early stage of treatment but ineffective at the late metastatic stages [6, 8]. However, the current therapies used in treating melanoma are the standard therapies which are often ineffective in making a substantial impact on the survival of the patients. It may be due to the intrinsic and acquired therapeutic resistance [9]. Therefore, there is a need of alternative treatment option in treating breast cancer to combat disease. There are significant problems in finding alternative treatments for meningiomas when assessing the beneficial impact on recurrent meningiomas with existing chemotherapies [10]. Therefore, there is urgent need of alternative treatment for lymphoma. Many literatures reported the use of medicinal mushroom because of the presence of bioactive constituents that may give the desired effect to treat these cancers, and further it was found that GA-DM is a potential candidate for acting as supplementary or alternative therapy with regard to the previously mentioned cancer types [11]. The *G. lucidum* has been tested in many studies against different cancers due to the presence of GA-DM.

3.1 GA-DM as an Alternative Chemotherapeutic Agent

There are mainly approaches to treat cancer including surgery, radiation, and chemotherapy [1]. There are many chemotherapeutics which may involve the use of drugs that kill cancer cells or ability of cancer cells to proliferate. Furthermore, the side effects associated with the most anticancer drugs are toxic [1]. GA-DM is capable of inducing cell death in various types of cancer cells without affecting the healthy cells [12–14]. There are many pathways which may lead to cell death like apoptosis, autophagy, and necrosis [15, 16]. Apoptosis is the main cell death pathway that is achieved through cellular processes and generally poses some beneficial effects to the organism [17]. On the contrary, necrosis, which uses various receptors, leads to loss of the cell membrane and causes leakage of cellular components. Autophagy is also implicated in cell death [17]. In which the cell itself is used to degrade the protein and autophagic molecules result in protein degradation and kill the cell. The current research reported to induce apoptosis and autophagy in various cancer cell types [17]. The use of GA-DM therefore plays a key role in cell death by apoptosis and autophagy, providing an alternative to traditional chemotherapy. Further, GA-DM produces the long-term effects to cancer patients by stimulating the immune system against various cancer types [17].

3.2 GA-DM as an Immune Stimulatory Agent in Cancer Therapy and Chemoimmunotherapy

By stimulation of the immune systems, GA-DM induces cell death through apoptosis and autophagy which enhances tumor antigen (Ag) presentation to CD4+ T cells. Furthermore, GA-DM upregulates the autophagic protein, Beclin-1, which may bind to the survival protein Bcl-2, provoking apoptosis via the activation of caspase 3 [18]. Apart from these, GA-DM found significant upregulation of HLA class II molecules as well as lysosomal LAMP-2 proteins. Overall, the autophagic process is conveyed by Beclin-1 expression in the melanoma cells [19]. The experimental data revealed that GA-DM administration increases HLA class II Ag presentation and CD+T cell identification of melanoma tumor cells [19]. Chemoimmunotherapy is the term used for the combination of conventional chemotherapy and immunotherapy [19]. Chemotherapy uses different drugs to kill or inhibit the growth of cancer cells, whereas immunotherapy uses treatments to stimulate or restore the ability of the immune system to fight against cancer [18, 19]. A common chemoimmunotherapy includes cyclophosphamide, doxorubicin, vincristine, and rituximab with immunomodulators. In the recent research, GA-DM is used as chemoimmunotherapy, by stimulating the immune response and inducing cell death against malignant tumor [19, 20].

3.3 GA-DM Administration in Cancer Therapy Via Various Routes and Emergence of Nanomedicine in Its Effective Delivery

GA-DM is administered in cancer therapeutics by usually two methods: systemic administration and nanoparticle-mediated delivery [2]. The different routes of administration have their own advantages and disadvantages in combating cancer progression. The systemic administration covers broader area of coverage when introduced into the patient [2, 3]. GA-DM does not easily enter into the tumor cells; therefore, larger quantity of GA-DM is required to establish a therapeutic effect [5]. Ultimately, the higher concentration leads to toxicity toward healthy cells also. To mitigate the problems, nanoparticles loaded with GA-DM are used to significantly reduce the concentration of GA-DM needed and are delivered directly to malignant tumor cells by using nanoparticles to deliver drugs [16]. From the preliminary studies, there is substantial change in dosage depending on the route of administration. Ultimately, it is inferred from animal experiments that a therapeutic impact can be detected with administration of GA-DM loaded nanomedicines [16]. Therefore, lowering in the concentration of GA-DM via nanoparticle drug delivery suggests less cytotoxic effects to the healthy cells. Furthermore, the nanoparticles also reduce metastasis, by delivery of drug to the tumor cells, from where cells prefer to metastasize [16]. In the latest study by Rahman et al. develops GA-loaded nanocarriers for treatment against hepatocellular carcinoma (HCC) in Wister rats. The GA-NLC was developed by double emulsion solvent displacement method and is used for the preparation of the said nanocarriers. The optimized formulation presented the particle size of 156 nm with higher than 85% encapsulation efficiency and showed initial burst release at the first 4 h and followed sustained drug release for up to next 24 h [21]. The said drug-loaded formulations showed better tolerant and antitumor efficacy against hepatic carcinoma. Furthermore, the histopathological demonstration found less alteration in the hepatic tissue via less inflamed vessels, less necrosis, and lower depositions of microdroplets [21]. Ultimately, the experimental data showed that enhanced in vivo antitumor activity by combining hepatic damage factors, biochemical markers, and antioxidants [21]. Another recent study develops self-assembled thermal gold nanorod-loaded thermosensitive liposome-encapsulated GA which has implicated against cancer in combination with near-infrared (NIR) irradiation [22]. The said nanoparticle-loaded GA was utilized as nanophotosensitizers to achieve synergistic photochemotherapy against antitumor action. Moreover, it exhibits anticancer activity against MCF-7 cells [22]. The said nanoparticle-loaded GA with NIR irradiation was found similar with cisplatin against MCF-7 cells. Furthermore, the in vivo toxicity studies for the said NPs revealed no toxicity in mice [22].

3.4 GA-DM in Treating Prostate Cancer

Prostate cancer is the most common type of cancer, which is diagnosed in men, and the second most common cause of cancer-related death in the Western world. There is therefore increasing interest in developing more effective therapeutic options for managing high mortality cancer. The current most common treatment options are including prostatectomy, radiation, and chemotherapy [23]. These treatment option is only effective in treating most local forms of prostate cancer. However, in some cases where castrate-resistant prostate cancer, there alternative therapies are required [23]. Yet, one of the major problems associated with prostate cancer is its ability to metastasize which penetrates particularly to the bone. In the bone, they start to initiate osteoclastogenesis. In the management of prostate cancer, immunotherapies provide an effective long-term treatment in the recent research studies [23]. GD-DM, proved as potential candidate of chemoimmunotherapy in the management of later stage of prostate cancer, causes cytotoxicity in both androgen-dependent and androgen-independent prostate cancer cells [24]. Furthermore, GA-DM administration inhibits both the activity of 5-α-reductase and conversion of testosterone to dihydrotestosterone (DHT) [24]. 5-α-Reductase is playing a key role in the conversion of testosterone to active metabolite called dihydrotestosterone (DHT). The said enzymes are to inhibit the conversion of testosterone to dihydrotestosterone (DHT). This might be due to conformational similarity in the structures of DHT and GA-DM [25]. Although GA-DM blocks the androgen receptors competitively, this inhibits DHT binding and obstructs the usual DHT-mediated signaling pathway, which leads to the cell survival [25, 26]. Furthermore, GA-DM inhibits osteoclastogenesis, which is a major challenge in prostate cancer metastasis.

3.5 GA in Treating Melanoma

It is the most common among skin cancers, which cause cancer-related deaths. There is need to search a novel therapy which can destroy the tumors while delivering the immune stimulating and chemotherapeutic drugs against the metastatic melanoma tumors [27]. GA-DM is a potential novel therapy that is utilized against metastatic melanoma via apoptosis. GA-DM also induces autophagic and apoptotic cell death and enhances tumor Ag presentation via HLA class II [28]. More importantly, GILT expression inhibits a tumorigenic molecule paired box-3 (PAX-3) protein. Furthermore, the in vivo experiments were conducted on B16 mouse melanoma model and found that enhanced Ag presentation led to greater T-cell infiltration of the tumor tissue and clearance of melanoma after treated with GA-DM [29].

3.6 GA in Treatment of Breast Cancer

It is the leading cause of cancer-related death in women. The chemotherapy includes surgery, radiation, chemotherapy, and hormonal therapy [4, 30]. Although chemotherapy treats breast cancer, its efficacy is restricted due to therapeutic resistance in the tumor cells. GA-DM is a potential candidate which treats breast cancer [30]. The literatures reported that GA-DM can effectively inhibit cell proliferation and colony formation in breast cancer cells. GA-DM shows G1 cell cycle arrest and decreases the protein level of CDK2, CDK6, cyclin D1, p-Rb, and c-Myc [4]. Furthermore, GA-DM induced apoptosis and decreases the mitochondrial membrane potential in breast cancer cells via causing DNA fragmentation and cleavage of PARP [4, 30]. Therefore, GA-DM suggests that G1 cell cycle arrest and apoptosis are result from the DNA damage. So, from the above study, it concludes that GA-DM could be a potential natural and better alternative therapeutic option for treating breast cancer [30].

3.7 GA in Treatment of Meningioma

It is the second most common tumor in adults which is found in the central nervous system. It arises from the arachnoid cells of the meninges which cover the brain. Usually, meningiomas are benign, localized, nonaggressive, and noninvasive [31]. However, the higher grade meningiomas are aggressive, malignant, and invasive which can cause multiple neurological and physiological complications [31]. Currently available medication has better therapeutic effect but is restricted with limitations. So, there is demands of developing novel therapy for the aggressive and nonresectable meningiomas. GA-DM works as to treat these aggressive meningiomas via the action on Wnt5/GSK3β/β-catenin signaling pathway [31, 32]. Further, the study reported that GA-DM inhibited the expression of Wnt5 α/β and catenin which enhanced the phosphorylation of GSK3β in IOMM-Lee and CH157MN cells, in which the phosphorylation of Ser 9 is a marker for inactivation of GSK3β [33, 34]. Furthermore, GA-DM has to suppress antiapoptotic proteins such as Akt, Bcl-XL, and Mcl-1 and upregulates the expression of apoptotic protein Bax and induced apoptosis. Moreover, all the above discussion of GA and GA-loaded nanocarriers is also illustrated in Fig. 1.

4 Conclusion

Ganoderic acid (GA) is very beneficial in the treatment of various cancer. A number of studies reported the anticancer activity as well in animal model. The most active constituents of *Ganoderma lucidum* is GA-DM. It is a potential therapeutic molecule to treat cancers. This novel compound has the capability to induce apoptosis in cancer cells to receive long-term protection from the malignant tumors. GA-DM is given to the patient from different routes of administration, and each route has its own qualities. While the study suggests that cancer research is highly successful as an additional or complementary treatment for different types of cancer, the latest research is very encouraging, and further studies into how mushroom-extracted

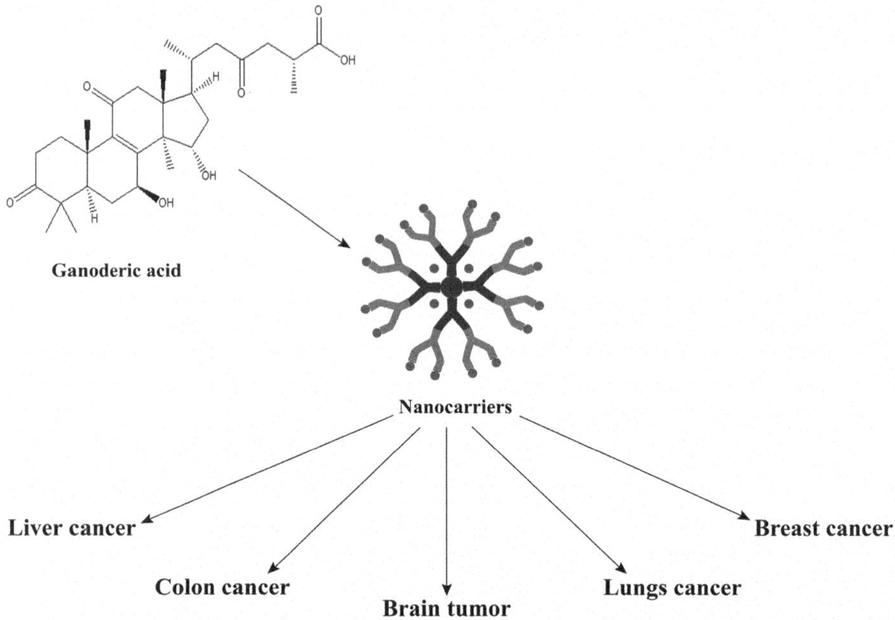

Fig. 1 Illustration of GA-loaded nanocarriers and its application in various cancers

bioactives can be incorporated into conventional medicines for the treatment of malignancies should be greatly encouraged.

References

1. Rahman M, Ahmad MZ, Kazmi I, Akhter S, Afzal M, Gupta G et al (2012) Advancement in multifunctional nanoparticles for the effective treatment of cancer. Expert Opin Drug Deliv 9(4):367–381
2. Lindequist U, Niedermeyer TH, Julich WD (2005) The pharmacological potential of mushrooms. Evid Based Complement Alternat Med 2:285–299
3. Wasser SP (2011) Current findings, future trends, and unsolved problems in studies of medicinal mushrooms. Appl Microbiol Biotechnol 89:132313–132332
4. Wu GS, Lu JJ, Guo JJ, Li YB, Tan W et al (2012) Ganoderic acid DM, a natural triterpenoid, induces DNA damage, G1 cell cycle arrest and apoptosis in human breast cancer cells. Fitoterapia 83:408–414
5. Radwan FF, Perez JM, Haque A (2011) Apoptotic and immune restoration effects of ganoderic acids define a new prospective for complementary treatment of cancer. J Clin Cell Immunol S3:4
6. Bishop KS, Kao CH, Xu Y, Glucina MP, Paterson RR et al (2015) From 2000 years of Ganoderma lucidum to recent developments in nutraceuticals. Phytochemistry 114:56–65
7. Wachtel-Galor S, Yuen J, Buswell JA, Benzie IFF (2011) Ganoderma lucidum (Lingzhi or Reishi): a medicinal mushroom (Chapter 9) In: Benzie IFF, Wachtel-Galor S (eds) Herbal medicine: biomolecular and clinical aspects, 2nd ed. CRC Press/Taylor & Francis, Boca Raton, FL

8. Radwan FF, Hossain A, God JM, Leaphart N, Elvington M et al (2015) Reduction of myeloid-derived suppressor cells and lymphoma growth by a natural triterpenoid. J Cell Biochem 116:102–114
9. Huang SZ, Ma QY, Kong FD, Guo ZK, Cai CH et al (2017) Lanostane-type triterpenoids from the fruiting body of Ganoderma calidophilum. Phytochemistry 143:104–110
10. Wei JC, Wang AH, Wei YL, Huo XK, Tian X et al (2017) Chemical characteristics of the fungus Ganoderma lucidum and their inhibitory effects on acetylcholinesterase. J Asian Nat Prod Res 20(10):1–10
11. Liu J, Shimizu K, Tanaka A, Shinobu W, Ohnuki K et al (2012) Target proteins of ganoderic acid DM provides clues to various pharmacological mechanisms. Sci Rep 2:905
12. Akihisa T, Nakamura Y, Tagata M, Tokuda H, Yasukawa K et al (2007) Anti-inflammatory and anti-tumor-promoting effects of triterpene acids and sterols from the fungus Ganoderma lucidum. Chem Biodivers 4:224–231
13. Shi L, Ren A, Mu D, Zhao M (2010) Current progress in the study on biosynthesis and regulation of ganoderic acids. Appl Microbiol Biotechnol 88:1243–1251
14. Xia Q, Zhang H, Sun X, Zhao H, Wu L et al (2014) A comprehensive review of the structure elucidation and biological activity of triterpenoids from Ganoderma spp. Molecules 19:17478–17535
15. Hossain A, Radwan FF, Doonan BP, God JM, Zhang L et al (2012) A possible cross-talk between autophagy and apoptosis in generating an immune response in melanoma. Apoptosis 17:1066–1078
16. Shafei A, El-Bakly W, Sobhy A, Wagdy O, Reda A et al (2017) A review on the efficacy and toxicity of different doxorubicin nanoparticles for targeted therapy in metastatic breast cancer. Biomed Pharmacother 95:1209–1218
17. Johnson BM, Doonan BP, Radwan FF, Haque A, Ganoderic Acid DM (2010) An alternative agent for the treatment of advanced prostate cancer. Open Prost Cancer J 3:78–85
18. Hoelzer D, Gökbuget N (2012) Chemoimmunotherapy in acute lymphoblastic leukemia. Blood Rev 26:25–32
19. Opat S, Hawkes EA (2017) Chemoimmunotherapy may not be dead yet in chronic lymphocytic leukemia, but fludarabine plus cyclophosphamide plus rituximab is potentially facing life support. J Clin Oncol 2017:4093–4094
20. Reddy NM, Thieblemont C (2017) Maintenance therapy following induction chemoimmunotherapy in patients with diffuse large B-cell lymphoma: current perspective. Ann Oncol 28:2680–2690
21. Rahman M, Al-Ghamdi SA, Sharma K, Beg S, Alharbi KS, Abbasi FA et al Ganoderic acid loaded nano-lipidic carriers improvise treatment of hepatocellular carcinoma. Drug Deliv. https://doi.org/10.1080/10717544.2019.1606865
22. Zhang W, Yu W, Ding X, Yin C, Yan J, Yang E et al (2019) Self-assembled thermal gold nanorod-loaded thermosensitive liposome-encapsulated ganoderic acid for antibacterial and cancer photochemotherapy. Artif Cells Nanomed Biotechnol 47(1):406–419
23. Sumanasuriya S, De Bono J (2018) Treatment of advanced prostate cancer-a review of current therapies and future promise. Cold Spring Harb Perspect Med. https://doi.org/10.1101/cshperspect.a030635
24. Bhutia SK, Dash R, Das SK, Azab B, Su ZZ et al (2010) Mechanism of autophagy to apoptosis switch triggered in prostate cancer cells by antitumor cytokine melanoma differentiation-associated gene 7/interleukin-24. Cancer Res 70:3667–3676
25. Aoun F, Bourgi A, Ayoub E, El Rassy E, van Velthoven R et al (2017) Androgen deprivation therapy in the treatment of locally advanced, nonmetastatic prostate cancer: practical experience and a review of the clinical trial evidence. Ther Adv Urol 9:73–80
26. Rucci N, Angelucci A (2014) Prostate cancer and bone: the elective affinities. Biomed Res Int 2014:167035
27. Lee N, Barthel SR, Schatton T (2014) Melanoma stem cells and metastasis: mimicking hematopoietic cell trafficking? Lab Investig 94:13–30

28. Haque MA, Li P, Jackson SK, Zarour HM, Hawes JW et al (2002) Absence of gamma-interferon-inducible lysosomal thiol reductase in melanomas disrupts T cell recognition of select immunodominant epitopes. J Exp Med 195:1267–1277
29. Hathaway-Schrader JD, Doonan BP, Hossain A, Radwan FFY, Zhang L et al (2018) Autophagy-dependent crosstalk between GILT and PAX-3 influences radiation sensitivity of human melanoma cells. J Cell Biochem. https://doi.org/10.1002/jcb.26383
30. Sotgia F, Fiorillo M, Lisanti MP (2017) Mitochondrial markers predict recurrence, metastasis and tamoxifen-resistance in breast cancer patients: early detection of treatment failure with companion diagnostics. Oncotarget 8:68730–68745
31. Das A, Miller R, Lee P, Holden CA, Lindhorst SM et al (2015) A novel component from citrus, ginger, and mushroom family exhibits antitumor activity on human meningioma cells through suppressing the Wnt/beta-catenin signaling pathway. Tumour Biol 36:7027–7034
32. Ragel BT, Jensen RL, Gillespie DL, Prescott SM, Couldwell WT et al (2005) Ubiquitous expression of cyclooxygenase-2 in meningiomas and decrease in cell growth following in vitro treatment with the inhibitor celecoxib: potential therapeutic application. J Neurosurg 103:508–517
33. Piscevic I, Villa A, Milićević M, Ilić R, Nikitović M et al (2015) The influence of adjuvant radiotherapy in atypical and anaplastic meningiomas: a series of 88 patients in a single institution. World Neurosurg 83:987–995
34. Preusser M, Berghoff AS, Hottinger AF (2013) High-grade meningiomas: new avenues for drug treatment? Curr Opin Neurol 26:708–715

Vitamin E-Based Nanomedicines for Anticancer Drug Delivery

Roya Karimi Alavijeh and Kamran Akhbari

Abbreviations

ABC	ATPγ
AMF	Alternative magnetic field
Ap	Apigenin
AUC	Area under the concentration−time curve
BBB	Blood–brain barrier
BCT	Breast conservation treatment
BMM	Bone marrow derived macrophages
BTB	Blood–tumor barrier
CA	Chitosan
CA	Cholic acid
CEHC	Carboxyethylhydroxychromans
CLSM	Confocal laser scanning microscopy
CMC	Critical micelle concentration
CPT	Camptothecin
Cremophor-EL	Polyoxyethylated castor oil
CSO	Chitosan oligosaccharide
DMAP	4-dimethylamioprydine
DNPs	Docetaxel loaded polymeric nanoparticles
DOPE	Dioleoyl phosphatidyl ethanolamine
DOX	Doxorobicin
DTX	Docetaxel
DTX	Marketed formulation of Docetaxel
EDC	(1-ethyl-3-(3-dimethylaminopropyl) carbodiimide hydrochloride
EE	Drug encapsulation efficiency

R. Karimi Alavijeh · K. Akhbari (✉)
School of Chemistry, College of Science, University of Tehran, Tehran, Iran
e-mail: akhbari.k@khayam.ut.ac.ir

© Springer Nature Singapore Pte Ltd. 2020

M. Rahman et al. (eds.), *Nanomedicine for Bioactives*,
https://doi.org/10.1007/978-981-15-1664-1_2

EGFR	Epidermal growth factor receptor
EPR	Enhanced permeability and retention effect
ER	Estrogen
FDA	Food and Drug Administration
GAR	Garcinol
GI	Gastrointestinal
GMO	Glyceryl monooleate
GSH	Glutathione
HCPT	10-Hydroxycamptothecin
IOs	Nano-sized iron oxides
LC	Drug-loading content
LUT	Luteolin
MDR	Multidrug resistance
MNPs	Magnetic nanoparticles
MPS	Mononuclear phagocyte system
MRP1	Multidrug resistance protein 1
mTHPC	Temoporfin
NIR	Near-infrared
NLC	Nanostructured lipid carriers
NO	Nitric oxide
NONOates	N-diazeniumdiolates
NP	Nanoparticle
PAA	Poly acrylic acid
PAMAM	Poly (amidoamine)
PBS	Phosphate-buffered saline
PCL	Poly (ε-caprolactone)
PDT	Photodynamic therapy
PEG	Poly (ethylene glycol)
PF-TOS micelles	TOS modified pluronic micelles
P-gp	P-Glycoprotein
PLA	Poly (lactide)
PLA–TPGS	Poly (L-lactide)-vitamin E TPGS
PLGA	Poly (lactide co-glycolide)
PR	Progesterone
PTX	Paclitaxel
QDs	Quantum dots
RES	Reticuloendothelial system
ROS	Reactive oxygen species
RSNOs	S-nitrosothiols
RT	Radiation treatment
SA	Succinic anhydride
SC	Solvent casting
SESD	Spontaneous emulsification solvent-diffusion method
Taxotere®	Clinical DTX formulation
TC	Terephthaloyl chloride

TCOsomes	α-Tocopherol-oligochitosan-based oligomersomes
THP	Pirarubicin
TNBC	Triple negative breast cancer
TOC	Tocopherols
TPGS	α-Tocopherol polyethylene glycol 1000 succinate
TPNs	TPGS-functionalized PLGA nanoparticles
TQR	Tariquidar
TT	Tocotrienols
VE	Vitamin E
VES	Vitamin E succinate
VES-g-CSO	TPGS and VES-grafted CSO
α-Fmoc	α-Fluorenylmethoxycarbonyl
α-TOS	α-Tocopheryl succinate

1 Introduction

Cancer is one of the leading causes of death throughout the world. It is defined as abnormal cell growth with the potential to invade or spread to other parts of the body [1]. Today, cancer is known as a dreadful disease with a death of 8.2 million people in cancer and 14.1 million diagnoses of new cancer cases worldwide in 2012. It is expected that the number of new cancer cases will reach 21.7 million in 2030. According to the food and nutrition report in 2003, in 30–40% cases, cancer is directly related to nutrition. Therefore, the use of chemopreventive agents such as vitamin E from nutritional source is an acceptable method to reduce the rate of cancer [2]. In most cases, conventional treatment methods, including chemo/surgical/radiation treatments, fail to target and treat the disease adequately. Recent progress in the field of tumor-directed nanotechnologies provides hope of treatment for cancer patients. Nanoparticle (NP)-based therapeutics can significantly improve the treatment outcomes of oncological diseases by increasing the tumor delivery of drugs and limiting off-target uptake [1, 3]. Nanomedicines are defined by the European Science Foundation as follow: "nanometer size scale complex systems, consisting of at least two components, one of which being the active ingredient." Nanomedicines can increase drug function including efficacy, specificity, tolerability, and therapeutic index [4]. In other words, the loading of drugs into nanocarriers can make some advantages:

- Increase in vivo stability
- Extend blood circulation time
- Allow for controlled drug release

Another potential of nanomedicines is combination therapy, which means the delivery of more than one therapeutic agent. Theranostic nanomedicines that contain imaging and therapeutic agents have also been developed to enable diagnosis and therapy together with monitoring of therapeutic response.

Nanomedicines can enable drugs to accumulate, preferably at the tumor site. This phenomenon is known as enhanced permeability and retention (EPR) effect [3]. Because of abnormal basement membranes and incomplete endothelial linings, the tumor vessels surrounding the tumors are leaky and allowing nanomedicines to reach tumor passively through the leaky vasculature. Therefore, the EPR effect enables the higher local concentration of the drug at the tumor site. There are two main strategies for delivery of NPs to the tumor cells: (1) passive and (2) active targeting. In passive targeting, NPs pass through leaky tumor vasculature, but in active targeting, receptor-mediated targeting is utilized for targeting nanomedicines to tumors. A schematic representation of passive and active targeting of nanosized drug delivery systems has been shown in Fig. 1.

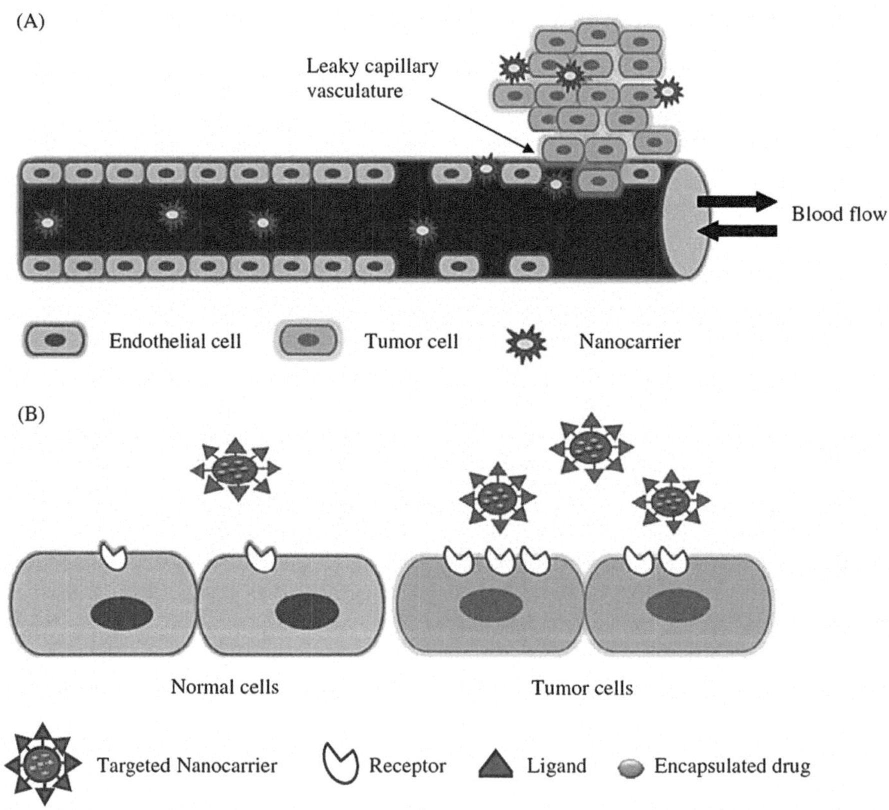

Fig. 1 Schematic of (**a**) passive (via the EPR effect) and (**b**) active (receptor-mediated) targeting utilized for targeting nanopreparations to tumors [6] (Reproduced with the permission of Taylor & Francis)

2 Vitamin E: Structure and Properties

Since its discovery in 1922, vitamin E attracted attention in the biomedical community. This vitamin is not a single compound but includes the family of lipid soluble tocochromanols. The term vitamin E (VE) refers to a family of neutral plant lipids including tocopherols and tocotrienols. These molecules possess structural similarities. All members of the vitamin E families have a common part, aromatic chromanol head, which determine the nomenclature of the isomers based on the number of methyl groups located on this part. There are eight naturally occurring tocols that are known as vitamin E. These compounds include four tocopherols (TOC) and four tocotrienols (TT) that are existing in alpha (α), beta (β), gamma (γ), and delta (δ) isomers (Fig. 2). Saturation of the 16-carbon hydrocarbon tail defines the analog names as tocopherol (saturated) or tocotrienol (unsaturated).

Vitamin E has many properties, including antioxidant, anti-inflammatory, antithrombotic, and other therapeutic effects. The tocopherols and tocotrienol isomers are known for their antioxidant and anticancer activity, respectively. This vitamin is a major lipid-soluble antioxidant with the ability to protect lipids and membranes from oxidative damage in vitro and in vivo. There is a firmly electrophilic hydroxyl group on the chroman ring of all tocochromanols that are capable efficiently for quenching carbon radicals. The generated oxidized tocopheroxyl radical during this reaction is regenerated back to tocopherol by the other water-soluble antioxidants such as ascorbic acid (vitamin C). As mentioned above, vitamin E also has anti-inflammatory effects. This feature is based on lipoxygenase inhibition. This enzyme is responsible for leukotriene production that causes inflammation [5, 7–9].

It has been found that tocol esters are more stable against oxidation than the nonester form of vitamin E. Vitamin E esters including vitamin E acetate, vitamin E succinate, and vitamin E polyethylene glycol-1000 succinate are commercially available [5].

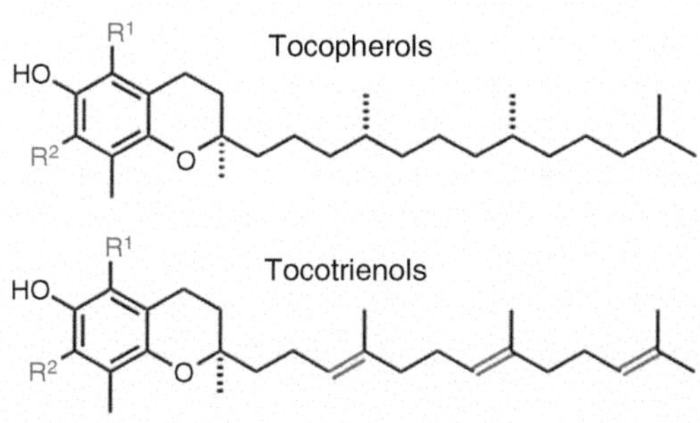

Tocochromanol type	R_1	R_2
α-Tocopherol	CH_3	CH_3
α-Tocotrienol	CH_3	CH_3
β-Tocopherol	CH_3	H
β-Tocotrienol	CH_3	H
γ-Tocopherol	H	CH_3
γ-Tocotrienol	H	CH_3
δ-Tocopherol	H	H
δ-Tocotrienol	H	H

Fig. 2 The vitamin E family [7] (Reproduced with the permission of Elsevier)

3 Vitamin E Sources and Metabolism

Generally, the recommended daily intake of vitamin E is about 2–4 mg/kg body weight, that is, about 20–30% of dietary vitamin E is absorbed. Humans and animals are not able to synthesize vitamin E. Therefore, obtain the eight isomers from plant sources [10]. Many studies indicate the importance of vitamin E in the diet. In one report, it was shown that the risk of developing colon cancer in people in the Mediterranean area who consume foods rich in vitamin E was lower than people in Northern Europe and the USA. Also, another study demonstrated that low vitamin E in diet be could responsible for increased prostate cancer risk [11, 12]. Vitamin E dietary sources include vegetable oils, particularly wheat germ and soybean oil, sprouted seeds, nuts, grains, green leafy vegetables, egg yolks, and legumes. Deficiency in vitamin E absorption is not detectable for a long time. However, when the damage is found, it is generally irreversible, particularly in nervous tissue. Muscular weakness and reproductive failure are other symptoms. Also there are several genetic diseases such as cystic fibrosis, cholestasis, and inflammatory bowel diseases which lead to the poor absorption of vitamin E. It was found that ingested vitamin E homologs are generally metabolized to yield carboxyethyl hydroxychroman (CEHC), a hydrosoluble compound which has been reported to be biologically active and exhibit pharmacological properties. This compound is excreted through the kidney. Vitamin E toxicity is very rare, and it is generally observed at very high doses (>3.2 g/kg/day). Interference with vitamin K metabolism and elevation in blood pressure are the reasons for vitamin E toxicity that is observed in particular people with sensitivities to this vitamin [13].

4 Vitamin E and Cancer

Vitamin E is a well-known antioxidant and acts as an essential component of biological membranes and circulating lipoproteins. Two primary analogs of vitamin E, α-tocopheryl succinate (TOS) and α-tocopherol polyethylene glycol 1000 succinate (TPGS), are exciting groups of novel anticancer agents. For example, TOS represents a potentially high pharmacological importance. This compound can inhibit the growth of breast cancer cells in vitro and in vivo [14]. It was initially thought that the ability of vitamin E to neutralize the free radicals generated by the fecal bacteria in the gut and thereby DNA damage prevention are the reason for chemopreventive properties of this vitamin. Nevertheless, accumulating evidence in the literature suggests that independent of the antioxidant functions of vitamin E, its signal transduction activities may also be responsible for its cancer chemopreventive effects [15].

One of the major problems for effective cancer treatment is intrinsic or acquired resistance to chemotherapeutic drugs. Multidrug resistance (MDR) is the most common form of resistance recognized in cancer patients. Alterations in cellular mechanisms are the primary reason for MDR that leads to decreased intracellular accumulation of drugs in the cells. Overexpression of the ATP-binding cassette (ABC) superfamily of transporters is one of the identified mechanisms for MDR that leads to efflux of the drug in cancer cells. P-glycoprotein (P-gp) is the most critical member of ABC superfamily of proteins that is ATP dependent. It means that P-gp can use the energy provided by ATP hydrolysis and pump drugs out of the tumor cells. Multidrug resistance protein 1 (MRP1) is another kind of ABC transporters that are frequently overexpressed in drug-resistant cancer cells. MRP1 can export a broad range of organic anions and neutral drugs that are conjugated with anions such as glutathione (GSH), glucuronide, and sulfide. Also, this protein can export unconjugated drugs together with GSH. In drug-resistant cancer cells, GSH biosynthesis enzymes are commonly overexpressed or activated together with MRP1. It has been shown that the efflux of drugs depends on intracellular GSH levels. TOS could lower the intracellular GSH concentration and easily incorporate into the liposomal layers. These results suggest that TOS may be a clinically useful adjuvant in cancer treatment [16–18]. Also, there are some reports about the application of TPGS for overcoming MDR phenomenon. Vitamin E TPGS can inhibit P-gp. For example, in a literature that is published in 2010, a polylactide-surfactant block copolymer, poly (L-lactide)-vitamin E TPGS (PLA–TPGS), was synthesized and used for doxorubicin delivery. Results showed that PLA–TPGS nanoparticles under conditions of various polymer-to-drug ratios produced particles ranging in size from 131.2 to 274.5 nm with drug encapsulation efficiency (EE %) ranging from 69.1 to 98.5. The best nanoparticles in terms of size and EE % were obtained from the polymer-to-drug ratio of 15:1. Synthesized nanoparticles with this condition were used for drug delivery investigations. Obtained results showed that this copolymer was able to reduce the activity of P-gp in breast cancer cells (MCF-7/ADR) and lead to an increase in the nuclear accumulation of doxorubicin [19].

5 Vitamin E-Based Nanomedicines

There are many reports for vitamin E grafted on various nanopolymers to efficient anticancer drug delivery. Many anticancer drugs are suffering from low solubility in aqueous media. Therefore, nanomaterials as a carrier can improve their performances. Reported carriers are in different forms including nanoparticles, micelles, liposomes, oligomersomes, prodrugs, and MOFs where vitamin E is in their structures or is used for modification of carriers, and in some cases such as MOFs, vitamin E acts as a cargo that is loaded on the pores of the MOFs. In the following, two kinds of vitamin E, TOS and TPGS, which are frequently used for anticancer drug delivery, are described.

5.1 TOS as an Anticancer Agent

Literature published in 2010 reported α-TOS as a physiologically acceptable lipid material used to prepare lipid nanospheres and for paclitaxel (Taxol®)[1] delivery. This study demonstrated that TOS has the potential for making lipid nanospheres.

[1] Taxol is a name for paclitaxel that is used clinically.

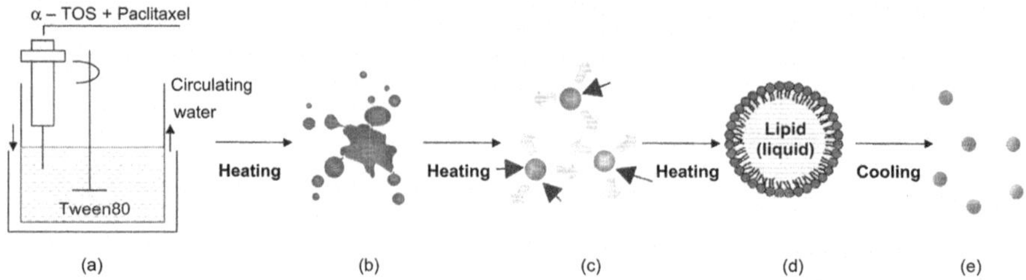

Fig. 3 Schematic representation of the formation of α-TOS nanospheres by the modified SESD process. (**a**) Solvent injection into aqueous solution containing surfactant, (**b**) self-emulsification by interfacial turbulences and rapid diffusion, (**c**) counterdiffusion and solvent evaporation, (**d**) emulsified liquid droplet (o/w), (**e**) solidified nanospheres obtained by cooling down of the liquid emulsion droplets [20] (Reproduced with the permission of Taylor & Francis)

A modified spontaneous emulsification solvent diffusion (SESD) method was used to obtain the nanosuspensions (Fig. 3) with small size (127 ± 48 nm) and satisfactory drug entrapment efficiency (90%). The in vitro release profile of paclitaxel from the prepared nanospheres indicated two stages: an initial fast release and a sustained release up to 72 h. These results suggested that paclitaxel release was controlled by both the diffusion of drug and also lipid matrix corrosion, and α-TOS-based lipid nanospheres have the potential for paclitaxel delivery [20].

Combination-based chemotherapy is a method for the treatment of multiple solid tumors. The combination of TOS with doxorubicin (DOX) is a method to increase antitumor efficacy that is investigated in literature published in 2018. In this report, nanostructured lipid carrier (NLC) loaded with α-tocopherol succinate (TS) and DOX (NLC-TS-DOX) was prepared by hot melting homogenization method and examined for antitumor activity in breast cancer models (MDA-MB-231 and 4T1 cell lines). In this study, NLC-TS-DOX was labeled with technetium-99m (99mTc-NLC-TS-DOX) to the evaluation of blood clearance, biodistribution, and scintigraphic images. The obtained results by scintigraphic images showed that uptake was mainly in the liver and spleen. It is expected for nanoparticles due to the recognition by mononuclear phagocyte system (MPS). The antitumor activity of prepared NLC-TS-DOX was investigated in 4T1-tumor-bearing BALB/c mice. This study represents more reduced tumor volume in the case of NLC-TS-DOX in comparison with other groups (blank-NLC as a control model, free DOX, and NLC-DOX). In vitro DOX-release data from NLC-TS-DOX was around 5% after 2 h and reached 10% in 24 h. In contrast, aqueous solution (free DOX) released around 90% after 4 h. Some unique features of NLC-TS-DOX including small particle size (85 nm), EPR effect, and the controlled release of DOX from the particle led to an increase of its antitumor efficacy. Its drug encapsulation efficiency was 98.97 ± 0.33. The reason for the high value of EE % was attributed to an ion-pare formation between DOX and TS. Therefore, NLC-TS-DOX has the potential to improve antitumor efficacy in DOX therapeutic regimens [21].

In addition to nanoparticles, there are many reports about micelles constructed from TOS that are used for delivery of anticancer drugs. Micelles are self-assembled core-shell structure with the ability of incorporating poorly water-soluble drugs into the core (hydrophobic part) and improving solubility, stability, and bioavailability of drugs. The outer shell of micelles is hydrophilic and prevents drug molecules from the inactivation process. The size of micelles is usually small (<200 nm); therefore, they indicate passive targeting into the tumor cells and consequently enhanced EPR effect [22].

One of the most reported anticancer drugs is paclitaxel (PTX). PTX is isolated from the bark of North American Pacific yew tree, *Taxus bravifolia*, with the ability to block the mitotic pathway of cell division by microtubule stabilization and apoptosis/cell death. This drug has been approved as a chemotherapeutic agent by the Food and Drug Administration (FDA). This chemotherapeutic drug has been used for the treatment of many cancers. However, its applications are restricted by its poor water solubility (~0.4 μg/mL). Its commercial formulation (Taxol®) consists of PTX, dehydrated alcohol, and polyoxyethylated castor oil (Cremophor EL). Cremophor EL can make several side effects, such as nephrotoxicity, hypersensitivity, and neurotoxicity reactions. Therefore, there is a need to use some strategies to solve these problems and improve its bioavailability [23].

One of the micelles that are constructed from TOS was reported in 2013. In this study, TOS-modified pluronic micelles (PF-TOS micelles) with an average size of about 58 nm were synthesized to paclitaxel delivery. Terminal NH_2 groups of PF were modified with TOS to improve the PTX encapsulation efficiency and enhance antitumor ability (Fig. 4). Bone marrow-derived macrophages (BMM) and human glioma U87 cells were examined to determination tumor-specific toxicity of PF-TOS-PTX micelles. The obtained results indicated that PTX-loaded PF-TOS micelles have high anticancer activity to U87 cells, longer systemic circulation time, and slower plasma elimination rate in comparison with that of the nonmodified controls. These obtained results indicated that PF-TOS micelles could provide efficient encapsulation and stabilization drug and, therefore, can improve the hydrophobic chemotherapeutic drug delivery [24].

Recently, in 2019, Biswas et al. reported a vitamin E-conjugated amphiphilic polymer to prepare a DOX-loaded nanomicellar system. Three polymeric components including TOS, polyethylene glycol (PEG), and dioleoylphosphatidylethanolamine (DOPE) were attached via lysine linker and make a conjugated amphiphilic polymer that was used to prepare micelles (VPM). DOX-loaded micelles (DOX-VPM) were developed by the dialysis method and examined in some cell lines with DOX-resistant ability, human colon carcinoma (SW 620-R), and human breast adenocarcinoma (MDA MB-231-R). High encapsulation efficiency and stable and sustained DOX release were observed in in vitro drug release experiments. The vitamin E-based polymer can inhibit the overexpressed P-gp in resistant cancer cells that was confirmed in this study with fluorescence images. DOX-VPM released the DOX in a sustained manner (~40 % of DOX was released in 20 h). This sustained release can solve one of the major drawbacks of conventional chemotherapy, that is, unwanted toxicity. This feature can reduce the side effects [25].

Another report of the application of TOS is attributed to the preparation of oligomersomes for siRNA delivery in 2011. α-Tocopherol-oligochitosan-based oligomersomes (TCOsomes) with the mean sizes of 220 and 377 nm were used for the

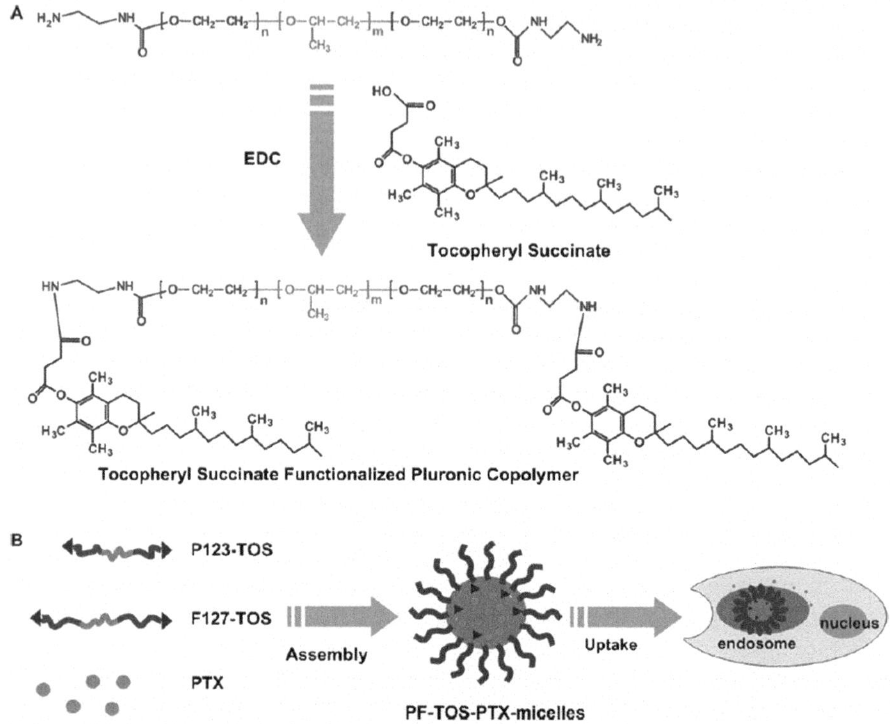

Fig. 4 Preparation scheme for tocopheryl succinate conjugated Pluronic block copolymers (PF-TOS) (**a**). Illustration on the preparation of PTX-loaded PF-TOS micelles (**b**) [24] (Reproduced with the permission of Elsevier)

efficient delivery of siRNA in vitro and in vivo conditions. The size of TCOsomes was controllable by chain lengths of oligochitosan. In this study, the antitumor activity of human Mcl-1 (hMcl-1)-specific siRNA (siMcl-1) delivered by prepared $TCOsome_{4K}$ was examined. Mcl-1 is a critical survival factor in various tumors, and siMcl-1 has anticancer activity by enhancing apoptosis of tumor cells. Because of ineffective siRNA delivery systems, the antitumor effects of siMcl-1 in vivo condition were rare. In this study, the obtained results showed silence of the expression of the target gene in the tumor tissues significantly. Therefore, TCOsomes could inhibit the growth of tumors by apoptosis of cancer cells in the tissues of the mice (treated with siMcl-1/$TCOsome_{4K}$ complexes), but siGL2/$TCOsome_{4K}$ complexes were not efficient (Fig. 5) (luciferase-specific siGL2 was used for comparison of its antitumor activity with siMcl-1/$TCOsome_{4K}$) [26].

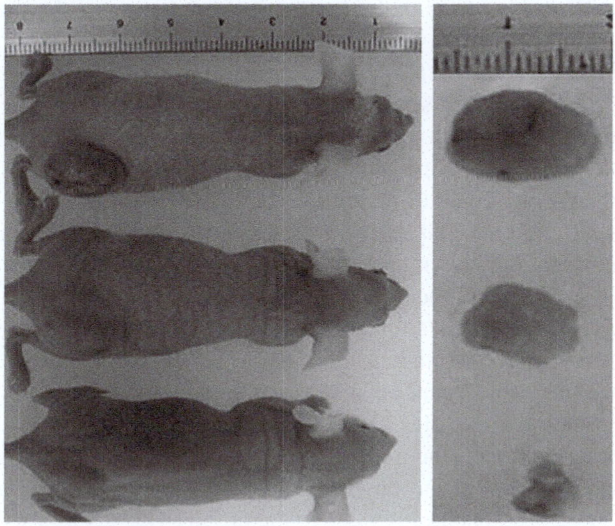

Fig. 5 Images of whole mice (left panel) and excised tumors (right panel) were taken at 22 days after tumor inoculation [26] (Reproduced with the permission of Elsevier)

5.2 TPGS as an Anticancer Agent

D-α-Tocopheryl polyethylene glycol 1000 succinate, which is represented with TPGS, is a water-soluble derivative of the natural vitamin E (α-tocopherol). TPGS is miscible in the water at room temperature and is prepared by esterification reaction of vitamin E succinate with PEG_{1000}. TPGS is a nonionic and high hydrophilic compound and used as a vehicle for lipid-based delivery formulations. For the first time, Mu and Feng introduced TPGS as an effective emulsifier for biodegradable nanoparticle formulations of drugs. They used paclitaxel as a model drug. There are many benefits for using TPGS in drug delivery because the lipophilic part of TPGS can facilitate solubilization of poorly soluble drugs in the micelles. TPGS on the surface of the polymeric nanoparticles can also increase the cellular uptake of the nanoparticle formulations and the oral bioavailability, therefore resulting in prolonged blood circulation time of drugs [27].

Based on reported studies, the performance of TPGS for drug delivery applications is related to the molecular weight of the PEG chain. For example, there is a study showing that TPGS with shorter PEG chains represents better cellular uptake and therapeutic effects. Another report revealed that TPGS with a longer PEG chain has better loading capacity and stability and lower critical micelle concentration (CMC) value [28–30].

As previously mentioned, in the clinical administration of paclitaxel, Cremophor EL is used as an adjuvant, and this compound is not very safe and causes severe side effects. In this regard, many kinds of literature reported methods to solve this problem. One solution is the nanoparticle formulation of the drug. One of them is related to a report published in 2003. In this report, TPGS as an emulsifier was blended with poly(lactic-co-glycolide) acid (PLGA) to prepare nanoparticles for achieving controlled PTX release. As the author mentioned, this report was the primary literature about the application of TPGS as a matrix material for preparing nanoparticles. The obtained nanoparticles with narrow size distribution and high encapsulation efficiency (100%) showed controlled drug release kinetics [31].

Nanoparticle colloidal delivery systems can resolve the limitations in conventional chemotherapy by improving drug delivery and tumor targeting. Incorporation of biodegradable and biocompatible polymers, such as poly(lactide) (PLA), PLGA, and surface functionalization, can provide target delivery of drugs directly into tumor microenvironments and reduce side effects (it is entirely explained in Sect. 9). Besides these advantages of nanoparticles, they suffer from problems such as acid degradation, rapid clearance via the liver, and autocatalytic effect. One method to solve these issues is the incorporation of TPGS and specific ligands in nanocarriers to facilitate active and passive transportation of chemotherapeutics drugs into the cancer cells and achieve effective therapy. TPGS acts as a stabilizer, solubilizer, and penetration enhancer.

Poly(lactide)–vitamin E TPGS (PLA–TPGS) was reported in 2006 as a copolymer to cancer chemotherapy. This copolymer was synthesized from lactide and $TPGS_{1000}$ and used for encapsulation of paclitaxel with modified solvent extraction/evaporation technique. The amount of drug-loading and drug encapsulation efficiency of this nanosized copolymer (300 nm) was 5% and 89%, respectively, and the results showed that this copolymer is able to sustain the release of drug in two stages: the first is burst release (17%) and then sustained release (51%) after 31 days [32].

Another report for encapsulation of paclitaxel was published in 2007 using a biodegradable nanoparticle PLGA. Solvent extraction/evaporation method with vitamin E TPGS as the emulsifier was used for capsulation of paclitaxel. This compound with 288 ± 11.67 nm size was tested on mortality of C6 glioma cells by colorimetric MTT assay method, and the results showed five times more effectiveness and more sustainability (168 h vs. 22 h) in comparison with Taxol[2]. Sustained release behavior, high encapsulation efficiency (92.63 ± 10.01%), low side effect, and biocompatibility of nanoparticle are advantages of this compound. The benefits of this nanoparticle formulation of paclitaxel vs. Taxol were confirmed by the xenograft model investigation (Fig. 6). This graph represents more obtained tumor growth inhibition by TPGS-emulsified PLGA nanoparticles [33].

Another report of vitamin E TPGS-emulsified PLGA nanoparticles was published in 2016 to garcinol loading as an anticancer drug with a sustained release manner for 6 days. Garcinol (GAR) is a polyisoprenylated benzophenone derivative, and because of its polyphenolic structure, it has antioxidative property. This compound is isolated from the fruit rind of *Garcinia indica* (known as kokum). This report was the first study about the encapsulation of GAR in polymeric nanoparticles to anticancer potency improvement. TPGS-emulsified PLGA

[2] Taxol is a brand name of paclitaxel (PTX).

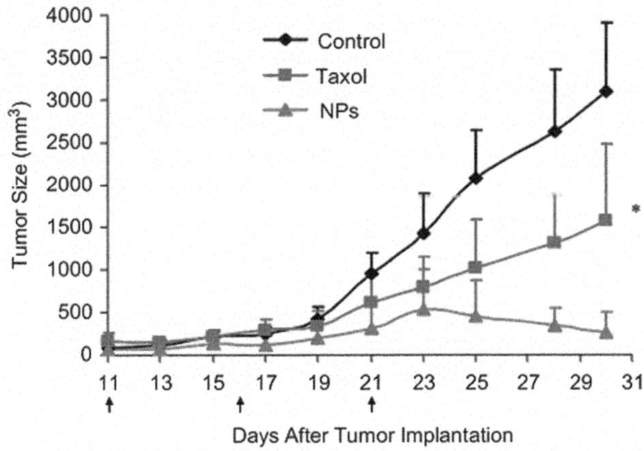

Fig. 6 Changes in the size of C6 tumor nodules after intratumoral injections on days 11, 16, and 21 with saline, Taxol, and Paclitaxel formulated in TPGS-emulsified PLGA nanoparticles at the same 10 mg/kg dose [33] (Reproduced with the permission of Elsevier)

nanoparticles were used to improve its water solubility and bioavailability and therefore enhance its potential therapeutic efficacy. Drug-loaded vitamin E TPGS-emulsified PLGA nanoparticles (GAR-NPs) were synthesized by the nanoprecipitation method. This method leads to the preparation of GAR-NPs with an average diameter of 88.05 ± 2.7 nm. Drug-loading amounts and encapsulation efficiencies were tested by the various amounts of GAR and PLGA. The highest LC %[3] and EE % were detected as 9.25 ± 2.8 % and 88 ± 3.3 %, respectively. The MTT assay was done on B16F10, HepG2, and KB cells, and the results indicated considerable cell apoptosis in B16f10 and KB cell lines. In vivo analysis was also done on B16F10 cells. Also, to the investigation of in vivo deposition at the tumor cells, these nanoparticles were radiolabeled with technetium-99m (99mTc-labeled GAR-NPs). These investigations represented the moderate accumulation of labeled nanoparticles in the tumor region. 99mTc-labeled GAR-NPs also showed high cell binding (B16F10) and internalization. This study revealed the antitumor effect of GAR-NPs and supported the ability of nanosystems for GAR delivery [34].

In spite of extensive investigation of PLGA NPs and the TPGS-b-PLGA NPs as drug delivery vehicles, most of the researches were about linear copolymers. In a literature reported in 2013, star-shaped PLGA-vitamin E TPGS copolymer that was functionalized with cholic acid (CA) was used for delivery of docetaxel to cervical cancer treatment. Star-shaped block polymers are a simple branched polymer. In this model, all branches extend from a single point (core) and show better properties such as lower solution viscosity, smaller hydrodynamic radius, higher drug-loading content, and drug encapsulation efficiency. In this study, CA was used because of its ability to improve biocompatibility for polymers. This acid with three hydroxyl groups and one carboxyl group is the primary bile acid in the body, and it is

[3] Drug-loading content

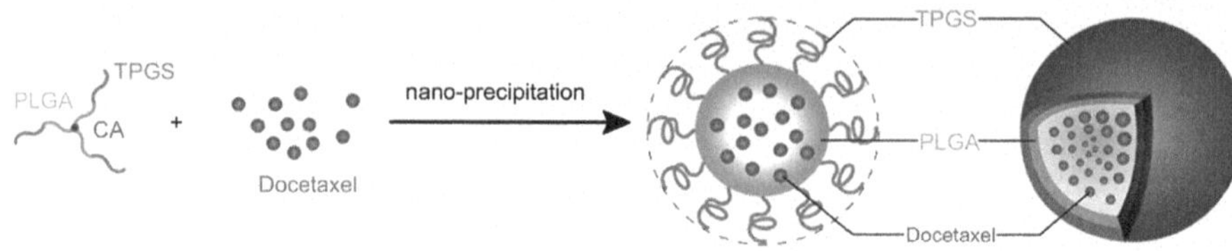

Fig. 7 Schematic representation of technique for preparation of the DTX-loaded CA-PLGA-b-TPGS NPs [35] (Reproduced with the permission of Elsevier)

composed of a steroid unit. The presence of CA moiety in materials could enhance both cell adherence and proliferation. The star-shaped block copolymer CAPLGA-b-TPGS was synthesized through the core-first approach, and DTX-loaded CA-PLGA-b-TPGS nanoparticles were prepared by a modified nanoprecipitation method (Fig. 7).

Cholic acid-functionalized star-shaped PLGA-b-TPGS (CA-PLGA-b-TPGS) NPs with small size (100 nm) were compared with PLGA NPs and the linear PLGA-b-TPGS NPs. These star-shaped NPs showed higher drug-loading concentration (10.08% vs. 8.34 and 8.75 for PLGA and PLGA-b-TPGS, respectively), drug encapsulation efficiency (97.85% vs. 78.36 and 85.91 for PLGA and PLGA-b-TPGS respectively), and also cellular uptake and cytotoxicity on HeLa cells. Altogether, the obtained results from in vitro and in vivo experiments indicated that the DTX-loaded CA-PLGA-b-TPGS NPs exhibit better antitumor effect than the linear PLGA-b-TPGS NP formulation and also the clinical DTX formulation (Taxotere®) [35].

A noticeable point here is a method for synthesis polymeric NPs. The most reported method for preparation of polymeric NPs, including TPGS-containing NPs, is the double emulsion or nanoprecipitation process. Each method has some drawbacks. For example, the disadvantages of double emulsion process are its several stages, low production rate, and large size of nanoparticles (>200 nm). Another method is nanoprecipitation; although it can produce particles with appropriate size (<200 nm) and high EE % (~80%), there are some problems such as the need of high concentrations of emulsifier (~5%). One method that was reported in 2013 is the inverse-phase nanoprecipitation method that was used for the preparation of TPGS-functionalized PLGA nanoparticles (TPNs). This method leads to the development of homogeneous particle size (in the range of 130–150 nm) and high EE % (>80%) for delivery of paclitaxel as a model drug for cancer treatment. Stages of NP preparation are shown in Fig. 8. The inverse-phase nanoprecipitation process consists of two steps (represented with I and II in the picture). In this method, the water phase is added slowly into the oil phase. At the first stage, the volume of the organic phase is larger than the aqueous phase. In this stage, soft nanoparticles are formed slowly, and the bigger nanoparticles would re-self-assemble into smaller size and homogenous nanoparticles by the external shear force. In the second stage, the volume of water is bigger than the oil phase, and the solidifying of the presolidified droplets is quickened. Finally, when the aqueous phase and organic phase volume rate reached five times, the nanoparticles with a stable uniform structure will be formed [36].

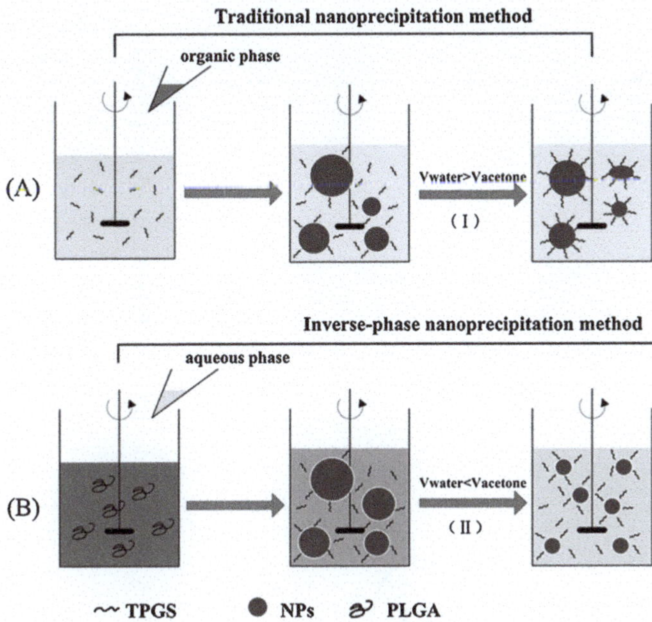

Fig. 8 The schematic diagram of nanoparticle forming process by (**a**) the traditional nanoprecipitation method and (**b**) the modified nanoprecipitation technique [36] (Reproduced with the permission of Elsevier)

Another application of NPs is their ability for co-delivery of drugs. Co-delivery of multiple drugs is a strategy that has been used for treatment of multidrug resistance. One of the examples was published in 2014. In this paper, porous PLGA nanoparticles were utilized for co-delivery of docetaxel (DTX) as an anticancer drug together with vitamin E TPGS. TPGS in this formulation plays a dual performance, that is, pore-forming agent in the nanoparticles and also as a bioactive agent with the potential to P-gp inhibition to overcome multidrug resistance. One of the reasons for difficulty in MDR to cancer cells is attributed to its complicated process; MDR to cancer cells with different mechanisms and at different stages requires different treatment concentration and drug exposure time. In this report, the DTX-loaded PLGA NPs with different amount of TPGS (0, 10, 20, and 40%) were synthesized by the nanoprecipitation method and investigated for their performance in drug delivery (Fig. 9).

The size of prepared nanoparticles with different amount of TPGS ranges from 100 to 120 nm with EE % ranging from 85 to 95% and drug loading around 10%. Results showed that DTX-loaded PLGA NPs with 10 or 20% TPGS were more (11.85- and 52.7-fold) effective than the DTX-loaded PLGA NPs without any TPGS. These results indicate advantages of co-delivery of DTX with TPGS by PLGA NPs [37].

Co-delivery of paclitaxel, as a chemotherapeutic agent, and tariquidar (TQR) as a drug resistance inhibitor in the TPGS nanoparticles to form TPGS/PTX/TQR nanoparticles was reported in 2016 (Fig. 10). These dual-drug-loaded nanoparticles

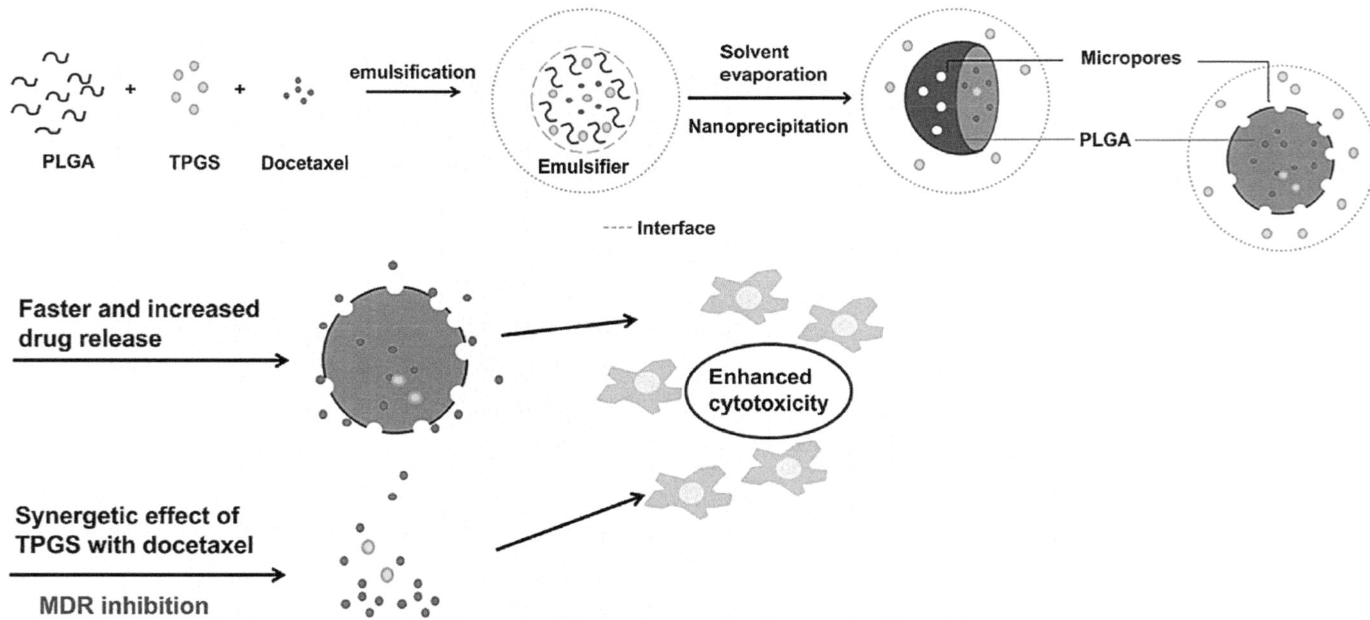

Fig. 9 Schematic graph of TPGS functions as a pore former and promotes antitumor activity of DTX-loaded NPs [37] (Reproduced with the permission of Elsevier)

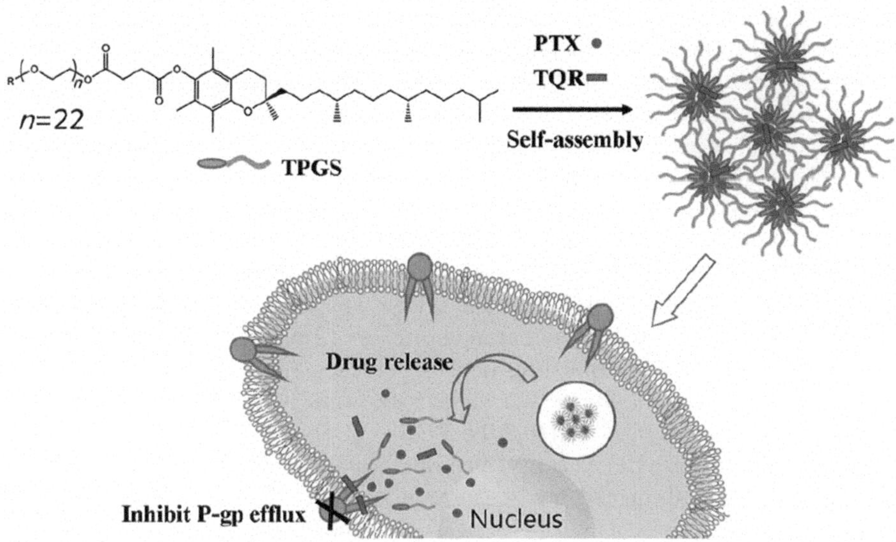

Fig. 10 Preparation of TPGS/PTX/TQR nanoparticles and schematic illustration of the drug delivery process [28] (Reproduced with the permission of Elsevier)

prepared by facile self-assembly method were with the appropriate size (below 150 nm) and high encapsulation efficiency (~80%) and used for the treatment of drug-resistant cancer cells (MCF-7/ADR). TQR, an anthranilamide-based compound, was used because this drug shows high potential to inhibit P-gp. A noticeable point in this study was the evaluation of the immunostimulatory effect. For this purpose, the effect of TPGS/PTX/TQR nanoparticles as a drug delivery system on the production of IL-10 was investigated. IL-10 is an anti-inflammatory cytokine, in the drug-resistant cells. The obtained results showed that TPGS/PTX/TQR NPs in comparison with TPGS/PTX NPs, mono-drug loaded nanoparticles, have significantly enhanced cellular inhibition effect. After the treatment by the dual-drug delivery systems, both P-gp expression and the concentration of IL-10 were decreased in drug-resistant cells. This study showed that the TPGS-based multiple drug delivery system has potential in tumor chemotherapy [28].

It has been shown that incorporation of a drug interactive domain such as 9-fluorenylmethoxycarbonyl (α-Fmoc) into a lipid surfactant can improve drug-loading capacity and formulation stability. This improved drug/carrier interaction was attributed to some interactions (hydrogen bonding, hydrophobic-hydrophobic interaction, and π–π stacking) between drug and Fmoc-containing carrier. Therefore, in a study in 2014, Fmoc-conjugated PEG-vitamin E2 micelles (PEG_{5K}-FVE2) were synthesized using lysine as a scaffold and used for delivery of paclitaxel. Evidence showed that PEG_{5K}-FVE2 micelles with small sizes (~60 nm) have higher PTX loading capacity than that of PEG_{5K}-VE2 micelles. Based on the experiments, PEG_{5K}-FVE2:PTX with ratio 1:1 leads to the preparation of micelles with 61.60 ± 3.55 nm size, LC = 11.6%, and EE = 96%. Investigation of toxicity was evaluated on breast cancer (4T1.2) cell line. Results from in vitro studies showed that PTX-loaded PEG_{5K}-FVE2 micelles have more cytotoxicity potential than free PTX or Taxol formulation. Therefore, these results indicated that incorporation of Fmoc as a drug-interactive motif into PEG5K-VE2 micelles resulted in improvement in the micelle formulation for the delivery of PTX [38].

In a published paper in 2016, modification of linear PEG chains with different molar ratios of vitamin E succinate (VES) has been reported. One of the drawbacks of esterification of vitamin E succinate with PEG_{1000} for preparation of TPGS is its high critical micelle concentration (CMC) of 0.2 mg/mL. Also, because of its short PEG chains, the circulation time is not long enough to avoid the accumulation of these micelles in organs such as liver and spleen. Therefore, in this study, the PEG-vitamin E amphiphilic with different molar ratios of PEG and vitamin E was prepared, and their stability and drug delivery efficacy were examined. Then, the developed micelles with an optimal molar ratio were used for drug loading and further investigations. This amphiphilic PEG_{5K}-VES6 (PAMV6) with wheat-like structure and small size (98.42 ± 1.38 nm) was used for encapsulation of pirarubicin (THP) as a model drug and breast cancer treatment (Fig. 11). These prepared PAMV6 micelles showed a LC and EE of 24.81% and 82.37%, respectively. In vitro and In vivo analysis on G2/M arrest in 4T1 cells showed that PAMV6/THP micelles had more efficiency and tumor growth inhibition than THP solution. The lowest critical micelle concentration (CMC), high stability in serum, and also highest tumor distribution and retention are other advantages of these micelles [39].

In the same year, these reported nanomicelles (PAMV6) with one-step pirarubicin (THP) loading were used for MDR breast cancer treatment. Based on this study, PAMV6/THP micelles were able to avoid P-gp-mediated efflux. Mechanistic studies showed that blank PAMV6 micelles inhibited P-gp activity by reducing mitochondrial membrane potential and decreasing intracellular ATP levels. Further studies showed that PAMV6/THP led to much slower tumor growth in comparison with free THP. These results represented that encapsulating THP as a drug into PAMV6/THP nanomicelles can overcome MDR in MCF-7/ADR cells [40].

6 The pH-Sensitive Vitamin E-Based Micelles/Nanoparticles

One of the exciting features of some drug delivery systems is their pH sensitivity. This feature is essential because of the acidic environment of cancer cells. A well-known anticancer drug, doxorubicin, was used with a carrier with pH sensitivity. In 2014, α-TOS containing diblock copolymer micellar system was prepared by conjugation of α-TOS molecules and mPEG-b-PHEMA hydrophilic diblock copolymer to delivery of DOX for colon cancer therapy (Fig. 12). The DOX loading process was done by the solvent exchange process. The size of DOX-loaded micelles was 116.3 ± 0.2 nm with DOX content of 2.2 ± 0.72 wt %. A noticeable property of this micelle is its pH sensitivity. In acidic pH, this DOX-loaded micelle deformed and released the drug. Therefore, these micelles showed cytotoxicity for human HCT116 colon cancer cells and high antitumor activity [41].

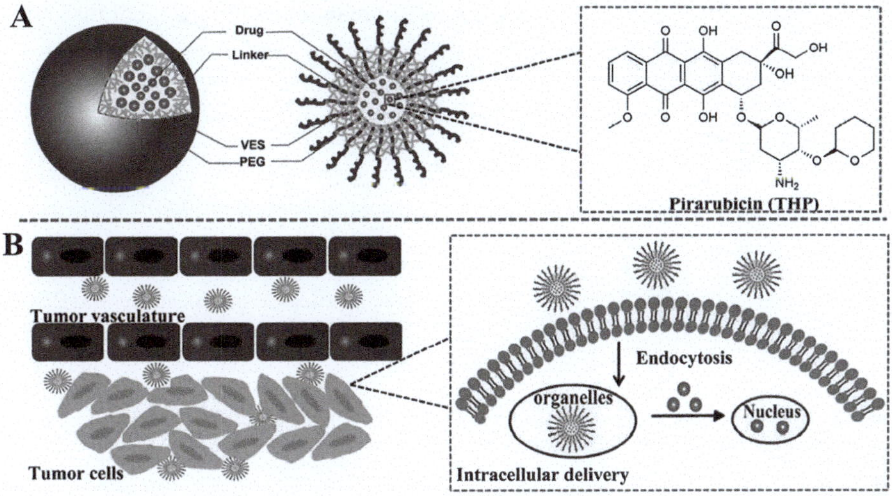

Fig. 11 Schematic diagram of $PAMV_6$ micelles for the tumor cell-specific delivery of THP. (**a**) Graphical elucidation of the $PAMV_6$/THP micelles; (**b**) graphical elucidation of tumor cell targeting delivery of $PAMV_6$/THP micelles [39] (Reproduced with the permission of ACS publications)

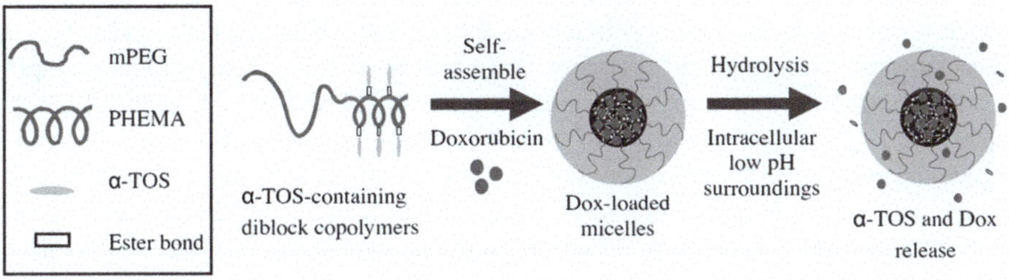

Fig. 12 Schematic diagram of DOX-loaded mPEG-b-P (HEMA-g-α-TOS) polymeric micelles and their drug (DOX and a-TOS) release at low pH surroundings [41] (Reproduced with the permission of Elsevier)

DOX-loaded pluronic-α-tocopheryl succinate polymeric micelles (P/TOS-DOX) are another reported polymeric micelle delivery system that was reported in 2015 to increase the therapeutic efficacy of DOX and reduce side effects. This micelle was synthesized with chemical conjugation of the amine group of pluronic123 with the carboxylic group of TOS and used to increase the intracellular concentration of DOX in the liver cancer cells (HepG2 hepatic cancer cell). These nanosized micelles were pH dependent and showed sustained release profile over 50 h without initial burst release of the drug. According to obtained results, higher amount of DOX released was achieved at lower pH (Fig. 13). The reason for this observation was related to the basic nature of DOX (pKa = 8.3) and its tendency to solubilize at lower pH [42].

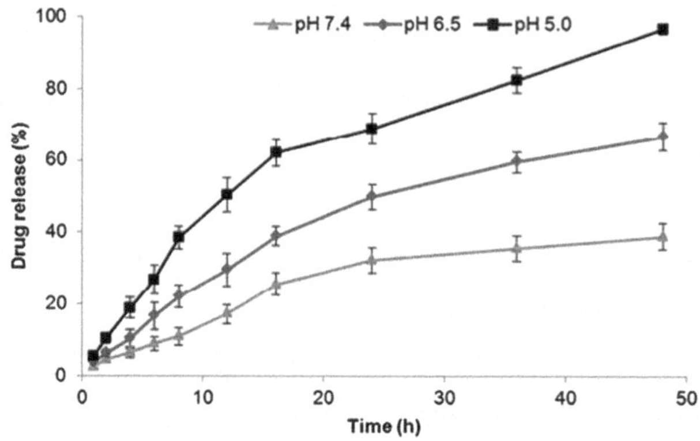

Fig. 13 In vitro release profile of DOX from P/TOS-DOX micelles [42] (Reproduced with the permission of Royal Society of Chemistry)

In the literature published in 2016, pH-sensitive nanoparticles composed of TPGS, chitosan (CA), polyacrylic acid (PAA), and PLGA were synthesized for delivering etoposide (VP-16) as an anticancer drug to the nucleus of MDR cells. The drug encapsulation efficiency of CS/PAA/VP-16 NPs and CS/PAA/VP-16@ TPGS/PLGA NPs was 88.9% and 91.7%, respectively. Because free VP-16 is a P-gp substrate, much of VP-16 undergoes P-gp-mediated efflux. Therefore, there is a need for preparation of support to deliver this drug to MDR cells. The pH value of tumor microenvironment differs from that of normal cells. The extracellular pH value of normal and most of solid tumor tissues is 7.4 and 6–7, respectively. The intracellular pH values are also different (in endosomes pH = 5.0 to 6.0, in lysosomes pH = 4.0 to 5.0, and in the nucleus pH = 7.4). This point was used in this report, and intracellular pH gradients were used to achieve a pH-sensitive release system. Therefore, chitosan-poly(acrylic acid) nanoparticles (CS/PAA NPs) as a type of pH-sensitive nanoparticles are a good candidate for the preparation of these nano-drug delivery systems (NDDS). In this report, NPs ((CS/PAA/VP-16@TPGS/ PLGA NPs) with S@L designation were prepared. Smaller NPs (CS/PAANPs containing the drug etoposide) were loaded on the larger nanoparticles (TPGS/PLGA NPs) to the preparation of final NDDS. Drug-loaded CS/PAA/VP-16@TPGS/ PLGA NPs were nearly spherical with the size of approximately 150 nm in radius. Therefore, they were small enough and can be internalized by endocytosis and were also large enough to be retained in blood circulation for a long time. CS/PAA/VP-16 NPs were released in the acid environment of cancer cells. Then, these NPs release the contained anticancer drug in the alkaline environment of the nucleus. The mechanism of its performance is shown in Fig. 14. As shown in this picture, the prepared CS/PAA/VP-16@TPGS/PLGA NPs were internalized into cancer cells via endocytosis. These nanoparticles are transported in vesicles from early endosomes to late endosomes and ultimately to lysosomes. CS/PAA/VP-16@TPGS/PLGA NPs or TPGS/PLGA NPs induced autophagy. The autophagosomes, by the ability to enwrap the nanoparticles, delivered the nanoparticles to lysosomes for degradation process. The small nanoparticles (CS/PAA/VP-16 NPs) released occurred in the

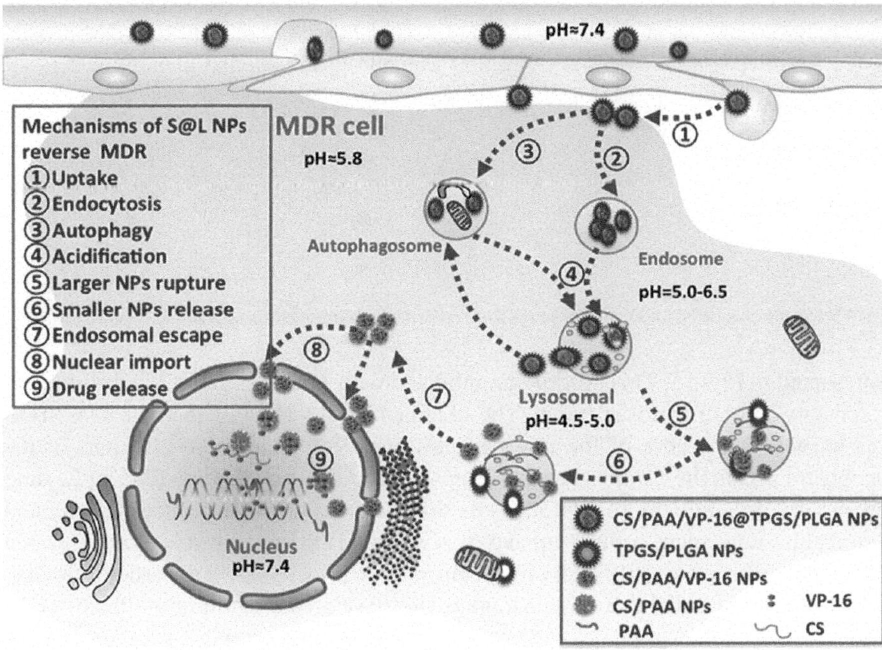

Fig. 14 Illustration of the mechanism by which CS/PAA/VP-16@TPGS/PLGA NPs reverse MDR [17] (Reproduced with the permission of Elsevier)

acidic environment of the cytoplasm or the lysosomes and are accumulated in the nucleus by transit through the nucleopores. This is the first pH-sensitive release stage, and the second pH-sensitive phase occurs in the alkaline environment of the nucleus that leads to VP-16 release. These results showed that S@L NPs could enhance the drug delivery to the nucleus and also increase the cytotoxicity (by test on A549/DDP cells) of anticancer drugs on MDR cells [17].

7 Vitamin E Conjugation to Drugs (Prodrugs)

A promising approach to increase the antitumor efficacy of anticancer drugs and reduce their toxicity is the preparation of self-assembled prodrugs forming nanoaggregates. The major strategy for drug modifications is polymer–drug conjugation. This strategy manipulates therapeutic agents (drugs) at the molecular level and leads to increase solubility, permeability, and stability of drugs. The main assumption in this method is a modification of drug's molecular structure and to make their analogous agents with distinct chemical properties from the original form of drugs and produce similar or even better biological effect [43, 44].

One reported paper in 2014 used this strategy for preparing a prodrug with DOX. In this study, doxorubicin was chemically conjugated to TOS, through an amide bond, for developing N-doxorubicin-α-D-tocopherol succinate (N-DOX–TOS) that is

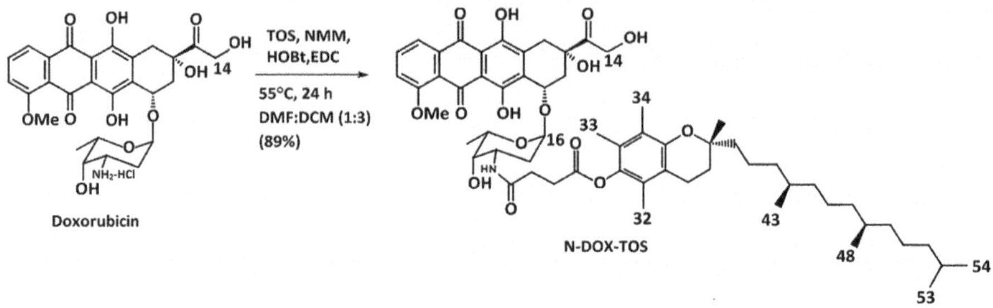

Fig. 15 Synthesis of N-DOX–TOS [44] (Reproduced with the permission of ACS publications)

represented in Fig. 15. This prodrug is stabilized with $TPGS_{2000}$ and self-assembled in water into 250 nm nanostructures. The obtained results showed that N-DOX–TOS was located in the core of the nanoparticles and PEG chains and also part of the tocopherol was in the corona. This prodrug showed high drug loading (34% w/w) and low in vitro drug release. In vitro analysis showed appropriate anticancer activity and temperature-dependent cellular uptake of N-DOX–TOS nanoparticle that suggested that active diffusion was the main mechanism of NDOX–TOS nanoparticles uptake. In vivo studies revealed that these nanoparticles have greater antitumor efficacy compared with free DOX [44].

Two camptothecin (CPT) prodrugs were reported in 2015. These two prodrugs were prepared by conjugating CPT to α-tocopherol in two ways: (1) carbonate ester bond (CPT-VE) or (2) disulfide linkage (CPT-S-S-VE). Both of them self-assembled into nanofibers and showed high stability. Results showed that during the self-assembly process, these prodrugs were successfully incorporated into PEG_{5K}-Fmoc-VE2, which was used as a nanomicellar carrier. Strong carrier/carrier and carrier/drug interactions lead to the formation of nanofiber structures (Fig. 16). In vitro analysis represented that PEG_{5K}Fmoc-VE2/CPT-S-S-VE had a higher level of cytotoxicity and tumor growth inhibition in a breast cancer model (4T1.2) compared to PEG_{5K}-Fmoc-VE2/CPT-VE. Size, EE %, and LC % of these two prodrugs are listed in Table 1. It is concluded that VE-conjugated prodrugs could represent an effective method for managing the therapeutic agents with complex structures and laborious formulation process [45].

In 2018, two kinds of prodrugs including vitamin E-based salinomycin (SAL) prodrug nanoparticles (TS NPs) and hyaluronic acid (HA)-coated TS NPs (HTS NPs) were prepared to deliver paclitaxel. These SAL prodrugs, denoted as TOS-ss-SAL (TS), were synthesized by covalent bonding between the carboxyl groups of TOS and SAL via cystamine linkages. In this synthesis, $TPGS_{1000}$ was used as a surfactant. Both prodrug NPs with small size (~200 nm) showed excellent drug-loading capacity (EE % and LC % of 92.61 ± 0.98 and 3.08 ± 0.12 for TS NPs and 87.64 ± 0.44 and 2.92 ± 0.05 for HTS NPs) and release profiles for PTX. However, the PTX-loaded HTS NPs represented more potent anticancer efficacy due to its more capability for tumor tissue permeation. Therefore, it is concluded that the TS prodrug NPs, especially the coated HTS NPs, have potential for cancer-targeted treatments [46].

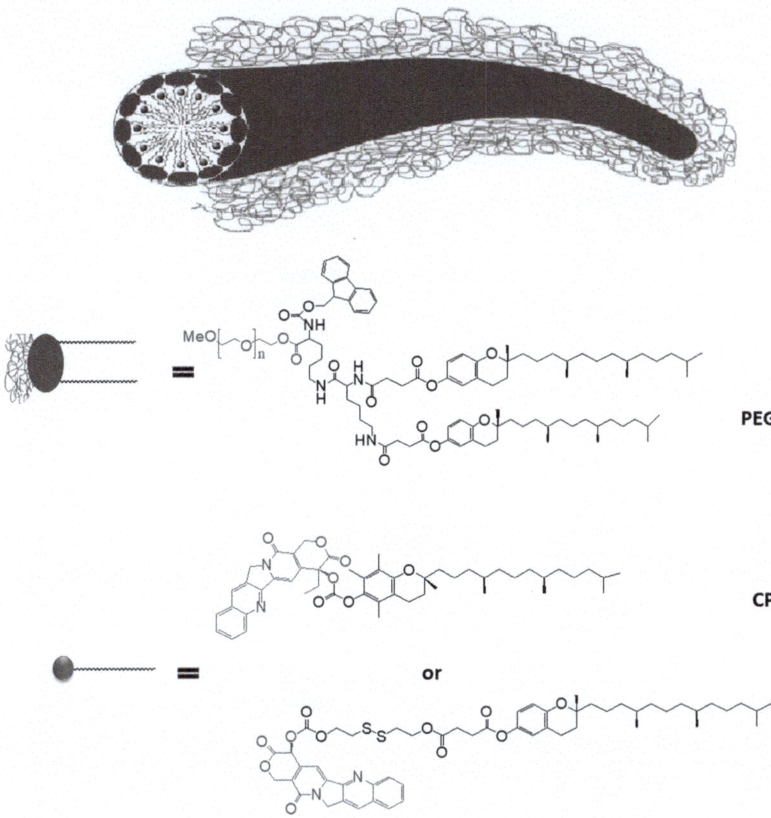

Fig. 16 Proposed mechanism for the interaction of PEG5K-Fmoc-VE2 with CPT-VE or CPT-S-S-VE and the formation of nanofiber structures [45] (Reproduced with the permission of Elsevier)

Table 1 Size and drug loading amount

Drug carrier	Drug	Size (nm)	Drug loading	Ref.
PEG$_{5K}$-Fmoc-VE$_2$/CPT-VE	Camptothecin (CPT)	87.4 ± 4.6	DLC % = 5.4 DLE % = 89.1	Lu et al. [45]
PEG$_{5K}$Fmoc-VE$_2$/ CPT-S-S-VE	Camptothecin (CPT)	49.7 ± 1.8	DLC % = 5.7 DLE % = 96.3	Lu et al. [45]

In addition to TOS prodrugs, there are some reports about TPGS application in the preparation of prodrugs for cancer treatment that is explained further.

In a study in 2008, TPGS was employed as a carrier of DOX. In this report, DOX was chemically conjugated to TPGS to form TPGS–DOX. For the preparation of TPGS–DOX, the activated hydroxyl group of TPGS (by succinic anhydride (SA)) interacted with the primary amine group of doxorubicin. Its synthetic procedure is shown in Fig. 17. The DOX content in this reported TPGS–DOX conjugate was 8.0 wt.%. MCF-7 breast cancer cells and C6 glioma cells were used as in vitro cell model. This reported system showed higher cellular uptake and also broader

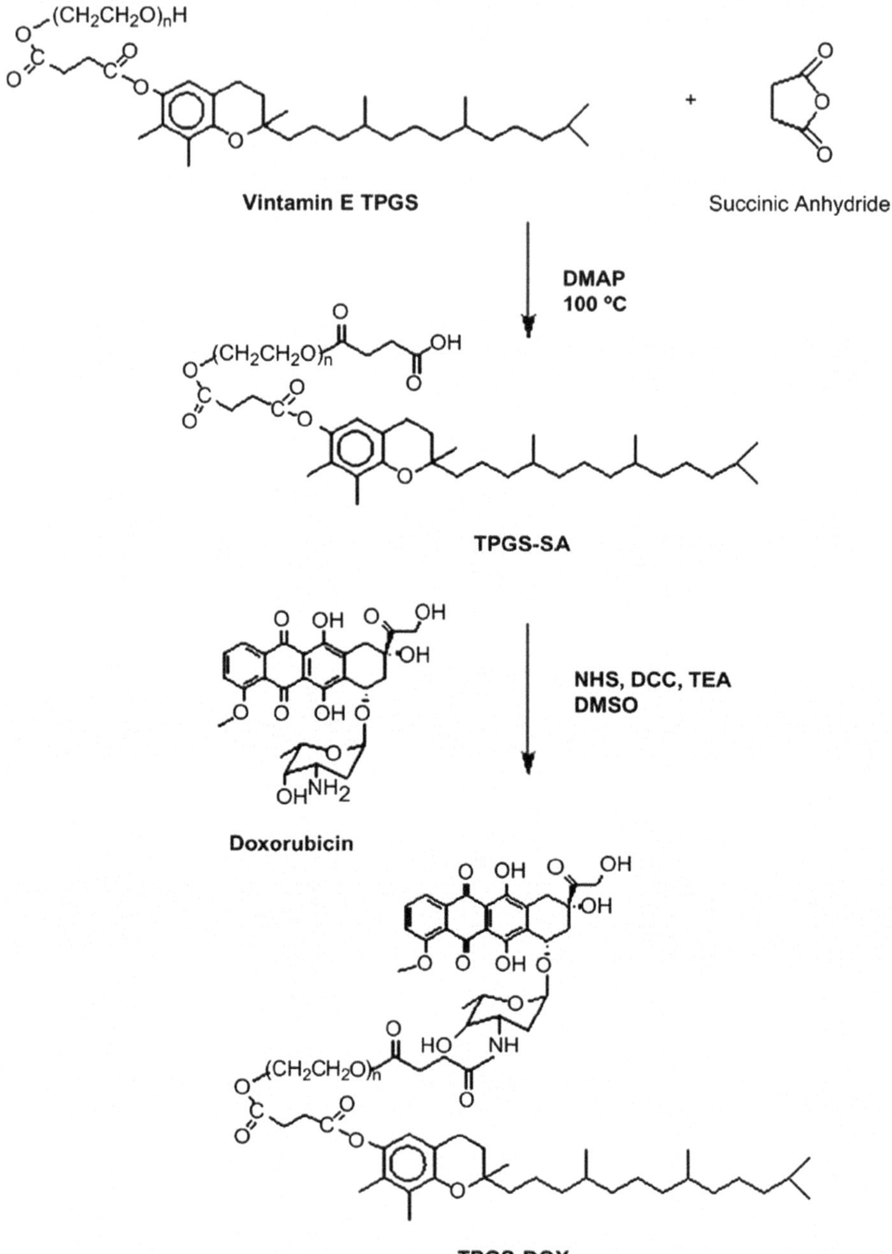

Fig. 17 Synthetic scheme of the TPGS–DOX conjugate [43] (Reproduced with the permission of Elsevier)

distribution within the cells. The drug level in some organs including the heart, gastric, and intestine was measured, and results indicated that the drug level was significantly reduced. These results showed that this TPGS–DOX conjugate had the potential to be a prodrug with higher therapeutic effects and fewer side effects in comparison with DOX itself [43].

A TPGS prodrug micelle system with cisplatin as a hydrophilic drug was reported in 2012. Some advantages, such as high drug loading of 4.95% (w/w), pH-responsive drug release kinetics, and also higher cellular uptake, were obtained in comparison with the original drug. Averaged size of these reported micelles was 12.56 ± 0.27 nm. The structure of TPGS-cisplatin prodrug is shown in Fig. 18. This reported prodrug micelle showed chemotherapy with neuroprotective effects due to increasing the IC_{50}[4] value from 1.75 μg/mL for the drug (cisplatin) to 3.02 μg/mL for the reported TPGS-cisplatin prodrug that was obtained after 24-h culture with the SH-SY5Y neuroblast cells [47].

Another TPGS-cisplatin prodrug was reported in 2013 to co-delivery of docetaxel, cisplatin, and Herceptin to multimodality treatment of breast cancer with HER2 overexpression. In this study, Herceptin-conjugated nanoparticles of TPGS-cisplatin prodrug (HTCP NPs) were synthesized. Two components including co-polymers poly(lactic acid)–TPGS (PLA–TPGS) and carboxyl group-terminated TPGS (TPGS-COOH) were added in the polymeric matrix to increase the stability of the prodrug nanoparticles and also facilitate Herceptin conjugation (Fig. 19). Nanoprecipitation method was used for the preparation of three models of HTCP NPs with high, moderate, and low docetaxel versus cisplatin ratio. Among these three models, the HTCP NPs (size of 187.4 ± 2.4 nm) with high docetaxel versus cisplatin ratio showed better efficacy than the others (drug loading for docetaxel, cisplatin, and Herceptin were 9.0 ± 0.5%, 3.5 ± 0.1%, and 73.1 ± 5.8%, respectively). These systems showed a pH-sensitive release profile for both anticancer drugs [48].

In 2014, a TPGS-based PTX prodrug (TPGS-S-S-PTX) was synthesized to overcome MDR in cancer cells (human ovarian cell line A2780 and PTX-resistant human ovarian cell line A2780/T). This prodrug self-assembled into stable micelles with a small size (diameter of ~140 nm) can disassociate in the reductive condition in cancer cells and release its active ingredients (PTX and TPGS) rapidly. In this study, the amounts of LC and EE were 30 % and 100 %, respectively. In vivo evaluation of this prodrug represents some advantages in comparison with Taxol (clinical formulation of PTX) and TPGS-C-C-PTX including an extended half-life, enhanced tumor distribution, increased AUC (area under the concentration-time curve), and reduced side effects. The main reason for the rapid release of PTX in cancer cells was attributed to disulfide bonds that are stable in human blood plasma but unstable under intracellular reductive conditions (Fig. 20). The slow release of PTX from TPGS-C-C-PTX has attributed to the presence of insensitive "–C-C–" conjugation. TPGS-S-S-PTX after endocytosis by tumor cells can be quickly cleaved and release the active drugs. TPGS can bind with P-gp and restrain its activity and therefore reduce the efflux of PTX. These reasons resulted in more activity of TPGS-S-S-PTX than Taxol and TPGS-C-C-PTX [49].

---

[4] The concentration of drug needed to kill half amount of the cells

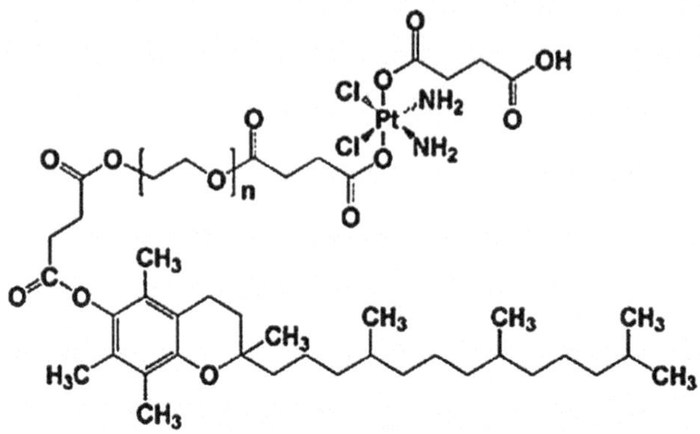

Fig. 18 Structure of TPGS-cisplatin prodrug [47] (Reproduced with the permission of Elsevier)

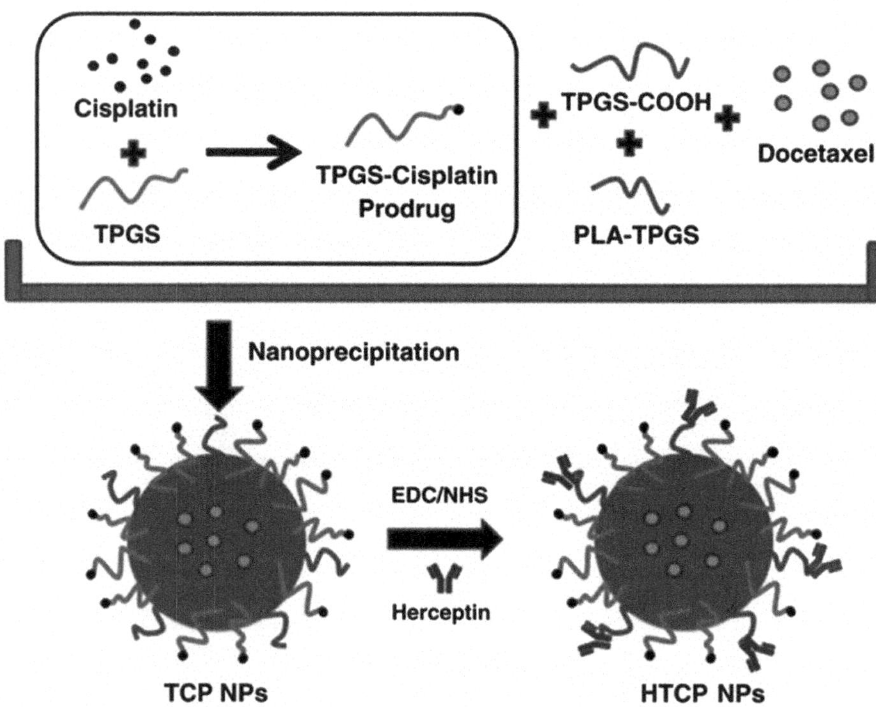

Fig. 19 Schematic illustration of the formulation of docetaxel-loaded TPGS-cisplatin prodrug nanoparticles (TCP NPs) and Herceptin-conjugated, docetaxel-loaded TPGS-cisplatin prodrug nanoparticles (HTCP NPs) [48] (Reproduced with the permission of Elsevier)

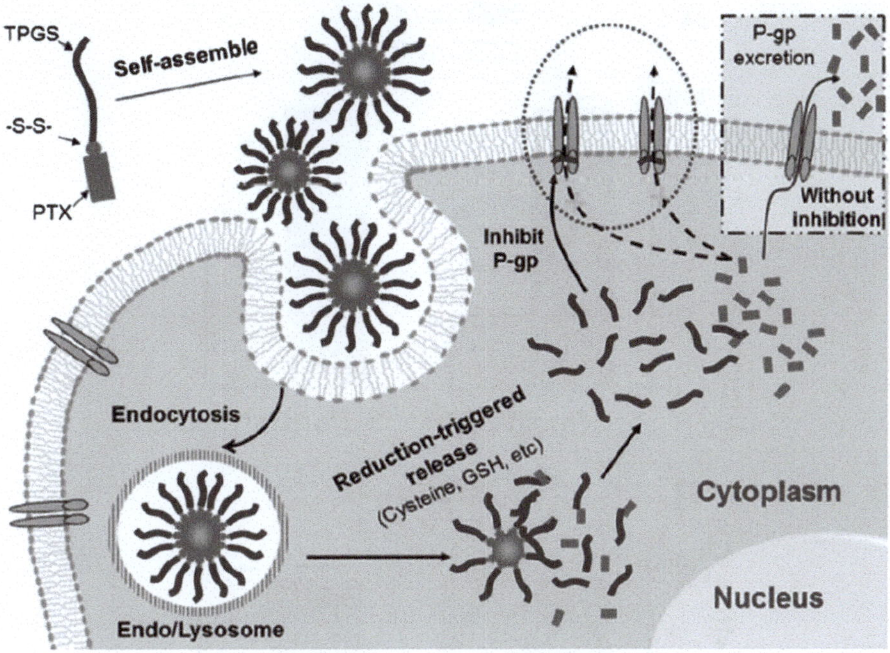

Fig. 20 Schematic illustration of redox-sensitive TPGSS-S-PTX prodrug for overcoming MDR of cancer cells [49] (Reproduced with the permission of ACS publications. Reproduced with the permission of Elsevier)

8 Vitamin E and Mixed Micelles

There are many reports about the application of mixed micelles. Micelles have much potential for antitumor drug delivery such as their solubilization properties and high biocompatibility, but they suffer from poor stability and low drug-loading capacity. The mixed micelle system was developed to solve these problems. Mixed micelles can provide a better hydrophobic microenvironment and therefore better encapsulation and stabilization of hydrophobic drugs in their inner core. This property leads to an increase in the bioavailability of drugs and circulation time. The mixed micelles can passively accumulate at the tumor site because of their small size. However, in most mixed micelle systems, low drug-loading coefficients remain a challenge. Selection of proper copolymers with compatible hydrophobic segments is a strategy for solving this problem [50]. Some reports used mixed micelles as drug delivery systems. These reports are summarized further.

PEG_{2000}-DSPE and vitamin E TPGS mixed micelles were reported in 2012. They have been used for combination loading of two drugs, including paclitaxel and parthenolide, to antitumor activity against non-small-cell lung cancer (NSCLC) cells. Drug loading stage was done by the film casting method that leads to high drug encapsulation efficiency of 95 %. TPGS was efficient in the stability of the micelles and prevents drug leakage (Fig. 21).

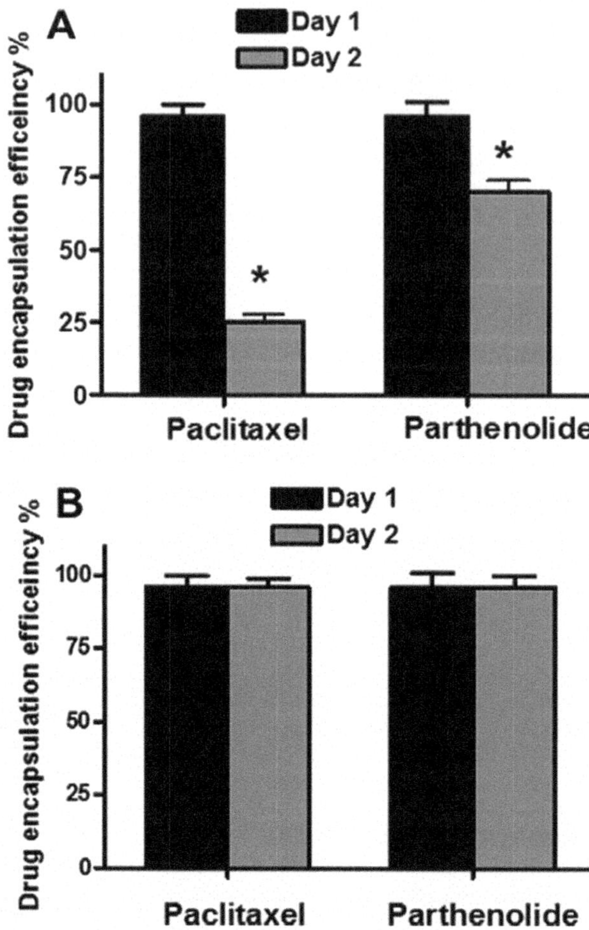

Fig. 21 Drug encapsulation efficiency of (**a**) paclitaxel and parthenolide loaded in PEG_{2000}-DSPE micelles and (**b**) paclitaxel and parthenolide loaded in PEG_{2000}-DSPE/vitamin E TPGS mixed micelles. Data presented in this figure corresponds to micelles and mixed micelles loaded with both drugs in the same formulation [51] (Reproduced with the permission of Elsevier)

The reason was explained based on the disability of PEG_{2000}-DSPE to maintain large hydrophobic and hydrophilic domains for two-drug loading. The reason for using mixed drug was solving drug resistance issue in cancer cells. This resistance is attributed to nuclear factor-κB (NF-κB). According to previous studies, the activity of this factor is responsible for cancerous cell protection against programmed cell death. Therefore, in this literature, co-administration of paclitaxel with parthenolide was studied. Parthenolide can suppress NF-κB activity by NF-κB DNA binding inhibition and also prevention of the degradation of IκB-α and IκB-β (inhibitor of nuclear factor-kB). The obtained results showed that this reported mixed micelles caused 79% cell death in Taxol-sensitive cell lines that were higher in comparison with 46 % cell death caused by the drugs in solution. This mixed micelle was also

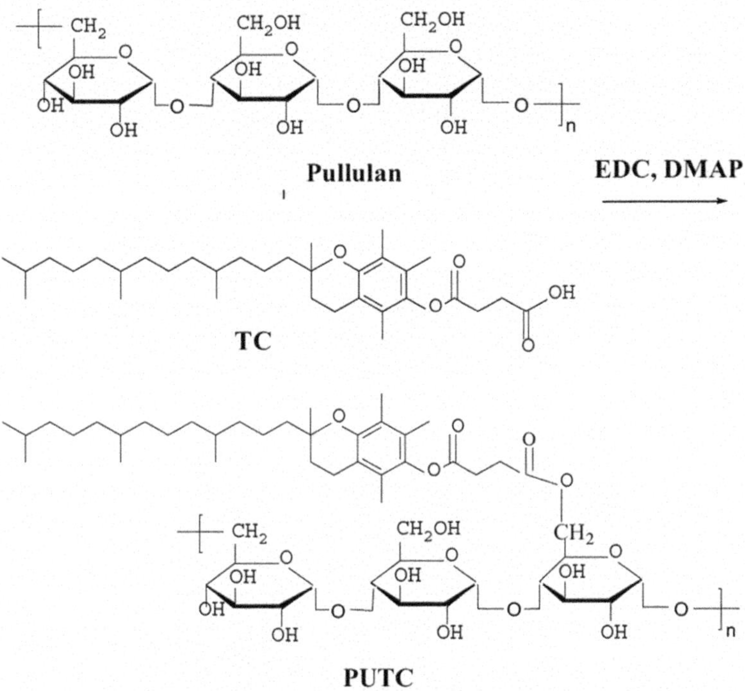

Fig. 22 Synthesis route of α-tocopherol succinate-modified pullulan polymers [53] (Reproduced with the permission of Elsevier)

examined on Taxol-resistant cell lines. The cell death induced by mixed micelles and un-encapsulated drugs was 70 % and 45 %, respectively. Therefore, co-encapsulation of paclitaxel and parthenolide in mixed micelles improve the anticancer activity of paclitaxel for treating lung cancer cells [51].

Based on the benefits of TOS and TPGS in a paper published in 2014, micelles consisting of a mixture of TOS and $TPGS_{2000}$ (1:1) were synthesized. In this paper, TOS was used in solubilization of DOX, and $TPGS_{2000}$ was used due to its amphiphilic character, therefore able to stabilize the molecular structuration of the particles. DOX-loaded TOS–TPGS micelles with nanosized structure (78 nm) and high drug loading (40 % w/w) were used for antitumor activity investigation on two tumor models (CT26 and MCF-7). A noticeable point of this study was the similarity of the survival time of mice treated with unloaded TOS–TPGS micelles and with free DOX. This result indicates the anticancer activity of vitamin E derivatives [52].

In the same year, α-tocopheryl succinate[5] conjugated linear homopolysaccharide, pullulan, for the preparation of three α-tocopherol pullulan polymers PUTC1, PUTC2, and PUTC3 was reported (Fig. 22). In the reported process for the synthesis of these micelles, EDC (1-ethyl-3-(3-dimethylaminopropyl) carbodiimide hydrochloride) and DMAP (4-dimethylamioprydine) were used as catalysts of the system. These

[5] TC was used in this study as an abbreviation for α-tocopheryl succinate.

polymers are different in substitution degrees of TC on the polysaccharide backbone. Therefore, because of the higher substitution degree of TC, PUTC3 is more hydrophobic than other mentioned polymers. Dialysis method was used to self-assemble these polymers into nanomicelles. 10-Hydroxycamptothecin (HCPT) as an anticancer drug was loaded into the micelles. These HCPT-loaded PUTC nanomicelles (HCPT–PUTC) with desirable particle size (171.5–257.8 nm) showed high encapsulation efficiency (98.3%) with sustained release of the drug (HCPT) over 80 h. The drug release rate was accelerated by low pH condition. The analysis illustrated that PUTC micelles could rapidly transport drug into the cell nuclei. In vitro antitumor activity studies were done on the MCF-7 cell line. The obtained results represented that HCPT-loaded PUTC2 was more cytotoxic than free drug (HCPT). The reason was attributed to the higher uptake of HCPT–PUTC2 than the free drug. Therefore, these results showed that PUTC micelles have the potential for hydrophobic antitumor drug delivery [53].

Oleanolic acid (OA) is another drug that was loaded on the mixed micelles constructed from TPGS and P105 in 2016. This drug with potential antitumor activity is a natural product which can be found in various fruits and vegetables. This drug, like most of the anticancer drugs, suffers from low water solubility and permeability. Therefore, nanodrug delivery systems are suitable for its performance improvement. In this report, OA-loaded mixed micelles were synthesized by single-step solvent evaporation method that led to the preparation of micelles with small size (95.7 ± 3.6 nm in aqueous phase) and high drug-loading encapsulation efficiency (93.6% ± 0.05%). The amount of drug loading was 3.5%. During the solvent evaporation process, drug (OA) and TPGS/P105 self-assembled into micelles and the drug were encapsulated in the core of micelles. The drug in this study was used to non-small-cell lung cancer (NSCLC) treatment. This is the most common lung cancer type. TPGS was chosen because it is an efflux inhibitor with the amphiphilic property. The P105, a PEO37–PPO56–PEO37 triblock copolymer, was used because of its ability to sensitize multidrug-resistant tumors and encapsulate hydrophobic drugs. The obtained results showed that OA was released slower from the mixed micelles than that from the free drug system. In vitro and in vivo analysis was done by examination of these OA-loaded mixed micelles on the NSCLC cell lines (A549 and PC-9). The obtained results indicated that OA micelles had higher antitumor efficacy than free OA in both cell lines which enhanced the OA sensitivity of NSCLC and significant inhibitory effect on the efflux. Therefore, these results revealed that this OA micelle could provide an effective method for the lung cancer treatment [54].

Vitamin E succinate (VES)-grafted chitosan oligosaccharide (CSO)/RGD-conjugated TPGS (TPGS-RGD) mixed micelles (VeC/T-RGD MM) were prepared and loaded with PTX (PTX@VeC/T-RGD MM) to U87MG tumor therapy. PTX@VeC/T-RGD MM with small size (150.2 nm) and 5.92% drug-loading coefficient was prepared by organic solvent emulsification-evaporation method. In this study, chitosan oligosaccharide (CSO) was used instead of chitosan because chitosan in the pH values above its pKa (pH 6.4) is insoluble, and its micelles would precipitate. Therefore, CSO with better solubility and many reaction sites for further modifications was chosen. Another component of this system is VES, which is a hydrophobic vitamin E analog. This compound is an excellent solvent for hydrophobic drugs. VES could be interconnected on the CSO backbone and form an amphiphilic copolymer (VES-g-CSO). This copolymer can self-assemble to form micelles. TPGS is another component of this system that was used to improve drug solubilization and cellular uptake enhancement of drugs and enhance colloidal stability (by reduction of the surface charge related to its negative charge). In this study, arginine-glycine-aspartic acid (RGD) peptide-conjugated TPGS (TPGS-RGD) was prepared and used as a component of mixed micelles. The reason for this choice was to decrease the toxicity of healthy cells and, therefore, increase the antitumor efficiency. The PTX@VeC/T-RGD MM indicated a superior therapeutic effect. This system showed a sustained-release profile with high cytotoxicity for human glioma U87MG cells. The reported tumor inhibitory rate in U87MG tumor spheroids and U87MG tumor-bearing mice was 49.3% and 88.4%, respectively. These results demonstrated that VeC/T-RGD MM is an ideal system for drug delivery [50].

Another report from the application of chitosan and TPGS was published in 2018 for colon cancer therapy. Bufalin-loaded vitamin E succinate-grafted-chitosan oligosaccharide/RGD conjugated TPGS mixed micelles (BU@VeC/T-RGD MM) were prepared with small size (140.3 nm) and EE % 2.24 and used to enhance the drug efficacy for colon cancer therapy. This amount of EE was higher than that of BU@VeC micelles. The possible reason was reported based on the solubility enhancement of TPGS. This multifunctional delivery system is constructed from vitamin E succinate-grafted-chitosan oligosaccharide (VES-CSO) and also cyclic arginine-glycine-aspartic acid (RGD) peptide-modified TPGS that was prepared by emulsion-solvent evaporation method. These bufalin-loaded micelles were tested on colon cancer LoVo/ADR and HCT116/LOHP cells. The BU@VeC/T-RGD MM exhibited good stability and sustained release property with the ability to enhance apoptosis rate and also P-gp efflux inhibition. Obtained results revealed that not only BU@VeC/T-RGD MM but also empty VeC/T-RGD MM was able to decrease the expression of P-gp in colon cancer cells. These results showed that BU-loaded VeC/T-RGD MM has the appropriate potential for antitumor activity against drug-resistant colon cancer cells [18].

9 Vitamin E and Target Drug Delivery

Anticancer drug delivery to the target site attracts many attentions because nontargeted drug delivery systems release the loaded drug directly into the plasma, therefore causing many side effects. There are many approaches to solve this problem, such as ligand-mediated targeting using carbohydrates, antibodies, peptides, and folic acid (FA). Cancer cells overexpress FA receptors; therefore, FA can act as an active targeting ligand.

One of the articles about the application of FA for targeted delivery of α-Tos is published in 2014. In this study, generation 5 poly(amidoamine) (PAMAM) dendrimers (with small size, 5.4 nm) were used as a platform for the preparation of multifunctional G5.NHAcn-α-Tos-FI-FA conjugate (Fig. 23). PAMAM can be used as a drug delivery system. This dendrimer can act as a platform in two ways: (1) physical encapsulation of hydrophobic drugs (within the highly branched internal cavity) or (2) a platform to covalently conjugate drugs onto its surface. In this report, PAMAM dendrimers were used for covalent conjugation of α-Tos (as a drug), FA (as targeting

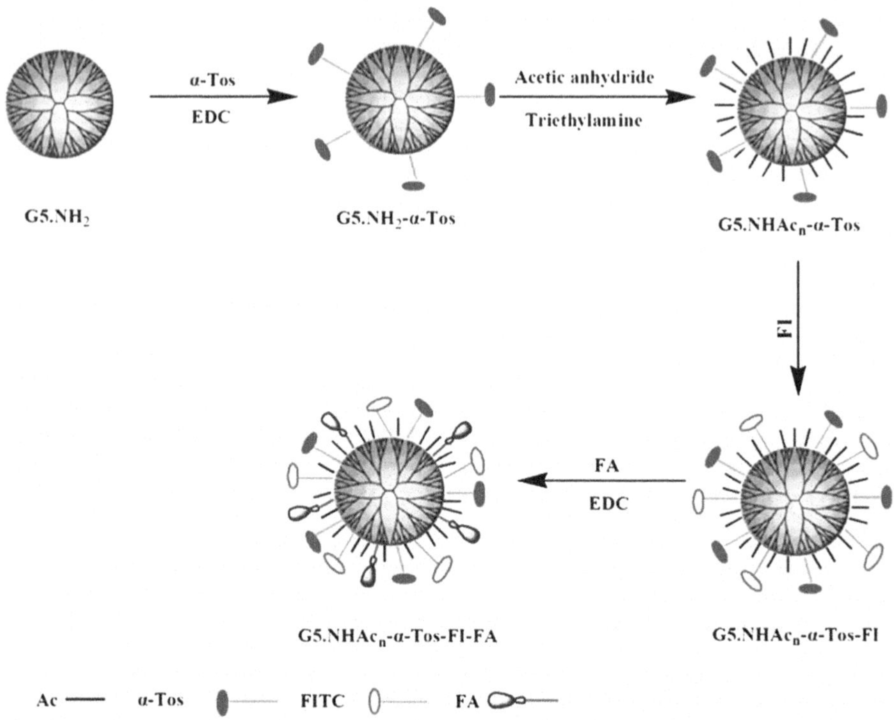

Fig. 23 Schematic illustration of the synthesis of G5.NHAcn-α-Tos-FI-FA conjugate [55] (Reproduced with the permission of Royal Society of Chemistry)

ligand), and FI (as an imaging dye). To test the therapeutic efficacy of G5.NHAcn-α-Tos-FI-FA conjugate, KB cells (human epithelial carcinoma cell) were used. The obtained results showed that this system had better therapeutic efficacy than the free α-Tos. Furthermre, this system showed target delivery of the drug to cancer cells [55].

Another study about target drug delivery is reported in 2015. In this study, folate-functionalized poloxamer 407-TPGS-mixed micelles were used for targeted delivery of DOX. In this study, P407 conjugated with folic acid was used for folate-mediated receptor targeting to cancer cells. The thin-film hydration method was used for P407-TPGS/FA-P407-TPGS-mixed micelle synthesis. The encapsulation efficiency of FA-P407-TPGS micelles was 73%. These micelles with small size (<200 nm) show potential for target drug delivery by minimal representation toxicity to healthy human cells and increased DOX–DNA binding in SKOV3 human ovarian cancer cells [56].

Target therapy is essential for brain cancer treatment because of two barriers, including the blood–brain barrier (BBB) and blood–tumor barrier (BTB). The first one limits the drug penetration of molecule into the brain cells, and the other one prevents drug uptake and accumulation. These reasons result in reduced drug accumulation in brain tumor cells. In the literature published in 2015, transferrin (β-1 glycopeptide) receptor-targeted vitamin E TPGS micelles were used to solve these problems. The expression of this receptor on cancer cells is up to 100-fold higher than the healthy cells. Thereby, transferrin has the potential to act as a targeting

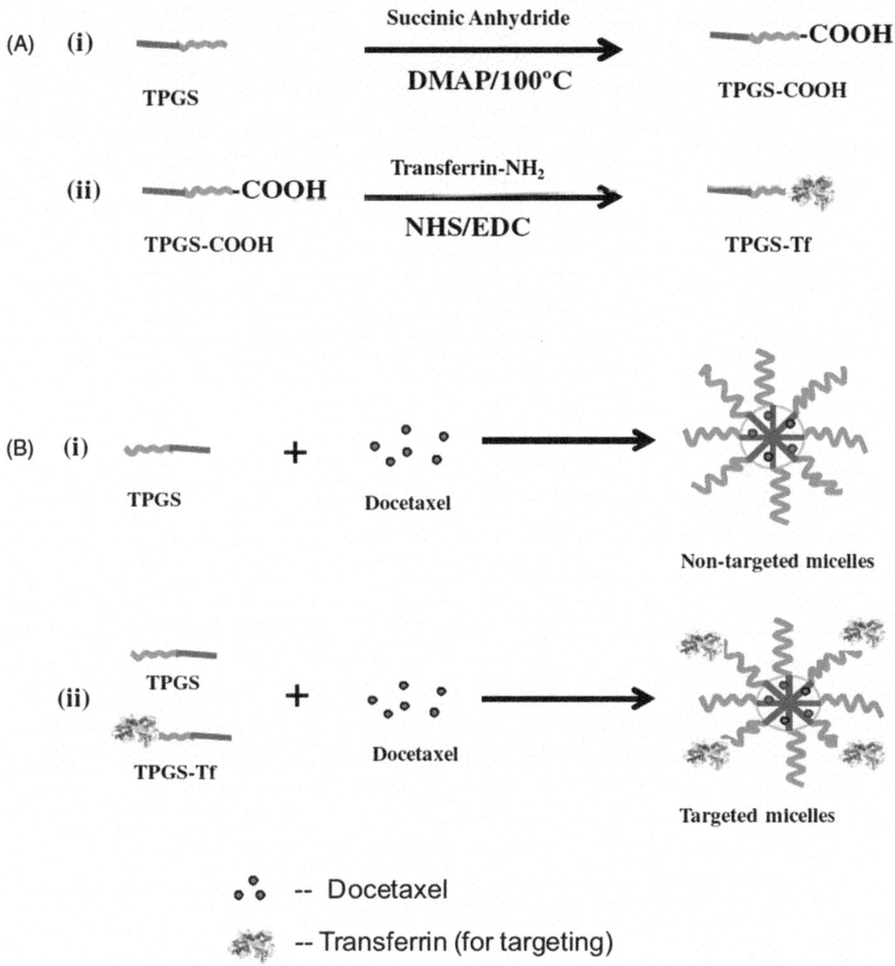

Fig. 24 Schematic diagram for (**a**) synthesis of (i) TPGS-COOH and (ii) TPGS-Tf and (**b**) preparation of (i) nontargeted (DTX-TPGS-150) and (ii) transferrin receptor-targeted micelles (DTX-TPGS-Tf). DTX-TPGS-150, DTX-loaded TPGS micelles; DTX-TPGS-Tf, transferrin-conjugated DTX-loaded TPGS micelles [57] (Reproduced with the permission of Taylor & Francis)

ligand for nanocarriers to therapeutic agent delivery into cancer cells. The preparation of transferrin receptor-targeted micelles (DTX-TPGS-Tf) was done in three steps. At the first step, acid functionalized TPGS (TPGS-COOH) was synthesized, and then it was conjugated covalently to the amino group of transferrin form TPGS-Tf. Finally, conjugated (TPGS-Tf) was assembled on the surface of the micelles. DTX molecules as a model drug were encapsulated in the core of the micelles by solvent casting (SC) method (Fig. 24).

These micelles with 520 nm size and ~85% of drug encapsulation efficiency were investigated for target drug delivery. In vitro analysis showed the sustained release of DOX and showed micelles with transferrin coating led to slower drug release profile. This effect was increased with increasing TPGS-Tf concentration.

Also, in vivo analysis was done to confirm the brain-targeting potential of DTX-TPGS-Tf. Because of the presence of transferrin in DTX-TPGS-Tf, results showed higher drug concentration in brain tissue in comparison with nontargeting micelles and the marketed formulation Docel™ [57].

Plumbagin (PLB) (5-hydroxy-2-methylnaphthalene-1,4-dione) is an anticancer drug and has been used for treatment of various cancers such as leukemia, ovarian, melanoma, and breast cancer; despite these therapeutic effects, this drug has short biological half-life (4 h) and suffers from low solubility that leads to its low oral bioavailability (39%). Therefore, there is a need to use carriers. Among various reported carriers, TPGS as an effective solubilizer, surfactant, and absorption enhancer is an appropriate candidate. In a literature published in 2016, $TPGS_{1k}$-folic acid conjugated nanomicelles (PTFM) were used for controlled and targeted delivery of PLB. FA in this study was applied to the maximum distribution of the drug in cancer cells. As mentioned previously, in most types of human cancer cells, folate receptors are overexpressed. Some benefits of this reported micelle (PTFM) was their small size (128.4 ± 0.8), good encapsulation efficiency (67.24% ± 2.22), sustained release efficiency, and enhanced biocompatibility. The in vitro drug release profile of PLB, PLB-loaded TPGS without folic acid conjugation (PTM), and PTFM are shown in Fig. 25. The obtained results showed that PTM and PTFM revealed the sustained release of the drug in 24 h. In PTM, hydrophilic part of TPGS is responsible for the water uptake and therefore acceleration of PLB diffusion through water channels. Thus, because of the presence of larger hydrophilic part of folic acid PTFM, this micelle showed the faster release of PLB than PTM. The obtained results showed that these developed micelles have some therapeutic effects and can reduce dose and cost of drugs [22].

Breast cancer classification is based on the receptor (hormone receptors and growth factor receptor) overexpression on the cancer cell membrane. Patients that have both the hormone and growth factor receptors such as estrogen (ER), progesterone (PR), and epidermal growth factor receptor 2 (HER2) are classified as triple

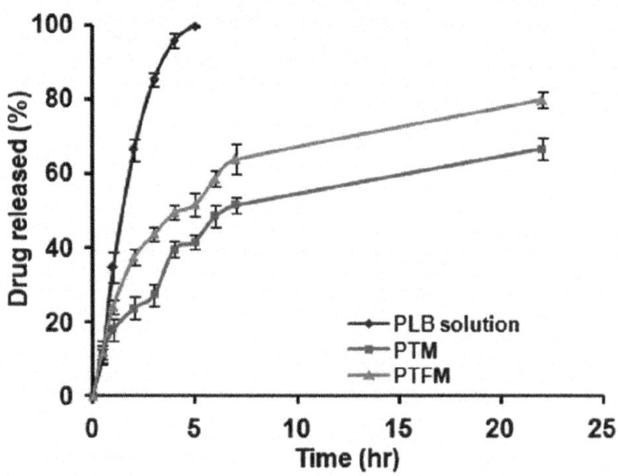

Fig. 25 In vitro PLB release from PLB solution, PTM, and PTFM in phosphate buffer saline pH 7.4 [22] (Reproduced with the permission of Royal Society of Chemistry)

positive breast cancer and the other that is not positive to any of these mentioned receptors is called the triple negative breast cancer (TNBC). In comparison with different types of cancers, TNBC is the most challenging type to cure. The reason for this difficulty is the high rate of recurrence and distant metastasis. Breast conservation treatment (BCT) followed by radiation treatment (RT) and chemotherapy is a common method for cure, but there is a need to more search to achieve target therapy. Some studies demonstrated that in triple negative breast cancers, epidermal growth factor receptor (EGFR) is overexpressed. In a literature published in 2013, cetuximab was used as a targeting ligand with the ability to specific binding to the extracellular region of EGFR. Another potential of cetuximab is its anti-angiogenic properties. Therefore, it can provide effective treatment of tumor xenografts. In this report, cetuximab conjugated vitamin E TPGS micelles were synthesized and used for loading of docetaxel (Fig. 26). The obtained results demonstrated that by the formulation of cetuximab conjugated TPGS micelles, the docetaxel therapeutic effect could be enhanced. The docetaxel-loaded cetuximab conjugated vitamin E TPGS micelles (TPCD) with small size (22.35 ± 2.170) and drug loading of 48.48% demonstrated lower cancer cell viability in comparison with TPD (without cetuximab conjugation) and free drug (Taxotere®) [58].

Further investigations of this micelle were published in 2015 to study the efficacy of this micellar system in xenograft models of TNBCs and the effect of the micelles on posttreatment tumors. Gene analysis and immunohistochemistry staining indicated that tumors treated with these micelles exhibit enhanced cell cycle arrest in comparison with the control and nontargeting micelle groups. The obtained results from in vivo and ex vivo study confirmed that reported micelles effectively targeted EGFR-overexpressing MDA-MB-231 breast cancer cells [59].

Another effort for targeted chemotherapy in the treatment of HER2 overexpressing breast cancer was done with a similar strategy in the same year. In this study, trastuzumab-conjugated vitamin ETPGS liposomes were synthesized and used for sustained and targeted delivery of docetaxel. Liposomes are lipid bilayer vesicles and can be seen as the simplest artificial biological cells. Nanoliposomes have many advantages because they are capable of increasing solubility of drugs, enhancing cellular uptake, and bioavailability. Because of their small size and unique biological properties, they play an essential role in cancer treatment by EPR effect. The presence of

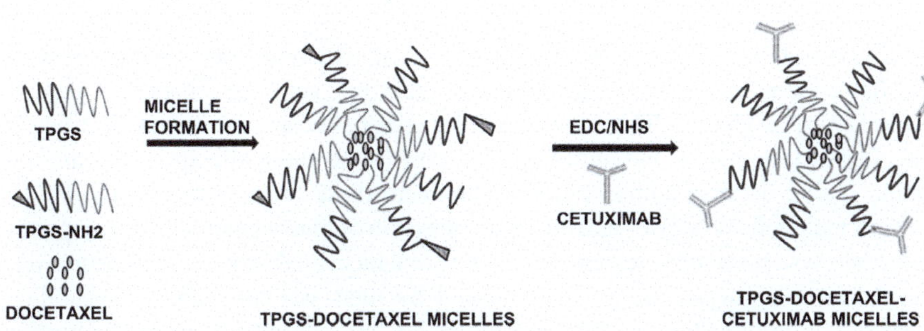

Fig. 26 Schematic illustration of formulation of the cetuximab-conjugated, docetaxel-loaded vitamin E TPGS micelles (TPCD) [58] (Reproduced with the permission of Elsevier)

functional materials in nanoliposomes can add potential for tumor-targeting drugs. For the first time, Wang et al. (2005) introduced TPGS-coated liposomes. They reported that doxorubicin loaded in TPGS-coated liposomes in comparison with PEG-coated liposomes has longer circulation time. TPGS-coated immunoliposomes were synthesized by the solvent injection method. The trastuzumab-conjugated liposomes (DTX-TP-HER) with small size (191.7 ± 0.6) and EE % 47.86 ± 0.6 showed appropriate potential application for targeted and sustained drug delivery.

In vitro studies on the SK-BR-3 breast cancer cells were done with the investigation of the amount of IC_{50} value. The obtained results showed that DTX-loaded trastuzumab-conjugated liposomes were 97% more efficient than the TPGS-coated liposomes after 24 h. In vivo analysis also showed better half-life of the drug in comparison with the marketed preparation of DTX and also PEG-coated liposomes [60, 61].

TPGS-modified liposomes were again reported in 2016 to targeted delivery of luteolin (LUT) in lung cancer treatment. LUT has a variety of pharmacological properties such as anti-inflammatory, antioxidant, and antimicrobial activities. Notably, its action in some cancers has attracted some attention. The mechanism of its antitumor activity was related to its ability to prevent the proliferation of cancer cells and inhibit angiogenesis and metastasis. However, some drawbacks limit its clinical application, such as poor solubility and high excretion. In this study, TPGS has been used as a modifier of liposomes. Passive tumor-targeting liposomes in this report were prepared by the film-dispersion method in a spherical shape with ~176.2 nm in size and the LC and EE 8.98 ± 1.32% and 71.55 ± 0.26%, respectively. In vivo imaging system, with a near-infrared fluorescence dye (DiR) as the probe, was used to investigate the biodistribution of prepared TPGS-coated liposomes in mice with A549 tumor cells. DiR-TPGS liposomes were prepared with luteolin replaced by DiR. The results represent in Fig. 27 indicated that DiR-loaded liposomes are accumulated in tumor sites with high fluorescence signals in tumor sites for more than 24 h. These results showed that TPGS-coated liposomes could deliver the drug directly into the tumor site and enhance their antitumor effect. Based on the obtained results from in vitro and in vivo studies on A549 cells, TPGS-coated liposomes with sustained drug release property were able to accumulate in tumor tissue and therefore enhanced tumor inhibition [60].

10 Vitamin E and Oral Drug Delivery

Another application of TPGS was investigated for oral chemotherapy. Oral delivery of drugs for cancer treatment is an essential step for achieving patients' dream: "Chemotherapy at Home." This method has some benefits such as enhanced efficacy and decreased side effects, but unfortunately, most of the chemotherapy drugs are not available orally, because some processes eliminate these drugs. Cytochrome P450-dependent metabolic processes and the overexpression of plasma membrane transporter P-gp are two systems that are responsible for low oral bioavailability of drugs. Moreover, the oral bioavailability of drugs is influenced by their solubility and permeability. Nanoparticles (NPs) with their small size and ability to modify the surface can improve the absorption of the NPs to drug transportation across the gastrointestinal (GI) barrier. Nanotechnology presents some benefits in oral drug delivery including (1) improvement

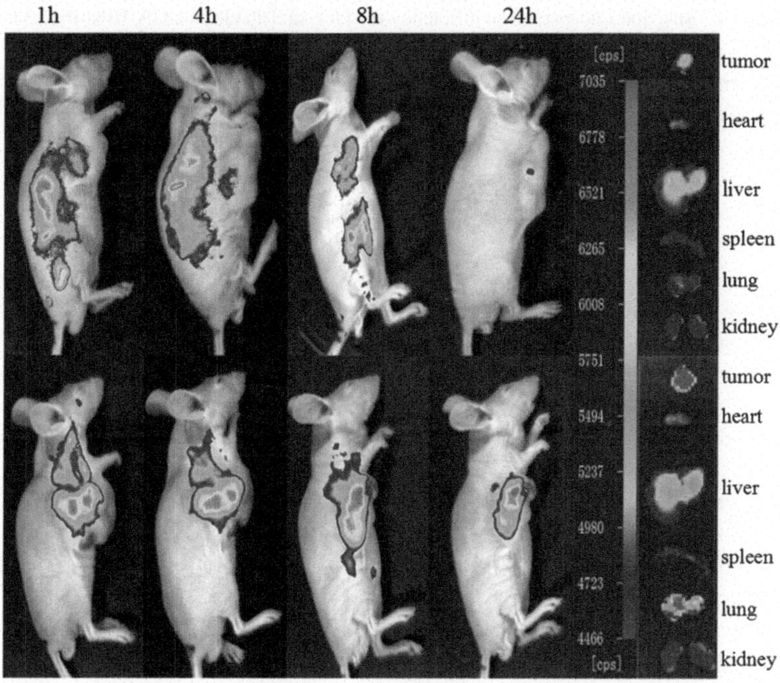

Fig. 27 Optical in vivo images of tumor-bearing mice after intravenous injection of DiR-TPGS liposomes and DiR [60] (Reproduced with the permission of Elsevier)

of water solubility of drugs, (2) target delivery of drugs to specific parts of the GI tract, and (3) ability to transcytosis of drugs across the GI tract barrier [62, 63].

There are some kinds of literature about oral drug delivery for cancer treatment. Some reported drugs for this purpose consist of docetaxel, paclitaxel, and rapamycin, which are further explained.

For oral drug delivery, the poly(lactide)–vitamin E TPGS copolymer incorporated with medical clay, montmorillonite, (PLA–TPGS/MMT NPs) was synthesized in 2009 for achieving oral delivery of docetaxel as a model anticancer drug. Three systems of biodegradable polymers including PLGA NPs[6], PLA–TPGS NPs[7], and PLGA/MMT NPs[8] were synthesized for comparison. The main reasons for using TPGS were its high emulsification effects, high drug encapsulation efficiency, high cellular adhesion, and adsorption. The drug-loading step was done by a modified solvent extraction/evaporation method. Gastrointestinal problem is the common side effect of cancer chemotherapy. This issue leads to some problems including mucositis and ulceration of the gastrointestinal tract and diarrhea. MMT was used for solving this problem. Because it can act as antidiarrhetic by absorbing excess water from feces, this compound is also able to provide the mucoadhesive capability

[6] Poly(lactic-co-glycolic acid) nanoparticles

[7] The poly(lactide)–vitamin E TPGS nanoparticles

[8] Poly(lactic-co-glycolic acid)–montmorillonite nanoparticles

Table 2 Size and encapsulation efficiency of docetaxel-loaded PLGA NPs, PLGA/MMT, PLA–TPGS NPs, and PLA–TPGS/MMT NP

Polymer	Size (nm)	Encapsulation efficiency
PLGA	293 ± 11.4	79.3 ± 7.5%
PLGA/MMT	333 ± 10.5	60.2 ± 3.2%
PLA–TPGS	201 ± 5.6	83.4 ± 3.6%
PLA–TPGS/MMT NPs	234 ± 16.8	80.4 ± 4.6%

of the NPs that leads to crossing from the GI barrier. The size and drug encapsulation efficiency of four synthesized nanoparticles are listed in Table 2.

Obtained results from in vitro cytotoxicity experiments showed that the amount of IC_{50} for PLA–TPGS/MMT NP and PLA–TPGS NP is more effective than the Taxotere®[9] formulation. In vivo experiment showed that the oral delivery of the PLA–TPGS/MMT NP and PLA–TPGS NP formulation was 26.4 and 20.6 times longer than half-life of docetaxel in the plasma in comparison with Taxotere® administration [63].

Another drug that was investigated for oral delivery is paclitaxel. Polymeric nanoparticles prepared by PLGA with vitamin E TPGS as emulsifier was used for oral chemotherapy in 2010. Prepared spherical shape nanoparticles with 200–300 nm diameter and EE % of 80.9% were used for in vivo and in vitro studies. In vitro studies were done on MCF-7 human breast cancer cells, and results showed that this NP formulation could be more effective than Taxol®. In vivo studies also confirmed the advantages of this nanoparticle versus Taxol® by 9.74-fold higher therapeutic effect and 12.56-fold longer sustainable therapeutic time [64].

Micelles of TPGS-modified apigenin phospholipid complex (TPGS-Ap-Ph) is another reported system in 2013 to increase the oral bioavailability of apigenin (Ap). Ap is one of the most bioactive plant flavones that is recognized in traditional medicines as anticancer. This drug is present in some fruits and vegetables like parsley, grapefruit, oranges, tea, etc. TPGS-Ap-Ph with small size (137.1 ± 3.4) and 87.35 % drug encapsulation and 12.6 % drug loading was prepared by thin-film method. These micelles showed more (2.4-fold) absorption of Ap compared to free drug. In vitro analysis of TPGS-Ap-Ph against A549 cancer cells showed a significant cytotoxic effect. In vivo analysis against S180 mice revealed an effective inhibition (72.9 %), while this amount for Ap-Ph complex was 19.5 %. Therefore, this study showed that TPGS-Ap-Ph could successfully protect apigenin and increase its intestinal absorption [65].

Another report for oral drug delivery was published in 2017 where paclitaxel (PTX) improves delivery to lung cancer cells. In this study, mixed micelle delivery system (PTX-TP-M) was prepared by ethanol thin-film dispersion method. TPGS

[9] Taxotere is the trade name for docetaxel.

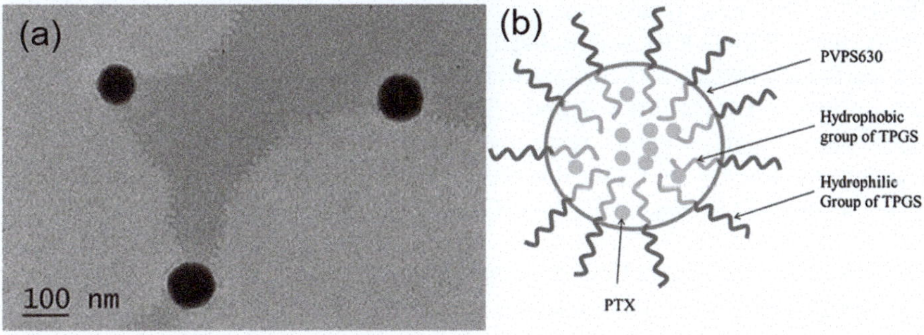

Fig. 28 TEM micrographs of PTX-TP-M (panel **a**); schematic illustration of the structure of PTX-TP-M (panel **b**). Scale bar = 100 nm [66]

and Plasdone®S-630 copovidone (PVPS630) were used in the synthesis of this mixed micelle. Its spherical morphology and the absence of any conglutination and adhesion were confirmed with TEM analysis (Fig. 28a). Schematic representation of the structure of PTX-TP-M is shown in Fig. 28b.

The average size of the prepared mixed micelles was 83.5 ± 1.8 nm with LC % and EE % of 3.09 ± 0.09% and 95.67 ± 2.84%, respectively, with the slow release of the drug over 48 h. Also, because of the presence of TPGS with the ability to modulate efflux pumps (such as P-gp or MDR-associated proteins) in the structure of the prepared micelles, the efflux of PTX was reduced effectively. The obtained results showed the improvement of the oral bioavailability of PTX. This result indicates that PTX-TP-M can promote the absorption of this drug in the gastrointestinal tract [66].

11 Vitamin E and Cancer Cell Diagnosis

Vitamin E derivatives have potential in the field of cancer cell diagnosis. For example, some reports indicate TPGS can play a role in cancer diagnosis field by encapsulation of luminescence probes such as quantum dots or the ability for surface coating of nanosized iron oxide (IO) suspensions.

The application of TPGS to the encapsulation of quantum dots (QDs), which is widely used as luminescence probes for cancer diagnosis, was reported in 2009. QDs have several advantages such as tunable emission by size and composition changing, the high quantum yield of fluorescence, high absorption coefficients, intense brightness, and high photobleaching resistance, but their application is limited because of its toxic effects. There are two main strategies for solving this problem: (1) surface modification and (2) targeted delivery of QDs for selective delivery to the cancer cells. In a reported literature by Feng et al., two copolymers of poly(lactide)–vitamin E TPGS (PLA–TPGS) and vitamin E TPGS carboxyl

(TPGS-COOH) were synthesized and blended and used for loading of QDs. In this report, the TPGS-COOH was utilized for folate-NH_2 conjugation. The obtained nanoparticles with size 280–300 nm were used for in vitro cellular uptake investigation on two cell lines: MCF-7 and NIH 3T3 fibroblast cells with high and low expression of folate receptors, respectively. The obtained result from confocal laser scanning microscopy (CLSM) demonstrated much higher internalization of the folate-decorated QD-loaded PLA–TPGS/TPGS-COOH NPs by MCF-7 cells. In fact, in the case of MCF-7 cell line, the CLSM analysis showed low internalization of QD-loaded NPs without FD[10] (Fig. 29a), while Fig. 29b shows higher internalization of QD-loaded FD NPs in this cell line. In the case of NIH 3T3 cell line, with low expression of folate receptors, the internalization for NPs with and without FD is almost similar (Fig. 29c, d). Therefore, these results represent the targeting effects of QD-loaded FD NPs for MCF-7 cells.

Cytotoxicity studies showed that QDs formulated in the folate-decorated PLA–TPGS/TPGS-COOH NPs have lower cytotoxicity for healthy cells in comparison with breast cancer cells such as MCF-7. This result is due to folate targeting effect. These results showed that QDs formulated in folate-decorated PLA–TPGS/TPGS-COOH nanoparticle can be used for targeted imaging with reduced side effects of QDs to healthy cells [67].

12 Vitamin E and Hyperthermia

Hyperthermia is defined as the treatment of malignant diseases by administering heat. This method has been used since the early 1970s as one method for cancer treatment. In spite of heat effects in the treatment of cancer, there are some challenges including the difficulty (1) providing uniform heating only on the tumor region and (2) creating uniform hyperthermia throughout the tumor cells. Combining nanotechnology and hyperthermia leads to nanothermotherapy. In nanothermotherapy, there is a possibility to design nanostructured devices with the ability of selective penetration into the cancer cells and reduce side effects [68].

In this regard, some kinds of literature are about nanotechnology application in thermotherapy. In the following, some reports about the use of TPGS for this method are noted.

One report is attributed to the application of TPGS as a surface coating material to the stability of nanosized iron oxide (IO) suspensions and was published in 2011. In this report, TPGS has been used for synthesis micellar formulation of IOs for their potential in nanothermotherapy and magnetic resonance imaging (MRI). In this report, vitamin E TPGS micelles were synthesized and used for IO formulation. At the first stage, the IOs were coated with oleic acid and oleylamine; then, they were formulated in the core of the TPGS micelles by the solvent-exchange method. The IO-loaded TPGS with the hydrodynamic size of 145.5 nm showed better properties in comparison with other commercial Resovist® and the Pluronic®F127

[10] Folate decorated

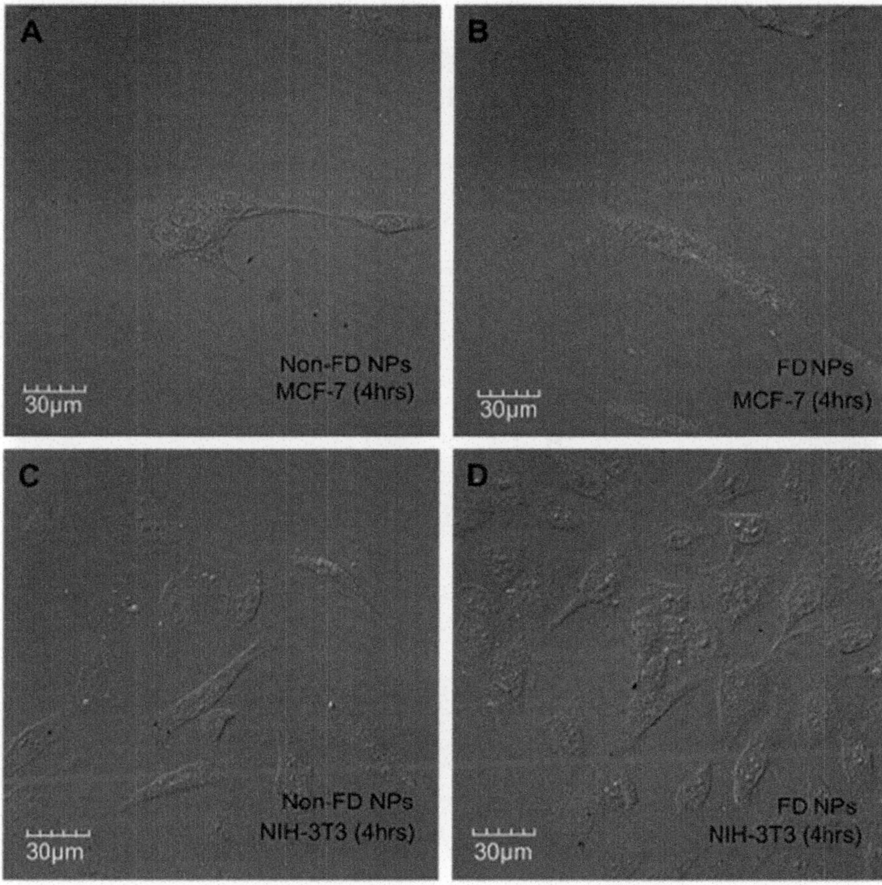

Fig. 29 Confocal laser scanning microscopy (CLSM) of (**a**) MCF-7 breast cancer cells treated with QD-loaded 11.1% TC NPs, (**b**) MCF-7 cells treated with QD loaded 11.1% FD NPs, (**c**) NIH 3T3 cells treated with QD-loaded 11.1% TC NPs, and (**d**) NIH 3T3 cells treated with QD-loaded 11.1% FD NPs at 0.38 nM QD concentration at 37 °C after 4 h incubation. Here green channel shows the transmission images, while the intensity-coded (red for QDs and blue for DAPI) channel shows the fluorescence [67] (Reproduced with the permission of Elsevier)

micelles. Better thermal and magnetic properties, lower cytotoxicity, and better in vivo therapeutic and imaging effects are some advantages of these IO-loaded TPGS micelles. The IO-loaded micelles showed enhanced uptake and reduced the toxic effect by the cancer cells (MCF-7). TEM images in Fig. 30, with two magnifications, represent the qualitative uptake of IO-loaded TPGS micelles in MCF-7 cells. Figure 30a shows the phagocytosis process. The IO-loaded TPGS micelles are distributed inside the cell within vesicles associated with lysosome, which is called phagolysosomes. A part of the cell's defense mechanism is the formation of these phagolysosomes. As shown in this picture, some IO-loaded TPGS micelles are attached to the cell membrane, and the membrane invagination resulted in the

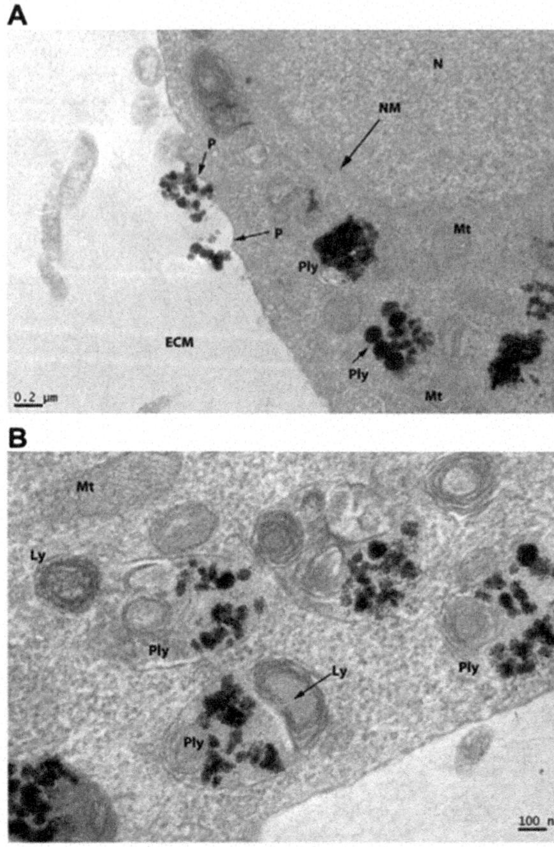

Fig. 30 Qualitative uptake of the IO-loaded TPGS micelles using transmission electron microscopy (TEM). (**a**) Shows the MCF-7 cancer cell line at lower magnification. The site at P shows the invagination of the cell membrane resulting in the uptake of the micelles (legend: N, nucleus; NM, nuclear membrane; P, site of phagocytosis; Mt, mitochondria; ECM, extracellular matrix; Ply, phagolysosomes containing TPGS IOs). (**b**) Shows the MCF-7 cancer cell at higher magnification. The image shows the xenophagic of the cells. Ly represents the lysosome associated with vesicles carrying the IO-loaded micelles; together they form the phagolysosome (Ply) [69] (Reproduced with the permission of Elsevier)

uptake of these micelles. Figure 30b shows the phagolysosomes with higher magnification. This picture shows the IO-loaded TPGS micelles with spherical shape. Therefore, TEM images indicate the stability of these micelles in physiological condition. Also, in vivo, MRI showed that this TPGS micelle formulation is more promising than Resovist® and F127 micelles because of increased IO uptake in the tumor cells [69].

Another report is attributed to polymeric nanoparticles composed of PLGA with vitamin E TPGS as an emulsifier for sustained docetaxel release. First, the amino silane-coated magnetic nanoparticles (MNPs) modified with amino groups were prepared by chemical precipitation method; then, these magnetic nanoparticles were covalently attached to the surface of carboxylic terminated docetaxel-loaded polymeric nanoparticles (DNPs) to develop the so-called solar-planet structured magnetic nanocomposites (MNCs). The diameter of DNPs was reported as 222.8 ± 10.5 nm with EE % as high as 85%. This size is close to the optimum amount for cellular uptake. In this report, inductive heating generation was done by stimulating the MNCs under alternative magnetic field (AMF). U251 human glioma cells were chosen for in vitro evaluation nanocomposite. Docetaxel-loaded MNCs were sustained release system. In vitro, docetaxel release profile demonstrated that 46 °C heating for 20 min could facilitate the drug release. The reason for this phenomenon was reported based on acceleration diffusion of the drug from the MNCs to the PBS buffer under AMF exposure. All the obtained results of in vitro and in vivo studies support that this reported solar-planet structured MNC is an effective medium for magnetic thermochemotherapy [68].

13 Vitamin E and NO Delivery

Nitric oxide (NO) is a signaling molecule that has shown great potential in inhibiting carcinogenesis and tumor growth. However, because of its short half-life (1–5 s) and chemical instability, its delivery to tumor cells is not sufficient. A number of NO donors with the ability to generate NO in situ have been developed, such as S-nitrosothiols (RSNOs), N-diazeniumdiolates (NONOates), etc. However, most of them are systems with low molecular weight which enhance the exposure to the reticuloendothelial system (RES) and therefore accelerate their clearance from the body. Macromolecular NO donors (such as dendrimers, liposomes, and nanoparticles) have been studied to solve these problems. In a study in 2014, a kind of nitrate-functionalized TPGS derivative (TNO_3) was synthesized. TNO_3 is a polymer modified with the nitrate group to achieve NO releasing ability, and it can self-assemble into stable micelles. The hydrodynamic diameter of TNO_3 micelles was 133.9 ± 0.9 nm. To enhance antitumor activity, doxorubicin was selected as the model drug for investigation of co-delivery of drugs. The in vitro analysis was done against hepatocarcinoma cell line (HepG2), and the in vivo tests were done on murine hepatic carcinoma H22 and sarcoma S180 tumor-bearing mice. The obtained results showed sustained NO release in tumor cells (~90% of NO content release in cancer cells for 96 h). TNO_3 combined with the anticancer drug DOX also exhibited an excellent capability of enhancing cancer cell cytotoxicity in vitro and synergistic antitumor efficiency [70].

14 Vitamin E and Cancer Phototherapy

In a study in 2017, TPGS-TOS micelles were used as nanoplatforms for IR780-mediated cancer phototherapy and imaging (Fig. 31). IR780 iodide is a near-infrared (NIR) dye. This dye has a high potential for cancer imaging and phototherapy. The main problem for the use of this dye is its lipophilic character. Therefore, an amphiphilic micelle is an appropriate candidate for IR780 loading. The IR780-loaded TPGS-TOS micelles (IR780-TTM) can induce a cytotoxic effect in cancer cells when exposed to NIR irradiation (photodynamic therapy). This ability is attributed to the generation of reactive oxygen species (ROS). IR780-loaded micelles have the potential to act as both photothermal and imaging agents. It was suggested that TPGS and TOS might have an essential role by displaying an intrinsic capacity to generate ROS [71].

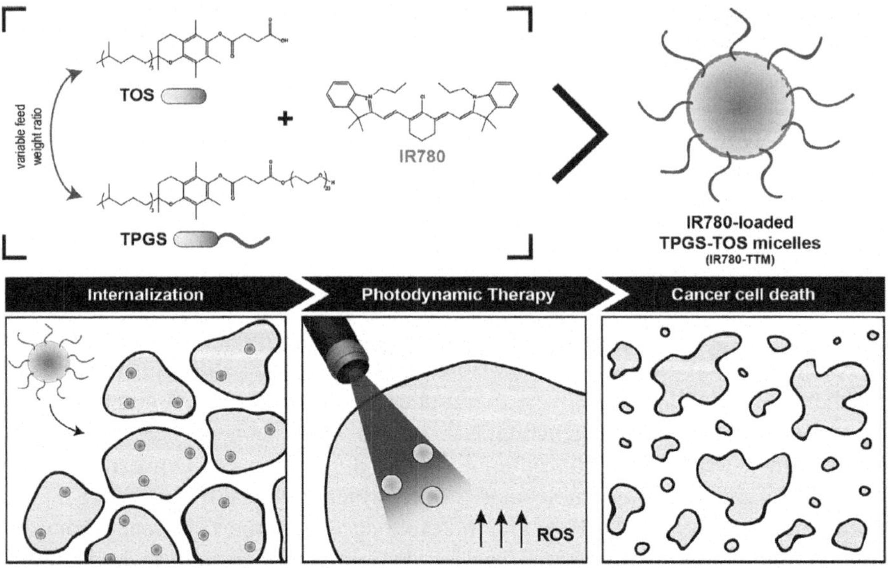

Fig. 31 Schematic of the experimental set-up used to produce and characterize IR780-loaded TPGS-TOS micelles suitability to be used for cancer photodynamic therapy [71] (Reproduced with the permission of Elsevier)

The poor tumor selectivity of the photosensitizer is a significant challenge in cancer photodynamic therapy (PDT). Therefore, in a study published in 2017, temoporfin-loaded vitamin E succinate-grafted-chitosan oligosaccharide/D-a-tocopheryl polyethylene glycol 1000 succinate nanoparticles (RGD-NPs) were prepared to overcome this issue and increase target therapy. These NPs are constructed from some components including vitamin E succinate (VES), chitosan oligosaccharide (CSO), TPGS, and VES-grafted CSO (VES-g-CSO).

VES is a vitamin E derivative with the ability to excellently encapsulate and deliver hydrophobic drugs. CSO is a natural polysaccharide with the pendant amine groups in its structure that allow further modification. TPGS is constructed from VES (as the lipophilic polar part) and PEG (as the hydrophilic part) which because of its bulky structure and its large surface area can improve the drug encapsulation efficiency. VES-g-CSO and TPGS can self-assemble and form nanoparticles. VES forms the inner core, PEG, and CSO forms the outer shell of NPs and forms a core-shell structure. This core-shell structure can act as a vehicle for hydrophobic photosensitizers. These RGD-NPs with small size (144.9 nm) and LC and EE of 2.24 ± 0.10% and 59.5 ± 2.95%, respectively, were prepared and displayed to enhance cellular uptake and also cytotoxicity in U87MG cells. These NPs were used for encapsulation of temoporfin (mTHPC). mTHPC is a photosensitizer. Photosensitizers are essential components of PDT. A cyclic RGD peptide, c(RGDfK), which is able to interact with $\alpha_v\beta_3$ integrin receptor-rich tumors was

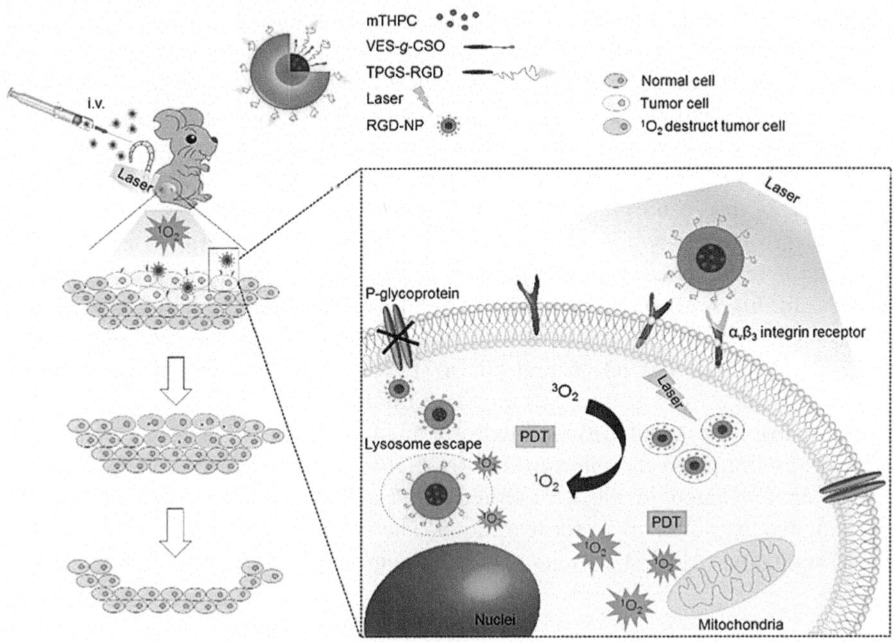

Fig. 32 Schematic illustration of mTHPC-loaded c(RGDfK)-modified VES-g-CSO/TPGS nanoparticles (RGD-NPs). RGD-NPs specifically bound integrin-$\alpha_v\beta_3$ receptor-rich U87MG tumors and then accumulated selectively and deeply in the tumor tissue, enhancing the efficiency of PDT [72] (Reproduced with the permission of Elsevier)

used for decorating the surface of VES-g-CSO/TPGS nanoparticles. The prepared RGD-NPs were internalized in the tumor cells by integrin $\alpha_v\beta_3$ receptor and enhance the efficiency of PDT (Fig. 32).

In comparison with unmodified nanoparticles, the RGD-NPs showed higher tumor-targeting ability and antitumor efficacy. Based on these results, it can be concluded that the RGD-NPs can enhance the efficiency of PDT [72].

15 Metal-Organic Frameworks and Vitamin E

Recently, porous materials as drug delivery systems have attracted attention. Metal-organic frameworks (MOFs) are a new class of coordination polymers that are built from metal ions or clusters and organic linkers [73–80]. MOFs have many advantages such as pore size tenability, ability to adjust the framework's functional groups, post-synthesis modification, and high surface area [81–83]. Because of these features, they were used in many fields, such as catalyst [84, 85], sensor [86], drug delivery [87, 88], antibacterial activity [89–91], gas storage [92–95], and preparation of nanomaterials with desirable morphologies [96–99]. Férey and coworkers reported the first report of the application of MOFs for drug delivery. Besides these advantages, MOFs are usually unstable in an aqueous solution. Many reports suggest methods for solving this problem. One of them is the application of biodegradable polyester polymers for controlling drug release.

In a study in 2016, poly(ε-caprolactone) (PCL) was used because of its biocompatibility and nontoxicity (Fig. 33). Carbonyl groups of PCL can functionalize with various hydrophilic materials (such as PEG). This functionalization leads to avoid the reticuloendothelial system and create long-circulating "stealth" nanocarriers. In this study, poly(ε-caprolactone)–tocopheryl polyethylene-glycol-succinate (PCL–TPGS) copolymer was prepared to engineer the surface of two kinds of Zr-based NMOFs (UiO-66 and UiO-67). The dimension of unloaded MOFs was in the range from 100 to 250 nm. Two models of the drug were used in this study: hydrophilic (cisplatin) and hydrophobic (Taxol) drugs. After adsorption of these drugs onto/into the nano-MOFs (NMOFs), the drug-loaded NMOFs were further encapsulated inside a modified poly(ε-caprolactone) with TPGS polymeric matrix by solid-oil-water technique. The in vitro cytotoxicity investigation was done against HSC-3 (human oral squamous carcinoma; head and neck) and

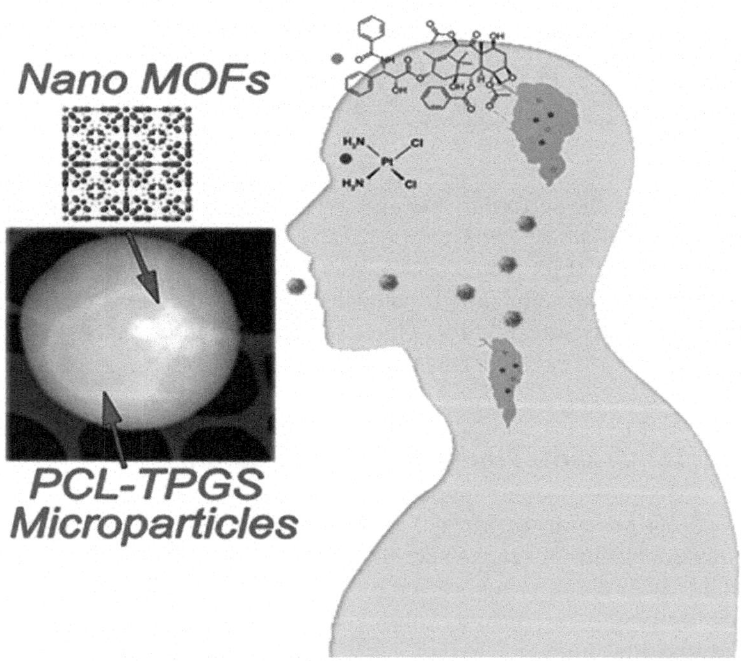

Fig. 33 Schematic illustration of application of Zr-based nanoscale MOFs coated with modified poly(ε-caprolactone) as anticancer drug carriers [100] (Reproduced with the permission of Elsevier)

U-87 MG (human glioblastoma grade IV; astrocytoma) cancer cells. The obtained results indicated low toxicity of polymer-coated MOFs. In this study, MOFs were selected because of their porous and crystalline nature that leads to delivering the drug molecules in a controlled manner. The in vitro drug release investigation was done for both drugs, and results showed a sustained release rate in the case of the cisplatin and high dissolution rate for the Taxol. The high dissolution rate of Taxol was related to the amorphization of the drug. This study indicated the effect of the biocompatible and biodegradable polymeric matrix to enhance the sustained release drug and decrease the burst effect. Therefore, this type of materials (the drug-loaded MOFs encapsulated inside the biocompatible polymeric matrix) displayed a good potential for drug delivery application because of the combination of MOFs with their porous structure and the polymer with biocompatibility nature [100].

Another application of MOF is the transportation of biomolecules, especially vitamins into the body. Recently, in 2019, modified PCN-100 (a mesoporous metal-organic framework with the amino group) with terephthaloyl chloride (TC) has been reported to loading and release of vitamin E and compared with pristine PCN-100 and activated carbon (AC) (Fig. 34). This study is the first report of postsynthetic modification of a mesoporous MOF (PCN-100) with TC at room temperature. Functionalization of PCN-100 is done to introduce extra-functional groups such as carboxylic, phenyl, and amide groups in the structure. PCN-100 is a mesoporous MOF that is composed of $Zn_4O(CO_2)_6$ as secondary building units. This MOF is constructed from nontoxic metal (Zn) and low-toxic ligand (carboxylic acids). TC-modified PCN-100 (TC-PCN-100) showed the slower release of vitamin E. This slow release is related to secondary hydrophobic, π–π [101, 102], and H-bonding

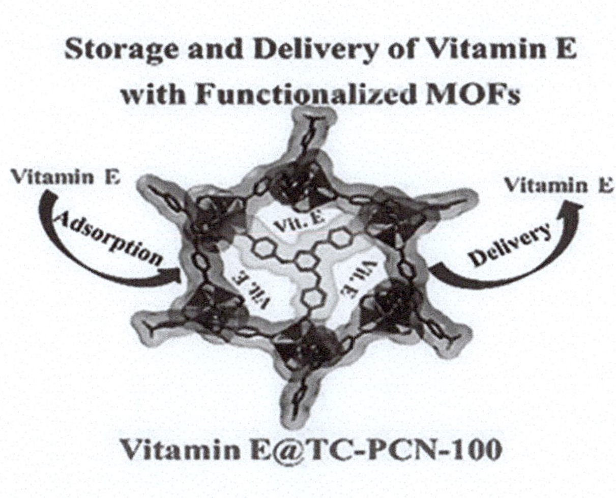

Fig. 34 Schematic illustration of functionalized PCN-100 as a carrier for vitamin E storage and delivery [106] (Reproduced with the permission of Elsevier)

[103–105] interactions. Furthermore, the obtained results indicate that vitamin E delivery rate might be controlled by pH. Decreasing pH from 7.4 to 3.0 leads to reduction of the release rate of vitamin E. It can be concluded that modified PCN-100 (TC-PCN-100) could be used for load and delivery of big molecules such as vitamin E [106].

16 Conclusion

Vitamin E, especially TOS and TPGS, has been known as antioxidant and anticancer agent with the ability to P-gp inhibition and, therefore, prevention of efflux of drugs from cancer cells and overcome MDR phenomenon. Succinate analogs of vitamin E are nontoxic to normal cells; therefore, vitamin E-based nanomedicines have many potentials for cancer treatment. These nanomedicines show high drug-loading efficiency, high cellular uptake, and therapeutic effect in vitro and in vivo. They can solve some problems such as low water solubility of cancer drugs that is a significant problem in efficient cancer treatment. There are many kinds of literature about the application of vitamin E in different forms such as nanoparticles, micelles, liposomes, oligomersomes, dendrimer, and nanoemulsion for anticancer drug delivery including some well-known anticancer drugs such as PTX, DOX, and rapamycin. Also, in some reports, vitamin E itself was used as an anticancer agent. Also, in some cases such as MOFs, this vitamin was used for surface modification of the platform to increase their bioavailability and therapeutic effects. This vitamin is a promising candidate for achieving patients' dream: "Chemotherapy at Home." Some paper reported the role of vitamin E nanomedicines for making this dream. Moreover, the therapeutic effect of vitamin E was used for cancer cell diagnosis application. Based on the mentioned benefits of this vitamin, it was concluded that vitamin E-based nanomedicine is a suitable candidate for the preparation of novel anticancer treatments in the future. Some of the reported nanomedicines based on vitamin E are listed in Table 3.

Table 3 Reported vitamin E-based nanomedicines for anticancer drug delivery

Carrier type	Composition	Drug/imaging probes	Model	Size (nm)	Studies	Ref.
Nanoparticles	α-TOS	Paclitaxel	–	100	In vitro	Shi et al. [20]
	NLC-TS-DOX	Doxorubicin	Breast cancer cell lines, MDA-MB-231 and 4T1	85	In vitro and in vivo	Fernandes et al. [21]
	PLGA/TPGS	Paclitaxel	-	300–1000	In vitro	Mu and Feng [31]
	PLA–TPGS	Paclitaxel	-	300	In vitro	Zhang and Feng [32]
	PLGA/TPGS	Paclitaxel	MCF-7 breast cancer cells C6 glioma cells	288 ± 11.67	In vitro and in vivo	Feng et al. [33]
	PLGA/TPGS	Garcinol	B16F10, HepG2, and KB cells	88.05 ± 2.7	In vitro and in vivo	Gaonkar et al. [34]
	CA-PLGA-b-TPGS	Docetaxel	HeLa cell	100	In vitro and in vivo	Zeng et al. [35]
	PLGA	Docetaxel, TPGS	HeLa cell, MCF-7/ADR cells	100–120	In vitro and in vivo	Zhu et al. [37]
	TPGS	Paclitaxel, Tariquidar	Drug-resistant cancer cells (MCF-7/ADR)	Below 150	In vitro	Liu et al. [28]
	PLA–TPGS/MMT NPs	Docetaxel	MCF-7 cells	234 ± 16.8	In vitro and in vivo	Feng et al. [63]
	TPGS-emulsified PLGA NPs	Paclitaxel	MCF-7 human breast cancer cell line	200–300	In vitro and in vivo	Zhao and Feng [64]

(continued)

Table 3 (continued)

Carrier type	Composition	Drug/imaging probes	Model	Size (nm)	Studies	Ref.
	TPGS-Ap-Ph	Apigenin (Ap)	A549 cancer cell lines, in vivo against S180 mice	137.1 ± 3.4	In vitro and in vivo	Munyendo et al. [65]
	Folate-decorated PLA–TPGS/TPGS-COOH NPs	Quantum dots as luminescence probes	MCF-7 cells and NIH 3T3 cells	280–300	In vitro	Pan and Feng [67]
	Solar-planet structured magnetic nanocomposites MNCs (constructed from TPGS/PLGA)	Docetaxel	U251 human glioma cells	222.8 ± 10.5	In vitro and in vivo	Zhao et al. [68]
	CS/PAA/VP-16@TPGS/PLGA NPs	Etoposide (VP-16)	A549/DDP cells (an MDR human lung adenocarcinoma cell line	150	In vitro	Wang et al. [17]
	RGD-NPs (constructed from VES-g-CSO and TPGS)	Temoporfin	U87MG tumor spheroids	144.9	In vitro and in vivo	Wu et al. [72]
	Poly(L-lactide)-vitamin E TPGS (PLA–TPGS)	Doxorubicin	Breast cancer MCF-7/ADR	100 (for P/D[a] ratio of 15:1)	In vitro	Li et al. [19]
	TPGS-functionalized PLGA nanoparticles (TPNs)	Paclitaxel	Human lung cancer A549 model cells	130–150	In vitro and in vivo	Wang et al. [36]
Micelles	TOS-modified pluronic micelles (PF-TOS micelles)	Paclitaxel	Human glioma U87 cells	58	In vitro and in vivo	Tao et al. [24]

	VPM (α-TOS, PEG and DOPE)	Doxorubicin	Human colon carcinoma, SW 620-R and human breast adenocarcinoma, MDA MB-231-R	141.2 ± 0.78	In vitro	Muddineti et al. [25]
	mPEG-b-P(HEMA-g-a-TOS)	Doxorubicin	Human HCT116 colon cancer cells	116.3 ± 0.2	In vitro and in vivo	Lee et al. [41]
	Pluronic-alpha-tocopheryl succinate polymeric micelles (P/TOS-DOX)	Doxorubicin	HepG2 cancer cell	100	In vitro and in vivo	Li et al. [42]
	PEG_{5K}-FVE_2	Paclitaxel	Breast cancer (4T1.2)	~60	In vitro and in vivo	Zhang et al. [38]
	PEG5K-VES6 (PAMV6)	Pirarubicin (THP)	L929 and HUVEC cells And 4T1 tumor spheroids	98.42 ± 1.38	In vitro and in vivo	Li et al. [39]
	DTX-TPGS-Tf	Docetaxel	-	<200	In vitro and in vivo	Sonali et al. [57]
	PTFM (TPGS/FA)	Plumbagin (PLB)	Human breast cancer MCF-7 cells	128.4 ± 0.8	In vitro and in vivo	Pawar et al. [22]
	TPCD (Cetuximab, TPGS)	Docetaxel	NIH3T3, SK-BR-3, MCF7, MDA MB 468, MDAMB 231	22.35 ± 2.170	In vitro	Kutty and Feng [58]

(continued)

Table 3 (continued)

Carrier type	Composition	Drug/imaging probes	Model	Size (nm)	Studies	Ref.
	TPGS	Iron oxides	and HCC38 cell lines MCF-7	145.5	In vitro and in vivo	Chandrasekharan et al. [69]
	TNO3	-	Hepatocarcinoma HepG2 cancer cells	133.9 ± 0.9	In vitro and in vivo	Song et al. [70]
Mixed micelle	VeC/T-RGD MM	Bufalin (BU)	Colon cancer LoVo/ADR and HCT116/LOHP cells	140.3	In vitro and in vivo	Yuan et al. [18]
	PUTC	10-Hydroxycamptothecin (HCPT)	MCF-7	171.5–257.8	In vitro	Wang et al. [53]
	VeC/T-RGD MM	Paclitaxel	Human glioma U87MG cells	150.2	In vitro and in vivo	Chen et al. [50]
	PEG2000-DSPE and vitamin-E TPGS	Paclitaxel and parthenolide	Taxol sensitive (A549) and resistant (A549-T24) NSCLC cell lines	12 ± 5	In vitro	Gill et al. [51]
	TOS, TPGS	Doxorubicin	CT26 and MCF-7	78	In vitro and in vivo	Danhier et al. [52]
	TPGS/P105	Oleanolic acid (OA)	A549 and PC-9	95.7 ± 3.6	In vitro and in vivo	Wu et al. [54]
	P407-TPGS/FA	Doxorubicin	SKOV3 human ovarian carcinoma cell lines	<200	In vitro	Butt et al. [56]
	PTX-TP-M (TPGS and Plasdone®S-630 Copovidone (PVPS630))	Paclitaxel	A549 and Lewis cell	83.5 ± 1.8	In vitro and in vivo	Hou et al. [66]

	TOS, TPGS	IR780 iodide	MCF-7 cells	123 (for 40% TPGS) 115 (for 33% TPGS) 121 (for 25% TPGS)	In vitro and in vivo	Pais-Silva et al. [71]
Liposome	TPGS, Trastuzumab)	Docetaxel	SK-BR-3 cells	191.7 ± 0.6	In vitro and in vivo	Raju et al. [61]
	LUT-TPGS	Luteolin	Adenocarcinomic human alveolar basal epithelial cell line A549 cells	~176.2	In vitro and In vivo	Li et al. [60]
Nanoemulsion	γ-T_3 and PEGylated γ-T_3	Paclitaxel	Bx-PC-3 and PANC-1 pancreatic tumor cells	<300	In vitro	Abu-Fayyad et al. [8]
Dendrimer	G5.NHAcn-α-Tos-FI-FA conjugate	A-TOS	KB (human epithelial carcinoma cell line)	5.4	In vitro	Zheng et al. [55]
Oligomersomes	$TCOsome_{4K}$ (α-tocopherol/ oligochitosan)0	siMcl-1	Human epidermal carcinoma cell line	218.1 ± 30.9	In vitro and In vivo	Noh et al. [26]
Prodrug	TS NPs salinomycin, TOS, TPGS	Paclitaxel	MCF-7	~ 200	In vitro	Liang et al. [46]
	N-DOX–TOS	Doxorubicin	MCF-7, CT26 cells	250	In vitro and in vivo	Duhem et al. [44]
	PEG_{5K}-Fmoc-VE_2/ CPT-VE	Camptothecin (CPT)	Breast cancer model (4T1.2)	87.4 ± 4.6	In vitro	Lu et al. [45]

(continued)

Table 3 (continued)

Carrier type	Composition	Drug/imaging probes	Model	Size (nm)	Studies	Ref.
	$PEG_{5K}Fmoc\text{-}VE_2$/ CPT-S-S-VE	Camptothecin (CPT)	Breast cancer model (4T1.2)	49.7 ± 1.8	In vitro	Lu et al. [45]
	HTS NPs (salinomycin, TOS, TPGS, hyaluronic acid)	Paclitaxel	MCF-7	~200	In vitro	Liang et al. [46]
	TPGS–DOX	Doxorubicin	MCF-7 breast cancer cells and C6 glioma cells as in vitro cell model	–	In vitro and in vivo	Cao and Feng [43]
	TPGS-cisplatin	Cisplatin	SH-SY5Y neuroblast Cells, HepG2 hepatocarcinoma cells	12.56 ± 0.27	In vitro	Mi et al. [47]
	HTCP NPs	Docetaxel, cisplatin, Herceptin	NIH3T3 cells and MCF7 cells SK-BR-3 cells	187.4 ± 2.4	In vitro	Mi et al. [48]
	TPGS-S-S-PTX	Paclitaxel	Human ovarian cell line A2780 and PTX-resistant human ovarian cell line A2780/T	~140	In vitro and in vivo	Bao et al. [49]
MOF	TPGS-block-PCL, Uio-66 and Uio-67	Paclitaxel and cisplatin	HSC-3 (human oral squamous carcinoma; head and neck) and U-87 MG (human glioblastoma grade IV; astrocytoma) cancer cells	100–250	In vitro	Filippousi et al. [100]
	TC-PCN-100	Vitamin E	–	Pore size ~2.5 nm	In vitro	Sarker et al. [106]

[a]P/D = polymer/drug ratio
[b]The TQR encapsulation efficiency should be similar to PTX due to their similar hydrophobicity

References

1. Zununi Vahed S, Fathi N, Samiei M, Maleki Dizaj S, Sharifi S (2019) Targeted cancer drug delivery with aptamer-functionalized polymeric nanoparticles. J Drug Target 27(3):292–299
2. Muddineti OS, Ghosh B, Biswas S (2017) Current trends in the use of vitamin E-based micellar nanocarriers for anticancer drug delivery. Expert Opin Drug Deliv 14(6):715–726
3. Wicki A, Witzigmann D, Balasubramanian V, Huwyler J (2015) Nanomedicine in cancer therapy: challenges, opportunities, and clinical applications. J Control Release 200:138–157
4. Plapied L, Duhem N, des Rieux A, Préat V (2011) Fate of polymeric nanocarriers for oral drug delivery. Curr Opin Colloid Interface Sci 16(3):228–237
5. Duhem N, Danhier F, Préat V (2014) Vitamin E-based nanomedicines for anti-cancer drug delivery. J Control Release 182:33–44
6. Kumari P, Ghosh B, Biswas S (2016) Nanocarriers for cancer-targeted drug delivery. J Drug Target 24(3):179–191
7. Ulatowski LM, Manor D (2015) Vitamin E and neurodegeneration. Neurobiol Dis 84:78–83
8. Abu-Fayyad A, Kamal MM, Carroll JL, Dragoi A-M, Cody R, Cardelli J et al (2018) Development and in-vitro characterization of nanoemulsions loaded with paclitaxel/γ--tocotrienol lipid conjugates. Int J Pharm 536(1):146–157
9. Aykin-Burns N, Pathak R, Boerma M, Kim T, Hauer-Jensen M (2019) Utilization of vitamin E analogs to protect normal tissues while enhancing antitumor effects. Semin Radiat Oncol 29(1):55–61
10. Constantinou C, Charalambous C, Kanakis D (2019) Vitamin E and cancer: an update on the emerging role of γ and δ tocotrienols. Eur J Nutr. https://doi.org/10.1007/s00394-019-01962-1
11. Berrino F, Muti P (1989) Mediterranean diet and cancer. Eur J Clin Nutr 43:49–55
12. Eichholzer M, Stähelin HB, Gey KF, Lüdin E, Bernasconi F (1996) Prediction of male cancer mortality by plasma levels of interacting vitamins: 17-year follow-up of the prospective basel study. Int J Cancer 66(2):145–150
13. Constantinides PP, Han J, Davis SS (2006) Advances in the use of tocols as drug delivery vehicles. Pharm Res 23(2):243–255
14. Zasońska BA, Pustovyy VI, Babinskiy AV, Palyvoda OM, Chekhun VF, Todor I et al (2019) Combined antitumor effect of surface-modified superparamagnetic maghemite nanoparticles and a vitamin E derivative on experimental Walker-256 mammary gland carcinosarcoma. J Magn Magn Mater 471:381–387
15. Constantinou C, Papas A, Constantinou AI (2008) Vitamin E and cancer: an insight into the anticancer activities of vitamin E isomers and analogs. Int J Cancer 123(4):739–752
16. Kang Y-H, Lee E, Youk H-J, Kim SH, Lee HJ, Park Y-G et al (2005) Potentiation by alpha-tocopheryl succinate of the etoposide response in multidrug resistance protein 1-expressing glioblastoma cells. Cancer Lett 217(2):181–190
17. Wang Y-Y, Zhang D-D, Kong Y-Y, Shao L-L, Zhang F-Y, Gao Y et al (2016) CS/PAA@TPGS/PLGA nanoparticles with intracellular pH-sensitive sequential release for delivering drug to the nucleus of MDR cells. Colloids Surf B Biointerfaces 145:716–727
18. Yuan Z, Yuan Y, Han L, Qiu Y, Huang X, Gao F et al (2018) Bufalin-loaded vitamin E succinate-grafted-chitosan oligosaccharide/RGD conjugated TPGS mixed micelles demonstrated improved antitumor activity against drug-resistant colon cancer. Int J Nanomedicine 13:7533
19. Li P-Y, Lai P-S, Hung W-C, Syu W-J (2010) Poly (L-lactide)-vitamin E TPGS nanoparticles enhanced the cytotoxicity of doxorubicin in drug-resistant MCF-7 breast cancer cells. Biomacromolecules 11(10):2576–2582
20. Shi K, Jiang Y, Zhang M, Wang Y, Cui F (2010) Tocopheryl succinate-based lipid nanospheres for paclitaxel delivery: preparation, characters, and in vitro release kinetics. Drug Deliv 17(1):1–10

21. Fernandes RS, Silva JO, Seabra HA, Oliveira MS, Carregal VM, Vilela JM et al (2018) α-Tocopherol succinate loaded nano-structured lipid carriers improves antitumor activity of doxorubicin in breast cancer models in vivo. Biomed Pharmacother 103:1348–1354
22. Pawar A, Patel R, Arulmozhi S, Bothiraja C (2016) D-α-Tocopheryl polyethylene glycol 1000 succinate conjugated folic acid nanomicelles: towards enhanced bioavailability, stability, safety, prolonged drug release and synergized anticancer effect of plumbagin. RSC Adv 6(81):78106–78121
23. Gorain B, Choudhury H, Pandey M, Kesharwani P (2018) Paclitaxel loaded vitamin E-TPGS nanoparticles for cancer therapy. Mater Sci Eng C 91:868–880
24. Tao Y, Han J, Wang X, Dou H (2013) Nano-formulation of paclitaxel by vitamin E succinate functionalized pluronic micelles for enhanced encapsulation, stability and cytotoxicity. Colloids Surf B Biointerfaces 102:604–610
25. Muddineti OS, Rompicharla SVK, Kumari P, Bhatt H, Ghosh B, Biswas S (2019) Vitamin-E/lipid based PEGylated polymeric micellar doxorubicin to sensitize doxorubicin-resistant cells towards treatment. React Funct Polym 134:49–57
26. Noh SM, Han SE, Shim G, Lee KE, Kim C-W, Han SS et al (2011) Tocopheryl oligochitosan-based self assembling oligomersomes for siRNA delivery. Biomaterials 32(3):849–857
27. Muthu MS, Avinash Kulkarni S, Liu Y, Feng S-S (2012) Development of docetaxel-loaded vitamin E TPGS micelles: formulation optimization, effects on brain cancer cells and biodistribution in rats. Nanomedicine 7(3):353–364
28. Liu B-Y, Wu C, He X-Y, Zhuo R-X, Cheng S-X (2016) Multi-drug loaded vitamin E-TPGS nanoparticles for synergistic drug delivery to overcome drug resistance in tumor treatment. Sci Bull 61(7):552–560
29. Lu J, Huang Y, Zhao W, Chen Y, Li J, Gao X et al (2013) Design and characterization of PEG-derivatized vitamin E as a nanomicellar formulation for delivery of paclitaxel. Mol Pharm 10(8):2880–2890
30. Zhao J, Feng S-S (2014) Effects of PEG tethering chain length of vitamin E TPGS with a Herceptin-functionalized nanoparticle formulation for targeted delivery of anticancer drugs. Biomaterials 35(10):3340–3347
31. Mu L, Feng S (2003) A novel controlled release formulation for the anticancer drug paclitaxel (Taxol®): PLGA nanoparticles containing vitamin E TPGS. J Control Release 86(1):33–48
32. Zhang Z, Feng S-S (2006) Nanoparticles of poly (lactide)/vitamin E TPGS copolymer for cancer chemotherapy: synthesis, formulation, characterization and in vitro drug release. Biomaterials 27(2):262–270
33. Feng S-S, Zhao L, Zhang Z, Bhakta G, Win KY, Dong Y et al (2007) Chemotherapeutic engineering: vitamin E TPGS-emulsified nanoparticles of biodegradable polymers realized sustainable paclitaxel chemotherapy for 168 h in vivo. Chem Eng Sci 62(23):6641–6648
34. Gaonkar RH, Ganguly S, Dewanjee S, Sinha S, Gupta A, Ganguly S et al (2017) Garcinol loaded vitamin E TPGS emulsified PLGA nanoparticles: preparation, physicochemical characterization, in vitro and in vivo studies. Sci Rep 7(1):530
35. Zeng X, Tao W, Mei L, Huang L, Tan C, Feng S-S (2013) Cholic acid-functionalized nanoparticles of star-shaped PLGA-vitamin E TPGS copolymer for docetaxel delivery to cervical cancer. Biomaterials 34(25):6058–6067
36. Wang G, Yu B, Wu Y, Huang B, Yuan Y, Liu CS (2013) Controlled preparation and antitumor efficacy of vitamin E TPGS-functionalized PLGA nanoparticles for delivery of paclitaxel. Int J Pharm 446(1–2):24–33
37. Zhu H, Chen H, Zeng X, Wang Z, Zhang X, Wu Y et al (2014) Co-delivery of chemotherapeutic drugs with vitamin E TPGS by porous PLGA nanoparticles for enhanced chemotherapy against multi-drug resistance. Biomaterials 35(7):2391–2400
38. Zhang Y, Huang Y, Zhao W, Lu J, Zhang P, Zhang X et al (2014) Fmoc-conjugated PEG-vitamin E 2 micelles for tumor-targeted delivery of paclitaxel: enhanced drug-carrier interaction and loading capacity. AAPS J 16(6):1282–1291
39. Li Y, Liu Q, Li W, Zhang T, Li H, Li R et al (2016a) Design and validation of PEG-derivatized vitamin E copolymer for drug delivery into breast cancer. Bioconjug Chem 27(8):1889–1899

40. Li Y, Li R, Liu Q, Li W, Zhang T, Zou M et al (2016b) One-step self-assembling nanomicelles for pirarubicin delivery to overcome multidrug resistance in breast cancer. Mol Pharm 13(11):3934–3944
41. Lee K-Y, Chiang Y-T, Hsu N-Y, Yang C-Y, Lo C-L, Ku C-A (2015) Vitamin E containing polymer micelles for reducing normal cell cytotoxicity and enhancing chemotherapy efficacy. Acta Biomater 24:286–296
42. Li Y, Dong M, Kong F, Zhou J (2015) Enhanced therapeutic efficacy and cytotoxicity of doxorubicin-loaded vitamin E–Pluronic micelles against liver cancer. RSC Adv 5(55):43965–43971
43. Cao N, Feng S-S (2008) Doxorubicin conjugated to D-α-tocopheryl polyethylene glycol 1000 succinate (TPGS): conjugation chemistry, characterization, in vitro and in vivo evaluation. Biomaterials 29(28):3856–3865
44. Duhem N, Danhier F, Pourcelle V, Schumers J-M, Bertrand O, LeDuff CS et al (2013) Self-assembling doxorubicin–tocopherol succinate prodrug as a new drug delivery system: synthesis, characterization, and in vitro and in vivo anticancer activity. Bioconjug Chem 25(1):72–81
45. Lu J, Liu C, Wang P, Ghazwani M, Xu J, Huang Y et al (2015) The self-assembling camptothecin-tocopherol prodrug: an effective approach for formulating camptothecin. Biomaterials 62:176–187
46. Liang D-S, Liu J, Peng T-X, Peng H, Guo F, Zhong H-J (2018) Vitamin E-based redox-sensitive salinomycin prodrug-nanosystem with paclitaxel loaded for cancer targeted and combined chemotherapy. Colloids Surf B Biointerfaces 172:506–516
47. Mi Y, Zhao J, Feng S-S (2012) Vitamin E TPGS prodrug micelles for hydrophilic drug delivery with neuroprotective effects. Int J Pharm 438(1–2):98–106
48. Mi Y, Zhao J, Feng S-S (2013) Targeted co-delivery of docetaxel, cisplatin and herceptin by vitamin E TPGS-cisplatin prodrug nanoparticles for multimodality treatment of cancer. J Control Release 169(3):185–192
49. Bao Y, Guo Y, Zhuang X, Li D, Cheng B, Tan S et al (2014) D-α-tocopherol polyethylene glycol succinate-based redox-sensitive paclitaxel prodrug for overcoming multidrug resistance in cancer cells. Mol Pharm 11(9):3196–3209
50. Chen Y, Feng S, Liu W, Yuan Z, Yin P, Gao F (2017) Vitamin E succinate-grafted-chitosan oligosaccharide/RGD-conjugated TPGS mixed micelles loaded with paclitaxel for U87MG tumor therapy. Mol Pharm 14(4):1190–1203
51. Gill KK, Kaddoumi A, Nazzal S (2012) Mixed micelles of PEG2000-DSPE and vitamin-E TPGS for concurrent delivery of paclitaxel and parthenolide: enhanced chemosensitization and antitumor efficacy against non-small cell lung cancer (NSCLC) cell lines. Eur J Pharm Sci 46(1–2):64–71
52. Danhier F, Kouhé TTB, Duhem N, Ucakar B, Staub A, Draoui N et al (2014) Vitamin E-based micelles enhance the anticancer activity of doxorubicin. Int J Pharm 476(1–2):9–15
53. Wang J, Cui S, Bao Y, Xing J, Hao W (2014) Tocopheryl pullulan-based self assembling nanomicelles for anti-cancer drug delivery. Mater Sci Eng C 43:614–621
54. Wu H, Zhong Q, Zhong R, Huang H, Xia Z, Ke Z et al (2016) Preparation and antitumor evaluation of self-assembling oleanolic acid-loaded Pluronic P105/d-α-tocopheryl polyethylene glycol succinate mixed micelles for non-small-cell lung cancer treatment. Int J Nanomedicine 11:6337
55. Zheng Y, Fu F, Zhang M, Shen M, Zhu M, Shi X (2014) Multifunctional dendrimers modified with alpha-tocopheryl succinate for targeted cancer therapy. Med Chem Commun 5(7):879–885
56. Butt AM, Amin MCIM, Katas H (2015) Synergistic effect of pH-responsive folate-functionalized poloxamer 407-TPGS-mixed micelles on targeted delivery of anticancer drugs. Int J Nanomedicine 10:1321
57. Sonali AP, Singh RP, Rajesh CV, Singh S, Vijayakumar MR et al (2016) Transferrin receptor-targeted vitamin E TPGS micelles for brain cancer therapy: preparation, characterization and brain distribution in rats. Drug Deliv 23(5):1788–1798

58. Kutty RV, Feng S-S (2013) Cetuximab conjugated vitamin E TPGS micelles for targeted delivery of docetaxel for treatment of triple negative breast cancers. Biomaterials 34(38):10160–10171
59. Kutty RV, Chia SL, Setyawati MI, Muthu MS, Feng S-S, Leong DT (2015) In vivo and ex vivo proofs of concept that cetuximab conjugated vitamin E TPGS micelles increases efficacy of delivered docetaxel against triple negative breast cancer. Biomaterials 63:58–69
60. Li J, Cheng X, Chen Y, He W, Ni L, Xiong P et al (2016c) Vitamin E TPGS modified liposomes enhance cellular uptake and targeted delivery of luteolin: an in vivo/in vitro evaluation. Int J Pharm 512(1):262–272
61. Raju A, Muthu MS, Feng S-S (2013) Trastuzumab-conjugated vitamin E TPGS liposomes for sustained and targeted delivery of docetaxel. Expert Opin Drug Deliv 10(6):747–760
62. Agrawal U, Sharma R, Gupta M, Vyas SP (2014) Is nanotechnology a boon for oral drug delivery? Drug Discov Today 19(10):1530–1546
63. Feng S-S, Mei L, Anitha P, Gan CW, Zhou W (2009) Poly (lactide)–vitamin E derivative/montmorillonite nanoparticle formulations for the oral delivery of Docetaxel. Biomaterials 30(19):3297–3306
64. Zhao L, Feng SS (2010) Enhanced oral bioavailability of paclitaxel formulated in vitamin E-TPGS emulsified nanoparticles of biodegradable polymers: in vitro and in vivo studies. J Pharm Sci 99(8):3552–3560
65. Munyendo WL, Zhang Z, Abbad S, Waddad AY, Lv H, Baraza LD et al (2013) Micelles of TPGS modified apigenin phospholipid complex for oral administration: preparation, in vitro and in vivo evaluation. J Biomed Nanotechnol 9(12):2034–2047
66. Hou J, Sun E, Zhang Z-H, Wang J, Yang L, Cui L et al (2017) Improved oral absorption and anti-lung cancer activity of paclitaxel-loaded mixed micelles. Drug Deliv 24(1):261–269
67. Pan J, Feng S-S (2009) Targeting and imaging cancer cells by folate-decorated, quantum dots (QDs)-loaded nanoparticles of biodegradable polymers. Biomaterials 30(6):1176–1183
68. Zhao L, Yang B, Wang Y, Yao Z, Wang X, Feng S-S et al (2012) Thermochemotherapy mediated by novel solar-planet structured magnetic nanocomposites for glioma treatment. J Nanosci Nanotechnol 12(2):1024–1031
69. Chandrasekharan P, Maity D, Yong CX, Chuang K-H, Ding J, Feng S-S (2011) Vitamin E (D-alpha-tocopheryl-co-poly (ethylene glycol) 1000 succinate) micelles-superparamagnetic iron oxide nanoparticles for enhanced thermotherapy and MRI. Biomaterials 32(24):5663–5672
70. Song Q, Tan S, Zhuang X, Guo Y, Zhao Y, Wu T et al (2014) Nitric oxide releasing D-α-tocopheryl polyethylene glycol succinate for enhancing antitumor activity of doxorubicin. Mol Pharm 11(11):4118–4129
71. Pais-Silva C, de Melo-Diogo D, Correia IJ (2017) IR780-loaded TPGS-TOS micelles for breast cancer photodynamic therapy. Eur J Pharm Biopharm 113:108–117
72. Wu J, Feng S, Liu W, Gao F, Chen Y (2017) Targeting integrin-rich tumors with temoporfin-loaded vitamin-E-succinate-grafted chitosan oligosaccharide/d-α-tocopheryl polyethylene glycol 1000 succinate nanoparticles to enhance photodynamic therapy efficiency. Int J Pharm 528(1–2):287–298
73. Akhbari K, Karami S, Phuruangrat A, Saedi Z (2018) Irreversible replacement of sodium with thallium in sodium coordination polymer nanostructures by solid-state mechanochemical cation exchange process. J Iran Chem Soc 15(6):1327–1335
74. Hasheminezhad M, Akhbari K, Phuruangrat A (2019) Solid-solid and solid-liquid conversion of sodium and silver nano coordination polymers. Polyhedron. https://doi.org/10.1016/j.poly.2019.03.032
75. Moeinian M, Akhbari K, Kawata S, Ishikawa R (2016a) Solid state conversion of a double helix thallium (I) coordination polymer to a corrugated tape silver (I) polymer. RSC Adv 6(85):82447–82449
76. Moeinian M, Akhbari K, Boonmak J, Youngme S (2016b) Similar to what occurs in biological systems; irreversible replacement of potassium with thallium in coordination polymer nanostructures. Polyhedron 118:6–11

77. Moghadam Z, Akhbari K, Phuruangrat A (2018a) Irreversible conversion of nanoporous lead (II) metal–organic framework to a nonporous coordination polymer upon thermal treatment. Polyhedron 156:48–53
78. Moradi Z, Akhbari K, Costantino F, Phuruangrat A (2017) Solid-state conversion of a three-dimensional sodium (I) coordination polymer with micro trigon morphology to two-dimensional silver (I) coordination polymer nanostructures. Polyhedron 121:33–40
79. Shirazi FS, Akhbari K (2015a) Solid-state thermal conversion of a nanoporous metal–organic framework to a nonporous coordination polymer. RSC Adv 5(63):50778–50782
80. Shirazi FS, Akhbari K, Kawata S, Ishikawa R (2016) Effects of different factors on the formation of nanorods and nanosheets of silver (I) coordination polymer. J Mol Struct 1123:206–212
81. Kang Y-S, Lu Y, Chen K, Zhao Y, Wang P, Sun W-Y (2019) Metal–organic frameworks with catalytic centers: from synthesis to catalytic application. Coord Chem Rev 378:262–280
82. Liu M, Wu J, Hou H (2019) Metal–organic framework (MOF)-based materials as heterogeneous catalysts for C– H bond activation. Chem A Eur J 25(12):2935–2948
83. Noori Y, Akhbari K (2017) Post-synthetic ion-exchange process in nanoporous metal–organic frameworks; an effective way for modulating their structures and properties. RSC Adv 7(4):1782–1808
84. Du M, He H, Zhu Q-Q, Zhao J-N, Sun H, Chen J et al (2019) Rational construction of an exceptionally stable MOF catalyst with metal-adeninate vertices toward CO2 cycloaddition under mild and co-catalyst free conditions. Chem Eur J. https://doi.org/10.1002/chem.201901471
85. Metzger ED, Comito RJ, Wu Z, Zhang G, Dubey RJ-C, Xu W et al (2019) Highly selective heterogeneous ethylene dimerization with a scalable and chemically robust MOF catalyst. ACS Sustain Chem Eng. https://doi.org/10.1021/acssuschemeng.8b05703
86. Tchalala M, Bhatt P, Chappanda K, Tavares S, Adil K, Belmabkhout Y et al (2019) Fluorinated MOF platform for selective removal and sensing of SO_2 from flue gas and air. Nat Commun 10(1):1328
87. Ke X, Song X, Qin N, Cai Y, Ke F (2019) Rational synthesis of magnetic Fe_3O_4@MOF nanoparticles for sustained drug delivery. J Porous Mater 26(3):813–818
88. Lazaro IA, Forgan RS (2019) Application of zirconium MOFs in drug delivery and biomedicine. Coord Chem Rev 380:230–259
89. Alavijeh RK, Beheshti S, Akhbari K, Morsali A (2018) Investigation of reasons for metal–organic framework's antibacterial activities. Polyhedron 156:257–278
90. Sava Gallis DF, Butler KS, Agola JO, Pearce C, McBride A (2019) Antibacterial countermeasures via MOF– supported sustained therapeutic release. ACS Appl Mater Interfaces 11(8):7782–7791
91. Usefi S, Akhbari K, White J (2019) Sonochemical synthesis, structural characterizations and antibacterial activities of biocompatible copper (II) coordination polymer nanostructures. J Solid State Chem 276:61–67
92. Akhbari K, Morsali A (2013) Modulating methane storage in anionic nano-porous MOF materials via post-synthetic cation exchange process. Dalton Trans 42(14):4786–4789
93. Akhbari K, Morsali A (2015) Needle-like hematite nano-structure prepared by directed thermolysis of MIL-53 nano-structure with enhanced methane storage capacity. Mater Lett 141:315–318
94. Gangu KK, Maddila S, Mukkamala SB, Jonnalagadda SB (2019) Characteristics of MOF, MWCNT and graphene containing materials for hydrogen storage: a review. J Energy Chem 30:132–144
95. Xue D-X, Wang Q, Bai J (2019) Amide-functionalized metal–organic frameworks: syntheses, structures and improved gas storage and separation properties. Coord Chem Rev 378:2–16
96. Mirzadeh E, Akhbari K (2016) Synthesis of nanomaterials with desirable morphologies from metal–organic frameworks for various applications. CrstEngComm 18(39):7410–7424

97. Moeinian M, Akhbari K (2015) How the guest molecules in nanoporous Zn (II) metal-organic framework can prevent agglomeration of ZnO nanoparticles. J Solid State Chem 225:459–463
98. Moeinian M, Akhbari K (2016) Solid-state synthesis of zinc oxide nano-structures with similar morphologies to their precursors and metal–organic frameworks topology. J Iran Chem Soc 13(3):547–552
99. Shirazi FS, Akhbari K (2015b) Preparation of zinc oxide nanoparticles from nanoporous metal–organic framework with one-dimensional channels occupied with guest water molecules. Inorg Chim Acta 436:1–6
100. Filippousi M, Turner S, Leus K, Siafaka PI, Tseligka ED, Vandichel M et al (2016) Biocompatible Zr-based nanoscale MOFs coated with modified poly(ε-caprolactone) as anticancer drug carriers. Int J Pharm 509(1):208–218
101. Akhbari K, Morsali A (2008) Spectroscopic, thermal, fluorescence and structural studies of new TlI pyridine dicarboxylate complexes,[Tl2 (py-2, 5-dc)] and [Tl2 (py-3, 5-dc)]. J Mol Struct 878(1–3):65–70
102. Noori Y, Akhbari K, Phuruangrat A, Costantino F (2017) Self-assembly through secondary interactions in formation of two-dimensional lead (II) supramolecular polymer with nanosheets morphology. J Mol Struct 1130:311–318
103. Mirzadeh E, Akhbari K, White J (2018) Mechanochemical conversion of nano potassium hydrogen terephthalate to thallium analogue nanoblocks with strong hydrogen bonding and straight chain metallophillic interactions. Appl Organomet Chem 32(5):e4313
104. Moghadam Z, Akhbari K, White J, Phuruangrat A (2018b) A new potassium-based coordination polymer with hydrogen bonding and zigzag metallophillic interactions. Appl Organomet Chem 32(12):e4613
105. Yusefi S, Akhbari K, White J, Phuruangrat A (2019) Sonochemical synthesis of a two-dimensional supramolecular polymer with nanoporous morphology, linear thallophilic and covalent hydrogen-bonding interactions. Appl Organomet Chem 33(3):e4747
106. Sarker M, Shin S, Jhung SH (2019) Functionalized mesoporous metal-organic framework PCN-100: an efficient carrier for vitamin E storage and delivery. J Ind Eng Chem 74:158–163

Rutin-Based Phytomedicines for Cancer Benefit

Mozhgan Parsaei and Kamran Akhbari

1 Introduction

For thousands of years, nature was considered an ideal source of pharmacological agents; as a result, the majority of medicines were composed or at least derived from natural products. In particular, medicinal plants provided a plethora of innovative chemotherapeutic agents that revolutionized the anticancer therapies clearly improving patient treatment outcome. Among different groups of natural products, flavonoids play an important role in the treatment of various disorders. Chemically, they are benzopyrone derivatives and widely located in various sources of vegetal origin (fruits, seeds, roots, flowers, tea, or wine). Their name has been derived from the Latin word *flavus* meaning yellow, and many of these compounds are responsible for the coloration of flowers, yolks, or leaves in autumn.

Natural compounds such as bioflavonoids have found application in healthcare system due to their wide biological activities, high safety margins, and lower cost. The flavonoids have an important role in successful medical treatments from past times to present. Flavonoids are also known as nature's tender drugs as they possess various biological/pharmacological activities including antioxidative activity, anti-inflammatory effects, antitumor effects, antithrombogenic effects, antibacterial activity, and antiviral activity. It is believed that many of the therapeutic effects of flavonoids result from their potent antioxidant and free radical-scavenging properties. Thus, flavonoids are getting increasing attention from the pharmaceutical and food industries. Medicines and food supplements containing flavonoids have been suggested to treat diseases.

M. Parsaei · K. Akhbari (✉)
School of Chemistry, College of Science, University of Tehran, Tehran, Iran
e-mail: akhbari.k@khayam.ut.ac.ir

© Springer Nature Singapore Pte Ltd. 2020

M. Rahman et al. (eds.), *Nanomedicine for Bioactives*,
https://doi.org/10.1007/978-981-15-1664-1_3

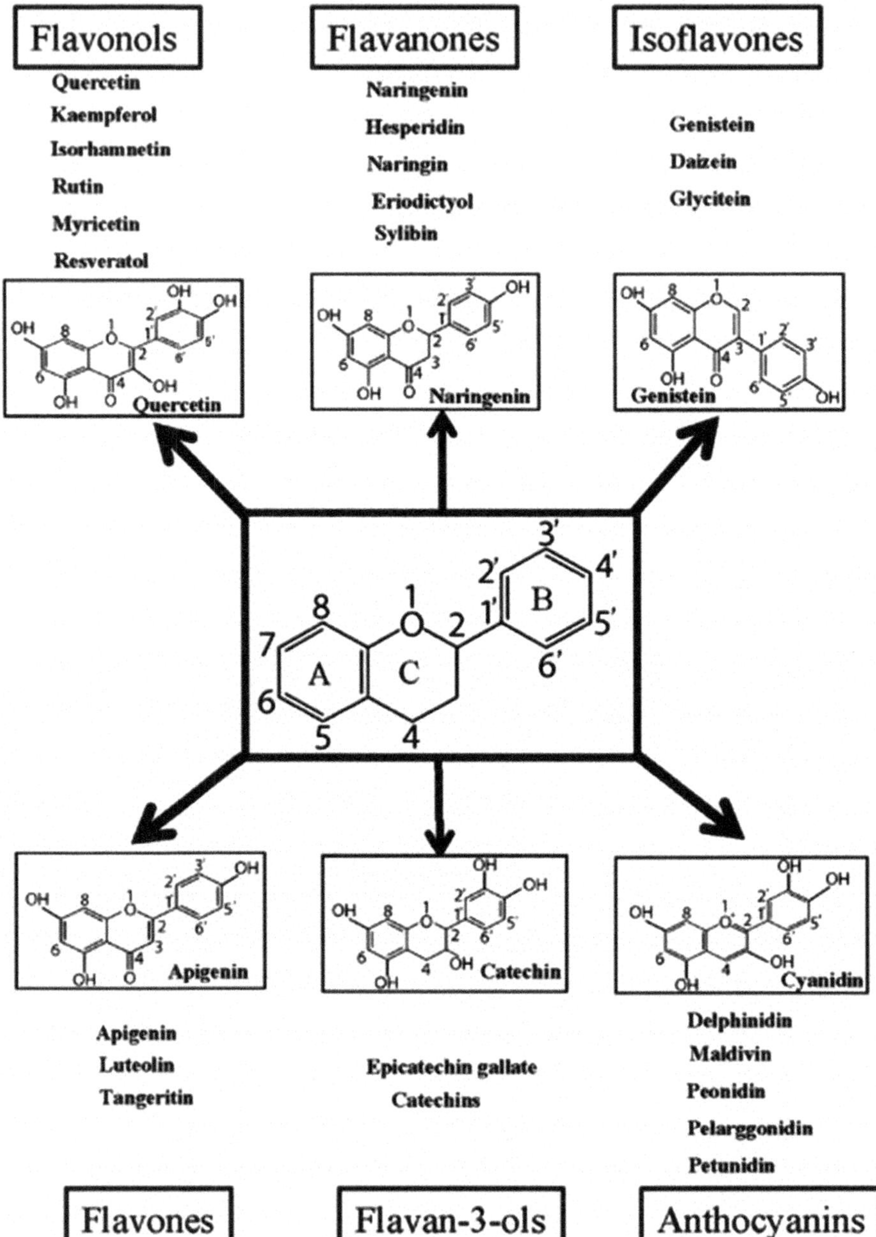

Fig. 1 Classification and example of flavonoids and their chemical structures (Reproduced from reference [1] under a Creative Commons License)

Flavonoids are secondary herbal products with numerous hydroxyl/phenolic groups in their molecular structure, which are responsible for their significant features. Based on the variation in the type of heterocycle involved, flavonoids are divided into six major subclasses (Fig. 1): flavonols, flavanones, flavanols, flavones, anthocyanins, and isoflavones. Individual differences within each group arise from the variation in number and arrangement of the hydroxyl groups and their extent of alkylation and/or glycosylation. Flavonols (such as quercetin and kaempferol) have a 3-hydroxy pyran-4-one group on the C ring. Flavanones (such as naringenin and taxifolin) have an unsaturated carbon-carbon bond in the C ring. Flavanols (such as the catechins) lack both a 3-hydroxyl group and the 4-one structure in the C ring. Flavones (such as luteolin) lack a hydroxyl group in the 3-position on the C ring. Anthocyanins (such as cyanidin) are characterized by the presence of an oxonium ion on the C ring and are highly colored as a consequence, and in isoflavones (such as genistein), the B ring is attached to the C ring in the 3-position, rather than the 2-position as is the case with the other flavonoids [1].

The use of flavonoids in therapy is broad and ancient, yet often empiric. Today, many efforts are underway to examine the possibility of cancer treatment by natural chemical compounds. Meanwhile, natural herbal compounds are more widely considered especially due to lower side effects (Fig. 2).

The effect of polyphenols on human cancer cell lines is most often protective and induces a reduction of the number of tumors or of their growth. These effects have been observed at various sites, including the mouth, stomach, duodenum, colon, liver, lung, mammary gland, or skin. Several mechanisms of action have been identified for chemoprevention effect of polyphenols; these include estrogenic/antiestrogenic activity, antiproliferation, induction of cell cycle arrest or apoptosis, prevention of oxidation, induction of detoxification enzymes, regulation of the host immune system, anti-inflammatory activity, and changes in cellular signaling [2].

Rutin (Fig. 3), also called rutoside, quercetin-3-O-rutinoside, and sophorin, is a flavonoid glycoside consisting of the aglycone form, quercetin bound at the C-3

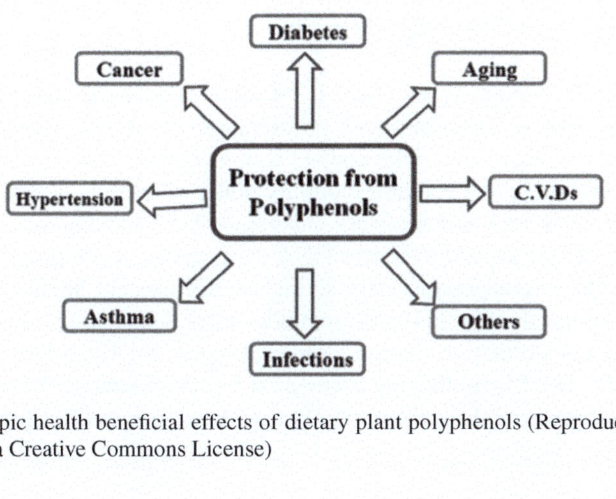

Fig. 2 Pleiotropic health beneficial effects of dietary plant polyphenols (Reproduced from reference [2] under a Creative Commons License)

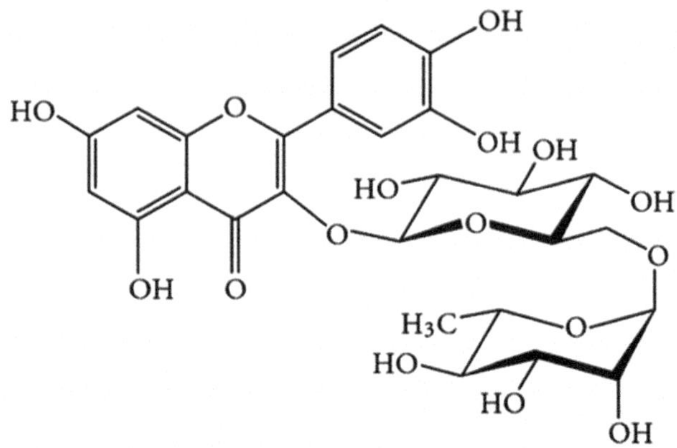

Fig. 3 Chemical structure of the flavonoid rutin

position (on ring C) to a disaccharide molecule, rutinose ($C_{12}H_{22}O_{10}$), which is composed of one molecule of rhamnose and one of glucose [3]. Rutin is found in many dietary sources such as onion, grape, apple, fruit drinks (especially citrus fruits, such as orange, grapefruit, lemon, and lime), red wine, black tea, buckwheat, red pepper, tomato, and berries such as mulberry, ash tree fruits, and cranberries. Rutin has significant scavenging properties on oxidizing species such as OH radical, superoxide radical, and peroxyl radical [4]. Therefore, it shows several pharmacological activities including antiallergic [5], anti-inflammatory and vasoactive [6], antitumor [7], antibacterial, antiviral, and antiprotozoal properties [8]. Moreover, it has also been reported that rutin has other therapeutic effects such as hypolipidemic [9], anticarcinogenic [10], and antidiabetic effect [11]. Rutin is also advantageous over aglycones whose use as pharmaceutical agents is restricted due to their mutagenic and cytotoxic activity. Therefore, it can be worth considering it being a potential but nontoxic and nonoxidizable molecule [12].

2 Anticancer Activity of Rutin

Cancer includes a group of disease that is characterized by the abnormal growth of cells and by the lack of the control cell proliferation and differentiation that latently assault or spread to other parts of the body [13]. It is a non-communicable disease that represents the second largest cause of mortality in the world [14]. The mechanisms by which cancer occurs are not completely understood [15]. Induction of cell proliferation, decreased apoptosis, and oxidative DNA damage have been proposed as predisposing factors to carcinogenesis [16]. Evidence from recent studies suggests that cancer cells, compared to normal cells, are under increased oxidative stress associated with oncogenic transformation, alterations in metabolic activity, and increased generation of ROS [17, 18]. Chemotherapy is essential for the treatment of a variety of cancers. Cancer is a major public health burden both in developed and developing countries.

Carcinogenesis is a complex process that involves multiple phases where molecular and cellular alterations, particularly of genetic origin, may occur [19]. The three phases involved in this process include the initiation stage in which the normal cell is exposed to the carcinogenic agent causing a genetic alteration, the promotion stage in which survival and replication of damaged cells occurs, and progression stage which is characterized by deregulation of cellular proliferation and differentiation, reduction of apoptosis of damaged cells (tumor growth), and increase of metastatic and angiogenic potential [19, 20]. Many anticancer agents from natural sources including plants, marine organisms, and microorganisms were tested for cytotoxic activity against cancer in cell lines grown in vitro or in vivo animal cancer models. Herbal-based dietary supplements contain many phytochemicals such as flavonoids which may contribute to cancer suppression. Polyphenols have been reported to interfere with cancer initiation, promotion, and progression, thereby acting as chemopreventive agents. These compounds act as anticancer agents by inhibiting cell growth, by inducing cell cycle arrest and/or apoptosis, or by inhibiting proliferation, angiogenesis, and/or metastasis and exhibit anti-inflammatory and/or antioxidant effects [21]. Cancer chemoprevention using bioactive compounds has attracted increasing attention in recent years [22]. Bioactive compounds with anticancer properties alone or in combination with other potent synthetic molecules are used to prevent or reverse the processes of carcinogenesis, furthermore to minimize the undesirable side effects, which are commonly associated with current therapies [23]. Different plant-derived compounds target different and diverse cancer-related signaling pathways. A plant derivative that is effective at a particular dosage for the treatment of one type of cancer may not be as efficient in the treatment of other forms of cancer owing to the complexity and diversity among different cancers. It is, therefore, necessary to identify a unified mechanism which can be attributed to and one which can be targeted for the treatment of many different cancers. It is also necessary that the active principles in crude plant extracts be clearly identified and, if possible, modified to increase their potency in the treatment of cancer. Additionally, the anticancer effects of plant derivatives were highly dose/concentration specific. While higher doses of a plant derivative may inhibit cancer signaling pathways, suppress tumor growth, and prove to be cytotoxic, lower concentrations of the same compound may support cell growth and proliferation. Therefore, detailed pharmacology/pharmacokinetic studies are warranted to determine the dosage for a plant-derived compound within the therapeutic window, the route of administration, and the required number of interventions to bring about the best possible anticancer effect with the least adverse side effects.

Over the past few years, many studies have provided evidence that rutin acts as an effective chemopreventive agent [24, 25]. Anticancer effects of rutin were investigated by Yang et al. in human leukemia HL-60 cells, and it was found that rutin inhibited tumor growth in a xenograft animal model [26]. For this purpose, human leukemia HL-60 cells (1×10^7/100 μl culture medium) were implanted into mice for 12-day incubation for solid tumor growth (Fig. 4). Animals bearing tumors were randomly assigned to three groups (six mice per group), and treatment was initiated when tumors reached volumes of about 200 mm^3. Group 1 was treated with 1% DMSO i.p. (intraperitoneally) only. Group 2 was treated with rutin i.p., and group 3 was treated

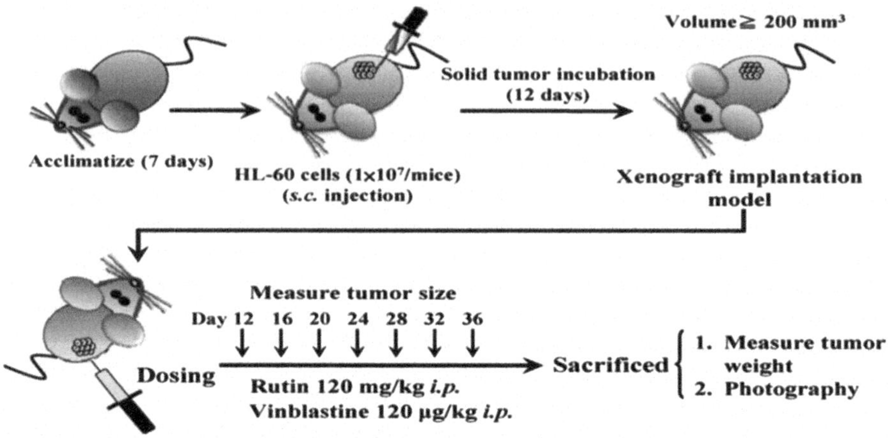

Fig. 4 Experimental design of rutin-suppressed HL-60 tumor in a xenograft animal model (Reproduced from reference [26] with permission from Wiley)

with vinblastine i.p. for 36 days. During the treatment, tumor size and weight of each animal were measured. The mice were injected i.p. once every 4 days with 30 μl of 1% DMSO control vehicle, rutin (120 mg/kg), and vinblastine (120 μg/kg).

It was found that treatment with 120 mg/kg of rutin or with 120 μg/kg of vinblastine resulted in a reduction of tumor weight and volume when compared with the control groups. Rutin decreased tumor weight by 62.99% when compared to control, and tumor size was also found to be significantly smaller in xenograft mice treated with 120 mg/kg of rutin than that in untreated control group (Fig. 5). These findings clearly indicated that rutin can also act as a leukemia preventive agent.

The antitumor effect of rutin was also investigated by Alonso-Castro et al. on a murine model [27]. In this study, human tumor-derived cell lines from cervical (HeLa), colorectal (SW-480), breast (MDA-MB-231), and liver (HepG2) carcinomas and leukemia (K562), nontumorigenic human immortalized keratinocytes (HaCaT), and ovarian (SKOV-3) and prostate (DU-145) carcinoma cell lines were used. Cells were seeded in 96-well microplates at a density of 5000 cells/well. After 24 h, cells were incubated with concentrations of RTN (rutin) or CDDP (cisplatin) between 0.1 and 300 μM. DMSO at a final concentration of 0.01% was used as vehicle control. After 48 h of treatment, MTT assay was performed, and relative viability was calculated as described. The concentration leading to 50% inhibition of viability (IC_{50}) was also calculated by regression analysis (percent survival vs. log concentration). Mice (weighing 18–23 g) were injected with SW480 cells (5 × 106), and 8 days after tumor implantation, mice were treated with doses of RTN between 1 and 20 mg/kg or CDDP 1 mg/kg daily over a 32-day period. Tumors were measured and weighed. RTN lacks toxic effects in nu/nu mice. Relative organ and body weight of untreated mice and mice treated with RTN or CDDP was calculated. The positive antitumoral drug CDDP 1 mg/kg reduced tumor growth by 63%, compared to untreated mice (Fig. 6a). RTN induced antitumoral effects in a dose-dependent manner. RTN administered at 1, 10, and 20 mg/kg inhibited significantly

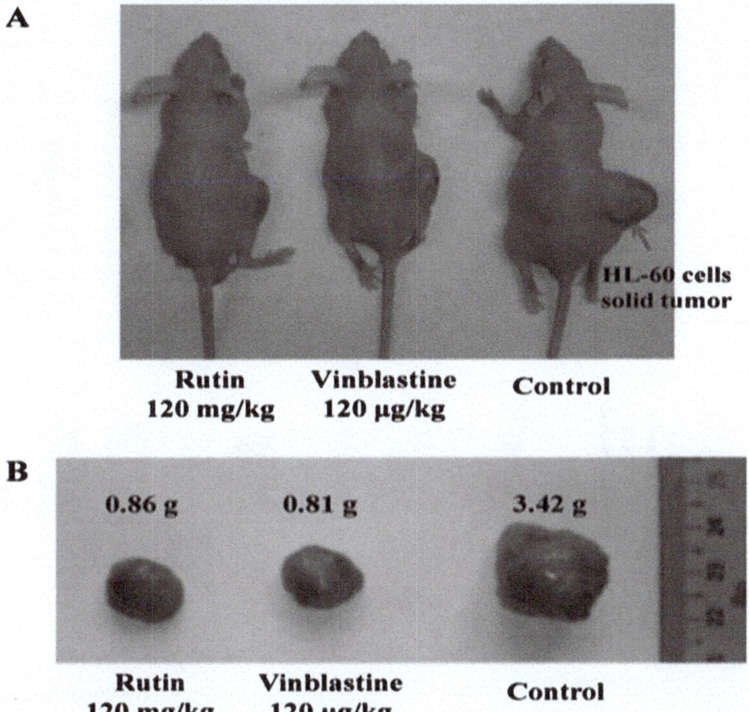

Fig. 5 Representative tumor on the xenograft animal model. (**a**) Representative animal with solid tumor; (**b**) solid tumor and tumor weight (Reproduced from reference [26] with permission from Wiley)

($p < 0.05$) the growth of the SW480 tumors by 16%, 38%, and 58%, respectively (Fig. 6a). Furthermore, CDDP 1 mg/kg decreased tumor weight by 55% (Fig. 6b). RTN also decreased tumor weight by 15% (1 mg/kg), 33% (10 mg/kg), and 53% (20 mg/kg). The toxic effects of RTN on mice were also evaluated. The results indicated that CDDP significantly decreased ($p < 0.05$) the relative organ weight of the spleen, liver, heart, and kidney and decreased body weight by 7% compared to untreated mice (Fig. 6c, d). In contrast, RTN at all concentrations (1e20 mg/kg) did not affect relative organ weight or body weight (Fig. 6c, d).

The results indicated that rutin induces high cytotoxic effects in vitro against SW480 cells. Also, rutin exerts in vivo antitumor and antiangiogenic activities. Thus, rutin may be a good option for cancer treatment due to its low toxicity in nontumorigenic cells and its antitumor effects. The results indicated that RTN showed the highest cytotoxic effects against SW480 cells ($IC_{50} = 125$ μM). Although RTN showed lower cytotoxic potency on SW480 cells compared to CDDP, RTN did not display toxic effects on nontumorigenic cells (HaCaT; $IC_{50} > 300$ μM). In contrast, CDDP showed high cytotoxic effects on nontumorigenic cells ($IC_{50} = 8.8$ μM). The results suggest that rutin exerts cytotoxic effects mainly against SW480 cancer cells with low cytotoxic effects on human nontumorigenic cells.

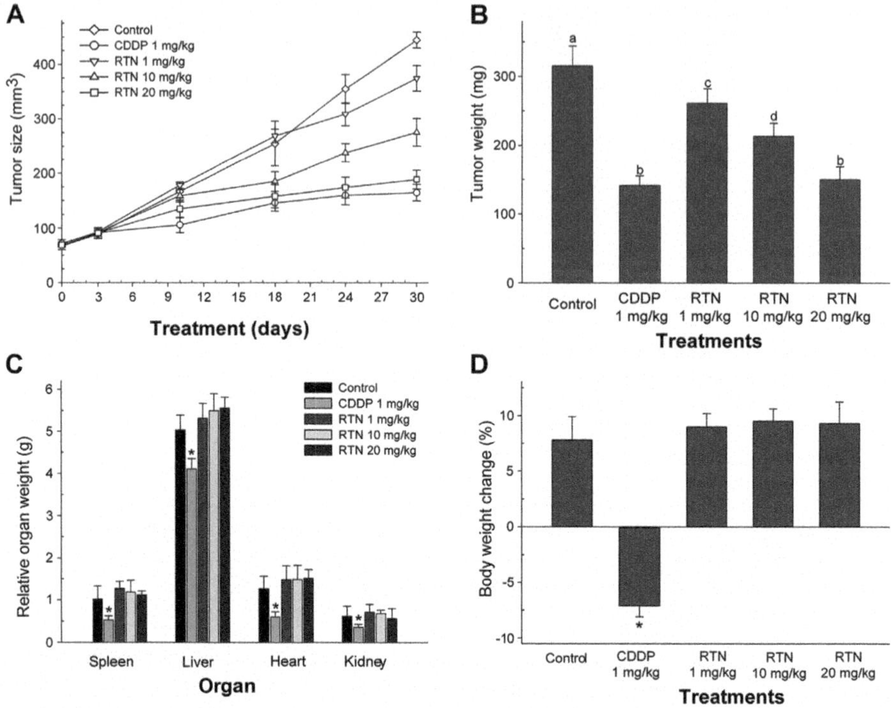

Fig. 6 RTN induces antitumor effects in nu/nu mice bearing SW480 tumor. Mice were injected with SW480 cells (5 × 10) and 8 days after tumor implantation, mice were treated with doses of RTN between 1 and 20 mg/kg or CDDP 1 mg/kg daily over a 32-day period. Tumors were measured (**a**) and weighed (**b**). RTN lacks toxic effects in nu/nu mice. Relative organ weight (**c**) and body weight (**d**) of untreated mice and mice treated with RTN or CDDP was calculated as described in Materials and Methods. Data are representative of three independent experiments in quintuplicate. Results represent the mean ± standard deviation (SD). Lowercase letters indicate significant differences according to ANOVA test ($p < 0.05$), followed by post hoc Tukey test (Reproduced from reference [27] with permission from Elsevier)

The anticancer effect of rutin and silibinin in combination on colon cancer cells, HT-29, was examined by Nafees et al. [28]. Cells were treated with increasing concentration of rutin, silibinin, and their combination, and cytotoxic effect was assessed by MTT assay. Results showed that rutin and silibinin induced cytotoxicity in HT-29 cell line in a dose-dependent manner; moreover, their combination exhibited higher sensitivity. IC_{50} values of rutin, silibinin, and their combination were estimated to be 78.025, 150.753, and 73.978 at 24 h and 39.16, 75.70, and 37.051 at 48 h respectively, in HT-29 colon cancer cell line. Results confirmed that rutin and silibinin in combination induced significantly higher toxicity in comparison to their individual treatment in HT-29 colon cancer cell line. Apoptosis is a programmed cell death with the involvement of diverse genetic pathways having a significant function in the progress and maintenance of cellular machinery in normal tissues. DNA breakage in colon cancer cells was tested (IC_{50}, 48 h) in the presence of rutin, in silibinin individually, and in their combination using comet assay. As shown in Fig. 7, DNA

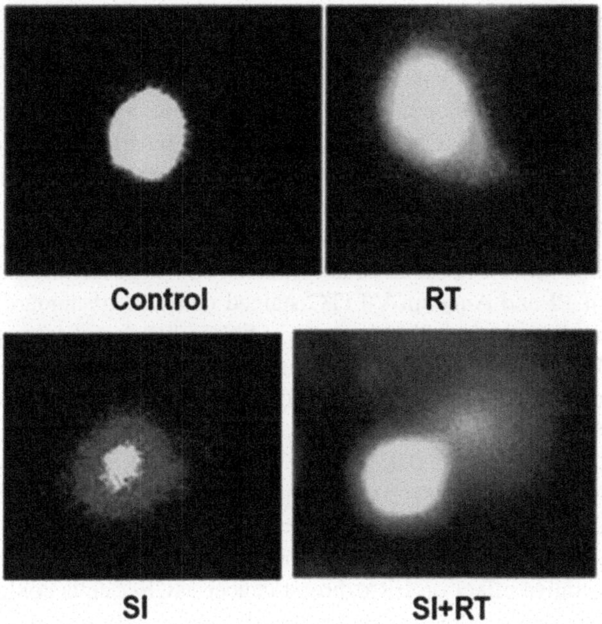

Fig. 7 Effect on DNA breakage. DNA damage by rutin and silibinin individually was observed as shown by increase in corresponding comet tail length (Reproduced from reference [28] with permission from Elsevier)

damage by rutin and silibinin individually was observed as evident by progressive increase in corresponding comet tail length. The image length from 2.116 ± 0.039 to 14.884 ± 0.664 μm (in case control no significant tail lengths) was observed as shown in Fig. 7. However, the extent of DNA damage was enhanced in their combination. Thus, it is established that rutin and silibinin in combination were capable of causing DNA breakage in colon cancer cells which is a marker of apoptotic cell death.

Results supported the viability of developing rutin and silibinin in combination as a novel therapeutic prodrug for colon cancer treatment.

In another study, p53- and cyclin b1-mediated apoptotic effects of apigenin and rutin in erα^+-breast cancer MCF-7 cells were investigated by Hasani et al. [29]. Tamoxifen is an effective treatment for estrogen receptor alpha-positive (ERα^+) breast cancer; however, patients who received the treatment for 5 years have greater mortality risk compared to those who did not receive tamoxifen. Furthermore, patients treated with tamoxifen developed resistance to the drug which is mediated through p53 and PTEN. Therefore, the study was undertaken to determine the potential adjuvant properties of flavonoids, apigenin, and rutin to promote the anticancer activity induced by tamoxifen using ERα^+-breast cancer MCF-7 cell lines. MCF-7 and nontransformed breast MCF-10A cells were treated separately with apigenin, rutin, tamoxifen, or the combination of each flavonoids with tamoxifen. Antiproliferative activity and respective IC_{50} concentrations were determined using MTT assay. The respective IC_{50} concentrations obtained were used in the subsequent experiments. The antiproliferative mechanism was determined using Annexin

V-FITC morphological staining and DNA fragmentation assays. The effect on tumor suppressor (p53 and PTEN) and cell cycle-related genes (p21, CDK1, and cyclin B1) was determined by QuantiGene Plex assay. Results indicate that the MCF-7 cells were more sensitive to apigenin and rutin compared to MCF-10A and tamoxifen. Antiproliferative effect on MCF-7 cells was enhanced by apigenin and rutin. Results showed that apigenin and rutin treatments lead to apoptosis. To examine whether cells undergo apoptosis, untreated, flavonoid-treated (apigenin or rutin), and tamoxifen-treated breast cancer MCF-7 and nontransformed breast MCF-10A cells were stained with Annexin V-FITC and PI at respected IC_{50} concentrations. Flow cytometry analysis of PI and Annexin V-FITC stained cells are a standard procedure to distinguish early apoptotic cells (Annexin V-FITC−/PI−) and late apoptotic cells (Annexin V-FITC+/PI+). Apigenin, rutin, and tamoxifen treatments in MCF-7 cells resulted in total apoptotic population of 48.40 ± 2.41% ($p < 0.01$, $n = 3$), 47.60 ± 1.67% ($p < 0.01$, $n = 3$), and 51.05 ± 0.62% ($p < 0.01$, $n = 3$), respectively, compared to untreated control cells (1.45 ± 0.03 at $p < 0.01$, $n = 3$). Rutin treatment only resulted in 13.07 ± 0.76% ($p < 0.01$, $n = 3$) of total apoptotic population in MCF-10A cells compared to tamoxifen, 45.73 ± 2.94% ($p < 0.01$, $n = 3$) even though rutin IC_{50} concentration was thrice higher than tamoxifen. This study revealed that apigenin and rutin antiproliferative effect on $ER\alpha^+$-breast cancer MCF-7 cells could be mediated through p53-dependent-G2/M phase arrest and apoptosis. It appeared that apigenin and rutin are able to synergistically enhance tamoxifen efficacy on MCF-7 cells and are potential candidates as adjuvant agents in reducing tamoxifen adverse effects.

Chemopreventive potential of rutin was also determined by ben Sghaier et al. in human lung (A549) and colon (HT29 and Caco-2) cancer cell lines [30]. In this study, they examined the effect of rutin on the viability, superoxide anion production, adhesion, and migration of human lung (A549) and colon (HT29 and Caco-2) cancer cell lines. They investigated the potential antiproliferative and antioxidant activities of rutin in human lung (A549) and colon (HT29 and Caco-2) cancer cell lines. In order to control the harmlessness of the tested concentrations of rutin, the viability of cancer cell lines was assessed using a 3-(4,5-dimethylthiazol-2-yl)-2,5-diphenyltetrazolium bromide (MTT) assay. ROS generation was measured by lucigenin chemiluminescence detecting superoxide ions. To investigate the effect of rutin on the behavior of human lung and colon cancer cell lines, we performed adhesion assays, using various purified extracellular matrix (ECM) proteins. Finally, in vitro cell migration assays were explored using modified Boyden chambers. The viability of cancerous cells was inhibited by rutin. It also significantly attenuated the superoxide production in HT29 cells. In addition, rutin affected adhesion and migration of A549 and HT29 cell. They evaluated the cytotoxicity of rutin (31.25–1000 mM) after 24 h of incubation on the different cancer cell lines (A549, HT29 and Caco-2) using MTT assay. Cells were plated at 5×10^3 cells per dish in the absence and the presence of rutin (31.25–1000 mM). As control, the cells were treated without or with 0.1% DMSO (vehicle). Rutin was able to inhibit the viability of A549, HT29, and Caco-2 cells in a dose-dependent manner. A significant difference in growth-inhibitory effects of this compound on A549, HT29, and Caco-2 cells was revealed. Rutin was most efficient in HT29 cell with an estimated IC_{50} value of 388.33 ± 4 μM after 72 h.

However, A549 and Caco-2 cells seemed to be more resistant to flavonoid treatment with IC_{50} values of 559.83 ± 3.5 μM and 710.59 ± 3.5 μM, respectively. They showed rutin inhibits adhesion of tumor cells. A549 cells were seeded at 31.2–1000 μM and were preincubated with various concentrations of rutin (25, 100, and 200 μM) for 30 min at room temperature. The cells were then added to 96-well microtiter plates coated with fibronectin (10 mg/ml), vitronectin (10 mg/ml), laminin 1 (10 mg/ml), fibrinogen (50 mg/ml), collagen types I and IV (10 mg/ml), and poly-L-lysine (10 and 50 mg/ml) and allowed to adhere for 1 h at 37 °C. HT29 cells were seeded at 10^6 cells/ml and were preincubated with various concentrations of rutin (25, 100, and 200 μM) for 30 min at room temperature. The cells were then added to 96-well microtiter plates coated with fibronectin (10 mg/ml), collagen types I and IV (10 mg/ml), and poly-L-lysine (10 mg/ml) and allowed to adhere for 2 h at 37 °C. After washing, adherent cells were stained with crystal violet and solubilized by SDS, and absorbance was measured at 600 nm. Rutin impaired attachment in a dose-dependent manner of both A549 and HT29 cells to fibronectin (inhibition of cell adhesion by 47 ± 2.5% for A549 cells and 38 ± 1.9% for HT29 cells in presence of rutin 200 μM). An impairment of attachment to collagen types I and IV was observed in A549 (inhibition of cell adhesion by 36.6 ± 2 and 50.2 ± 3%, respectively). In contrast, no effect of rutin in attachment of HT29 cells to collagen types I and IV was observed In addition, rutin inhibited human lung A549 cell migration and colon (HT29) cancer cell lines. Cell migration can be considered as a finely regulated process including successive steps of cell adhesion and deadhesion. Results showed that rutin at 100 and 200 μM blocked A549 and HT29 cell migration. This inhibition was dose dependent with migration percentage of 47.9 ± 1.5 and 38.61 ± 1% in the presence of type IV collagen and 52.75 ± 1 and 9.48 ± 0.5% in the presence of fibronectin toward A549 cells with rutin at 100 and 200 μM, respectively. For HT29 cells, the migration percentage values were 55.19 ± 2.5 and 37.63 ± 1.5% toward type IV collagen and 54.49 ± 3 and 37.82 ± 2% toward fibronectin with rutin at 100 and 200 μM, respectively.

ROS generation was measured by lucigenin chemiluminescence detecting superoxide ions. After incubation of the cells for the desired time with drugs in 96-well plates (50 × 10^3 cells/well), luminescence was detected by a Fluoroscan Ascent FL fluorimeter. The detected signal was assessed at each minute over the course of 45 min. The results represent the integration of the signal for 45 min which is associated with the intermediate time of measurement. All measurements were performed at 37 °C, and the results were expressed as total reactive oxygen species measurements which represent the difference of reactive oxygen species production measured in untreated control. As high level of ROS production exists in tumor cells, ROS production has been linked to either the cell proliferation or cell death. The effect of rutin was investigated on superoxide anion production in the different cell lines. It appears that the superoxide anion-reducing effect was dose dependent in the presence of rutin. Therefore, the maximum reducing effect of rutin was observed at a concentration of 100 μM against HT29 cells with a reduction percentage of 64.41 ± 2%. However, at this concentration, rutin was able to reduce superoxide anion production with a reduction percentage of 56.65 ± 3 and 31.12 ± 1.5%

against A549 and Caco-2 cells, respectively. This result indicates that rutin is more effective in HT29 cells.

In another study, protective effect of rutin supplementation against cisplatin-induced nephrotoxicity in rats was investigated by Alhoshani et al. [31]. Cisplatin (CP) is commonly used in the treatment of different types of cancer, but nephrotoxicity has been a major limiting factor. This study investigated the possible protective effects of rutin against cisplatin-induced nephrotoxicity in rats. 40 male Wistar albino rats were randomly divided into 4 groups. Rats of group 1 control group intraperitoneal (i.p.) received 2.5 ml/kg, group 2 CP group received single-dose 5 mg/kg cisplatin i.p., group 3 rutin group orally received 30 mg/kg rutin, and group 4 (CP plus rutin) received CP and rutin as in groups 2 and 3. Kidneys were harvested for histopathology and for the study of the gene expression of c-Jun N-terminal kinases (JNK), mitogen-activated protein kinase 4 (MKK4), MKK7, P38 mitogen-activated protein kinases (P38), tumor necrosis factor alpha (TNF-α), TNF receptor-associated factor 2 (TRAF2), and interleukin-1 alpha (IL-1-α). Results showed that the cisplatin single-dose administration to rats induced nephrotoxicity associated with a significant increase in blood urea nitrogen (BUN) and serum creatinine and significantly increase malondialdehyde (MDA) in kidney tissues by 230 ± 5.5 nmol/g compared to control group. The animal treated with cisplatin showed a significant increase in the expression levels of the IL-1α (260%), TRFA2 (491%), P38 (410%), MKK4 (263%), MKK7 (412%), JNK (680%), and TNF-α (300%) genes compared to control group. Additionally, histopathological examination showed cisplatin-induced interstitial congestion, focal mononuclear cell inflammatory, cell infiltrate, acute tubular injury with reactive atypia, and apoptotic cells. Rutin administration attenuated cisplatin-induced alteration in gene expression and structural and functional changes in the kidney. Additionally, histopathological examination of kidney tissues confirmed gene expression data. Nephrotoxicity induced by CP is characterized by a reduction in renal function that leads to increase in serum creatinine and blood urea levels. Blood urea nitrogen (Fig. 8a) and serum creatinine (Fig. 8b) were used as biochemical markers for the nephrotoxicity. CP significantly increased the levels of BUN (128.6 ± 44 mg/dl) and serum creatinine (4.6 ± 0.34 mg/dl), compared to control group 37 ± 2.4 mg/dl and 1.2 ± 0.1 mg/dl, respectively, ($p < 0.001$). The rutin group showed no significant changes in BUN and serum creatinine compared to control group. However, administration of rutin in combination with CP resulted in complete reversal of CP-induced increase in BUN and serum creatinine to their normal values as in control group.

To investigate the effect of CP on oxidative stress genes, expression levels of IL-1α were measured in kidney tissues by using real-time PCR (Fig. 9). CP alone significantly increased the expression level of IL-1α in kidney tissues by 260% ($p < 0.05$) and 164% ($p < 0.001$) compared to control and rutin groups, respectively. Interestingly, administration of rutin to CP-treated rats resulted in a complete reversal of the reduction of IL-1α expression level induced by CP to control values. This reversal change results in significant decrease in IL-α expression level by 63% ($p < 0.007$) compared to CP group and by 73% compared to control group.

For histopathology examination, the kidneys harvested from each groups were fixed in 10% neutral buffered formaldehyde. Tissue dehydration, clearing in xylene,

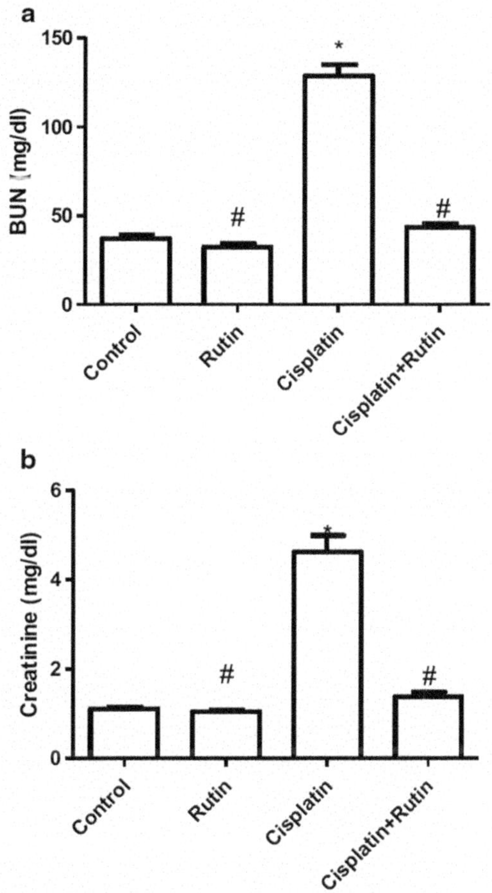

Fig. 8 Represents the changes in the levels of serum BUN (**a**) and creatinine (**b**) in rats. Data were presented as mean ± SEM ($n = 10$). Asterisk indicates significant change from control. Hash indicates significant changes from rutin (Reproduced from reference [31] under a Creative Commons License)

and paraffin embedding were done according to the standard method. Sections were cut by a rotary microtome at 5–7 μm thick and were stained by hematoxylin and eosin and periodic acid Schiff (PAS). Sections were examined under a light microscope and findings documented by two certified histopathologists. The MDA and GSH were measured as biomarkers for the oxidative stress. In the kidney tissue, the MDA level was significantly increased, and GSH level was decreased by the effect of cisplatin. However, rutin administration caused significant decreases in lipid peroxidation and promoted increases in GSH content in the kidney. Therefore, rutin can protect the kidney from CP-induced injury via improvement in oxidant status. At the end of the experiment, CP-treated animals significantly lost weight compared to control group ($p < 0.05$). However, administration of rutin alone resulted in an increase in the body weight compared to both control and cisplatin group. Interestingly, administration of rutin in combination with CP resulted in a significant increase in the body weight compared to CP group ($p < 0.05$). This study suggested that the antioxidant and anti-inflammatory effect of rutin may prevent CP-induced nephrotoxicity via decreasing the oxidative stress, inhibiting the

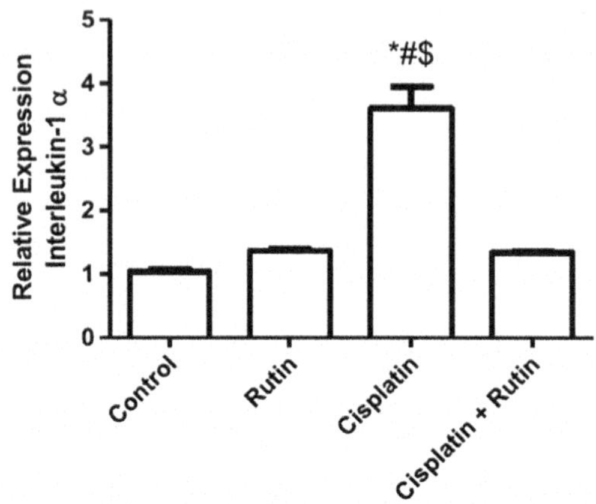

Fig. 9 Represents the changes in the expression level of IL-1α in rat kidney tissues induced by CP. Data were presented as mean ± SEM ($n = 10$). Asterisk indicates significant change from control, hash indicates significant changes from rutin, and dollar indicates a significant change in cisplatin plus rutin (Reproduced from reference [31] under a Creative Commons License)

interconnected ROS/JNK/TNF/P38 MAPK signaling pathways, and repairing the histopathological changes against cisplatin administration.

Azevedo et al. investigated the neuroprotective effects of rutin and quercetin on oxaliplatin-induced neurotoxicity [32]. Painful peripheral neuropathy appears to be caused by all platinum compounds used in cancer chemotherapy. Among these compounds, the third-generation drug oxaliplatin or the oxaliplatin-based regiments are one of the first choices in the treatment of colorectal cancer, but it is associated with the development of acute and chronic painful neuropathy that is difficult to treat. The acute phase affects almost every patient who receives this drug, and the chronic effects of oxaliplatin are cumulative and limit treatment. Data suggest that oxaliplatin's initial neurotoxic effect is peripheral and oxidative stress dependent. A spinal target is also suggested in its mechanism of action. Based on this background evidence on the biological effects of flavonoid compounds and the possible oxidative-related mechanism of neuronal injury due to oxaliplatin injection, the present study investigated the neuroprotective effects of rutin and quercetin on oxaliplatin-induced neurotoxicity. Oxaliplatin (1 mg/kg, Intravenous) was injected in male Swiss mice, twice a week (total of nine injections). The development of sensory alterations, such as thermal and mechanical allodynia, was evaluated using the tail immersion test in cold water (10 °C) and the von Frey test. Rutin and quercetin (25–100 mg/kg, i.p) were injected 30 min before each oxaliplatin injection. The animals' spinal cords were removed for histopathological and immunohistochemical evaluation and malondialdehyde assay. Mechanical nociceptive threshold was evaluated in the hind paws of mice that received an intravenous injection. The mice received two intravenous injections of oxaliplatin (OXL; 1 mg/kg) per week, for 4.5 weeks for a total of nine injections. Rutin (A) and quercetin (B) were injected intraperitoneally 30 min before every OXL administration. The control group received saline instead of rutin or quercetin. Mechanical threshold was assessed before and every 7 days after each treatment (Fig. 10).

The histopathological analysis revealed differences between naive and oxaliplatin-treated mice. The mice received two intravenous injections of OXL (1 mg/kg) per week for 4.5 weeks for a total of nine OXL injections. Rutin and quercetin (50 mg/kg) were injected intraperitoneally 30 min before every OXL administration. The control group received saline instead of rutin and quercetin. 24 hours after oxaliplatin injection, the presence of discrete lacunar spaces was observed between neuronal cells, suggesting nervous tissue edema. They also observed shrinkage of the size of dorsal horn neurons, mainly from the 14th experimental day, suggesting the atrophy of neurons and increase in glial cell number. Both rutin and quercetin prevented these changes, maintaining the tissue with aspects similar to the naive group (Fig. 11).

This study showed that oxaliplatin induced painful peripheral neuropathy in mice, an effect prevented by rutin and quercetin. The oxaliplatin-related neurotoxic effect appears to occur at least partially through oxidative stress-induced damage in dorsal horn neurons, reflected by lipid peroxidation and protein nitrosylation. Lipid peroxidation and tyrosine nitrosylation were prevented by flavonoid treatment. The decreased iNOS expression in the flavonoid-treated groups seems to involve the inhibition of peroxynitrite-associated neuronal damage due to the direct antioxidant effect of these compounds. The data may lead to a better understanding of the pathogenesis of oxaliplatin-induced neuropathy and the development of new approaches to treat this condition, which includes the prevention of oxidative damage.

Antitumor effect of rutin through inducing G2/M cell cycle arrest and promoting apoptosis was determined by Chen et al. on human neuroblastoma cell lines [33]. The antiproliferation of rutin in human neuroblastoma cells LAN-5 were detected by 3-(4,5-dimethylthiazol-2-yl)-2,5 diphenyltetrazolium bromide (MTT) assay. Chemotaxis of LAN-5 cells was assessed using transwell migration chambers and scratch wound migration assay. The cell cycle arrest and apoptosis in a dose-dependent manner were measured by flow cytometry and fluorescent microscopy analyses. The apoptosis-related proteins BAX and BCL2 as well as MYCN mRNA express were determined by RT-PCR analysis. Secreted TNF-α level was determined using specific enzyme-linked immunosorbent assay kits. Rutin induced cell cycle arrest in G2/M phase in LAN-5 cells. The effects of rutin on cell cycle progress in LAN-5 cells were investigated by flow cytometry. They examined the effects of rutin on the viability of LAN-5 cells. The MTT assay showed that rutin treatment significantly inhibited the viability of LAN-5 cells in a dose-dependent manner, OD values from 0.92 ± 0.06 to 0.83 ± 0.07 and 0.79 ± 0.06 as well as 0.71 ± 0.08, respectively. The effects of rutin on cell cycle progress in LAN-5 cells were investigated by flow cytometry. Cells incubated with different concentrations (0, 25, 50, and 100 μM) of rutin for 24 h were analyzed for the distribution of G0/G1, S, and G2/M phases of cell cycle. Compared with control, rutin treatment resulted in a significant accumulation of cells in the G2/M phase; the percentage of LAN-5 cells in the G2/M fraction increased from 8.8% to 10.05%, 14.46%, and 20.05% after

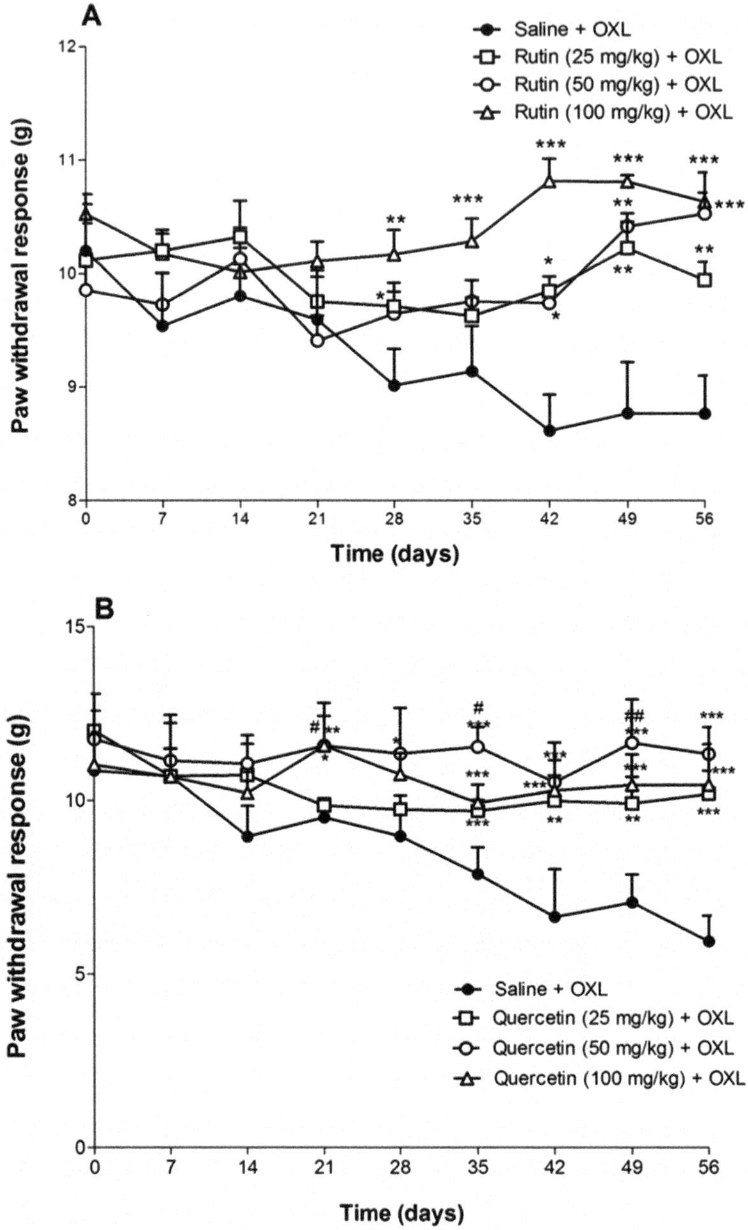

Fig. 10 Antinociceptive effects of rutin (**a**) and quercetin (**b**) on oxaliplatin-induced mechanical nociceptive threshold decrease (von Frey). (Reproduced from reference [32] under a Creative Commons License)

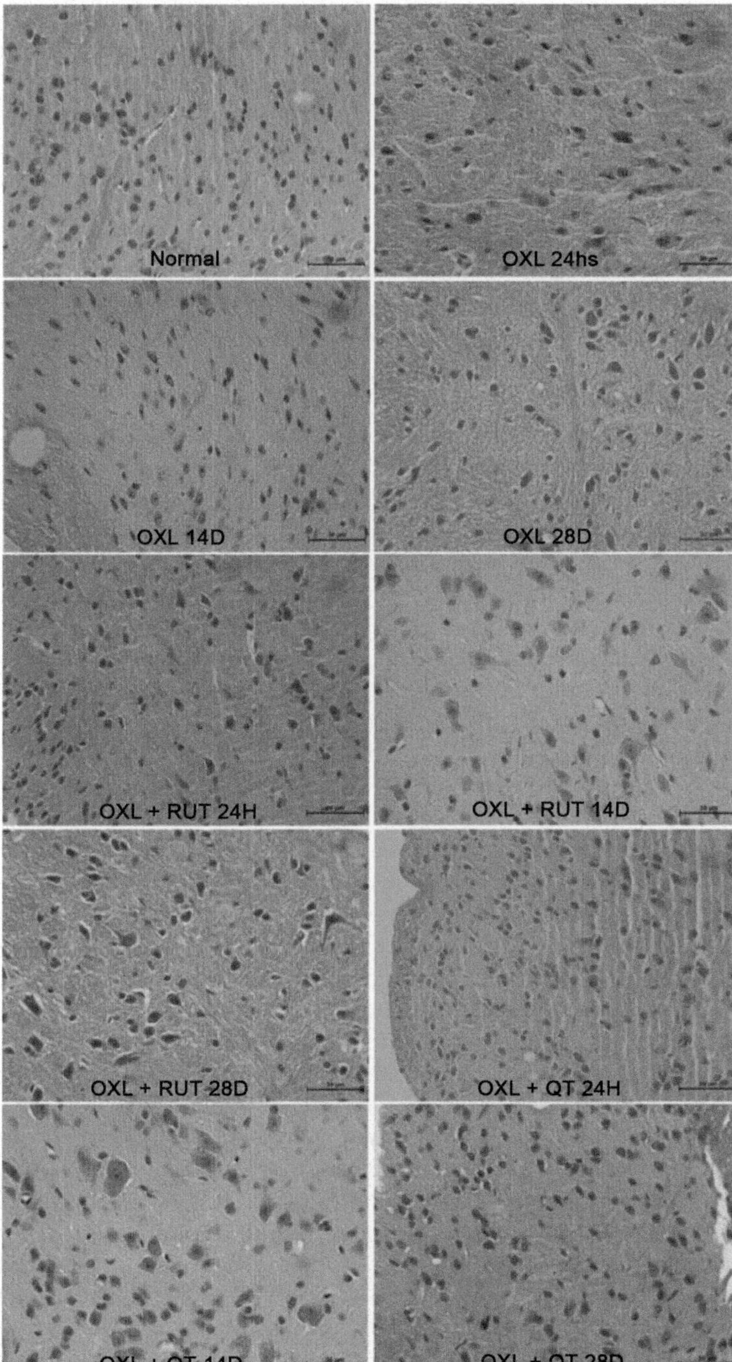

Fig. 11 Photomicrographs of the dorsal horn of the spinal cord of mice subjected to oxaliplatin (OXL)-induced neurotoxicity and treated with rutin (RUT) or quercetin (QT) (Reproduced from reference [32] under a Creative Commons License)

treated with 25, 50, and 100 μM of rutin, which underwent G2/M arrest in a dose-dependent way. Rutin-induced LAN-5 apoptosis is mediated by the induction of BCL2/BAX imbalance. Annexin V-FITC/PI apoptosis assay kit was used to detect cell apoptosis. Normal cells were used as the control group. It seems that the addition of different dose rutin induced apoptosis of LAN-5 cells with percentages of apoptosis cells increasing from 4.48% to 7.50%, 11.26%, and 20.37%. Meanwhile, RT-PCR analysis showed that the expression of BCL2 was downregulated by rutin and BCL2/BAX ratios were downregulated. Cell monolayer wounding and migration test were performed to examine the role of rutin on cell migration. LAN-5 cells were seeded in 6-well plates; wounds were created by scrapping the cell monolayer by P200 pipette tips. As shown in Fig. 12a, different dose rutin could decrease the migration rate of LAN-5 cells, as compared to that of control LAN-5 cells, indicating that rutin is involved in migration capability of LAN-5 cells.

Matrigel invasion assay was performed to examine the role of rutin in LAN-5 cell invasion. LAN-5 cells were seeded on the upper chamber of transwells coated with matrigel. After 24 h, the number of cells that invaded the lower chamber is significantly lower in different dose rutin groups compared to the control group ($p < 0.05$, $p < 0.01$; Fig. 14b), indicating that rutin suppressed the invasion abilities. These results indicated that the rutin plays an important role in the invasion and migration capability of LAN-5 cells. Overexpression of MYCN in neuroblastoma has been correlated with treatment failure; the inhibition of MYCN expression may prove to be a useful target for treatment of neuroblastoma. High MYCN mRNA expression was observed in LAN-5 cells (MYCN/GAPDH ratio was 1.06 ± 0.10). The different doses of rutin treatment significantly reduced MYCN mRNA expression in the LAN-5 cells (MYCN/GAPDH ratios were 0.96 ± 0.06, 0.69 ± 0.09, and 0.48 ± 0.08). They found that rutin could downregulate the expression of MYCN in a dose-dependent manner. Their study demonstrated that rutin plays the antiproliferation of human neuroblastoma cells' role through inducing G2/M phase cell cycle arrest and triggering apoptosis; this event is a BCL2-independent process. Moreover, rutin could inhibit MYCN expression. Therefore, rutin is a potential therapeutic drug for the treatment of neuroblastoma, particularly in MYCN expression tumor.

A study was undertaken by Gautam et al. to corroborate the effect of rutin against MTX-induced intestinal toxicity in experimental animals [35]. Methotrexate (MTX) is recognized as an antimetabolite in cancer chemotherapy and is associated with various toxicities assigned to inflammation and oxidative stress. Rutin has been reported to have significant anti-inflammatory and antioxidant along with antiulcer properties. Six groups of rats ($n = 6$) were dosed with normal saline (3 ml/kg, i.p.), MTX (2.5 mg/kg, i.p.), rutin (50 and 100 mg/kg, i.p.), rutin + MTX (50 mg/kg + 2.5 mg/kg, i.p.), and rutin + MTX (100 mg/kg + 2.5 mg/kg, i.p.) for 7 consecutive days and sacrificed on the 8th day. The intestinal contents were scrutinized physiologically (pH, total acidity, free acidity, CMDI), biochemically (TBARS, protein carbonyl, SOD, catalase, and GSH) and for immunoregulatory cytokines (IL-2, IL-4, and IL-10). The administration of rutin manifested significant protection against intestinal lesions damaged by MTX. The treatment with rutin elicited

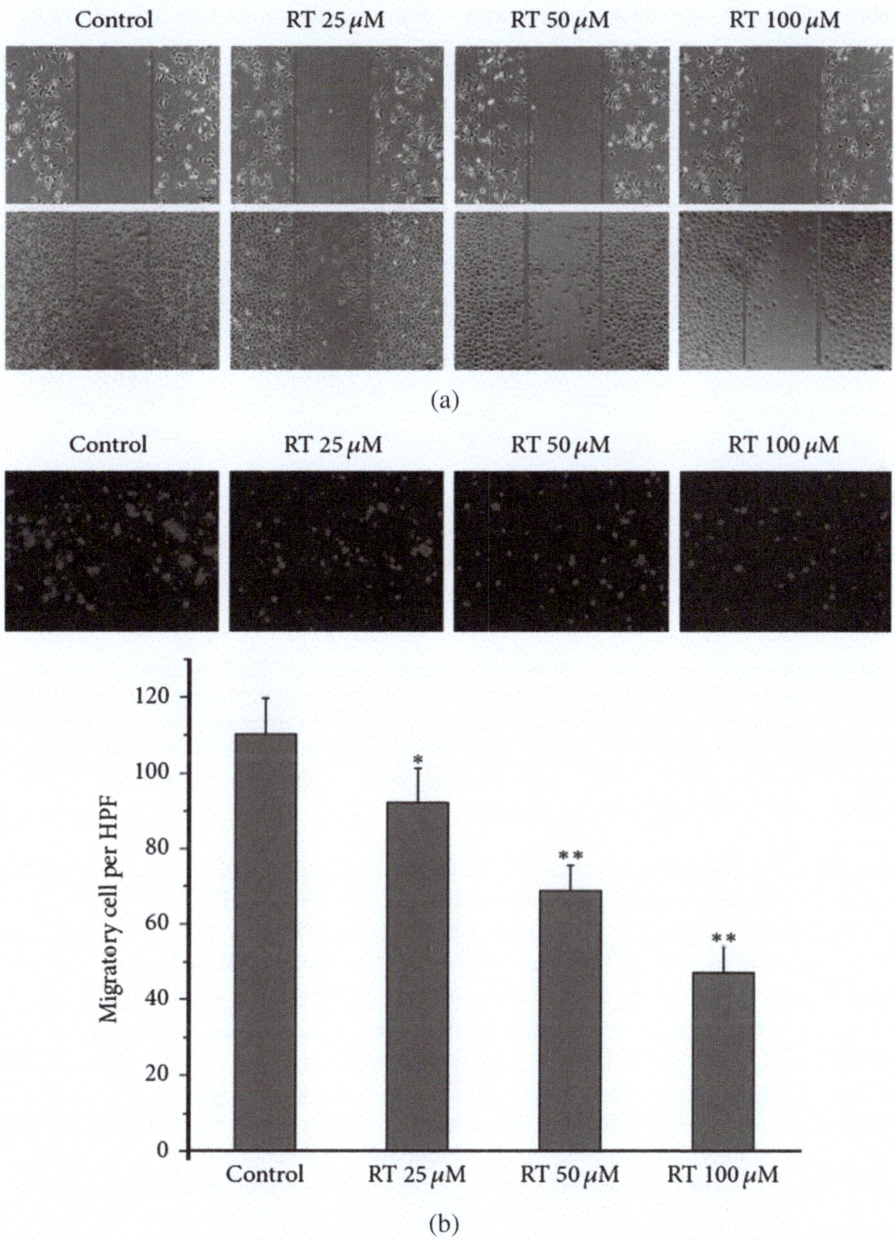

Fig. 12 The effect of rutin on cell migration and invasion capability. The representative pictures in cell wounding and migration assay. (**a**)The pictures show the migration of LAN-5 cells in the matrigel surface after 48 h of scrape in different groups. The dot line shows the initial start point where the cell began the migration process. (**b**) The pictures show the transwell assay. The cells that have migrated through the pores to the lower surface of the membrane were counted under microscope. All experiment data are expressed as means ± SEM of the three measurements $^{*}p < 0.05$; $^{**}p < 0.01$ compared to control (Reproduced from reference [33] under a Creative Commons License)

Table 1 Effects of rutin on intestinal pH, total acidity, free acidity, and CMDI induced by MTX in rats

Groups	Treatment	pH (mEq/L)	Total acidity	Free acidity (mEq/L)	CMDI
Group-I	Sham control (normal saline, 3.0 mL/kg)	7.15 ± 0.01	11.02 ± 0.61	7.47 ± 0.61	0.33 ± 0.15
Group-II	Toxic control (MTX, 25 mg/kg)	6.32 ± 0.04[a]	15.06 ± 0.80[a]	11.46 ± 058[a]	4.08 ± 005[a]
Group-III	Rutin (50 mg/kg)	7.67 ± 003[***a]	11.74 ± 0.69** (22.05%)	8.45 ± 0.57*** (26.20%)	1.17 ± 0.41***c (58.25%)
Group-IV	Rutin + MTX(50 + 25 mg/kg)	7.44 ± 0.02[***a]	12.68 ± 0.87*** (15.74%)	11.14 ± 0.60[a] (2.70%)	1.67 ± 0.52[***a] (58.32%)
Group-V	Rutin (100 mg/kg)	7.81 ± 001[***a]	11.06 ± 0.38*** (26.57%)	8.70 ± 0.31*** (24.01%)	1.17 ± 0.52[***c] (58.25%)
Group-VI	Rutin + MTX (100 + 2.5 mg/kg)	7.21 ± 0.01***	12.75 ± 0.85*** (15.28%)	10.13 ± 0.71[b] (11.52%)	2.50 ± 0.55[***a] (37.50%)

Each group contains six animals. Values are represented as mean ± SEM. Statistical significance compared to toxic control using one-way ANOVA followed by Bonferroni test (*p < 0.05, **p < 0.01, and ***p < 0.001). Values in parenthesis represent percentage inhibition. Statistical significance compared to sham control using one-way ANOVA followed by Bonferroni test [c]p < 0.05, [b]p < 0.01, and [a]p < 0.001 was considered statistically significant (Reproduced from reference [35] under a Creative Commons License)

noticeable inhibition of free acidity (26.20%), total acidity (22.05%), and CMDI (1.16%) in the experimental animals similar to control (Table 1).

They evaluated the effects of rutin on intestinal oxidative damage induced by methotrexate in rats. In MTX-treated toxic group, the level of MDA increased, which was subsequently restored after rutin treatment. As for protein oxidation, the MTX treatment is validated in a comparable increase in protein carbonyl level in the toxic group (42.38 ± 0.35 nmol/ml) in comparison to normal (38.41 ± 1.61 nmol/ml). The groups treated with rutin restored the level of protein carbonyl to a significant level. Similarly, a decrease in SOD level was observed in MTX-treated toxic group (0.29 ± 0.035 SOD/mg of protein) in resemblance to normal control (1.82 ± 0.035 SOD/mg of protein). Rutin restored the levels of SOD in treatment groups. A significant decrease in the enzymatic activity of catalase was observed in toxic control (0.45 ± 0.03 nM of H_2O_2/min/mg of protein). It is noteworthy that treatment with rutin restored the catalase activity in treatment groups as well. The tissue GSH level was upsurged in MTX-treated toxic group (117.17 ± 1.12 mg %) in context to normal control (30.63 ± 2.38 mg %). When rutin was given, the GSH level was restored. Effects of rutin on intestinal cytokines induced by methotrexate in rats were studied. The tissues were also contemplated for the presence of inflammatory markers like IL-2, IL-4, and IL-10. IL-2 is a proinflammatory marker, whereas IL-4 and IL-10 are

Table 2 Effects of rutin on intestinal cytokines induced by methotrexate in rats

Groups	Treatment	IL-2 (pg/mL)	IL-4 (pg/mL)	IL-10 (pg/mL)
Group-I	Sham control (normal saline, 3.0 mL/kg)	398.10 ± 5.32	162.18 ± 8.79	63.09 ± 5.51
Group-II	Toxic control (MTX, 2.5 mg/kg)	671.93 ± 56.65^{a}	213.79 ± 7.64^{a}	88.57 ± 3.45^{a}
Group-III	Rutin (50 mg/kg)	174.15 ± 8.20^{***a}	193.50 ± 1.47^{***a}	104.05 ± 0.65$_{**a}$
Group-IV	Rutin + MTX(50 + 25 mg/kg)	233.52 ± 14.69^{***a}	117.29 ± 0.023$_{***a}$	63.21 ± 2.96
Group-V	Rutin (100 mg/kg)	178.72 ± 3.28^{***a}	198.26 ± 7.76^{***a}	122.11 ± 1.88^{***a}
Group-VI	Rutin + MTX (100 + 25 mg/kg)	272.02 ± 15.00^{***a}	179.24 ± 1.57***	96.99 ± 1.50^{***a}

Each group contains six animals. Values are represented as mean ± SEM. Statistical significance compared to toxic control using one-way ANOVA followed by Bonferroni test $*p < 0.05$, $**p < 0.01$, and $***p < 0.001$ was considered statistically significant. Statistical significance compared to normal control (Group-I) using one-way ANOVA followed by Bonferroni test $^{c}p < 0.05$, $^{b}p < 0.01$, and $^{a}p < 0.001$ was considered statistically significant (Reproduced from reference [35] under a Creative Commons License) anti-inflammatory cytokines. The MTX-treated toxic group perceived in a comparable upsurge in the levels of interleukins, and rutin restored the same (Table 2).

They also examined the effects of rutin on activities of COX-1, COX-2, and 5-LOX. The rutin was found to inhibit the enzymatic activity of COX-1, COX-2, and 15 LOX with 75.63, 81.00, and 80.43% inhibition, respectively. Their findings demonstrate that rutin exhibits significant physiological, biochemical, and immunological protection against MTX-induced intestinal toxicity in a dose-dependent manner and can be advocated as an adjuvant therapy in MTX chemotherapy to countervail the deleterious effects. The free radical scavenging, immunoregulatory, and dual inhibitory potential of AA metabolism could be considered as the possible mechanisms for the actions of rutin as observed in the present experiment.

In an in vivo study, Lin et al. focused on the in vivo effects of rutin on leukemia WEHI-3 cells and also examined the effects of rutin on WEHI-3 in BALB/c mice in vivo [36]. These experiments were divided into two parts. In Part I, 60 BALB/c mice were divided into three groups: Group I was control (10 animals), Group II was treated with olive oil (vehicle), and Group III was treated with rutin in olive oil. In Part II, the 60 mice from Part I experiment were divided into three groups: Group I was injected with WEHI-3 cells for control (10 animals) and Groups II and III were injected with WEHI-3 cells and then treated with rutin (6 and 12 mg/kg) in olive oil. All animals were given orally the above dose daily for up to 3 weeks before being weighed. Rutin was dissolved in olive oil to treat mice. The BALB/c mice were randomly divided into six groups receiving different treatments.

First, blood samples were collected (about 1ml) from each animal from each group at the end of the experiments. Second, the collected blood was immediately treated with ammonium chloride for lysing of the red blood cells, followed by centrifugation for 15 min at 1500 rpm at 4 °C. Third, the isolated white blood cells were examined for cell markers, such as CD3, CD19, CD11b, and Mac-3 based on the staining with anti-CD3, CD19, CD11b, and Mac-3 antibodies, respectively. These cells were then stained with the second fluorescent antibody before being analyzed to determine the cell marker levels by flow cytometry. Figure 13a shows representative pictures of the animals after injection withWEHI-3 cells for 3 weeks that also showed the leukemia spleen tumors. The size of the spleen showed that rutin decreased the spleen tumor size and had dose-dependent effects (Fig. 13b). Liver and spleen tissues were isolated from animals and were photographed, weighed, and histopathologically examined. Representative results are presented in Fig. 13c, d. The results indicated that rutin affected the morphology, weight, and cells of spleen (Fig. 13) and liver. The histopathological examination indicated that after treatment with rutin, spleen demonstrated a pattern ranging from minimal histopathological change to scanty small neoplastic cell nests present in the sinusoid with apoptosis (Fig. 13d). Either the spleens disclosed markedly decreased numbers of neoplastic cells or the cells were hard to find in the red pulp. Also, the number of megakaryocytes increased (Fig. 13d). The marked expansion of red pulp was seen, but the white pulp was only changed a little in examined spleen tissues. The neoplastic cells in examined spleen tissues presented large irregular nuclei with clumped chromatin, prominent nucleoli, and abundant clear and light eosinophilic cytoplasm.

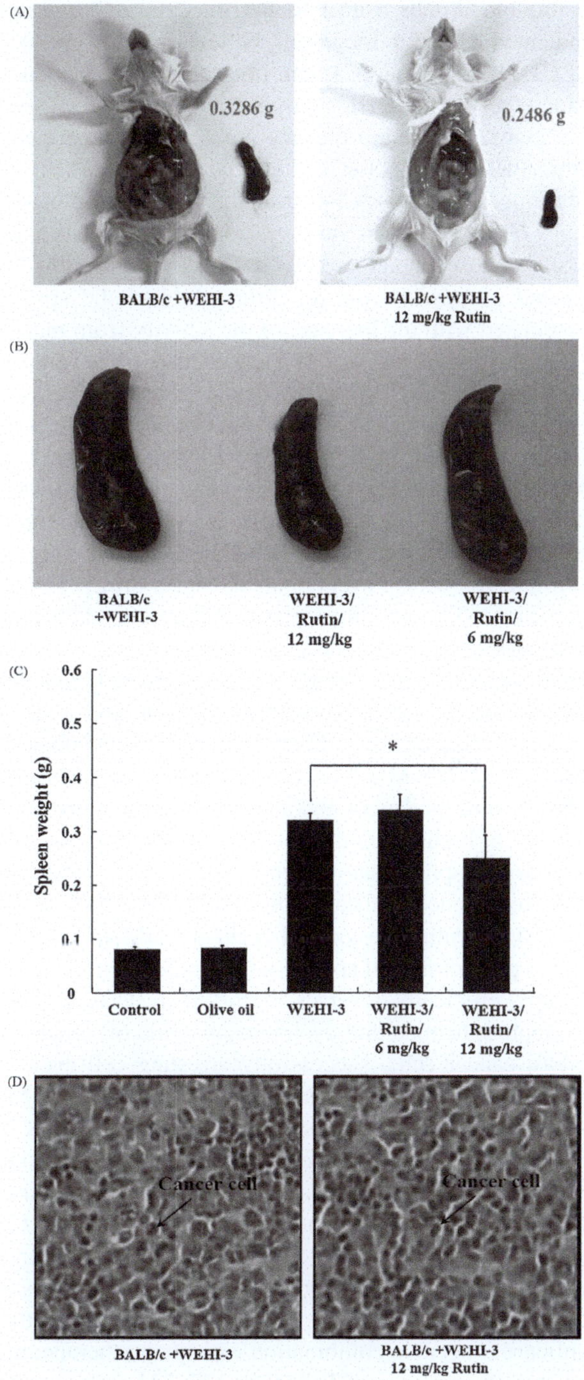

Fig. 13 Representative pictures of BALB/c mice after injection with WEHI-3 cells and were treated with rutin over 3 weeks. After animals were injected with WEHI-3 cells (1 × 10 cells/100 μL) in RPMI 1640 medium for 3 weeks and/or co-treated with rutin, blood was collected and animals were sacrificed (**a**) for examination of spleens, which showed morphologic changes (**b**), weighed (**c**), and histopathologically examined (**d**) (Reproduced from reference [36] with permission from Elsevier)

White blood cells from BALB/c mice are exposed to rutin in olive oil. The results indicated that rutin increased the levels of CD3 and CD19 but decreased the levels of CD11b and Mac-3. It can be seen that rutin induced a significant difference of Mac-3 between WEHI-3 only treated groups ($p < 0.05$). Results from flow cytometric analysis indicated that the activities of macrophage phagocytosis which come from PBMC and peritoneal in BALB/c mice after exposes to rutin in olive oil are presented. The data demonstrated that rutin promoted the activities of phagocytosis. They observed that in the rutin-treated groups, the activity of macrophage phagocytosis increased, and there was a significant difference between the control and rutin-treated groups. Therefore, rutin inhibited the spleen tumor which may also be associated with the increase of the activity from macrophage phagocytosis.

In another work, results showed that rutin restores chemosensitivity in human breast cancer cells [34]. In this study, rutin was investigated as chemosensitizer toward two different human epithelial breast cancer cell lines: (1) MB-MDA-231 selected as representative for triple-negative breast cancer and (2) MCF-7 used as a well-characterized model of HER2-negative breast cancer. To assess the cytocompatibility of rutin against non-cancer cells, primary human mammary fibroblasts were used as control and nontarget cells. In particular, the use of rutin as anticancer adjuvant agent in combination with conventional anticancer drugs cyclophosphamide (CYC) and methotrexate (MTX) was investigated. After rutin addition, half-dosage chemotherapeutic administration caused the same effects in cancer cells in comparison with the therapeutic doses of CYC and MTX, though strongly reducing their dose-dependent toxicity on nontarget cells. Also, MDR reversal, apoptosis induction, and cell cycle arrest were investigated to identify the mechanisms involved in rutin-induced chemosensitization. In MDA-MB-231 cells, 20 μM rutin enhanced cytotoxicity related to cyclophosphamide and methotrexate. Rutin significantly ($p < 0.05$) increased the anticancer activity of both chemotherapeutics, at 24–48–72 h, and decreased the activity of the adenosine triphosphate-binding cassette transporters, namely, P-glycoprotein (P-gp) and breast cancer resistance protein (BCRP). Flow cytometry analysis showed 20 μM and 50 μM rutin arrested cell cycle at G2/M and G0/G1 phases, respectively, significantly promoting cell apoptosis. Rutin, via nonselective inhibition of P-gp and BCRP pumps, efficiently reverses multidrug resistance and restores chemosensitivity to cyclophosphamide and cyclophosphamide of human chemoresistant, triple-negative breast cancer cells, successfully arresting cell cycle progression. In ER (estrogen receptor)- and PR (progesterone receptor)-positive MCF-7 cells, although at 24 h 20 μM rutin alone did not significantly affect 20 μM CYC and 2 μM MTX cytotoxicity (Fig. 14a), at 48 and 72 h chemotherapeutics, in the presence or not of rutin, significantly decreased cell viability (Fig. 14b, c, respectively) ($p < 0.05$, indicated by ∗). Rutin did not further enhanced efficacies of 20 μM CYC and 2 μM MTX alone, at each time point (Fig. 14a–c); thus, in terms of anticancer efficacy, no difference was observed between drugs alone and their combinations with this flavonoid.

In order to explore the mechanisms involved in the activity of rutin as adjuvant anticancer agent in combination with chemotherapeutics, the effects of this flavonoid on the efflux pumps of TN MDA-MB-231 cells were investigated. Resulting effects were compared with verapamil, inhibitor of P-gp, and Ko143, inhibitor of both P-gp and BCRP. Rutin showed similar inhibitory effect on P-gp and BCRP efflux pumps

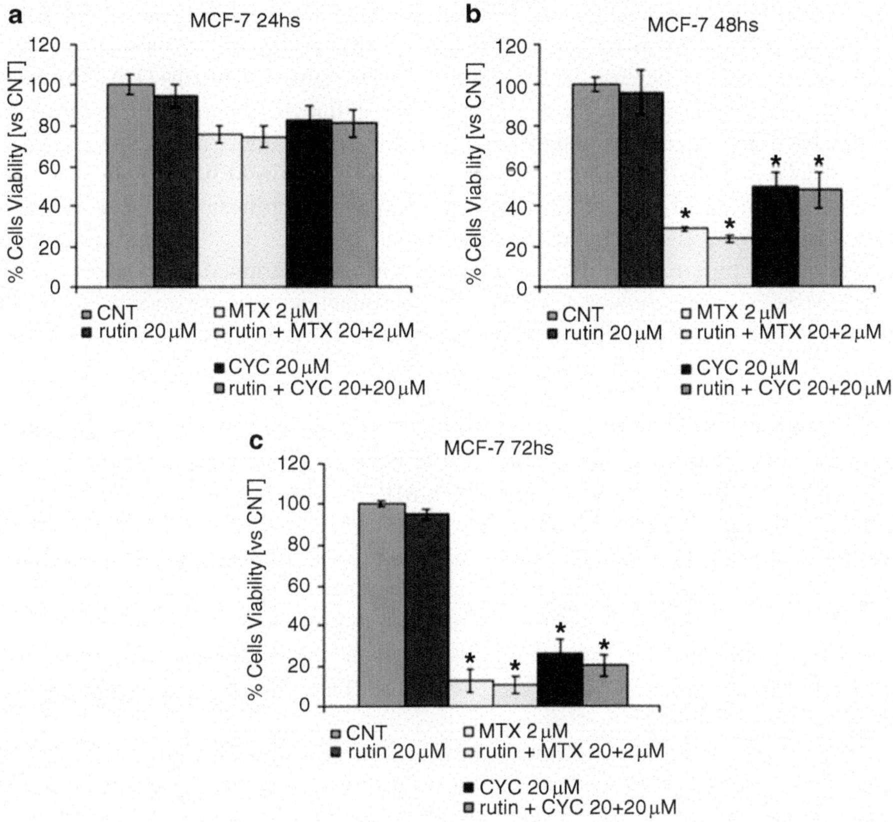

Fig. 14 Rutin as adjuvant agent towards MCF-7 cells. Rutin 20 μM was combined with half-dose cyclophosphamide (CYC) and methotrexate (MTX) to improve their efficacy. Cells were tolerant to the tested drugs (CYC or MTX) or combinations (rutin + CYC/MTX) after 24 h (**a**) as no difference was observed in comparison with untreated cells (CNT, $p > 0.05$). On the opposite, results were significant after 48 h (**b**) and 72 h (**c**) for all the tested formulations ($p < 0.05$, indicated by the asterisk, one-way ANOVA and Sheffè test). No significant difference was reported between rutin-chemotherapeutic combinations and chemotherapeutics alone ($p > 0.05$). Bars represent mean and standard deviation (Reproduced from reference [34] with permission from Wiley)

as the inhibitors are used as positive controls, that is, verapamil and Ko143. At fluorescence analyses of calcein AM (specific substrate of P-gp efflux pump, thus marker of its functionality), results revealed that, after 24 h of incubation with 20 μM rutin, 50 μM rutin, or 20 μM Verapamil, a statistically not significant difference among treatments could be observed. Similar trends were observed for calcein assay after 48 and 72 h. At fluorescence analyses of rhodamine 123 (specific substrate of P-gp efflux pump, thus marker of its functionality), at 72 h, the positive control, that is, 10 μM Ko143 in combination with 20 μM verapamil, was still not significantly different from the treatments with 20 and 50 μM rutin ($p > 0.05$). Also, apoptosis was detected after rutin treatment. After exposure to 20 and 50 μM rutin, MDA-MB-231 and MCF-7 cells were stained with Annexin V and 7-AAD; the former is a marker of phosphatidylserine externalization at cell surface, typical feature of apoptosis, and the latter is a DNA intercalator in dead cells. Cells were, then, analyzed by flow cytometry to assess the percentage of apoptotic and necrotic cells. Following 24 h of incubation, a slight increase in apoptotic cell phenotype (Annexin+/7-AAD+ and Annexin+/7-AAD) was observed in both cancer cell lines, compared with the control group, but without statistical significance (Fig. 15). The most noticeable changes were detected after 48 h of incubation with 50 μM rutin in MDA-MB-231. In this case, the number of dead cells (Annexin/7-AAD+) significantly increased over untreated controls ($p < 0.05$, indicated by ∗) (Fig. 15b).

Cell cycle progression after rutin treatment was examined. Considering TN MDA-MB-231, data showed that the 24-h treatment with both concentrations of rutin did not produce any change in cell cycle over controls. After 48-h incubation, instead, both 20 and 50 μM rutin significantly increased the percentage of MDA-MB-231 cells in G0/G1 ($p < 0.05$) and G2/M ($p < 0.05$) phases, respectively. Noteworthy, at this time point, MDA-MB-231 cells in S phase significantly decreased after treatment with 50 and 20 μM rutin ($p < 0.05$). The effect of 50 μM rutin in MCF-7 appeared earlier than MDA-MB-231: after 24 h, the flavonoid significantly increased the percentage of MCF-7 cells at G0/G1 phase ($p < 0.05$), decreasing cells in G2/M ($p < 0.05$). Results demonstrated that rutin reverses MDR in human breast cancer cells by overcoming the drug efflux from the cell by ABC transporters and promoting PCD with the arrest of cell cycle. These events impair the cancer cell machinery involved in drug-resistance development. In particular, rutin nonselectively counteracted P-gp and BCRP pumps, thus increasing sensitivity to CYC and MTX.

Elsayed et al. reported the bioassay-guided discovery of rutin among other flavonoids, as a potential antibreast cancer hit [37]. Rutin inhibited the proliferation of seven breast adenocarcinoma cell lines, with utmost activity toward the invasive c-Met-dependent TNBC cell lines. Moreover, rutin inhibited the migration and invasion of the metastatic TNBC MDA-MB-231 cells in a dose-response mode. Treatment effects were mediated via a significant inhibition of c-Met phosphorylation and its downstream mitogenic signaling involved in cellular proliferation and motility. Rutin showed the best antiproliferative activity, inhibiting the HGF-dependent growth across the highly c-Met expressing TNBC cell lines in a dose-responsive manner as compared to their vehicle-treated control groups and c-Met-scant ones. These results suggested c-Met inhibition can be a potential molecular mechanism mediating the in vitro anticancer effects of the flavonoid rutin on c-Met overexpressing TNBC cell lines. In addition to the modulation of cell proliferation, activation of the HGF/c-Met axis promotes cell migration and invasion, which contributes to the metastatic characteristics of malignant cells. Different concentrations of rutin were evaluated for ability to inhibit the migration of MDA-MB-231 cells across a wound-inflicted monolayer compared to the vehicle control treatment. Results showed that rutin exhibited a significant antimigratory activity against the invasive TNBC MDA-MB-231 cells in a concentration-dependent manner. For instance, 100, 200, and 300 μM of rutin concentrations inhibited the MDA-MB-231 cell migration by 43.4%, 59.3%, and 78.8%, respectively, compared to the DMSO vehicle control (Fig. 16). Cells were plated in sterile

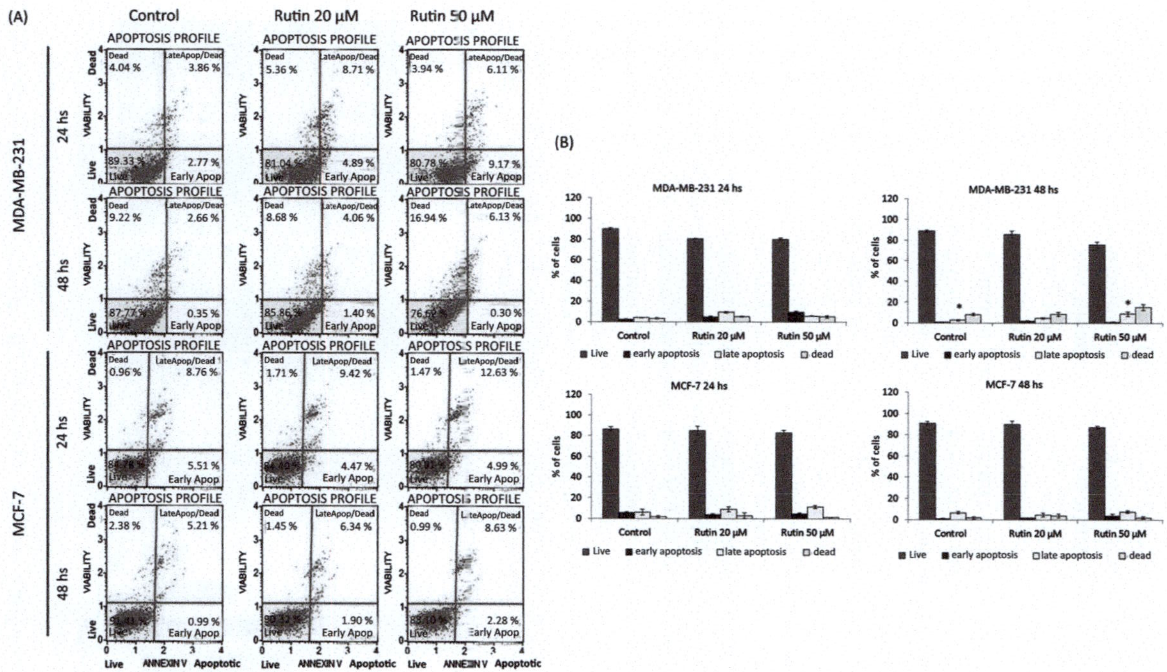

Fig. 15 Apoptosis detection after rutin treatment. MDA-MB-231 and MCF7 cells were treated with 20 or 50 μM rutin for 24 and 48 h and subjected to the combined Annexin V/7-AAD (7-aminoactinomycin D) staining and flow cytometry. (**a**) Annexin V/7-AAD assay divided cells into four groups: early apoptotic cells [Annexin V (+) and 7-AAD ()], late apoptotic or dead cells [Annexin V (+) and 7-AAD (+)], dead cells [Annexin V () and 7-AAD (+)], and live cells [Annexin V () and 7-AAD (). Untreated cells (left panels) were used as control. Experiments were carried out in triplicate, and representative results are shown. (**b**) Percentages of live, early apoptotic, late apoptotic, or dead cells. Experiments were carried out in triplicate and results expressed as mean; bars represent the standard deviation. Statistical differences were assessed by Friedman's ANOVA variance analysis followed by post hoc Dunn's test and Wilcoxon's test. The value of $p < 0.05$ was considered to be significant (∗). MDA-MB-231, triple-negative human breast cancer cells; MCF-7, HER2/neu negative human breast cancer cells (Reproduced from reference [34] with permission from Wiley)

24-well plate and allowed to be attached overnight; then, a scratch wound was inflicted in the confluent monolayers using a sterile 200 mL pipette tip. Media were aspirated; cells were washed twice with PBS and then reincubated in serum-free media for 5 h. Subsequently, media were replaced by fresh serum-free media supplemented with the scattering factor (100 ng/mL HGF). Percentage of migrated cells was determined relative to zero time control wells, and IC_{50} value was calculated using GraphPad Prism version 5.01. $*P$-value of <0.05 was considered to be significant.

The cytotoxicity of rutin was further probed using lactate dehydrogenase (LDH) release assay. Cayman LDH cytotoxicity kit was manipulated to evaluate cancerous cell death in response to rutin treatments. Cellular death was correlated to the loss of cell membrane integrity and release of cytoplasmic LDH to the cultural media, which was later quantified using a reaction buffer (NADC, lactic acid, INT, diaphorase). The analyzed data showed that rutin displayed minimal cytotoxicity of 1.95% at an effective anticancer concentration (500 mM) and even at higher doses as demonstrated by 9.95% cell death at 1000 mM. This clearly indicates that the anticancer effect of rutin is mediated though obstructing cell proliferation rather than nonselective toxicity induction. They investigated the effect of rutin on viability of MCF-10A mammary epithelial cells. The results clearly supported rutin's high degree of selectivity toward malignant cells coupled with the lack of potential toxicity to normal cells. The aforementioned in vitro results motivated the study of tumor genesis inhibitory effects of rutin in vivo. A tumor xenograft model generated by orthotopically implanting 1×10^6 green fluorescent protein-reporter stable MDA-MB-231/GFP cells into the mammary fat pad of nude mice was selected for this investigation. Latency period of 5 days postinoculation was encountered after which tumor size becomes palpable. Rutin treatment was initiated with 30 mg/kg (3×/week/mice/i.p.) and continued as described in the Experimental Section for additional 28 days. At the end of the experiment, rutin's treatment revealed a significant diminution in tumor growth by 53.2%, with a mean tumor volume of 812.0 ± 104 mm^3, compared to vehicle-treated control group (1737.0 ± 281.8 mm^3, Fig. 17a).

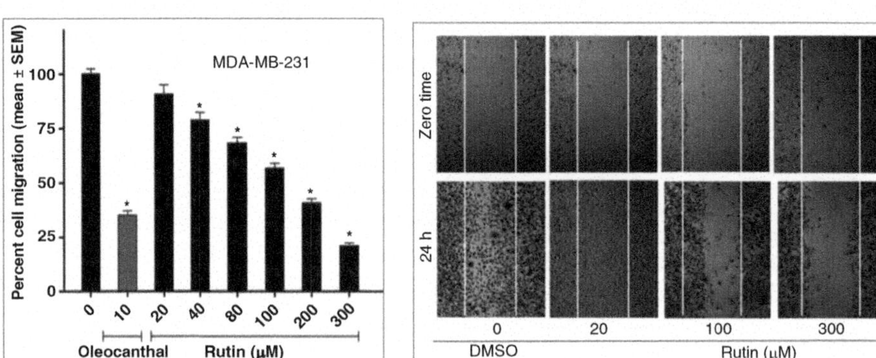

Fig. 16 Effect of rutin treatments on TNBC MDA-MB-231 cell migration (Reproduced from reference [37] with permission from Taylor & Francis)

Photographs of representative vehicle control and rutin-treated mice at the experiment end are shown in Fig. 17b. The mean tumor weights, obtained after treatment group euthanization, showed significant tumor reduction with an average tumor weight 1.27 ± 0.17 g, compared to 2.45 ± 0.72 g in vehicle-treated control group. Mean mice body weight changes in both groups were statistically nonsignificant all over the course of treatment (Fig. 17d).

Molecular modeling and docking studies further delineated the association between structural characteristics and c-Met kinase inhibitory activity of rutin, compared to the structurally related flavonoids. The cell-free Z0-LYTE kinase assay, Western blot, and in silico docking experiments uncovered c-Met kinase as a potential mechanistic target for rutin-mediated anticancer effects on TNBC cell lines. Likewise, the intraperitoneal injection of rutin at 30 mg/kg, 3×/week, significantly

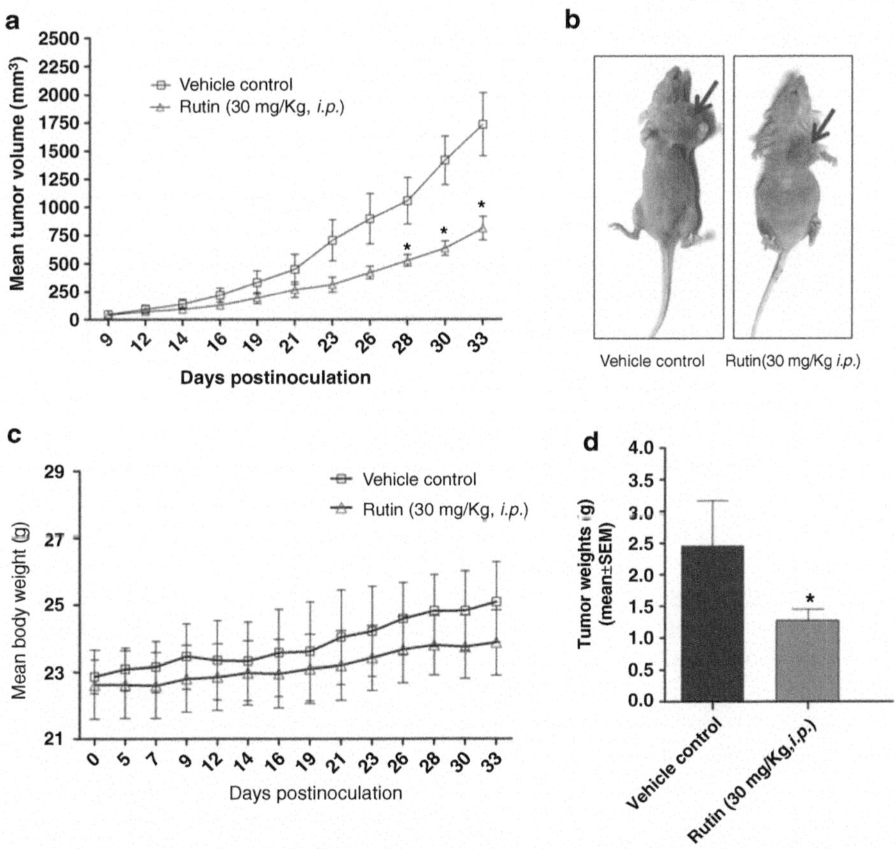

Fig. 17 (**a**) Effect of rutin treatment in nude mice breast cancer xenograft model. (**b**) Representative rutin treatment and vehicle-treated control mice. (**c**) Effect of rutin treatment on animals' body weight compared to the vehicle control treatment. (**d**) Animal tumor weight (mean ± SEM) of vehicle-treated control and rutin-treated groups at the end of study (Reproduced from reference [37] with permission from Taylor & Francis)

reduced the growth of the TNBC MDA-MB-231/GFP orthotopic xenograft in nude mouse model. Rutin significantly reduced the tumorigenic capacity of the invasive human TNBC MDA-MB-231 cells in orthotopic nude mice xenograft model, increasing the interest in this common lead natural product.

In another study, Saleh et al. investigated the possible anticancer activity of rutin and orlistat which are both safely used clinically in humans against two breast cancer models (in vivo EAC and in vitro MCF7) and the pancreatic cancer cell line (PANC-1) [38]. Mice were randomly distributed into five groups (n = 20) as follows:

- Group (1): Control
- Group (2): EAC group
- Group (3): Rutin IP, 20 mg/kg suspended in saline using 0.5% CMC daily for 21 days after EAC implantation
- Group (4): Orlistat IP, 240 mg/kg daily for 21 days after EAC implantation
- Group (5): Doxorubicin IP, 2 mg/kg on days 14, 16, 18, and m20 after EAC implantation. This dosing regimen for doxorubicin was chosen to avoid its toxicity.

The effect of cancer and treatment with rutin, orlistat, and doxorubicin on tumor volume and tissue histopathology was examined. At the end of the study, mice were sacrificed; tumor was excised, placed on a graph paper, and photographed; and then tumor volume was measured using the Image J program®. The following formula was used to calculate the volume of the developed tumor mass.

$$\text{Tumor volume}\left(\text{mm}^3\right) = 4\,\pi\left(A/2\right)^2 \times \left(B/2\right)$$

where A is the minor tumor axis, B is the major tumor axis, and π equals 3.14. Rutin, orlistat, and doxorubicin induced a significant decrease in tumor volume compared with the cancer group, reaching 1186, 1451, and 917.4, vs. 2418 mm^3, respectively (Fig. 18a). The isolated tumor was cut into sections (7–12 μm thick, frozen at −20 °C) and stained by hematoxylin and eosin stain. The sections were examined under light microscope and photographed with a digital camera. Tumors induced by subcutaneous implantation of EAC cells showed large numbers of malignant rounded to polygonal cells with hyperchromatic nuclei, prominent nuclei, and frequent mitosis. This picture was significantly improved in mice treated with rutin, orlistat, and doxorubicin as evidenced by decreasing numbers of malignant rounded to polygonal cells with hyperchromatic nuclei, prominent nuclei, and frequent mitosis. On the other hand, thigh sections from control mice showed no malignant cells (Fig. 18b).

The effect of cancer and treatment with rutin, orlistat, and doxorubicin on carcinoembryonic antigen, malondialdehyde, glutathione, fatty acid synthase, and tissue cholesterol was studied. Results showed that tumor-bearing mice had a significant increase in CEA compared to the control group (3.2 vs. 2.4 ng/ml). On the other hand, rutin, orlistat, and doxorubicin induced a significant decrease in CEA

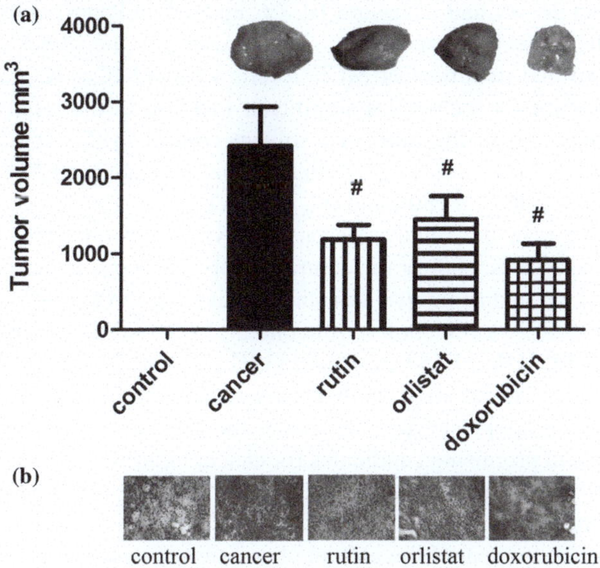

Fig. 18 The effect of cancer and treatment with rutin, orlistat, and doxorubicin on: (a) tumor volume and (b) tissue histopathology (H&E ×400). Data are expressed as mean ± S.D. $n = 20$. Number sign indicates significantly different from the cancer group at $p < 0.05$ using ANOVA followed by Tukey's post hoc test (Reproduced from reference [38] with permission from Springer)

compared with the cancer group (2.6, 2.6, and 2.5 vs. 3.2 ng/ml, respectively). EAC mice had a significant increase in tissue MDA compared to the control group (30.7 vs. 7.7 nmol/g) and a significant decrease in tissue GSH compared to the control group (9.4 vs. 22.7 mg/g). On the one hand, rutin, orlistat, and doxorubicin induced a significant decrease in tissue MDA compared with the cancer group (13.7, 19.9, and 12.3 vs. 30.7 nmol/g, respectively), and on the other hand, rutin, orlistat, and doxorubicin induced a significant decrease in tissue MDA compared with the cancer group (13.7, 19.9, and 12.3 vs. 30.7 nmol/g, respectively) and a significant increase in tissue GSH compared with the cancer group (16, 15.1, and 16.1 vs. 9.4 mg/g, respectively). Subcutaneous implantation of Ehrlich tumor cells resulted in a significant increase in FAS compared to the control group (4.3 vs. 1 ng/g). On the other hand, rutin, orlistat, and doxorubicin induced a significant decrease in FAS compared with the cancer group (1.6, 2.9, and 1.9 vs. 4.3, respectively). In addition, the EAC group showed a significant increase in tissue cholesterol compared to the control group (36.8 vs. 7.8 mg/g), while rutin, orlistat, and doxorubicin induced a significant decrease in tissue cholesterol compared with the cancer group (14.1, 23, and 11.7 vs. 36.8 mg/g, respectively). Also, the growth-inhibitory effects of rutin, orlistat, and doxorubicin on MCF-7 cells and Panc-1 cells after 24, 48, and 72 h were studied. IC_{50} values (μg/ml) for rutin after 24, 48, and 72 h were >100, 90, and 26 on Panc-1 cells and >100, 82, and 19.4 on MCF-7 cells, respectively. Similarly, IC_{50} values (μg/ml) for orlistat after 24, 48, and 72 h were >100, 58.8, and 14.5 on Panc-1 cells and 99.8, 74.7, and 18.7 on MCF-7 cells, respectively. IC_{50} values for doxorubicin after 24, 48, and 72 h were 10, 7.6, and 4.9 on Panc-1 cells and 9.5, 5.1, and 3.5 on MCF-7 cells, respectively. They examined the effect on apoptosis. Findings showed that treatment of MCF-7 cells with 1/2 IC_{50}, IC_{50}, and 2 IC_{50} of rutin and orlistat for 24 h caused a significant increase in the relative activity of caspase 3/7 compared to the untreated solvent control (195, 193, 228.5, 151.6, 274.7, and 274.8, respectively, vs. 100). In addition, treatment of PANC-1 cells with 1/2 IC_{50}, IC_{50}, and 2 IC_{50} of rutin and IC_{50} and 2 IC_{50} of orlistat for 24 h caused a significant increase in the relative caspase activity reaching 158, 174.8, 216.4, 134.7, 288.8, and 375 vs. 100 for the control group. Their results revealed that both rutin and orlistat exerted an in vivo anticancer activity as evidenced by the decrease in tumor volume, CEA level, cholesterol content, FAS, and the exerted antioxidant action (reduced MDA level and increased GSH content) and through histopathological examination. In addition, both were cytotoxic to MCF-7 and Panc-1 cell lines by promoting apoptosis. In conclusion, the anticancer activity of rutin and orlistat makes them promising candidates for cancer treatment alone or in combination with other anticancer drugs specially that they are used clinically with an acceptable safety profile.

Schwingel et al. compared the effects of a series of antioxidant compounds on simultaneous development of oxaliplatin-induced hepato- and neurotoxicity in mice [39]. Oxaliplatin (OXA) is a platinum compound widely used in the treatment of some solid tumors, especially colorectal cancer. Despite its usefulness, oxaliplatin-associated neurotoxicity represents the main dose-limiting factor of this drug. Chemotherapy with oxaliplatin also increases the rate of developing hepatic damages with inflammatory activity, termed chemotherapy-associated steatohepatitis (CASH). They administered the following antioxidant compounds—rutin (RT) (20 mg/kg), resveratrol (RVS) (100 mg/kg), quercetin (QT) (20 mg/kg), and quercetin nanoemulsion (NQT) (20 mg/kg)—daily by gavage to BALB/c, and N-acetylcysteine (NAC) was used as positive control. Treatments with RSV, RUT, or NQT were able to prevent mechanical allodynia when compared to the OXA group, and this effect was associated with decreased c-Fos immune positivity in the lumbar spinal cord. Regarding the effects on steatohepatitis, RVS, QT, and NQT almost completely reversed the mean liver weight increase induced by OXA. In accordance with these previous data, histological evaluation indicated attenuation of all features of hepatic steatosis evaluated in RSV, RUT, QT, and NQT groups. These compounds were able to reduce the immune positivity for the apoptosis marker caspase-3. On the other hand, only QT and NQT treatments were able to reduce neutrophil migration measured by myeloperoxidase (MPO) activity. The data presented in this study showed the positive effects of polyphenol compounds in diminishing pain, inflammation, and apoptosis drawn out by oxaliplatin. Their results indicate the use of the compounds such as resveratrol, rutin, quercetin, and, for the first time, quercetin nanoemulsion, as effective contributors to avoid oxaliplatin-induced hepato- and neurotoxicity in a rodent model.

Attenuation of genotoxicity, oxidative stress, apoptosis, and inflammation by rutin in benzo (a) pyrene exposed lungs of mice was reported by Shahid et al. [40]. They designed a study to evaluate the protective effects of rutin targeting genotoxicity, apoptosis, oxidative stress, and inflammatory markers like NF-κB, IL-6, TNF-α, COX-2, and Bcl-2. Benzo(a)pyrene [B(a)P] is an environmental contaminant and potential carcinogenic agent that causes lung injuries which lead to lung cancer. Pretreatment of rutin was given by oral gavage at doses of 40 and 80 mg/kg body weight (b.wt.) for 7 days before the administration of a single oral dose of B(a)P (125 mg/kg b.wt.). The ameliorative effect of rutin on oxidative stress, apoptotic and inflammatory markers in lung tissues, and genotoxicity was studied using an alkaline unwinding assay and DNA fragmentation. B(a)P enhanced lipid peroxidation, xanthine oxidase, H_2O_2 generation, and lactate dehydrogenase (LDH) activity; depleted activities of antioxidant enzymes and glutathione content; induced DNA strand breaks and fragmentation; disrupted normal histopathological architecture; and also showed abnormal expression of NF-κB, COX-2, IL-6, TNF-α, and Bcl-2. Rutin pretreatment caused a significant reduction in lipid peroxidation and LDH activity, increased glutathione content, restored antioxidant enzyme activity, reduced DNA strand breaks and fragmentation, modulated the expression of inflammatory and apoptotic markers, and restored the histopathological structure. The findings of this study supported the protective effect of rutin against B(a)P-induced lung toxicity and genotoxicity.

Vijay et al. evaluated radiosensitizing potential of rutin against human colon adenocarcinoma HT-29 cells [41]. Their study was intended to investigate the effect of rutin in combination with radiation on the proliferation of HT-29 colon cancer cells. To elucidate the radiosensitizing effects of rutin, HT-29 cells were treated with rutin (80 μM) for 24 h and then post-irradiated with 4 Gy of X-ray radiations. The experimental groups are represented as follows:

- Group 1: Sham control
- Group 2: Rutin treated (80 μM)
- Group 3: Irradiated (4 Gy)
- Group 4: Rutin-treated (80 μM) + irradiated (4 Gy) HT-29 cancer cells

The cytotoxic effects of rutin were examined on cultured HT-29 cells by exposing the cells to concentrations ranging from 10 to 100 μM for 24 h, 48 h, and 72 h. The reduced formazan was dissolved in DMSO, and the absorbance was read in a 96-well plate microtiter reader. Rutin induces cell death in a dose- and duration-dependent manner. Morphological changes in HT-29 cells treated and untreated with rutin and 4-Gy radiation are shown in Fig. 19.

They also investigated the effect of rutin and radiation on cell morphology. Cells treated with IC_{50} concentration of rutin for 24 h and exposed to 4 Gy radiation were stained with AO/EtBr. AO/EtBr staining of cells treated with both rutin and 4 Gy radiation showed a higher percentage of cell death, whereas AO/EtBr staining of cells treated with rutin alone and 4 Gy radiation alone showed a lower percentage of cell death (Fig. 20a). Quantitative data revealed that rutin alone at 80 μM concentration exhibits death in 30% of cells, whereas on treatment with 4 Gy radiation alone, the cell death occurred in 50% of cells, and on treatment with rutin (80 μM) along with 4 Gy radiation, the cell death occurred in 70% of cells with larger number of

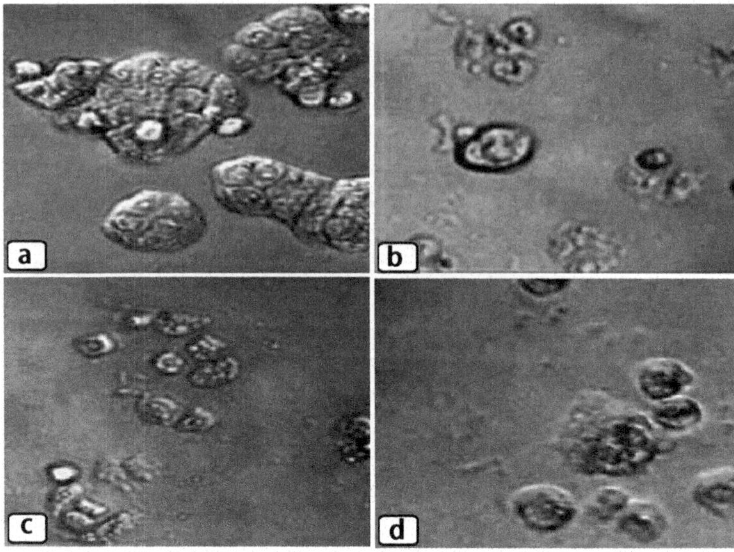

Fig. 19 Morphological changes in HT-29 cells after rutin-alone and radiation-alone treatments and in combination (light microscopy); (**a**) sham control cells, (**b**) cells treated with rutin (80 μM), (**c**) cells treated with 4 Gy radiation, (**d**) cells treated with both rutin and 4 Gy radiation (Reproduced from reference [41] with permission from AEPress)

apoptotic bodies (Fig. 20b). The effect of rutin and radiation on mitochondrial membrane potential was studied. HT-29 sham control cells showed intense green fluorescence due to high mitochondrial transmembrane potential, whereas cells treated with both rutin and 4-Gy radiation stained with rhodamine 123 showed a progressive loss of green fluorescence at 12 h and highly diminished fluorescence at 12 h due to a mild and complete loss of mitochondrial transmembrane potential, respectively. The loss in mitochondrial transmembrane potential can be correlated with the enhanced apoptosis encountered on rutin and/or radiation treatment. They also evaluated the effect of rutin and radiation on DNA damage. Treatment with both rutin and 4 Gy radiation caused severe DNA damage at 24 h (comet assay), whereas the cells treated with rutin alone and 4 Gy radiation alone exhibited lower percentage of DNA damage, as evidenced by the appearance of a prominent comet with a tail. An increased DNA damage (comet assay) was noted on simultaneous treatment of HT-29 cells with rutin and X radiation as evidenced by the increased comet tail length and density. This could contribute to the increased rate of cell deaths in cancer cells, and these effects may be due to the generation of ROS in HT-29 cells. Results illustrated that the treatment of HT-29 cells with rutin along with radiation exposure caused a pronounced chemoradiotherapeutic effect as confirmed by (1) the alterations in nuclear morphology; (2) loss of mitochondrial transmembrane potential; (3) marked induction of DNA damage, apoptosis, lipid peroxidation; and (4) reduction in enzymatic antioxidants. The abovementioned

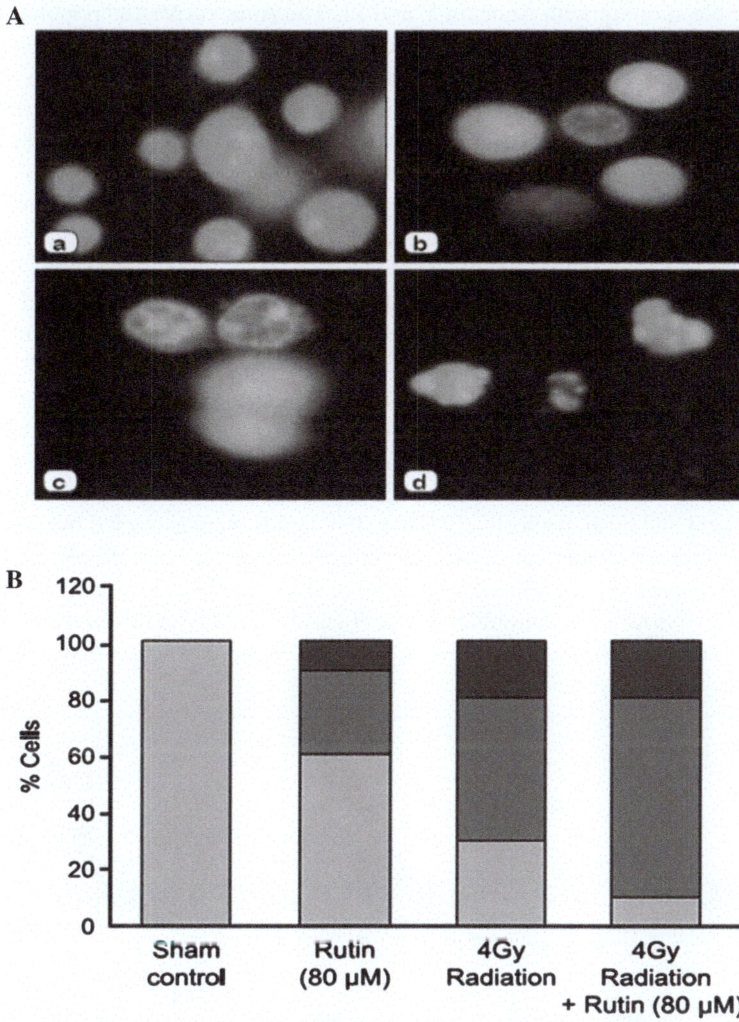

Fig. 20 (**A**) The effect of rutin-alone and radiation-alone treatments or in combination on HT-29 cells (AO/EtBr dual staining); (**a**) sham control cells, (**b**) cells treated with rutin (80 μM), (**c**) cells treated with 4 Gy radiation, (**d**) cells treated with both rutin and 4 Gy radiation. (**B**) Quantitative data on the effect of rutin-alone and radiation-alone treatments or in combination on HT-29 cells reveal apoptosis and necrosis on staining with AO/EtBr (Reproduced from reference [41] with permission from AEPress)

effects confirm not only the anticancer property of rutin but also its ability to act as a radiosensitizer.

Guon and Chung demonstrated the mechanism of two flavonol glycosides, hyperoside and rutin, in the induction of apoptosis in HT-29 human colon cancer cells through the bioactivity-guided fractionation and isolation method [42]. In order to quantify the percentage of apoptotic cells, flow cytometry analysis was performed using double staining with Annexin V and PI. The Annexin V⁻/PI⁻ population was considered to account for apoptosis, Annexin V⁺/PI⁺ for late apoptosis, and Annexin V⁻/PI⁺ for necrosis. The results showed that the treatment of the cells with hyperoside and rutin significantly increased the percentage of apoptotic cells compared with untreated control cells (Fig. 21). The total apoptotic cell populations of hyperoside and rutin were increased by 52.4 and 56.5% at 200 μm, respectively, compared with the control. These results indicate that hyperoside and rutin may induce apoptosis in HT-29 human colon cancer cells. The findings of this study indicated that hyperoside and rutin induce apoptosis in HT-29 human colon cancer cells and that this phenomenon is mediated via the death receptor-mediated and mitochondria-mediated apoptotic pathways. These results suggest that hyperoside and rutin may be useful in the development of a colon cancer therapy protocol.

In another study, the antioxidative and anticancer effects of extracted flavonoids from Pink Lady apples on human colon cancer LoVo cells and breast cancer MCF-7 cells were evaluated by Yang et al. [43]. In this study, Pink Lady apples were selected as the matter for investigation, because of their high content of flavonoids in extracts from the peel and flesh, respectively. Then, flavonoids were extracted from peel and flesh, and their contents were also determined. Furthermore, multiple evaluation systems are adopted to conduct a comparative study of the in vitro antioxidant and antitumor activities of flavonoids in the peel and flesh. Flavonoids in the peel were 2.2-fold to those in the flesh with the value of 24.05 mg/g. The extraction yields of

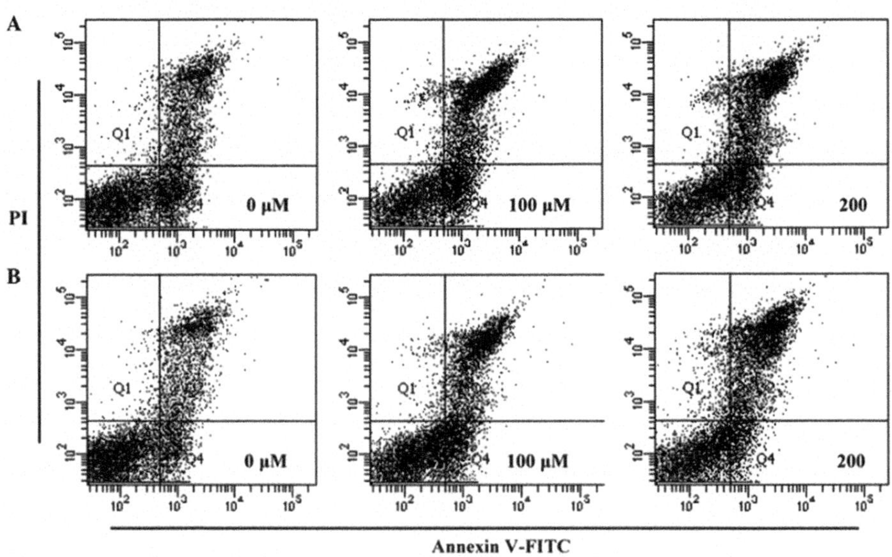

Fig. 21 Induction of apoptosis by (**a**) hyperoside and (**b**) rutin on HT-29 human colon cancer cells. Flow cytometric analysis was performed on HT-29 human colon cancer cells that were incubated with hyperoside and rutin for 24 h. The right bottom quadrant represents Annexin V-stained cells (early-phase apoptotic cells). The top right quadrant represents PI- and Annexin V-stained cells (late-phase apoptotic cells). Bcl-2, B-cell lymphoma 2; Bax, Bcl-2-associated X protein (Reproduced from reference [42] with permission from Spandidos Publications)

the peel flavonoids (Peel-F) and flesh flavonoids (Flesh-F) were 2.4% and 1.1%, respectively, which led the conclusion that the content of flavonoids in the peel was higher than that in flesh. To assess the bioactivity of the compound, antioxidant activity analysis is a significant method to reveal the potential of a compound to inhibit a free radical oxidation reaction by a direct effect on free radicals or the indirect blocking of intermediate material that easily generates free radicals. DPPH• is a nitrogen-centered organic synthetic radical, which is more stable in organic solvents, with the capacity for evaluating the antioxidation of matter. It has a single electron and can accept hydrogen ions or electrons. Its methanol solution has an obvious absorption peak at 517 nm. So when there is a free radical scavenger, it will be paired with DPPH• lone pair electrons, with the absorbance value decreased. The scavenging rate is calculated according to the changes in absorbance value. The higher scavenging rate indicates a stronger antioxidant capacity of the matter. Results showed that the apple flavonoids' scavenging effect of DPPH• was in a concentration-dependent manner. The scavenging effect enhanced with the increased treated concentration. The peel flavonoids' scavenging ability is higher than that of the flesh flavonoids. When the sample concentration was 0.8 mmL^{-1}, the DPPH• scavenging rate of Peel-F and Flesh-F reached 64.38% and 17.41%, respectively. Figure 22 shows the effect of flavonoids from different parts of a Pink Lady apple on the viability of LoVo cells and MCF-7 cells for 24 h and 48 h. Figure 22a shows that after 24 h of the treatment, the inhibition rates of 25 mg/mL, 50 mg/mL, and 100 mg/mL of the peel flavonoids (Peel-F) on LoVo cell proliferation were 2.5%, 12.32%, and 19.17%, respectively, while the inhibition rate of the 100 mg/mL flesh flavonoids (Flesh-F) on LoVo cell proliferation was 6.83%. After 48 h of the treatment, the inhibition rate of the 100 mg/mL Peel-F on LoVo cell proliferation reached 47.97%, while the inhibition rate of the Flesh-F on LoVo cell proliferation increased 20.1%. Figure 22b shows that after 48 h of treatment, the inhibition rates of the 50 mg/mL Peel-F and Flesh-F on MCF-7 cell proliferation were 48.77% and 17.35%, which were in contrast to the control group ($p < 0.01$). When the concentration of the Peel-F and Flesh-F flesh was increased to 100 mg/mL and after 48 h of treatment, further increased to 1118.76 U L and 909.50 U L. At this point, the inhibition rate of the Peel-F on the cell proliferation rate was slightly higher than that after 24 h of treatment of the anticancer agent 5-Fu, which was 54.25%. The IC_{50} values produced by the Peel-F and Flesh-F acting on LoVo cells were 110.33 ± 2.52 mg/mL and 378.14 ± 1.64 mg/mL, respectively, and the IC_{50} values for MCF-7 cells were 58.42 ± 1.39 mg/mL and 296.06 ± 3.71 mg/mL, respectively, which showed that the inhibitory effect of the Peel-F on cancer cells was significantly higher than that of the Flesh-F and that the Peel-F demonstrated a more pronounced antitumor effect on MCF-7 cells.

The effect of different concentrations of the Peel-F on cell morphology was also in a concentration-dependent manner, which means that the higher the drug concentration exposed to the cells, the greater the destructive changes in the cell morphology. This indicates that the drug in low concentrations causes no significant damage to cells, while high doses lead to significant cell damage and significant reduction in the number of cells, which is consistent with the cell viability assay results (Fig. 23).

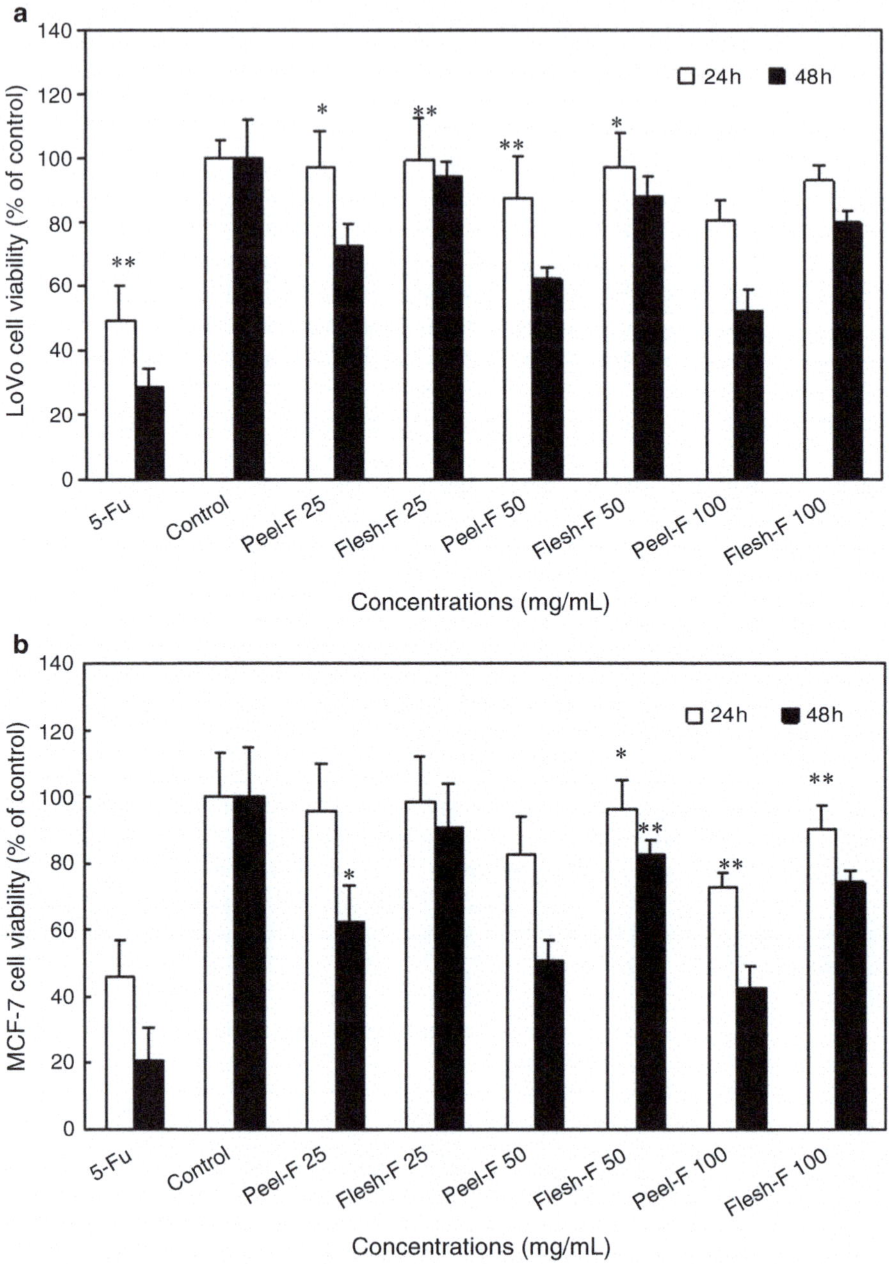

Fig. 22 The effects of flavonoids from different parts of an apple on cell viability in LoVo (**a**) and MCF-7 (**b**) human cancer cells $*p < 0.05$, $**p < 0.01$ compared with control group (Reproduced from reference [43] with permission from Royal Society of Chemistry)

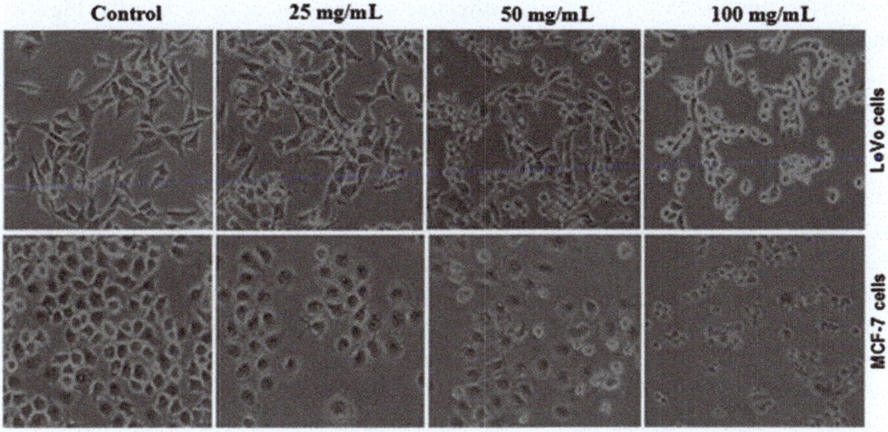

Fig. 23 The morphological changes of LoVo and MCF-7 human cancer cells on exposure to the peel flavonoids. These were observed by an inverted optic microscope (original magnification ×100) (Reproduced from reference [43] with permission from Royal Society of Chemistry)

For the study of antitumor efficiency, the release of LDH is usually regarded as an index to assess the integrity of cell membranes. In this research, the LDH release was tested after incubation for 24 h and 48 h of the different treated concentrations of the drug. The LDH activities produced by treating LoVo cells with the 50 mg/mL peel flavonoids (Peel-F) and 50 mg/mL flesh flavonoids (Flesh-F) for 24 h were 466.34 U/L and 149.18 U/L, respectively. However, after 48 h of treatment, the LDH activities significantly increased to 653.07 U/L and 380.26 U/L, respectively. When the concentration was increased to 100 mg/mL, the LDH activities produced by the peel flavonoids and flesh flavonoids acting on cells for 48 h rose to 751.52 U/L and 478.79 U/L, respectively. The LDH activities of MCF-7 cells treated with the 50 mg/mL Peel-F and 50 mg/mL Flesh-F for 48 h were 678.57 U/L and 457.14 U/L, respectively. When the concentration was increased to 100 mg/mL, the LDH activities produced by the Peel-F and Flesh-F acting on the cells for 24 h significantly increased to 963.83 U/L and 789.50 U/L, respectively, and after 48 h of treatment further increased to 1118.76 U/L and 909.50 U/L. The results revealed that the LDH activities with peel flavonoids were significantly higher than those with the positive control group after 5-Fu treatment. The LDH activity assay results showed that the toxicity of Peel-F and Flesh-F to cancer cells occurs in a concentration- or time-dependent manner, suggesting that with increased concentration of flavonoids, LDH activity is gradually enhanced and the toxicity to cells is also exacerbated. The results of toxicity to two cell lines LoVo and MCF-7 and the cell viability of flavonoids from different parts of the apples were consistent. The peel flavonoid effects were significantly better than the flesh flavonoids, and by comparison, the peel flavonoids were more sensitive to MCF-7 cells. The apoptosis detection results suggested that flavonoids from different parts of an apple can induce apoptosis of tumor cells, which is directly correlated with dosage. With the increase of dosage, more cells will enter the apoptosis period, and the Peel-F causes greater cell apoptosis than the Flesh-F. These results led to the conclusion that the Peel-F was more effective against cancer cells than the Flesh-F and that the scavenging ROS effects were significantly higher in the Peel-F in vitro.

3 Solubility and Bioavailability Enhancement of Rutin

The major problem associated with rutin is its poor solubility in aqueous media (0.8 mg/ml) [44], which is the main reason for its poor bioavailability. To overcome these limitations in addition to widening the scope of bioactive therapy, preparing stable bioactive complexes by integrating the distinct functions of individual natural compounds is a feasible alternative to the single compound-based therapy [45, 46]. A number of drug delivery systems were proposed for the entrapment of this active compound with the aim of improving its pharmacological activity. Common approaches used to enhance bioavailability of rutin are particle diminution to the submicron range and by complexation of rutin with cyclodextrins and various metals. Also, various carrier systems for delivery of rutin such as microemulsions, microencapsulation, nanocrystals, and nanosuspensions have been developed. Nanosystems and nanomaterials are general terms assigned to any presence having sizes ranging in the nanometric scale with different properties, such as nanoemulsions, nanoparticles (NPs), polyplexes, dendritic structures, micelles, and liposomes, among others [47]. Various approaches such as complexation using cyclodextrin and phospholipids have been researched and documented.

4 Cyclodextrin Complex

Cyclodextrins are cyclic oligosaccharides with a somewhat truncated, cone-like structure that are divided into three types, a, b and g, depending on whether they contain six-, seven-, or eight-membered sugar ring molecules, respectively. Cyclodextrins are known for their ability to form inclusion complexes with many size-suitable guest molecules owing to the unique molecular structure with hydrophobic cavity and hydrophilic external faces, which increases the bioavailability of poorly soluble drugs by enhancing their aqueous solubility and acts as a driving force for diffusion across the biological membrane. Cyclodextrins have both stabilizing and destabilizing effects on drugs from photolytic and hydrolytic degradation [48, 49]. Rutin has been encapsulated in different substituents of cyclodextrin, such as 2-hydroxypropyl beta-cyclodextrin (HP-β-CD), hydroxypropyl-γ-cyclodextrin (HP-γ-CD), β-cyclodextrin (β-CD), and G2-β-cyclodextrin (G2-β-CyD) for increasing the solubility and stability of the drug [48]. The order of stability constants calculated from the phase solubility method increased as HP-γ-CyD $<$ G2-β-CyD $<$ β-CyD $<$ HP-β-CyD. Spectroscopic studies also revealed that HP-β-CD and β-CD formed relatively more stable inclusion complexes with rutin. The dissolution rates of rutin increased by the complexation with CyDs in the order of rutin alone $<$ HP-β-CyD $\leq$ β-CyD. HP-β-CyD may also stabilize rutin in a gastrointestinal tract after oral administration as it inhibited the hydrolysis of rutin in alkaline buffer solution and small intestinal homogenates of rats. In vivo absorption studies suggested that HP-β-CyD increased the oral bioavailability of rutin from the gastrointestinal tracts of beagle dogs because of the increase in solubility, faster dissolution rate, and gastrointestinal stability. Oral bioavailability was determined by measuring the plasma levels of homovanillic acid (HVA) (a major stable metabolite of rutin) in beagle dogs, after oral administration of HP-β-CyD complex. HP-β-CyD also had a significant advantage with respect to providing high aqueous solubility while maintaining a lack of toxicity in oral pharmaceutical preparations of rutin. In another study, Sri et al. prepared and characterized cyclodextrin inclusion complexes of quercetin and rutin to improve their aqueous solubility and dissolution properties [50]. In another study, the effects of the complexation on the stability and antioxidant activity of rutin were studied. The complexes formed with HP-γ-CD and HP-β-CD had the greatest stability constants, followed by β-CD and HP-α-CD. The calculated IC_{50} (the antioxidant concentration required to scavenge 50% of the free radicals present) of rutin for DPPH· radicals was 0.022 mg/ml, while for rutin/HP-β-CD was only 0.014 mg/ml. After storage at 45 °C for 18 days, the amount of rutin remaining was 50.7 and 65.8% for free and complex forms, respectively. This study concluded that among all of the cyclodextrin-rutin complexes, complexes formed with HP-β-CD and HP-γ-CD had the greatest stability constants. The presence of hydroxypropyl groups in HP-β-CD was found to significantly enhance its capacity to entrap rutin. During complex formation, there was no covalent bond formed or broken, and the rutin molecules in complex were in rapid equilibrium with free molecules in the solution. The formation of such an inclusion complex conferred moderate degrees of protection to rutin from thermal and UV radiation degradation during storage and significantly enhanced its antioxidant capacity [12].

5 Phytosomes/Phospholipid Complex

Phytosome is an amphiphilic molecular complex in which bioactive molecules (mainly polyphenolics) are complexed with phospholipids, in particular phosphatidylcholine, resulting in the formation of supramolecular adducts having a definite stoichiometry. The design and construction of supramolecular arrays utilizing noncovalent bonding of tectons would be called supramolecular synthesis [51]. Thus, the supramolecular synthesis successfully exploits hydrogen-bonding [52–54] and other types of noncovalent interactions, in building supramolecular systems [55–59] such as intermolecular coordination bonds, which lead to formation of coordination polymers [60–66] and metal-organic frameworks [67–70], and metallophilic [52–54], polyhapto, or agostic interactions [71–77]. The advantage of phytosomes over other formulations is that it improves gastrointestinal absorption upon oral administration by enhancing the rate and the extent of solubilization of bioactive into aqueous intestinal fluids, thus enhancing the ability to cross the lipid-rich biomembranes, yielding higher plasma levels and reducing kinetic elimination, and resulting in enhanced systemic bioavailability [78, 79]. Rutin phytosomes (RN-P) were developed and characterized to establish its feasibility for transdermal application in inflammatory conditions. Phytosomes were prepared in five molar ratios of rutin (0.5–1.0) to phosphatidylcholine (1.0–0.5). All RN-Ps showed aqueous solubility higher than pure rutin [80]. Jain et al. developed rutin-phospholipid complex for improving the lipophilicity of rutin [81]. Rutin-phospholipid complex was prepared by anhydrous cosolvent lyophilization method. For the preparation of complex in the ratio of 1:1, 100 mg of rutin was dissolved in 10 ml of methanol separately, and 100 mg phospholipid was dissolved in 10 ml chloroform, separately. The authors examined the free radical-scavenging property of this molecule through these complexes. The complexes were tested for its physicochemical properties and in vitro antioxidant activity. Among all prepared complexes, 1:3 exhibited maximum free radical-scavenging activity, followed by 1:2 and then 1:1 in comparison to rutin. The results obtained, proved better efficiency in the form of complexes as compared to the molecules itself.

Singh et al. also prepared rutin-phospholipid complex in order to enhance the delivery of poorly soluble rutin and found that the water/n-octanol solubility of rutin was improved from 2.88 to 45.71 μg/ml and 68.17 to 245.18 μg/ml, respectively, in the complex [82]. The antioxidant activity indicated that the bioactivity of rutin was maintained even after being complexed with the phospholipid. In vitro drug release kinetics was studied to predict the in vivo performance of the drug (rutin) in the phospholipid complex. Unlike the free rutin which showed a total of 30.05% and 54.04% drug release, the rutin complex showed 60.05% and 80.89% drug release at the end of 24 h in the pH 1.2 HCl and 6.8 phosphate buffer saline, respectively. The complex showed a pH-dependent release profile as that of rutin, but the release rate was found to be of first-order kinetics. The improved release of rutin from the phospholipid complex may be explained by the improved solubility (due to the amorphous characteristics) following complexation.

6 Nanoparticulate System

For poorly water-soluble drugs, particle size often plays an important role in achieving adequate oral bioavailability. Particle diminution to the nanosize range is a useful formulation approach to increase the dissolution rate and the saturation solubility and in turn to enhance the oral bioavailability of poorly water-soluble drugs. Recently, nanoparticle-based drug-delivery systems have emerged as a solution for increasing the bioavailability of potential therapeutic agents. Nanoparticulate drug delivery systems such as nanoemulsion, liposomes, self-emulsifying drug-delivery systems (SMEDDS), niosomes, nanoparticles, and dendrimers of submicron size are stabilized with surfactants or polymers in nanosuspensions, which can be further processed into standard dosage forms for oral, parental, or topical delivery to improve the bioavailability of the active ingredients and introduce controlled/target release [83]. Mauludin et al. prepared nanocrystals of rutin for improving the dissolution behavior and bioavailability [4]. For this purpose, they prepared dried rutin nanocrystals by lyophilization technique and investigated their physicochemical properties such as redispersability, particle size, morphology, and dissolution behavior. Lyophilized rutin nanocrystals were then incorporated into tablets using direct compression method, and the dissolution behavior of the tablets was evaluated. It was observed that very fine particles of lyophilized rutin could be completely redispersed in the water. The PCS size average and polydispersity index (PI) of lyophilized rutin were of 721 nm and of 0.288 after redispersion. The dissolution velocity of the rutin nanocrystal-loaded tablet was compared to rutin microcrystal-loaded and a marketed tablet. It was noticed that rutin was released after 30 min and dissolved completely from the nanocrystal tablets in water. In contrast, only 71 and 55% of the total amount of rutin were dissolved from the microcrystal tablets and the marketed tablet, respectively. In another work, Almeida et al. studied the improvement of the rutin photostability and prolongation of its in vitro antioxidant activity by its association with nanostructured aqueous dispersions [84]. The in vitro antioxidant activity was evaluated by the formation of free radical ·OH after the exposure of hydrogen peroxide to a UV irradiation system. It was observed that rutin-loaded nanostructures showed lower rutin decay rates [$(6.1 \pm 0.6)\ 10^{-3}$ and $(5.1 \pm 0.4)\ 10^{-3}$ for nanocapsules and nanoemulsion, respectively] when compared to the ethanolic solution [$(35.0 \pm 3.7)\ 10^{-3}\ \text{min}^{-1}$] and exposed solution [$(40.1 \pm 1.7)\ 10^{-3}\ \text{min}^{-1}$] as well as the exposed nanostructured dispersions [$(19.5 \pm 0.5)\ 10^{-3}$ and $(26.6 \pm 2.6)\ 10^{-3}$, for nanocapsules and nanoemulsion, respectively]. SMEDDS has recently come into existence as one of the most interesting approaches to improve the solubility, dissolution, and bioavailability for poorly water-soluble drugs. An example of this technology is the Neoral formulation of cyclosporine A, which is commercially marketed by Novartis Pharmaceuticals (Basel, Switzerland). Self-emulsifying drug delivery systems (SEDDS) of rutin were prepared in order to improve the release profile of the drug and hence the bioavailability [85]. SEDDS were prepared by using coprecipitation and melting methods at different drug-polymer ratios. Polyethylene glycol (PEG 6000), polyvinyl pyrrolidone (PVP K30 and PVP K17), and sodium deoxycholate were used to prepare SEDDS by coprecipitation method. The results indicated that maximum amount of drug release was obtained after 30 min (93%).

Both natural polymers (collagen, chitin, chitosan, keratin, silk and elastin, starch, cellulose, pectin, etc.) [86] and synthetic polymers (poly(lactic acid), poly(glycolic acid), poly(lactic acid coglycolic acid), and poly(ε-caprolactone)) [87] are used as biomaterials. Among preformed polymers, poly(lactic-coglycolic acid) (PLGA) is beneficial due to its features [88]. It has been the most appealing copolymer used for drug delivery systems [89], nanovaccines, and nanoparticles (NPs) [90]. It is a biodegradable, biocompatible, nontoxic, and FDA-approved polymer [91]. In the literature, emulsification–diffusion [92], solvent emulsion–evaporation [92], interfacial deposition [93], and nanoprecipitation methods have been used to prepare PLGA nanoparticles [90]. The rutin-loaded PLGA nanoparticles in optimum formulation by oil-in-water (o/w) single- solvent emulsion–evaporation method were synthesized and characterized in details [94]. In accordance with this purpose, different amounts of rutin and PLGA, stabilizer concentration in aqueous phase, and oil/aqueous phase ratio were examined to determine the optimum NP formulation. The effects of variables were analyzed by reaction yield (RY), encapsulation efficiency (EE), hydrodynamic particle size, polydispersity index (PDI), and zeta (ζ)-potential. An optimized NP was determined after evaluation of analysis results, and then in vitro release of rutin from optimum NP was examined. Different amounts of PLGA and rutin were dissolved into the organic phases consisting of DCM and propylene glycol, respectively; then, they were mixed. These organic solutions were emulsified with aqueous solution of PVA with diverse concentrations and volumes by sonication (output power of 70 W, power of 80%, and 2 min) using a microtip probe sonicator over an ice bath. When high-energy ultrasound interacts with liquids, cyclic alternating areas of compression (high pressure) and rarefaction (low pressure) are formed. In the low pressure region, the pressure drops below the vapor pressure of the solvent, and/or the reactants and small bubbles, i.e., cavities, are formed. The bubbles grow under the alternating pressure through the diffusion of solute vapor into the volume of the bubble. Thus, ultrasonic energy is accumulated. Once the bubbles reach their maximum size, they become unstable and collapse [95, 96]. The free nanoparticles were prepared by the same method without rutin. They all were stirred overnight on a magnetic stirrer at room temperature for evaporation of organic phase. After centrifugation at 9.000 rpm (4 °C) for 40 min, collapsed particles were collected, washed three times with ultrapure water to remove excess PVA, and then lyophilized. All lyophilized nanoparticles were stored at −80 °C. The nanoparticle NP4 with high reaction yield (47%) and encapsulation efficiency (81%), small size (252.6 ± 2.854 nm), and PDI value (0.209 ± 0.008) was determined as optimized NP. In vitro % rutin release for two different concentrations of NP4 was subsequently investigated. NP4 with low concentration (0.5 mg/mL) successfully had 100% releases on the 5th day (120 h). Results showed that encapsulation with PLGA could be useful for poorly soluble but beneficial-to-health extracts like rutin as delivery systems to enhance their biological activities. Pandey et al. prepared and optimized rutin (RT)-loaded poly(lactic co-glycolic acid) (PLGA) nanoparticles (NPs) and explore their therapeutic efficacy against diethylnitrosamine (DEN)-induced hepatocellular carcinoma (HCC) in rats [97]. As per the Box–Behnken design, NPs of RT were rationally optimized for the critical factors, viz. concentration of PLGA and RT as the product parameters and sonication time as the process parameter, while other factors like ethyl acetate and 1% (w/v) polyvinyl alcohol were fixed as constant. Response surface methodology was employed for estimation of factor–response relationship and identifying the critical formulation factors influencing the responses of NPs. Optimized PLGA NPs showed particle size of 211 nm with polydispersity index of 0.24, ζ-potential in the range of −21.3 to −22.8 mV, drug-loading capacity of 6.39%, entrapment efficiency of 77.83%, and gastric stability in various pH media along with 71% in vitro drug release. During pharmacodynamic study, rats treated with RT group showed less hepatic nodules with tumor incidence varying between 90 and 54.5%, while RT-PLGA-NP group showed significant reduction in tumor incidence to 16.6%. RT-PLGA-NP-treated rats have 0.58 times lower relative liver weight as compared with rats treated with DEN group. RT treatment showed boosting level of total protein, albumin, and reduced the total bilirubin and γ-glutamyl transferase level at doses 5 and 50 mg/kg, whereas the RT-PLGA-PNP showed the similar pattern for increase in total protein, albumin, and decrease in the total bilirubin and γ-glutamyl transferase level. Dose-dependent alteration of red blood cells, hemoglobin, and upregulation of erythrocyte sedimentation rate and white blood cells by RT were observed. Moreover, RT-PLGA-NPs significantly ($p < 0.001$) altered hematological profile by upregulation of red blood cells, hemoglobin, and downregulation of erythrocyte sedimentation rate and white blood cells. Administration of RT-PLGA-NPs demonstrated the modulation of Phase I antioxidant enzymes including NADH-cytochrome-b5-reductase, NADPH-cytochrome P450 reductase, cytochrome P450, and cytochrome-b5 in DEN-induced group rats. Enhanced level of NF-κB was observed in DEN-administered rats, whereas its level reduces up to a significant level by using RT-PLGA-NPs. RT-PLGA-NPs showed less change in histopathology observation in terms of less necrosis, inflammatory cells, and enlargement of karyomegaly as compared with rats in DEN group. Results showed that the administration of RT-PLGA-NPs significantly reduced the production of free radical, oxidative stress and stabilized the membrane-bound enzyme activity via multiple mechanisms to provide therapeutic benefit against hepatocellular carcinoma.

In another study, rutin encapsulated in non-amidated low methoxyl pectin bead in various formulations with or without sorbitol and/or sodium bicarbonate was prepared; in vitro anticancer activities in three human cancer cell lines including human colon adenocarcinoma (HT-29) cells, human mouth epidermal carcinoma (KB) cells, and human hepatocellular carcinoma (HepG2) cells were investigated in order to evaluate their potential for further human cancer treatment formulation [98]. LMP (low methoxyl pectin) aqueous formulations including 3NA (3% non-amidated LMP), 3NA15Sor (3% non-amidated LMP with 15% sorbitol), 3NA1Bica (3% non-amidated LMP with 1% $NaHCO_3$), and 3NA15Sor1Bica (3% non-amidated LMP with 15% sorbitol and 1% $NaHCO_3$) were prepared followed by 2% w/v of rutin dispersed in the solution and stirred until a uniform dispersion was obtained. The pH and viscosity of the solution were measured. The beads were made using encapsulator UNIT VAR1 with a nozzle of 0.7 mm inner diameter. The slurry was dropped into 50 ml of a gently agitated solution of the crosslinking agent (2% w/v $CaCl_2$) at flow rate 100 mL/h with falling distance of 4 cm. The gelled beads were formed immediately and allowed to stand in the cross-linking solution for 10 min. Then, beads were separated by filtration, washed with deionized water, and dried at 37 ± 2 °C for 24 h in a drying room. All beads were oblong or spherical shape with around 600 μm size. The rutin which was encapsulated in 3NA15Sor1Bica (3% non-amidated LMP with 15% sorbitol and 1% $NaHCO_3$) bead showed anticancer activity higher than non-encapsulated rutin on human cancer cell lines especially on human colon adenocarcinoma (HT-29) and human mouth epidermal carcinoma (KB) cells. 3NA15Sor1Bica rutin bead composing both sorbitol and $NaHCO_3$ exhibited higher cytotoxic activity with cancer cell viability being 51.41 ± 7.01 and 55.12 ± 3.96% on HT-29 and KB cells, respectively. There was no significant different ($p > 0.05$) of cell viability between all bead formulation for HepG2 cells.

Banjare et al. developed sustained release nanoparticles of rutin for topical delivery. They were prepared by nanoprecipitation technique using ethylcellulose as polymer [99]. This colloidal carrier has an advantage of enhanced drug penetration into the skin due to its submicron size, and also because of its lipoidal nature, it formed a drug concentrate in the skin and remained localized for a longer period of time, thus enabling drug targeting to the skin. The ex vivo studies were done using goat skin which showed sustained release of the drug from the formulation. The cumulative amount of drug permeated at the end of 24 h was found to be 76.24 μg/cm^2 with skin permeation rate constants of 0.415 %/cm^2/h. The release of drug from these formulations followed a Fickian pattern with n value 0.418–0.423. The results demonstrated that rutin could possibly permeate through the human skin maintaining the sustained release. Kamel et al. developed novel vesicular system in the form of proniosomes using a binary mixture of sorbitan monostearate and polyethylene glycol fatty acid esters for rectal delivery of rutin [100]. The proniosomes were prepared using Span 60 or Span 60/Gelucire combination. These proniosomes were compared to conventional niosomes for encapsulation efficiency (EE%), particle size, zeta potential (Z), in vitro release, and ex vivo permeation using excised rat rectum. Span 60/Gelucire proniosomes exhibited lower particle size as well as better release and permeation results, compared to Span 60 proniosomes. Span 60/Gelucire 44/14 (2:1) proniosomes attained the highest percentage of drug release (98.39%) and drug deposition in the rectal walls (86.26%) after 8 h, indicating achievement of targeted therapeutic efficacy.

A rutin-fucoidan complex-based phytotherapy for cervical cancer through achieving enhanced bioavailability and cancer cell apoptosis was reported by Deepika et al. [101]. Rutin (Ru) is noncovalently complexed with fucoidan (Fu) using the functional groups to obtain a therapeutic polymeric complex. As tumor regions have slightly acidic extracellular environment, Ru-Fu release rate was evaluated employing the different pH conditions: pH 5.5 (simulates the pH in mature endosomes of cancer cells), pH 6.8 (the pH of cancer tissue), and pH 7.4 (corresponds to the circulatory environment of blood) at 37 °C, the physiological temperature. Observations showed that the acidic pH enhanced the rate of molecule release from Ru-Fu complex. The percentage of cumulative release of rutin and fucoidan is as follows: 45.2% and 43.8% at pH 7.4, 58.8% and 57.2% at pH 6.8, and 78.5% and 72.5% at pH 5.5, respectively. No initial burst release was observed using the formulation. The results demonstrated that the release process of compounds at pH 7.4 was much slower than that of pH 5.5. The carboxylic and hydroxyl linkages enable Ru-Fu complex to remain slightly stable for a considerable period of time during circulation at pH 7.4 and thereby eliminate the premature burst release. The stability of the complex might reduce the risk of adverse effects of compounds against normal cells due to their premature release by disintegration in circulation. Although at pH 6.8 the drug release behavior was higher than that at pH 7.4, the effect was lesser than at pH 5.5, and the slight change in the behavior could also be due to the possible changes in the degree of protonation. At lower pH 5.5, the destabilization of weaker hydrogen bonds between Ru-Fu contributes to the release of the compounds. To evaluate the effects of cytotoxicity induced by rutin, fucoidan, and complex, HeLa and HaCaT cells (control) were treated with various concentrations of rutin, fucoidan, and complex for 24 h, and the MTT reduction assay was performed. The IC_{50} values of rutin, fucoidan, and Ru-Fu complex in Hela cells were 30 μg/ml, 200 μg/ml, and 20 μg/ml, respectively, whereas they were 60 μg/ml, 400 μg/ml, and 80 μg/ml for HaCaT cells. The observed IC_{50} values of HeLa were much lower than HaCaT cells, indicating the intrinsic toxic nature of compounds and complex against HeLa cancer cells but less toxic to normal HaCaT cells (Fig. 24a–c). Moreover, Ru-Fu complex showed higher growth-inhibitory effects on HeLa cells than HaCaT cells in a concentration-dependent manner, and a drastic reduction in the IC_{50} value of Ru-Fu for HeLa cells was observed when compared to individual compounds (rutin and fucoidan) (Fig. 24c). The treatment by mixtures of the compounds (Ru+Fu) at lower concentrations also increased the antiproliferative effect in Hela cells than individual treatment (Fig. 24d). The percentages of cell viability at the selected sub-IC_{50} were found to be 75.3% (rutin), 81.6% (fucoidan), and 41% (Ru+Fu mixture) compared to control (Fig. 24d). However, the enhanced effect was observed with Ru-Fu complex, and the HeLa cell growth was found to be reduced to 14.8% by the treatment with the complex (Fig. 24d).

Further, the chemopreventive potential of Ru-Fu complex is investigated against HeLa cells by cellular apoptotic assays and flow cytometric analysis. The apoptotic extent was higher in Ru-Fu complex (33%) treated cells than the cells treated with rutin (25%), fucoidan (21%), and their mixtures (27%). Ru-Fu complex treated cells showed low population of necrotic cells (31%), while the control group did not significantly show either the apoptotic or necrotic population ($p \leq 0.01$). It showed that the complex is able to disrupt cell cycle regulation and has the ability to induce cellular apoptosis via nuclear fragmentation, ROS generation, and mitochondrial potential loss. In vitro cell viability assay with Ru-Fu complex shows that the complex is biocompatible on normal cells. The hemolysis assay also reveals that the complex does not release hemoglobin from human red blood cells (RBCs). This study provided scope for developing an anticancer complex integrating both structural and functional features of rutin (a low molecular weight phytochemical) and fucoidan (a large molecular weight sulfated polysaccharide) as an interacting complex.

A drug-polymer complex, rutin-Pluronic SD, was prepared by Lee et al. [102]. The solid dispersion (SD) method called amorphization, which is the formation of a molecular mixture in which poorly water-soluble drugs are encased in hydrophilic polymers, led to an increase in bioavailability and improvement in drug release profile [103]. Among materials employed for SD, Pluronic has been extensively used as copolymer-based excipient for many drug molecules with poor water solubility. Electrospinning loads the prepared polymer solution with various drugs in a thin fiber formation, in the range of micrometers to nanometers, and the electrospun fibers eventually obtain a large surface area with high porosity [104]. SDs were produced using a facile mixing method, and nanofibrous structures containing pluronic SD-loaded rutin were made by electrospinning. The optimization of the loading efficiency of the rutin-loaded SDs was performed by examining the effect of pluronic addition and various initial amounts of rutin. Furthermore, all samples

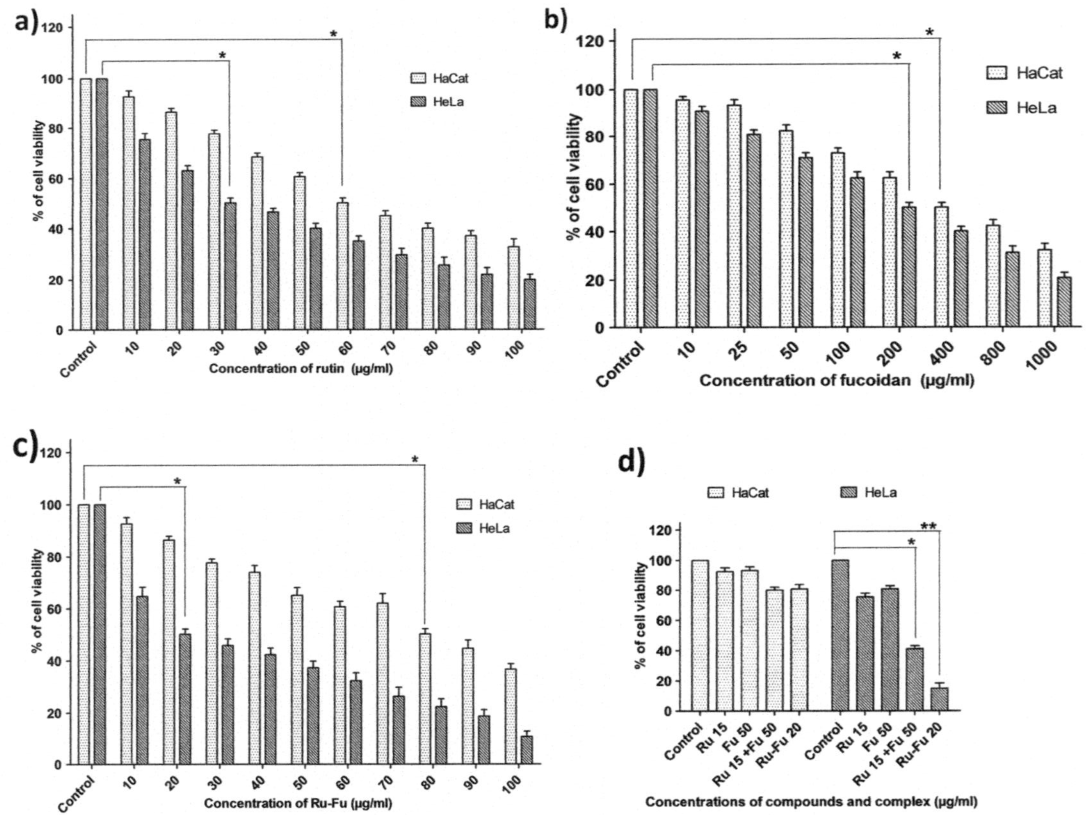

Fig. 24 (**a**) Cytotoxicity of rutin treatment in HeLa and HaCaT cells. (**b**) Cytotoxic effects of fucoidan in HeLa and HaCaT cells. (**c**) Cytotoxic effects of Ru-Fu (rutin-fucoidan complex) in HeLa and HaCaT cells. (**d**) The effect of rutin (Ru 15 μg/ml), fucoidan (Fu 50 μg/ml), rutin-fucoidan mixtures (Ru 15 + Fu 50), and rutin-fucoidan complex (Ru-Fu 20 μg/ml) on HaCaT and HeLa cell proliferation (Reproduced from [101] under a Creative Commons License)

containing rutin were evaluated through antioxidant assay as well as in vitro release test. Also, ABTS radical scavenging activity was investigated. The ABTS assay indicated that the presence of pluronic in PUL&RU@PF affected the increase in the amount of rutin and contributed to the enhanced antioxidant effect of rutin. An in vitro release test was conducted to assess the loading and release performance of rutin from the SD or nanofiber formation. The release profiles of raw rutin, RU@PF, PUL&RU, PUL&RU@PF, PUL&RU, and PUL&RU@PF films are shown in Fig. 25. The rutin release in RU@PF displayed an initial burst release pattern compared to that of raw rutin. The amount of raw rutin released during the predetermined period (6 h) was about 10%, which was largely attributed to the low water solubility of rutin. In case of film formations of PUL&RU and PUL&RU@PF, rutin release showed gradually increasing profiles compared to other formations. In contrast, rutin-loaded SD (RU@PF) and nanofibers (PUL&RU and PUL&RU@PF) displayed a burst release and reached a maximum level of release in 6 h. Notably, 90% of the rutin was released in 15 min. The release from PUL&RU@PF was faster than that from PUL&RU and RU@PF in SSF. At the start, a slight difference in release rate between PUL&RU and PUL&RU@PF was due to the presence of pluronic SD, which facilitated the fast release of rutin after PUL was dissolved in SSF. The fastest release was exhibited in the presence of both PUL matrix and PF SDs, which displayed a synergistic effect on the release of rutin. The pluronic, an amphiphilic surfactant, released rutin in SSF. To prepare the SD, RU@PF was fabricated by freeze dryer, and water molecules were removed from RU@PF to make more space. This porous formation was prone to swelling and dissolution. Thus, the prepared formulation obtained the advantage that rutin was delivered to the targeted site within a short time.

Therefore, PUL&RU@PF may be a promising alternative for enhanced solubility and UV stability with the remaining rutin's inherent antioxidant ability.

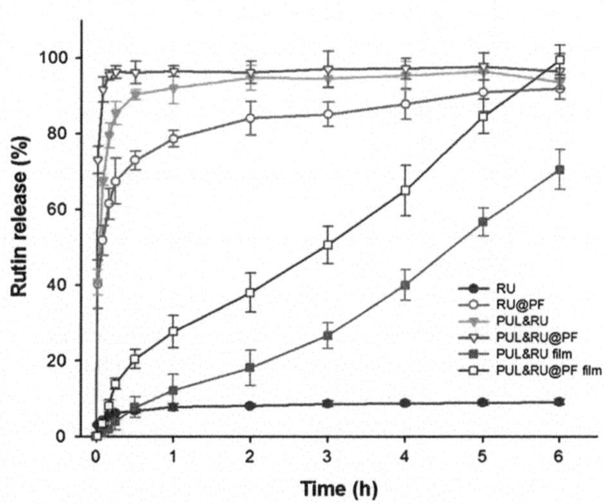

Fig. 25 In vitro rutin release profiles of RU, RU@PF, PUL&RU, and PUL&RU@PF in simulated salivary fluid (SSF) (Reproduced from reference [102] with permission from Wiley)

In another study, chitosan-encapsulated phytoconstituents of curcumin and rutin nanoparticles were formulated and characterized by Shanmugam et al. [105]. This formulation was prepared by solvent evaporation technique, and the optimized formulation was further subjected to in vitro and in vivo evaluations. The prepared nanoparticles were in the size range of 25–100 nm, and the release profile was found to be non-Fickian transport. In vivo pharmacokinetic studies were carried in rabbits, and the pharmacokinetic profile was studied. The results indicate that oral bioavailability of curcumin and rutin has been increased to 3.06- and 4.24-fold, respectively, when compared to their pure drugs. The data suggests that the prepared nanoparticles can be an excellent tool with these combinational drugs which can be used in treating multidrug resistance cancers. In the last few decades, remarkable developments occurred in the field of design of chemotherapeutics. However, most of the chemotherapeutics suffer from high side effects, poor adsorption, and lower solubility, high drug dosing, and uncontrolled and nonspecific delivery followed by high cytotoxicity, which restrict their uses [106]. To overcome these problems, an ideal drug delivery system, with attributes like (1) biocompatible, (2) biodegradable, (3) able to release medication in a controlled way, (4) high drug loading capacity, etc., is required [107–110]. To address these factors, several drug delivery systems, e.g., liposomes, dendrimer, polymers nanoparticles, etc., have been developed over the last few years, but none of them could address the issues described above independently [111–113]. So, there is a burgeoning interest to develop new kind of drug delivery system which can diminish the problems and can be used as an ideal delivery system. Metal-organic frameworks (MOFs) are porous crystalline materials made up from self-assembly of metal ions or clusters of metal ions and organic ligands or linkers that have porous structures and can be one, two, or three dimensional. These compounds can be used in catalysis [114], sensor [115], drug delivery [116], energy storage [117, 118], preparation of nanomaterials [119], etc. Because MOFs can exhibit exceptionally high surface areas with large pore sizes, loading is relatively high, which provided a brilliant basic material for the further loading of multicomponent drugs and controlling their release. However, nano-MOFs have also been considered effective carriers for chemotherapeutic agent delivery [120, 121]. Some investigation has been taken on the application of these compounds as carrier for other flavonoid [122]. However, this novel strategy has not yet been evaluated for delivery of rutin. So, evaluation of rutin-based metal-organic framework can be an excellent topic as a drug delivery system.

7 Conclusion

Rutin has been known as a therapeutic agent for fighting cancer because of its antioxidant and anti-inflammatory effects. It also has been shown to induce apoptosis, or cancer cell death, and display antitumor effects. It has been used to sensitize cancer cells to anticancer drugs and reverse drug resistance for patients undergoing cancer treatment. There is a great deal of research involving humans, rodents, and cells suggesting that this antioxidant works to reduce tumor size, increase cancer cell death, and increase survival time. Hence, rutin can be regarded as a "vital phytochemical" which is needed to be studied extensively to establish effective safety profile in human to get therapeutic benefits.

References

1. Hossain MK, Dayem AA, Han J, Yin Y, Kim K, Saha SK et al (2016) Molecular mechanisms of the anti-obesity and anti-diabetic properties of flavonoids. Int J Mol Sci 17(4):569
2. Pandey KB, Rizvi SI (2009) Plant polyphenols as dietary antioxidants in human health and disease. Oxid Med Cell Longev 2:270–278
3. Aherne SA, O'Brien NM (2002) Dietary flavonols: chemistry, food content, and metabolism. Nutrition 18:75–81
4. Mauludin R, Müller RH, Keck CM (2009) Development of an oral rutin nanocrystal formulation. Int J Pharm 370:202–209
5. Chen S-S, Gong J, Liu F-T, Mohammed U (2000) Naturally occurring polyphenolic antioxidants modulate IgE-mediated mast cell activation. Immunology 100:471–480
6. Ihme N, Kiesewetter H, Jung F, Hoffmann KH, Birk A, Müller A et al (1996) Leg oedema protection from a buckwheat herb tea in patients with chronic venous insufficiency: a single-centre, randomised, double-blind, placebo-controlled clinical trial. Eur J Clin Pharmacol 50:443–447
7. Deschner EE, Ruperto J, Wong G, Newmark HL (1991) Quercetin and rutin as inhibitors of azoxymethanol-induced colonic neoplasia. Carcinogenesis 12:1193–1196
8. Panasiak W, Wleklik M, Oraczewska A, Luczak M (1989) Influence of flavonoids on combined experimental infections with EMC virus and Staphylococcus aureus in mice. Acta Microbiol Pol 38:185–188
9. Park SY, Bok SH, Jeon SM, Parkc YB, Leed SJ, Jeong TS et al (2002) Effect of rutin and tannic acid supplements on cholesterol metabolism in rats. Nutr Res 22:283–295
10. Webster RP, Gawde MD, Bhattacharya RK (1996) Protective effect of rutin, a flavonol glycoside, on the carcinogen-induced DNA damage and repair enzymes in rats. Cancer Lett 109:185–191
11. Ahmed OM, Moneim AA, Yazid IA, Mahmoud AM (2010) Antihyperglycemic, antihyperlipidemic and antioxidant effects and the probable mechanisms of action of ruta graveolens infusion and rutin in nicotinamide-Streptozotocin-induced diabetic rats. Diabetol Croat 39:15–35
12. Sharma S, Ali A, Ali J, Sahni JK, Baboota S (2013) Rutin: therapeutic potential and recent advances in drug delivery. Expert Opin Investig Drugs 22(8):1063–1079
13. Markert C, Markert C (1968) Neoplasia: a disease of differentiation. Cancer Res 28:1908–1914
14. Islam F, Khatun H, Khatun M, Ali SM, Khanam JA (2014) Growth inhibition and apoptosis of Ehrlich ascites carcinoma cells by the methanol extract of Eucalyptus camaldulensis. Pharm Biol 52(3):281–290
15. Posey LM (2005) Cancer treatment and chemotherapy, pharmacotherapy: a pathophysiologic approach. McGraw-Hill, New York
16. Matés JM, Segura JA, Alonso FJ, Márquez J (2008) Intracellular redox status and oxidative stress: implications for cell proliferation, apoptosis, and carcinogenesis. Arch Toxicol 82(5):273–299
17. Hileman EO, Liu J, Albitar M, Keating MJ, Huang P (2004) Intrinsic oxidative stress in cancer cells: a biochemical basis for therapeutic selectivity. Cancer Chemother Pharmacol 53(3):209–219
18. Kang D, Hamasaki N (2003) Mitochondrial oxidative stress and mitochondrial DNA. Clin Chem Lab Med 41(10):1281–1288

19. Ramos S (2008) Cancer chemoprevention and chemotherapy: dietary polyphenols and signaling pathways. Mol Nutr Food Res 52:507–526
20. Jemal A, Siegel R, Ward E, Hao Y, Xu J, Murray T et al (2008) Cancer statistics. J Clin 58:71–96
21. Soobrattee MA, Bahorun T, Aruoma OI (2006) Chemopreventive actions of polyphenolic compounds in cancer. Biofactors 27:19–35
22. Liu RH (2003) Health benefits of fruit and vegetables are from additive and synergistic combinations of phytochemicals. Am J Clin Nutr 78:517S–520S
23. Pan MH, Ho CT (2008) Chemopreventive effects of natural dietary compounds on cancer development. Chem Soc Rev 37:2558–2574
24. Gong G, Qin Y, Huang W, Zhou S, Yang X, Li D (2009) Rutin inhibits hydrogen peroxide-induced apoptosis through regulating reactive oxygen species mediated mitochondrial dysfunction pathway in human umbilical vein endothelial cells. Eur J Pharmacol 628:27–35
25. Matsukawa Y, Nishino H, Okuyama Y, Matsui T, Matsumoto T, Matsumura S et al (1997) Effects of quercetin and/or restraint stress on formation of aberrant crypt foci induced by azoxymethane in rat colons. Oncology 54:118–121
26. Lin JP, Yang JS, Lin JJ, Lai KC, Lu HF, Ma CY et al (2011) Rutin inhibits human leukemia tumor growth in a murine xenograft model in vivo. Environ Toxicol 27:480–484
27. Alonso-Castro AJ, Dominguez F, García-Carrancá A (2013) Rutin exerts antitumor effects on nude mice bearing SW480 tumor. Arch Med Res 44(5):346–351
28. Nafees S, Mehdi SH, Zafaryab M, Zeya B, Sarwar T, Rizvi MA (2018) Synergistic interaction of rutin and silibinin on human colon cancer cell line. Arch Med Res 49(4):226–234
29. Hasani NAH, Amin IM, Kamaludin R, Nik Mohd Rosdyd NMM, Johari Ibahim M, Sheikh Abdul Kadir SH (2018) P53 and cyclin b1 mediate apoptotic effects of apigenin and rutin in $er\alpha^{+}$-breast cancer mcf-7 cells. J Teknol 80:133–140
30. Sghaier MB, Pagano A, Mousslim M, Ammari Y, Kovacic H, Luis J (2016) Rutin inhibits proliferation, attenuates superoxide production and decreases adhesion and migration of human cancerous cells. Biomed Pharmacother 84:1972–1978
31. Alhoshani AR, Hafez MM, Husain S, Al-sheikh AM, Alotaibi MR, Al Rejaie SS et al (2017) Protective effect of rutin supplementation against cisplatin-induced nephrotoxicity in rats. BMC Nephrol 18:194
32. Azevedo MI, Pereira AF, Nogueira RB, Rolim E, Brito GA, Wong D et al (2013) The antioxidant effects of the flavonoids rutin and quercetin inhibit oxaliplatin-induced chronic painful peripheral neuropathy. Mol Pain 9:53
33. Chen H, Miao Q, Geng M, Liu J, Hu Y, Tian L et al (2013) Anti-tumor effect of rutin on human neuroblastoma cell lines through inducing G2/M cell cycle arrest and promoting apoptosis. ScientificWorldJournal 269165:2013
34. Iriti M, Kubina R, Cochis A, Sorrentino R, Varoni E, Kabała-Dzik A et al (2017) Rutin, a quercetin glycoside, restores chemosensitivity in human breast cancer cells. Phytother Res 31(10):1529–1538
35. Gautam R, Singh M, Gautam S, Rawat JK, Saraf SA, Kaithwas G (2016) Rutin attenuates intestinal toxicity induced by methotrexate linked with anti-oxidative and anti-inflammatory effects. BMC Complement Altern Med 16:99
36. Lin JP, Yang JS, Lu CC, Chiang JH, Wu CL, Line JJ et al (2009) Rutin inhibits the proliferation of murine leukemia WEHI-3 cells in vivo and promotes immune response in vivo. Leuk Res 33:823–828
37. Elsayed HE, Ebrahim HY, Mohyeldin M, Siddique AB, Kamal AM, Haggag EG et al (2017) Rutin as a novel c-met inhibitory lead for the control of triple negative breast malignancies. Nutr Cancer 69(8):1256–1271
38. Saleh A, ElFayoumi HM, Youns M, Barakat W (2018) Rutin and orlistat produce antitumor effects via antioxidant and apoptotic actions. Naunyn Schmiedebergs Arch Pharmacol 392(2):165–175
39. Schwingel TE, Klein CP, Nicoletti NF, Dora CL, Hadrich G, Bica CG et al (2014) Effects of the compounds resveratrol, rutin, quercetin, and quercetin nanoemulsion on oxaliplatin-

induced hepatotoxicity and neurotoxicity in mice. Naunyn Schmiedebergs Arch Pharmacol 387(9):837–848
40. Shahid A, Ali R, Ali N, Hasan SK, Rashid S, Majed F et al (2016) Attenuation of genotoxicity, oxidative stress, apoptosis and inflammation by rutin in benzo(a) pyrene exposed lungs of mice: plausible role of NF-κB, TNF-α and Bcl-2. J Complement Integr Med 13(1):17–29
41. Vijay M, Sivagami G, Thayalan K, Nalini N (2016) Radiosensitizing potential of rutin against human colon adenocarcinoma HT-29 cells. Bratisl Lek Listy 117(3):171–178
42. Guon TE, Chung HS (2016) Hyperoside and rutin of Nelumbo nucifera induce mitochondrial apoptosis through a caspase-dependent mechanism in HT-29 human colon cancer cells. Oncol Lett 11(4):2463–2470
43. Yang S, Zhang H, Yang X, Zhu Y, Zhang M (2015) Evaluation of antioxidative and antitumor activities of extracted flavonoids from Pink Lady apples in human colon and breast cancer cell lines. Food Funct 6(12):3789–3798
44. Lauro MR, Torre ML, Maggi L, de Simone F, Ubaldo C, Rita Patrizia A (2002) Fast- and slow-release tablets for oral administration of flavonoids: rutin and quercetin. Drug Dev Ind Pharm 28:371–379
45. Carn F, Guyot S, Baron A, Pérez J, Buhler E, Zanchi DE (2012) Structural properties of colloidal complexes between condensed tannins and polysaccharide hyaluronan. Biomacromolecules 13(3):751–759
46. Lu KY, Li R, Hsu CH, Lin CW, Chou SC, Tsai ML, Mi FL (2017) Development of a new type of multifunctional fucoidan-based nanoparticles for anticancer drug delivery. Carbohydr Polym 165:410–420
47. Fornaguera C, Solans C (2017) Methods for the in vitro characterization of nanomedicines-biological component interaction. J Pers Med 7:2
48. Lin HS, Chean CS, Nug YY, Chan SY, Ho PC (2000) 2- Hydroxypropyl-beta-cyclodextrin increases aqueous solubility and photostability of all-trans-retinoic acid. J Clin Pharm Ther 25:265–269
49. Loftsson T, Brewster ME (1996) Pharmaceutical applications of cyclodextrins. Drug solubilization and stabilization. J Pharm Sci 85:1017–1024
50. Sri KV, Kondaiah A, Ratna JV, Annapurna A (2007) Preparation and characterization of quercetin and rutin cyclodextrin inclusion complexes. Drug Dev Ind Pharm 33:245–253
51. Akhbari K, Morsali A (2010a) Thallium(I) supramolecular compounds: structural and properties consideration. Coord Chem Rev 254:1977–2006
52. Mirzadeh E, Akhbari K, White J (2018a) A Phuruangrat mechanochemical conversion of nano potassium hydrogen terephthalate to thallium analogue nanoblocks with strong hydrogen bonding and straight chain metalophillic interactions. Appl Organomet Chem 32:4402
53. Moghadam Z, Akhbari K, White J, Phuruangrat A (2018a) A new potassium-based coordination polymer with hydrogen bonding and zigzag metallophilic interactions. Appl Organomet Chem 32:4613
54. Yusefi S, Akhbari K, White J, Phuruangrat A (2019) Sonochemical synthesis of a two-dimensional supramolecular polymer with nanoporous morphology, linear thallophilic and covalent hydrogen-bonding interactions. Appl Organomet Chem 33:4747
55. Akhbari K, Morsali A (2013a) Solid-state structural transformations of two AgI supramolecular polymorphs to another polymer upon absorption of HNO3 vapors. Inorg Chem 52:2787
56. Akhbari K, Morsali A (2013b) Solid- and solution-state structural transformations in flexible lead (II) supramolecular polymers. CrstEngComm 15:8915–8918
57. Noori Y, Akhbari K, Phuruangrat A, Costantino F (2017a) Self-assembly through secondary interactions in formation of two-dimensional lead(II) supramolecular polymer with nanosheets morphology. JMolStruct 1130:311–318
58. Noori Y, Akhbari K, Phuruangrat A, Costantino F (2017b) Studies the effects of ultrasonic irradiation and dielectric constants of solvents on formation of lead(II) supramolecular polymer; new precursors for synthesis of lead(II) oxide nanoparticles. Ultrason Sonochem 35:36–44

59. Noori Y, Akhbari K, Phuruangrat A, Costantino F (2018) What can only occur in supramolecular systems; first solid-state conversion of micro to nanostructures without any treatment in environmental conditions. Ultrason Sonochem 40:17–20
60. Hazrati Z, Akhbari K, Phuruangrat A (2017) The effects of altering reaction conditions in green sonochemical synthesis of a thallium(I) coordination polymer and in achieving to different morphologies of thallium(III) oxide nanostructures via solid-state process. Ultrason Sonochem 39:662–668
61. Mirzadeh E, Akhbari K, White J (2018b) Irreversible exchange of potassium with thallium in coordination polymer nanostructures upon solid-state mechanochemical process. Inorg Chem Commun 92:95–100
62. Moghadam Z, Akhbari K, White J, Phuruangrat A (2018b) Reversible desorption and absorption of water in a zinc-based coordination polymer nanostructure. Polyhedron 153:286–291
63. Mohaghegh F, Akhbari K, Phuruangrat A (2018) Preparation of thallium nanomaterials from thallium(I) coordination polymers precursors synthesized by green sonochemical and mechanochemical processes. Ultrason Sonochem 40:594–600
64. Moradi ZH, Akhbari K, Phuruangrat A, Costantino F (2017) Studies on the relation between the size and dispersion of metallic silver nanoparticles and morphologies of initial silver(I) coordination polymer precursor. J Mol Struct 1133:172–178
65. Shahangi Shirazi F, Akhbari K (2017) Sodium chloride nanorods from nanorods of one-dimensional pencil-shaped sodium coordination polymer. J Iran Chem Soc 14:1519–1525
66. Shahangi Shirazi F, Akhbari K, Boonmak J, Youngme S (2017) Solid-state mechanochemical conversion of one-dimensional pencil shaped sodium coordination polymer nanorods to corrugated tape silver coordination polymer nanoparticles. Inorg Chem Commun 82:6–10
67. Moeinian M, Akhbari K (2015) How the guest molecules in nanoporous Zn(II) metal-organic framework can prevent agglomeration of ZnO nanoparticles. J Solid State Chem 225:459
68. Moeinian M, Akhbari K (2016) Solid-state synthesis of zinc oxide nano-structures with similar morphologies to their precursors and metal–organic frameworks topology. Iran Chem Soc 13:547–552
69. Shahangi Shirazi F, Akhbari K (2015a) Solid-state thermal conversion of a nanoporous metal-organic framework to a nonporous coordination polymer. RSC Adv 5:50778
70. Shahangi Shirazi F, Akhbari K (2015b) Preparation of zinc oxide nanoparticles from nanoporous metal–organic framework with one-dimensional channels occupied with guest water molecules. Inorg Chem Acta 436:1
71. Akhbari K, Morsali A (2007) Highly polyhapto–aromatic interactions in thallium(I) coordination. J Organomet Chem 692:5109–5112
72. Akhbari K, Morsali A (2010b) Silver nanofibers from the nanorods of one-dimensional organometallic coordination polymers. CrstEngComm 12:3394–3396
73. Akhbari K, Morsali A (2012) Thallium(I) supramolecular polymer with Tl…π secondary interactions; preparation of nano-structures by sonochemical method as a new precursor for preparation of Tl2O3 with nano-structural surface. J Organomet Chem 700:125–131
74. Akhbari K, Morsali A, Zeller M (2007) Unique Ag–C bonds, thermal, fluorescence, structural and solution studies of two-dimensional silver(I) coordination polymer. J Organomet Chem 692:3788–3795
75. Moeinian M, Akhbari K, Boonmak J, Youngme S (2018) Three-dimensional organometallic thallium(I) supramolecular polymer nanostructures synthesized with sonochemical process. Ultrason Sonochem 41:11–16
76. Moghadam Z, Akhbari K, Phuruangrat A (2018c) Solid-state conversion of thallium(I) coordination polymer nanoparticles with cubic cage units to an organometallic silver(I) coordination polymer. J Organomet Chem 861:105–111
77. Shahangi Shirazi F, Akhbari K, Kawata S, Kanazashi K (2016) Reversible liquid assisted mechanochemical conversion of sodium coordination polymer nanorods to organosilver coordination polymer nanosheets. Inorg Chem Commun 74:31–34
78. Bombardelli E, Curri SB, Loggia RD et al (1989) Complex between phospholipids and vegetal derivatives of biological interest. Fitoterapia 60:1–9

79. Kidd P, Head K (2005) A review of the bioavailability and clinical efficacy of milk thistle phytosome: a silybin-phosphatidylcholine complex (Siliphos). Altern Med Rev 10:193–203
80. Das MK, Kalita B (2014) Design and evaluation of phyto-phospholipid complexes (phytosomes) of rutin for transdermal application. J Appl Pharm Sci 4(10):51–57
81. Jain S, Dhanotiya C, Malviya N (2012) Physicochemical characterization and determination of free radical scavenging activity of rutin-phospholipid complex. Int J Pharm Sci Res 3:909–913
82. Singh D, Rawat MSM, Semalty A, Semalty M (2012) Rutin-phospholipid complex: an innovative technique in novel drug delivery system-NDDS. Curr Drug Deliv 9:305–314
83. Kumar A, Ahuja A, Ali J, Baboota S (2010) Conundrum and therapeutic potential of curcumin in drug delivery. Crit Rev Ther Drug Carrier Syst 27:279–312
84. Almeida JS, Lima F, Ros SD, Bulhões LO, de Carvalho LM, Beck RC (2010) Nanostructured systems containing rutin: in vitro antioxidant activity and photostability studies. Nanoscale Res Lett 5:1603–1610
85. Kumar VP, Khan A (2012) Formulation design and evaluation of rutin loaded self-emulsifying drug delivery system (SEDDs) using edible oil. Asia J Pharm Clin Res 5:76–78
86. Sionkowska A (2011) Current research on the blends of natural and synthetic polymers as new biomaterials. Prog Polym Sci 36:1254–1276
87. Okamoto M, John B (2013) Synthetic biopolymer nanocomposites for tissue engineering scaffolds. Prog Polym Sci 38:1487–1503
88. Fornaguera C, Dols-Perez A, Caldero G, Garcia-Celma M, Camarasa J, Solans C (2015) PLGA nanoparticles prepared by nanoemulsion templating using low-energy methods as efficient nanocarriers for drug delivery across the blood–brain barrier. J Control Release 211:134–143
89. Makadia HK, Siegel SJ (2011) Poly lactic-co-glycolic acid (PLGA) as biodegradable controlled drug delivery carrier. Polymers 3:1377–1397
90. Kumari A, Yadav SK, Yadav SC (2010) Biodegradable polymeric nanoparticles based drug delivery systems. Colloids Surf B Biointerfaces 75:1–18
91. Masood F (2016) Polymeric nanoparticles for targeted drug delivery system for cancer therapy. Mater Sci Eng C 60:569–578
92. Sahana D, Mittal G, Bhardwaj V, Kumar M (2008) PLGA nanoparticles for oral delivery of hydrophobic drugs: influence of organic solvent on nanoparticle formation and release behavior in vitro and in vivo using estradiol as a model drug. J Pharm Sci 97:1530–1542
93. Reis CP, Neufeld RJ, Ribeiro AJ, Veiga F, Nanoencapsulation I (2006) Methods for preparation of drug-loaded polymeric nanoparticles. Nanomed: Nanotechnol Biol Med 2:8-–21
94. Kızılbey K (2019) Optimization of rutin-loaded PLGA nanoparticles synthesized by single-emulsion solvent evaporation method. ACS Omega 4:555–562
95. Akhbari K, Karami S, Saedi Z (2018) The effects of solvent and ultrasonic irradiation in synthesis of thallium(I) nano supramolecular polymers and use them as template for synthesis of thallium(III) oxide nanostructures with desirable morphology. Ultrason Sonochem 42:640–646
96. Shahangi Shirazi F, Akhbari K (2016) Sonochemical procedures; the main synthetic method for synthesis of coinage metal ion supramolecular polymer nano structures. Ultrason Sonochem 31:51–61
97. Pandey P, Rahman M, Bhatt PC, Beg S, Paul B, Hafeez A et al (2018) Implication of nano-antioxidant therapy for treatment of hepatocellular carcinoma using PLGA nanoparticles of rutin. Nanomedicine 13(6):849–870
98. Jantrawut P, Chambin O, Ruksiriwanich W (2015) Anti-cancer activity of rutin encapsulated in low methoxyl pectin beads. Cell Chem Technol 49(1):51–54
99. Banjare L, Ghillare N (2012) Development of biocompatible nanoparticles for sustained topical delivery of Rutin. Int J Pharm Bio Arch 3:326–332
100. Kamel R, Basha M, Abd El-Alim SH (2013) Development of a novel vesicular system using a binary mixture of sorbitan monostearate and polyethylene glycol fatty acid esters for rectal delivery of rutin. Liposome Res 23:28–36

101. Deepika S, Thangam R, Sheena TS, Rajendran S, Sivasubramanian S, Babu MD et al (2018) A novel rutin-fucoidan complex based phytotherapy for cervical cancer through achieving enhanced bioavailability and cancer cell apoptosis. Biomed Pharmacother 109(1):1181–1195
102. Lee IW, li J, Chen X, Park HJ (2017) Fabrication of electrospun antioxidant nanofibers by rutin-pluronic solid dispersions for enhanced solubility. J Appl Polym Sci 134:44859
103. Vasconcelos T, Sarmento B, Costa P (2007) Solid dispersions as strategy to improve oral bioavailability of poor water soluble drugs. Drug Discov Today 12:1068
104. Bhardwaj N, Kundu SC (2010) Electrospinning: a fascinating fiber fabrication technique. Biotechnol Adv 28:325
105. Ramaswamy S, Dwarampudi L, Kadiyala M, Kuppuswamy G, Karri V, Kumar C et al (2017) Formulation and characterization of chitosan encapsulated phytoconstituents of curcumin and rutin nanoparticles. Int J Biol Macromol 104:1807–1812
106. Kamaly N, Yameen B, Wu J, Farokhzad OC (2016) Degradable controlled-release polymers and polymeric nanoparticles: mechanisms of controlling drug release. Chem Rev 116:2602–2663
107. Brus J, Albrecht W, Lehmann F, Geier J, Czernek J, Urbanova M et al (2017) Exploring the molecular-level architecture of the active compounds in liquisolid drug delivery systems based on mesoporous silica particles: old tricks for new challenges. Mol Pharm 14:2070–2078
108. Madden TD, Adlakha-Hutcheon G, Bally MB, Shew CR (1999) Controlled destabilization of a liposomal drug delivery system enhances mitoxantrone antitumor activity. Nat Biotechnol 17:775–779
109. Uhrich KE, Cannizzaro SM, Langer RS, Shakesheff KM (1999) Polymeric systems for controlled drug release. Chem Rev 99:3181–3198
110. Zhang Y, Yu J, Bomba HN, Zhu Y, Gu Z (2016) Mechanical force-triggered drug delivery. Chem Rev 116:12536–12563
111. Huang X, Wu S, Ke X, Li X, Du X (2017) Phosphonated Pillar[5]arene-valved mesoporous silica drug delivery systems. ACS Appl Mater Interfaces 9:19638–19645
112. Mekuria SL, Debele TA, Chou HY, Tsai HC (2016) IL-6 antibody and RGD peptide conjugated poly(amidoamine) dendrimer for targeted drug delivery of HeLa cells. J Phys Chem B 120:123–130
113. Otla NG, Chandrasekar B, Rooney P, Sivaraman G, Larrañaga A, Krishna KV et al (2017) Biomimetic lipid-based nanosystems for enhanced dermal delivery of drugs and bioactive agents. ACS Biomater Sci Eng 3:1262–1272
114. Xu C, Fang R, Luque R, Chen L, Li Y (2019) Functional metal–organic frameworks for catalytic applications. Coord Chem Rev 388:268–292
115. Noori Y, Akhbari K (2017) Post-synthetic ion-exchange process in nanoporous metal-organic frameworks; an effective way for modulating their structures and properties. RSC Adv 7:1782–1808
116. Karimi Alavijeh R, Beheshti S, Akhbari K, Morsali A (2018) Investigation of reasons for metal–organic framework's antibacterial activities. Polyhedron 156:257–278
117. Akhbari K, Morsali A (2013c) Modulating methane storage in anionic nano-porous MOF materials via post-synthetic cation exchange process. Dalton Trans 42:4786–4789
118. Akhbari K, Morsali A (2015) Needle-like hematite nano-structure prepared by directed thermolysis of MIL-53 nano-structure with enhanced methane storage capacity. Mater Lett 141:315–318
119. Mirzadeh E, Akhbari K (2016) Synthesis of nanomaterials with desirable morphologies from metal–organic frameworks for various applications. CrstEngComm 18:7410–7424
120. Della Rocca J, Liu D, Lin W (2011) Nanoscale metal-organic frameworks for biomedical imaging and drug delivery. Acc Chem Res 44:957–968
121. Kundu T, Mitra S, Patra P, Goswami A, Díaz Díaz D, Banerjee R (2014) Mechanical downsizing of a gadolinium(III)-based metal-organic framework for anticancer drug delivery. Chemistry 20:10514–10518
122. Laha D, Pal K, Chowdhuri AR, Parida PK, Sahu SK, Jana K et al (2019) Fabrication of curcumin-loaded folic acid-tagged metal organic framework for triple negative breast cancer therapy in in vitro and in vivo systems. New J Chem 43:217–229

Nanoformulated Drug Delivery of Potential Betulinic Acid Derivatives: A Promising Approach Toward Cancer Therapy

Biswajit Mukherjee, Ashique Al Hoque, Debasmita Dutta, Brahamacharry Paul, Alankar Mukherjee, and Sahajit Mallick

1 Introduction

In last few decades, there has been a tremendous interest worldwide in the design of analogue libraries based on plant-derived product for anticancer drug development. The main objective in cancer drug development is to find new drugs that are cytotoxic to cancer cells but not to healthy cells. However, adverse side effects to normal cells, drug resistance property of conventional chemotherapeutic drugs, and non-specific tissue distribution are the primary obstacles to combat the disease. So potential plant-derived natural products are in search as an alternative to toxic drugs. This approach has received a positive response obtained from some natural products due to the efficacious anticancer properties and minimum side effects toward normal cells. It is worthy to mention here that around 50% of the currently used anti-cancer drugs suggested were either natural products or natural product-based conjugates [1].

An increasing number of triterpenoids, naturally occurring terpenes having potential for the neoplastic disease prevention and treatment properties, are ubiquitously distributed throughout the plant kingdom. Many of them have been reported to exhibit cytotoxicity against a variety of cancer cells without manifesting considerable toxicity toward normal cells [1]. Notable among them is the pentacyclic lupane type of triterpenoids that represent a very important class of natural products. Betulinic acid [3β-hydroxy-lup-20(29)-en-28-oic acid] and its reduced congener betulin [lup-20(29)-en-3β, 28-diol] belonging to this class are widely distributed in the plant kingdom. Betulinic acid was found to show significant antiproliferative activity on a broad panel of cancer cells [2, 3], and it was found to cause cancer cell death by induction of apoptosis involving caspases with no toxic effects [4]. The

B. Mukherjee (✉) · A. Al Hoque · D. Dutta · B. Paul · A. Mukherjee · S. Mallick
Department of Pharmaceutical Technology, Jadavpur University, Kolkata, India
e-mail: biswajit.mukherjee@jadavpuruniversity.in

© Springer Nature Singapore Pte Ltd. 2020

M. Rahman et al. (eds.), *Nanomedicine for Bioactives*,
https://doi.org/10.1007/978-981-15-1664-1_4

similar compounds were tested for anticancer activity on various human cancer and healthy normal cells, and the molecular mechanisms were studied. Betulinic acid is also cytotoxic against primary cancer cells isolated from tumor specimens obtained from neuroblastoma, glioblastoma, and leukemia [5]. It acted as a cytotoxic agent in different models of drug resistance, for example, primary pediatric acute leukemia samples that were refractory to standard chemotherapeutic agents. It promises to be an alternative and effective product against certain forms of drug resistance cases. Furthermore, there is evidence that betulinic acid is preferentially cytotoxic against metastatic over nonmetastatic melanoma cells.

Despite its distinguished biological activity, betulinic acid was not found to act successfully in vivo as a useful drug because of its poor solubility and low bioavailability. To resolve this difficulty, preparation of analogue library via synthetic modification of functional groups has been tried as an alternative option. A fascinating feature of this approach may provide innumerable possibilities for generating a diverse array of new types of molecules for application in biology. A large number of analogues have been synthesized by structural modification of betulinic acid for optimization of bioactivity, and some of these semisynthetic analogues are considered to be the most potent anti-inflammatory and anticarcinogenic triterpenoids known to mankind [6]. The antitumor efficacy of several such triterpenoids is currently being evaluated in phase I clinical trials [7].

Another alternative approach to reduce the side effects of drug is their nature of the administration so that it can act only on the target cells and not to normal functioning cells, thereby enhancing dosage efficacy. In this regard, nanotechnology-mediated drug delivery strategies have made a potential impact on clinical therapeutics in the last two decades. Advances in nanoscale drug carriers such as liposomes and polymeric nanoparticles have enabled more efficient and safer delivery of a myriad of drugs. Long circulating polymeric nanoparticles prepared with poly(lactic-co-glycolic) acid (PLGA) are particularly well accepted as effective drug carriers. Advantages in nanoparticle drug delivery include longer circulation half-lives, long shelf life, and high carrier capacity [8]. Thus, by using nano-sized PLGA drug delivery carriers, it is possible to deliver the drug directly to the specific site and also release it at a controlled rate and in a sustained manner for a prolonged period of time, avoiding predominant degradation in the circulation, thereby reducing the residual toxicity of drug that will be effective with a small dose. FDA and European Medicine Agency have already approved the polymer to prepare a drug delivery system for potential systemic administration. In the present chapter, efforts have been given to describe antineoplastic potential of various betulinic acid derivatives and their efficacy when administered through synthetic drug nanocarriers.

2 Overview on Betulinic Acid and Its Medicinal Use

Pentacyclic lupane types of triterpenoids are considered one of the very important classes of natural products. A plant-derived pentacyclic lupane-type triterpenoid, betulin, the reduced form of betulinic acid (BA), was first isolated from plants in 1788 by Johann Tobias Lowitz [4] and found to be a predominant constituent of the outer-bark of white-barked birch trees. Both BA and BN are widely distributed throughout the plant kingdom. BA has been extracted from a wide range of diverse plants, ranging from meat-eating plants like *Sarracenia flava* (Sarraceniaceae) [9] to trees and shrubs like *Diospyros* spp. (Ebenaceae), *Inga punctata* (Fabaceae) [10], *Ziziphus* spp. (Rhamnaceae), *Vauquelinia corymbosa* (Rosaceae) [11], and *Syzygium* spp. (Myrtaceae) [12–14]. Owing to the high betulin content (up to 22%) in the bark of the white birch tree (Betula alba), the most convenient source for BA is via a simple oxidation process from betulin isolated from this tree [13, 15].

In 1995, betulinic acid was reported as a selective inhibitor of human melanoma [4]. Then, it was demonstrated to induce apoptosis in human neuroblastoma in vitro and in vivo in different models [16]. Also, betulinic acid was found active against neuroectodermal neuroblastoma, medulloblastoma, Ewing's sarcoma, and malignant brain tumors [17], and ovarian carcinoma [3] and in human leukemia HL-60 cells [18] and malignant head and neck squamous cell carcinoma SCC25 and SCC9 cells [19].

BA is an inducer of apoptosis in human neuroblastoma cells. BA can directly activate mitochondria and triggers apoptosis pathway Tezuka et al. [20] focused on its different extracts of the aerial parts of the plant on murine colon carcinoma cell 26-L5 [20]. Kwon et al. [21] reported that antiangiogenic activity of BA occurs through a modulation of mitochondrial function rather than APN (aminopeptidase N) activity in endothelial cells [21]. Thurnher et al. [19] proclaimed betulinic acid as a new cytotoxic compound against malignant head and neck cancer cells. Mukherjee et al. [22, 23] investigated the potential of betulinic acid in angiogenesis and have developed more potent BA derivatives through bioactivity-guided structure activity studies [22, 23]. Liu et al. [24] determined the anticancer potential of different extracts of the plant on human WI-38, VA-13, and HepG2 cells [24]. Wada and Tanaka [25] tested the ethanolic extract of *E. serrata* on human K562 cells for cytotoxic and apoptosis-inducing activity and found BA and its derivatives as potent DNA topoisomerase II inhibitors [25]. Fu et al. [26] studied the chloroform ($CHCl_3$) extract of the bark of *Bruce javanica* on A549 cell for assessing the activity of betulinic acid on DNA topoisomerase II activity [26]. Abdille et al. [27] reported that the total phenolic content and antioxidant activity (DPPH assay (2,2-diphenyl-1-picryl-hydrazyl-hydrate assay)) of *Dillenia indica* fruit extracts. The highest phenolic content as well as highest antioxidant capacity was found for methanolic extracts (34.1%) [27]. The effects of BA on the VEGF expression and the growth of colorectal cell SW480 xenografts in nude mice were investigated [28]. Kumar et al. [29] assessed the antileukemic activity of the methanolic extract of *Dillenia indica* L. fruits on human U937, HL60, and K562 cells [29]. Gao et al. [30] investigated the antiproliferative activity of natural product BA and its underlying mechanisms of actions in pancreatic cancer models. Besides its expected antiproliferation activity, BA exhibited strong antiangiogenesis and anti-invasion abilities in pancreatic cancer. Chintharlapalli et al. [31] determined the effects of BA on colon cancer cell proliferation and apoptosis and tumor growth in vivo and analyzed the effects of BA on specificity proteins (Sp) and Sp-regulated gene products and also assessed microRNA-27a (miR-27a) and ZBTB10 mRNA expression [31]. Antioxidant activity of BA synthesized from the leaf extract of Dillenia indica Linn was observed. Yang et al. [32, 33] synthesized novel derivatives of BA and assessed for antitumor activity on MGC-803, PC3, A375, Bcap-37, and A431 cells.

Reiner et al. [34] found out that BA blocked multiple deubiquitinases (DUBs), which resulted in the accumulation of poly-ubiquitinated proteins, reduced levels of oncoproteins, and enhanced apoptotic cell death [34]. Gomathi et al. [35] reported that BA and its derivatives act on cancerous cells selectively without harming the normal cells, and this makes them preferable than chemotherapeutic agents.

3 Mechanism of Action of BA in Different Cancer Cells

The compounds of this class which includes betulin and BA were found to show significant antitumor activity on a broad panel of cancers. Different mechanisms are reported about the mode of action of betulinic acid. According to Thurnher et al. [19], betulinic acid exerts its cytotoxic action through induction of apoptosis and chromatin condensation. Apoptosis induction by BA is mediated in several ways out of which caspase activation and mitochondrial membrane depolarization are two major pathways. In a study on neuroectodermal cells, it was found that on exposure to BA caspases were activated by several proteolytic enzymes causing initiation of caspase cascade. The activated caspases can also activate other proteolytic enzymes like caspase-activated DNase (CAD) which triggers DNA fragmentation. In a later stage, chromatin condensation initiates and cells progress in apoptotic pathway of cell death. BA directly affects mitochondria and causes release of cytochrome-c through formation of membrane permeability transition pore (MPTP). Cytochrome-c once outside mitochondria (i.e., in cytosol) activates caspases (mainly caspase-3 alone and caspase-8 with the help of some other mediators like AIF, apoptosis inducing factor). This caspase activation regulates other "downstream" events of apoptosis; thus, the formation of MPTP is sufficient to trigger apoptosis by BA. The activation of apoptotic pathway by cytochrome-c release is naturally inhibited by antiapoptotic proteins of Bcl-2 family (like Bcl-2, Bcl-XL, and Mel-1) while it is alleviated by apoptotic proteins like Bax and Bad. Treatment with BA and its analogues were found to decrease level of Bcl-2 and to upregulate Bax, suggesting their role in apoptosis induction and regulation [36].

The role of p53 in BA-induced apoptosis is controversial. According to Raisova et al. [37], treatment with BA does not accumulate wild-type p53, while doxorubicin can accumulate wild-type p53. So, anticancer activity of BA is supposed to be p53 independent, while that exerted by doxorubicin is mediated by upregulation of p53. However, in a different study on C8161 melanoma cells, partial involvement of p53 in apoptosis regulation and DNA fragmentation was observed. In a comparative study in melanoma and nonmelanoma cells, apoptosis induction by BA is also independent of death receptors and also of Fas. Zuco et al. [3] demonstrated that BA is quite less potent (10%) than doxorubicin but also less toxic to nonmelanoma cells (around one fifth) and has a less chance of resistance development by the cancer cells against it.

BA was also found to be cytotoxic in different drug resistance cells, e.g., primary pediatric acute leukemia samples which were refractory to standard chemotherapeutic agents. This implicates BA can overcome drug resistance. BA bypasses resistance to CD95 and doxorubicin-mediated apoptosis, due to different molecular mechanism of BA-induced apoptosis. In a study reported by Selzer et al. [38] combination of BA with γ-irradiation has clearly additive effects and indicated that the combination of treatment differed in their mode of action.

Many of the synthetic analogues of natural products have been found to exhibit unusual properties and biological activity as the different molecular segments act cooperatively to control and modulate conformation, recognition, communication, transport, and solubility among other properties. Chemical modification leads to development of chemical entities capable of controlling solubility, conformational conservancy, recognition of binding sites, avoidance of unwanted degradation, safe transport to desired site, etc. Therefore, the motive behind the preparation of analogues should be to find out lead compound(s) possessing higher solubility and enhanced bioavailability as well as bioactivity compared to the parent compound. An appealing feature of this approach is that it may provide innumerable possibilities for generating a diverse array of new types of molecules for application in biology.

Toward this goal, different analogues of BA have been prepared and tested. Some of the significant results appeared in the literature on the structural modifications of BA, and its structure-activity relationship (SAR) studies are illustrated and briefly described here.

Betulinic acid (3β hydroxy-lup-20(29)-en-28-oic acid) is a lupine-type pentacyclic triterpene (Fig. 1) and holds a great promise as an antitumor agent. Betulinic acid (BA) and its analogue betulin (BN) are available widely in plant kingdom mainly in the genera *Ziziphus*, *Syzygium*, and *Betula*. White-barked birch trees (*Betula* spp.) are the highest source of betulin (around 20% of dry weight). The major drawback to use BA as an anticancer agent in vivo is its poor solubility in aqueous solutions. Modifications of BA are possible at numerous positions; C-3, C-28, and C-20 are the prime locations to chemically modify BA in search of BA derivatives. Modifications at C-20 did not enhance cytotoxicity in several cancer cells [39], but derivatives at the C-3 and C-28 positions were found to be promising. Amino acid conjugates at the C-28 position enhanced water solubility as well as cytotoxicity [2]. Hydroxylation at the C-3 position gave promising results when

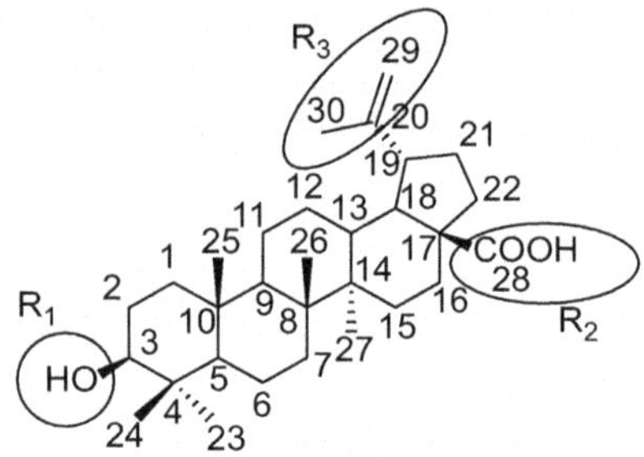

Fig. 1 Betulinic acid and its possible sites of chemical modification

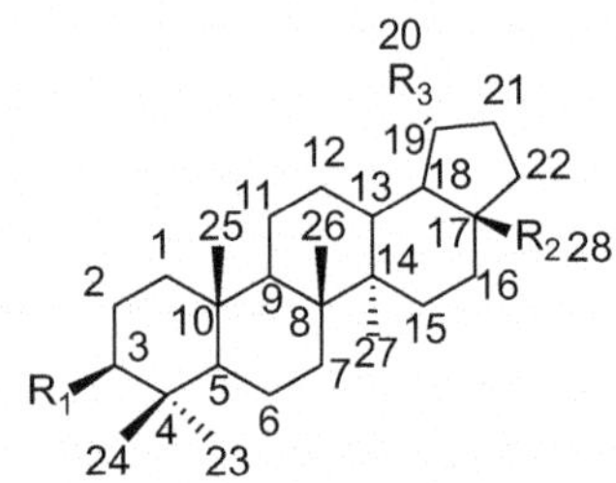

Fig. 2 Backbone structure of betulinic acid along with the substantial groups R_1, R_2, and R_3. Chemical moieties for R_1, R_2, and R_3 and corresponding cells on which they were tested have been given in Table 1

tested on murine melanoma cells [16]. Another C-3 modification gave better antitumor efficacy in a colon cancer xenograft mouse model when compared with BA [40].

In subsequent sections, we mainly discussed on modification at C-3, C-28, or C-20 positions, solely or in a combination of them. Structures of the compounds mentioned in text as "bold numbers" correspond to Fig. 2, and the substituted groups are listed in the table as R_1, R_2, and R_3.

The C-3 hydroxyl group of BA is easily replaceable by acylation, amination, or esterification. Ahmad et al. [41] reported a number of O-acyl derivatives produced from esterification of C-3 position by aliphatic or aromatic anhydrides, effective against human lung cancer (A549) and ovarian cancer (CAOV3) cells. Some of such notable compounds with low IC_{50} value against A549 cells are **1**, **2**, **3**, **4**, and **5**. Faujan et al. [42] also reported compounds **1** and **3** to be effective against some other cells. Additionally, they have reported compound **6** to be effective against HL-60, HeLa, and DBTRG0.5MG cells. Chloroacyl substitution at C-3 (**7**) was found to be effective against lung, colon, and breast cancer cells by Kommera et al. [43].

Introduction of amine or oxime group at C-3 (compound **8–12**) well retains cytotoxicity against MEL-2 melanoma cells but lower than parent compound as reported by Kim et al. [44]. However, the oxime derivatives **9** and **12** showed higher activity than parent compound against ovarian cancer cell PA-1 [22, 23]. C3-oxime derivative **9** is also active against KB cells. C-3-oxime derivative with C-20, C-29 dihydro modification (compound **11**) showed interesting activity against lung cancer (A549, L132) and prostate cancer cells (DU145) [45]. Rajendran et al. [46] modified C-3 position by insertion of nitro- or fluoro-substituted benzyl or benzoyl derivatives, and some of

Table 1 Betulinic acid derivatives formed by R_1, R_2, and R_3 substitutions on betulinic acid backbone (Fig. 2) are found efficacious on various cells

No.	R_1	R_2	R_3	Cells active against
1	CH_3COO–	–COOH	–C(=CH_2)CH_3	A549, Hl-60, CEM-SS, WEHI-3B, HeLa, HCT116
2	$CH_3CH_2CH_2COO$–	–COOH	–C(=CH_2)CH_3	A549
3	$HOOCCH_2CH_2COO$–	–COOH	–C(=CH_2)CH_3	A549, HL-60, CEM-SS, HeLa
4	$HOOCCH_2CH_2CH_2COO$–	–COOH	–C(=CH_2)CH_3	A549
5	$CH_3CH(CH_3)COO$–	–COOH	–C(=CH_2)CH_3	A549
6	$HOOCCH_2C(CH_3)_2COO$–	–COOH	–C(=CH_2)CH_3	HL-60, HeLa, DBTRG0.5MG
7	$ClCH_2COO$–	–COOH	–C(=CH_2)CH_3	A549, MCF-7, HT-29, FADU, 518A2
8	–NH_2	–COOH	–C(=CH_2)CH_3	Daoy, MEL-2
9	=NOH	–COOH	–C(=CH_2)CH_3	MEL-2, PA-1, KB
10	=NOH	–CHNOH	–C(=CH_2)CH_3	MEL-2
11	=NOH	–COOH	–$CH(CH_3)_2$	MEL-2, A549, L132, DU145
12	=$NOCOCH_3$	–COOH	–C(=CH_2)CH_3	PA-1
13	–$NOCH_2C_6H_4NO_2$ [4]	–COOH	–C(=CH_2)CH_3	MOLT-4, U-937, DU145, L132, PA-1, HT-29
14	–$OCOC_6H_3F_2$ [2, 4]	–COOH	–C(=CH_2)CH_3	MOLT-4, U-937, DU145, L132, PA-1, HT-29
15	–$NCHC_6H_3F_2$ [2, 4]	–COOH	–C(=CH_2)CH_3	MOLT-4, U-937, DU145, L132, PA-1, HT-29
16	–$OCOC_6H_4(CF_3)$ [3]	–COOH	–C(=CH_2)CH_3	ECV304
17	–$OCOC_6H_3F_2$ [3,5]	–COOH	–$CH(CH_3)_2$	ECV304
18	–NHNHPh	–COOH	–$CH(CH_3)_2$	PA-1, ECV304
19	–NHNH-(p-methoxy phenyl)	–COOH	–$CH(CH_3)_2$	ECV304
20	–N=CH-(2,4-difluorophenyl)	–COOH	–$CH(CH_3)_2$	PA-1
21	–NH-O-CH_2-(p-nitrophenyl)	–COOH	–$CH(CH_3)_2$	PA-1
22	=$NNHC_6H_5$	–COOH	–$CH(CH_3)_2$	PA-1, DU145
23	=$NNHC_6H_2Cl_3$ (2,4,6)	–COOH	–$CH(CH_3)_2$	ECV304, A549
24	=$NNHC_6H_4F$ (4)	–COOH	–$CH(CH_3)_2$	ECV304
25	=(2,4-DNP)	–COOH	–$CH(CH_3)CHO$	A2780, 8505C, 518A2, MCF-7, A549

(continued)

Table 1 (continued)

No.	R_1	R_2	R_3	Cells active against
26	=(2,4-DNP)	–COOH	–CO(CH_3)	A2780, 8505C, 518A2, MCF-7, A549
27	–OH	–COOH	–C(CH_3)(=2,4-DNP)	A2780, 8505C, 518A2, MCF-7, A549
28	=O	–COOH	–C(=CH_2)CH_3	MCF-7, MEL-2
29	–O-α-D-mannopyranose	–CH_2OH	–C(=CH_2)CH_3	A549, DLD-1, WS1
30	–O-α-L-rhamnopyranose	–COOH	–C(=CH_2)CH_3	A549, DLD-1, WS1
31	–O-α-D-arabinopyranose	–COOH	–C(=CH_2)CH_3	A549, DLD-1, WS1
32	–O-β-D-xylopyranose	–COOH	–C(=CH_2)CH_3	A549, DLD-1, WS1
33	Methyl phthalate	–COOH	–C(=CH_2)CH_3	CEM, K562, HT-29
34	–$OCOCH_3$	–CH_2O-phthalate	–C(=CH_2)CH_3	K562-tax
35	–$OCONHC_2H_5$	–$COOCH_3$	–C(=CH_2)CH_3	8505C, A549, DLD-1, FaDu, HT-29, MCF-7
36	–$OCONHC_2H_5$	–$CH_2OCONHC_2H_5$	–C(=CH_2)CH_3	518A2, A253, A431, HCT-8, HCT-116, SW480, A2780
37	–$OCONHC_2H_5$	–CH_2OCOCH_3	–C(=CH_2)CH_3	A253, A549, HCT-116, SW480, SW1736
38	–OCONH-S(O)$_2$-Ph	–CONH-S(O)$_2$-Ph	–C(=CH_2)CH_3	MDA-MB-231
39	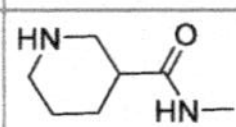	–COOH	–C(=CH_2)CH_3	HCT-116, HT-29, PC3, DU145, MDA-MB-231, 4T1, MCF-7
40	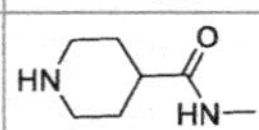	–COOH	–C(=CH_2)CH_3	HCT-116, HT-29, PC3, DU145, MDA-MB-231, 4 T1, MCF-7
41	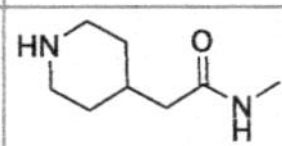	–COOH	–C(=CH_2)CH_3	HCT-116, HT-29, PC3, DU145, MDA-MB-231, 4T1, MCF-7
42	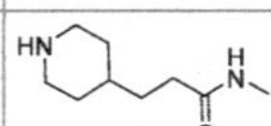	–COOH	–C(=CH_2)CH_3	HCT-116, HT-29, PC3, DU145, MDA-MB-231, 4T1, MCF-7
43	-OCO(CH_2)$_2$CONHCH$_2$- 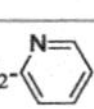	–COOH	–C(=CH_2)CH_3	CEM, MCF-7, HeLa, G-361

44	-OCO(CH$_2$)$_2$CONHCH$_2$- 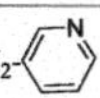	–COOH	–C(=CH$_2$)CH$_3$	CEM, MCF-7, HeLa, G-361
45	–OH, –C≡CH$_2$-pyrrolidinyl	–COOH	–C(=CH$_2$)CH$_3$	MCF-7, SW1736, A549, A253, 518A2, NiH 3 T3
46	–OH, –C≡CH$_2$-diisopropyl	–COOH	–C(=CH$_2$)CH$_3$	MCF-7, SW1736, A549, A253, 518A2, NiH 3T3, DLD1, 8505C
47	–OH, –C≡CH$_2$-pyrrolidinyl	–COMe	–C(=CH$_2$)CH$_3$	MCF-7, SW1736, A549, A253, 518A2, NiH 3T3, DLD1, 8505C
48	–OH, –C≡CH$_2$-piperidinyl	–COMe	–C(=CH$_2$)CH$_3$	MCF-7, SW1736, A549, A253, 518A2, NiH 3T3, DLD1, 8505C
49	–OH, –C≡CH$_2$-azepanyl	–COMe	–C(=CH$_2$)CH$_3$	MCF-7, SW1736, A549, A253, 518A2, NiH 3T3, DLD1, 8505C
50	–OH, –C≡CH$_2$-N-methyl-piperidinyl	–COMe	–C(=CH$_2$)CH$_3$	MCF-7, SW1736, A549, A253, 518A2, NiH 3T3, DLD1, 8505C
51	–OH, –C≡CH$_2$-dimethylamino	–COMe	–C(=CH$_2$)CH$_3$	MCF-7, SW1736, A549, A253, 518A2, NiH 3T3, DLD1, 8505C
52	–OH	–CH(OH)C≡CCH$_2$N- (isoPr)$_2$	–C(=CH$_2$)CH$_3$	SW1736, MCF-7, DLD-1, A549, A2780, A253, 8505C, 518A2, NiH 3 T3
53	–OH	–CH(OH) C≡CCH$_2$- pyrrolidinyl	–C(=CH$_2$)CH$_3$	SW1736, MCF-7, DLD-1, A549, A2780, A253, 8505C, 518A2, NiH 3T3
54	–OH	–CH(OH) C≡CCH$_2$-morpholinyl	–C(=CH$_2$)CH$_3$	SW1736, MCF-7, DLD-1, A549, A2780, A253, 8505C, 518A2, NiH 3 T3
55	–OAc	–CH(OH) C≡CCH$_2$N-(n-Pr)$_2$•MeI	–C(=CH$_2$)CH$_3$	SW1736, MCF-7, DLD-1, A549, A2780, A253, 8505C, 518A2, NiH 3 T3
56	–OAc	–CH(OH) C≡CCH$_2$N-(isoPr)$_2$•MeI	–C(=CH$_2$)CH$_3$	SW1736, MCF-7, DLD-1, A549, A2780, A253, 8505C, 518A2, NiH 3T3
57	–OH	–CH(OH) CH$_2$CH$_2$CH$_2$N-(isoPr)$_2$	–C(=CH$_2$)CH$_3$	SW1736, MCF-7, DLD-1, A549, A2780, A253, 8505C, 518A2, NiH 3T3

(continued)

Table 1 (continued)

No.	R_1	R_2	R_3	Cells active against
58	–OH	$–CH_2O$-L-Ala-NH_2	$–C(=CH_2)CH_3$	Me-45
59	–OH	$–CH_2O$-L-Lys-NH_2	$–C(=CH_2)CH_3$	Me-45
60	–OH	$–CH_2O$-L-Orn-NH_2	$–C(=CH_2)CH_3$	Me-45
61	–OH	$–CH_2O$-L-Dab-NH_2	$–C(=CH_2)CH_3$	Me-45
62	–OH	$–CH_2O$-L-Dap-NH_2	$–C(=CH_2)CH_3$	Me-45
63	–O-Ala	$–CH_2OCOCH$(Ala)-NH_3^+	$–C(=CH_2)CH_3$	EPP85-181P, EPP85-181RDB, EPG85-257P
64	–O-Met	$–CH_2OCOCH$(Met)-NH_3^+	$–C(=CH_2)CH_3$	EPP85-181P, EPG85-257P
65	–O-Lys	$–CH_2OCOCH$(Lys)-NH_3^+	$–C(=CH_2)CH_3$	EPP85-181P, EPP85-181RDB, EPG85-257P
66	–OCOCH(Lys)-NH_3^+	–COOH	$–C(=CH_2)CH_3$	EPP85-181P, EPG85-257P
67	–OCOCH(Abu)-NH_3^+	–COOH	$–C(=CH_2)CH_3$	EPP85-181P, EPG85-257P
68	–OH	$–CONHCH(CH_3)COOH$	$–C(=CH_2)CH_3$	KB
69	–OH	–CONHCH(isoPr)COOH	$–C(=CH_2)CH_3$	KB
70	–OH	$–CONHCH_2COOH$	$–C(=CH_2)CH_3$	MEL-2
71	–OH	$–CONHCH(CH_3)COMe$	$–C(=CH_2)CH_3$	MEL-2
72	–OH	–CONHCH(isoPr)COMe	$–C(=CH_2)CH_3$	MEL-2
73	–OH	$–COOCH_2CH{\equiv}CH$	$–C(=CH_2)CH_3$	T47D, CCRF/CEM, SW707, P388
74	=O	$–COOCH_2CH{\equiv}CH$	$–C(=CH_2)CH_3$	T47D, CCRF/CEM, SW707, P388
75	–OH	$–COOCH_2CH{\equiv}CHCH_3$	$–C(=CH_2)CH_3$	T47D, CCRF/CEM, SW707, P388
76	–OH	$–COOCH_2CH_2CH{\equiv}CH$	$–C(=CH_2)CH_3$	T47D, CCRF/CEM, SW707, P388
77	–OH	$–CONHCH_2CH{\equiv}CH$	$–C(=CH_2)CH_3$	T47D, SW707, P388
78	–OH	$–CH_2OCOCH{\equiv}CH$	$–C(=CH_2)CH_3$	G-361
79	–OH	$–COOCH_2CH_2N(C_2H_5)_2$	$–C(=CH_2)CH_3$	MGC-803, PC3, A375, A431, Bcap-37
80	–OH	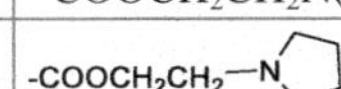	$–C(=CH_2)CH_3$	MGC-803, PC3, A375, A431, Bcap-37
81	–OH	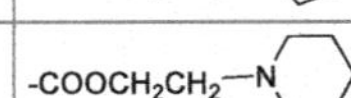	$–C(=CH_2)CH_3$	MGC-803, PC3, A375, A431, Bcap-37

82	–OH	$–COO(CH_2)_3N(C_2H_5)_2$	$–C(=CH_2)CH_3$	MGC-803, PC3, A375, A431, Bcap-37
83	–OH	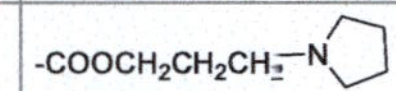	$–C(=CH_2)CH_3$	MGC-803, PC3, A375, A431, Bcap-37
84	=O	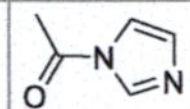	$–C(=CH_2)CH_3$	HepG2, HeLa, Jurkat
85	=O, C_2=	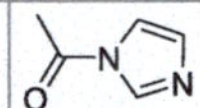	$–C(=CH_2)CH_3$	HepG2, HeLa, Jurkat
86	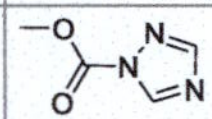	$–CH_2OH$	$–C(=CH_2)CH_3$	HepG2, HeLa, Jurkat
87	–OH	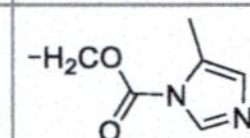	$–C(=CH_2)CH_3$	HepG2
88	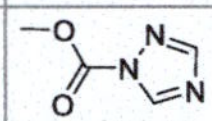	$–CH_2OH$	$–C(=CH_2)CH_3$	HepG2, Hela
89	–OH	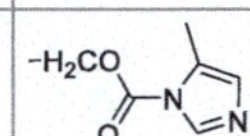	$CH_3OCH_2CH=CH2$	HeLa, HT-29
90	–OH	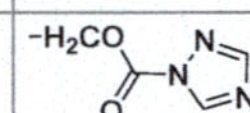	$CH_3OCH_2CH=CH2$	HepG2, Jurkat, HT-29
91	–OH	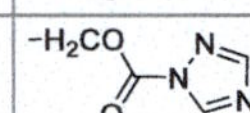	$–C(=CH_2)CH_3$	PC3, HT-29
92	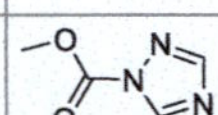	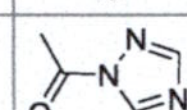	$–C(=CH_2)CH_3$	PC3
93	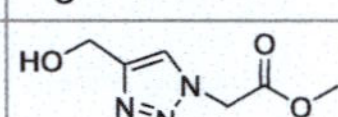	–COOH	$–C(=CH_2)CH_3$	HT29, Jurkat

(continued)

Table 1 (continued)

No.	R_1	R_2	R_3	Cells active against
94	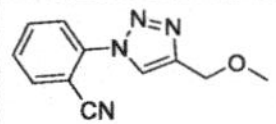	–COOH	–C(=CH_2)CH_3	HL-60, A549, MCF-7, THP-1
95	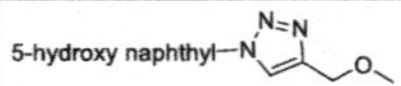	–COOH	–C(=CH_2)CH_3	HL-60, THP-1, MCF-7
96	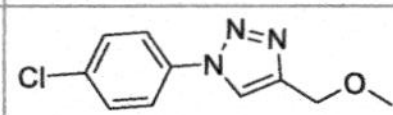	–COOH	–C(=CH_2)CH_3	HL-60, A549
97	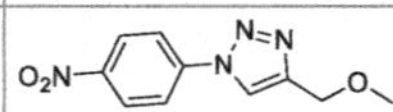	–COOH	–C(=CH_2)CH_3	THP-1, HL-60, SF-295
98	–OH	–CO-[Gly]$^-$ [benzalkonium]$^+$	–C(=CH_2)CH_3	A3375, A431, SH-SY5Y, MCF-7
99	–OH	–CO-[Gly]$^-$ [cholinium]$^+$	–C(=CH_2)CH_3	A3375, A431, SH
100	–OH	–COO$^-$ [benzalkonium]$^+$	–C(=CH_2)CH_3	A3375, A431, SH
101	–OH	–COO$^-$ [cholinium]$^+$	–C(=CH_2)CH_3	A3375, A431, SH
102	–OCOCHCl_2	–COO-(n-butyl)-PPh_3^+ Br^-	–C(=CH_2)CH_3	MCF-7, TET21N
103	–OCOCH_3	–COO-(triethyleneglycol)-PPh_3^+ Br^-	–C(=CH_2)CH_3	MCF
104	–OH	–COO-(triethyleneglycol)-PPh_3^+ Br^-	–C(=CH_2)CH_3	MCF
105	–OCOCHCl_2	–COO-(triethyleneglycol)-PPh_3^+ Br^-	–C(=CH_2)CH_3	MCF
106	–OH	–COO$CH_2CH_2N^+(CH_3)_3$• Br^-	–C(=CH_2)CH_3	CEM
107	–OH	–COOH	CH_3CHCH_3	L132
108	–OH	–COOH	CH_3CH=NOH	MCF-7
109	–OH	–COOH	CH_3CH=NOCH_3	MCF-7
110	–OH	–COOH	C(=CH_2)CH_2OH	M14-MEL, HCT-116
111	–OH	–COOH	CH_3CH=NNH-(2,4-dinitrophenyl)	MCF-7

112	–OH	–COOH	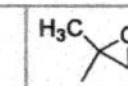	HCT116
113	–OH	$–CH_2OH$	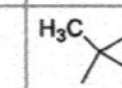	Colo-38
114	Artesunic succinate ester	$–CH_2OAc$	$–C(=CH_2)CH_3$	CEM/ADR5000
115	–OH	$–CH_2$-ASC ester	$–C(=CH_2)CH_3$	CEM/ADR5000
116	–ASC ester	$–CH_2$-ASC ester	$–C(=CH_2)CH_3$	CEM/ADR5000
117	–OH	$–COOCH_2$-AZT	$–C(=CH_2)CH_3$	KB, HepG2
118	$–OCO(CH_2)_2COOH$	$–COOCH_2$-AZT	$–C(=CH_2)CH_3$	KB, HepG2
119	–OH	–C(S)OH	$–C(=CH_2)CH_3$	HCT116
120	–OH	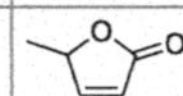	$–C(=CH_2)CH_3$	HCT116

these compounds (**13–15**) were found to be effective against multiple types of cancer cells. Mukherjee et al. [45] also developed fluorinated benzoyl derivatives (**16**) by modifying C-3 position and suggested that the saturation at C-20 position (**17**) improves activity against spontaneously transformed immortal endothelial cell (ECV304). Mukherjee et al. [45] in another report described a series of C-3 bezylidene, hydrazine, and hydrazine derivatives (**18–24**) with C-20 saturation. Some of them are highly toxic against specific cells, e.g., PA-1. Baratto et al. [47] also tried substitution of 2,4-dinitrophenyl hydrazine (2,4-DNP) at both C-3 and C-20 positions and found that monosubstituted derivatives (**25–27**) were selectively cytotoxic while disubstitution at both C-3 and C-20 positions was not advantageous anyway. Oxidation of C-3 hydroxyl group to keto group (**28**) improved cytotoxicity of parent compound (BA) against MCF-7 and MEL-2 cells [44, 47]. Insertion of sugar molecules at C-3 position also improved aqueous solubility and bioavailability of BA (or BN). Thibeault et al. [48] produced a number of C-3-glycosylated BA derivatives; some of them are mentioned here (**29–32**) as they have low IC_{50} values.

Kvasnica et al. [49] reported insertion of phthalic ester at C-3 position of BA and esterification of C-3 and C-28 OH group of BN by phthalic acid. They found that the methyl-phthalate derivative (**33**) is more active than BA against CEM, K562, and HT-29 cells. The phthalate derivative of BN (**34**) was highly active against K562-tax cells compared to BN. Kommera et al. [50] developed a series of C-3 and C-28 carbamate derivatives of BA (and BN) among which compound (**35–37**) showed high toxicity against a number of cancer cells. In another study, a thiocarbamate derivative of BA **38** possessing high toxicity against MDA-MB-231 cells was developed by Bar et al. [51]. Cui et al. [52] tested insertion of heterocyclic groups at C-3 position of BA with the help of amide linker and found substitution by unsaturated six-membered heterocycles (e.g., pyrazine carboxamides and nicotinamides) does not improve cytotoxicity, but insertion of saturated six-membered heterocycles (e.g., piperidine carboxamides (**39–42**)) improves cytotoxicity much more than BA alone. Bildziukevich et al. [53] introduced heterocyclic amides at C-3 and C-28 positions of BA, and some of their compounds (**43, 44**) showed high toxicity against multiple types of cancer cells. A number of Mannich base of BA produced by Csuk et al. [54], after insertion of ethynyl group at C-3 position, were found to be effective against many cancer cells with considerably low IC_{50} values. Some of them are shown in **45–51**.

The same group also reported [55] introduction of Mannich base at C-28 position, and these compounds were effective against many cancer cells. Some of them with low IC_{50} values are shown in **51–57**. Attempts to couple amino acids with BA (or BN) to improve its aqueous solubility produced a number of cytotoxic analogues. Drag-Zalesinska et al. [56] produced a number of C-28 amino acid derivatives (**58–62**) of BN effective against Me-45 human melanoma cells. The same group in another study reported a number of amino acid substituted derivatives of BN (**63–67**) effective against human gastric and pancreatic cancer cells [57]. Jeong et al. [58] coupled amino acids at C28 position of BA and found that alanine (**68**) and valine (**69**) derivatives are effective against mouth carcinoma cells (KB), while glycine (**70**), methyl alanine (**71**), and methyl valine (**72**) derivatives against MEL-2 melanoma cells. Bebenek et al. [59] reported a number of acetylenic derivatives of BA by acetenylation of C-28 COOH group. Some of those cytotoxic compounds are shown in **73–77**. They found that increment in chain length of acetylenic groups did not improve cytotoxicity rather increased IC50 values. Orchel et al. [60] found compound **78** produced by propynoylation of C28 of BN which was active against human melanoma cell, G-361. A series of N-alkyl esters produced by esterification of C-28 COOH group of BA was reported by Yang et al. [32, 33]. They found many of these compounds (**79–83**) possess moderate to high toxicity against multiple types of cancer cells. Insertion of azole functionality at C-28 and/or C-3 position was found to remarkably improve cytotoxicity of the parent compound (BA or BN). Santos et al. reported that pyrazole and triazole [61, 62] derivatives (**84–90**) are effective against HepG2, HeLa, and Jurkat cells. Compounds **91** and **92** were found to be effective against PC3 cells. Chakraborty et al. [36] also reported a number of C-3 azole derivatives out of which compound **93** was found to be effective against HT-29 and Jurkat cells. Regarding substituted triazole derivatives reported by Majeed et al. [63], compounds **94–97** showed significant cytotoxicity against multiple types of cancer cells. Suresh et al. [64] reported a number of ionic derivatives of BA by salt formation between BA and C-28-amino acid BA and tertiary amines, mainly cholinium and benzalkonium (**98–101**). They have shown more cytotoxicity against A375, A431, SH-SY5Y, and MCF-7 cells compared to the parent compound. In a different way, Nedopekina et al. [65] developed a number of triphenylphosphonium salts of BA (**102–105**) by inserting PPh_3 with the help of hydrophobic linker (n-butane) or hydrophilic linker (triethylene glycol) and found them to be effective against MCF-7 and TET21N cells. Introduction of quaternary ammonium group at C-28 position (**106**) also improves water solubility and cytotoxicity against selected cells.

While considering modification at C-20 position alone, a number of compounds were reported (**107–113**) including spiro compounds (**112, 113**), but in most cases, cytotoxicity became lower than the parent compound. Preserving cytotoxicity by C20 substitution is highly speculative and depends on size and charge density of the substituents [45]. Attempts to couple other cytotoxic agents with BA (or BN) produced some hybrid antitumor agents; some of them are more effective than their parent analogues. For example, coupling of BN with artesunic succinic acid (ASC) produced compounds **114–116** [5]; and they were found to be active against multidrug-resistant leukemia cell, CEM/ADR5000. In another study, Dang et al. [66] coupled BA with azidothymidine (AZT) at C28 position, which produced compounds **117** and **118** which were effective against KB and HepG2 cells.

Some other types of compounds produced by modification at C-28 position are also described [5] such as compounds **119** and **120** showing high activity against HCT116, a human colon adenocarcinoma.

Apart from these aforelisted compounds, some other compounds with C-2 modification, ring modification by insertion of heterocyclic atoms, ring expansion, or opening (Fig. 3), were also demonstrated by several researchers [67, 68].

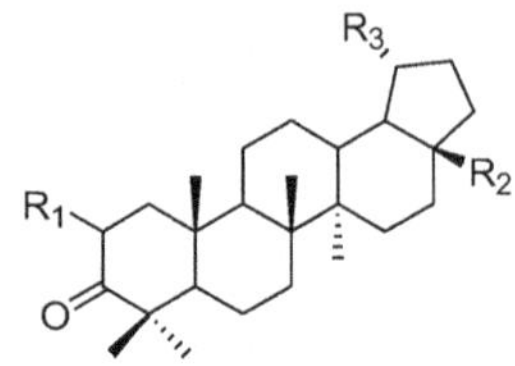

121 R_1 = (=CHOH), R_2 = COOH, R_3 = -CH(CH_3)$_2$

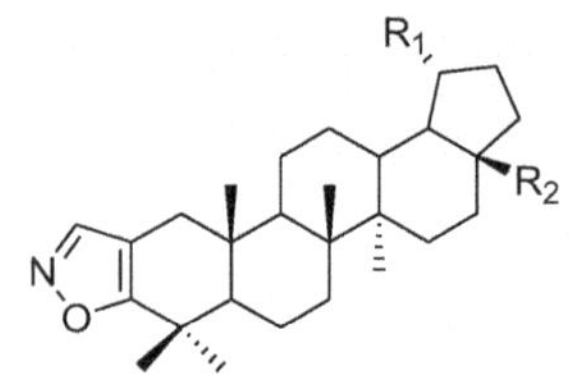

128 R_1 = -C(=CH_2)CH_3, R_2 = COOH
129 R_1 = -C(CH_3)$_2$, R_2 = COOH

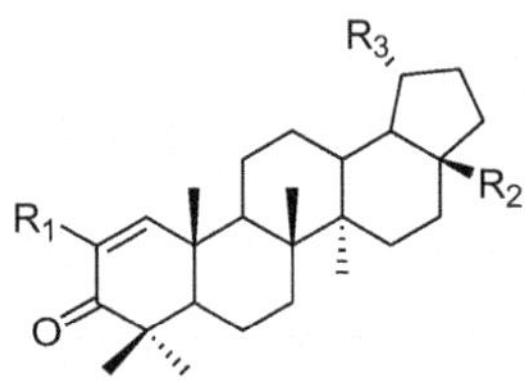

122 R_1 = CN, R_2 = $COOCH_3$, R_3 = -C(=CH_2)CH_3
123 R_1 = CN, R_2 = COOH, R_3 = -CH(CH_3)$_2$
124 R_1 = Cl, R_2 = COOH, R_3 = -CH(CH_3)$_2$
125 R_1 = CHO, R_2 = $COOCH_3$, R_3 = -CH(CH_3)$_2$
126 R_1 = OH, R_2 = COOH, R_3 = -C(=CH_2)CH_3
127 R_1 = OAc, R_2 = COOH, R_3 = -C(=CH_2)CH_3

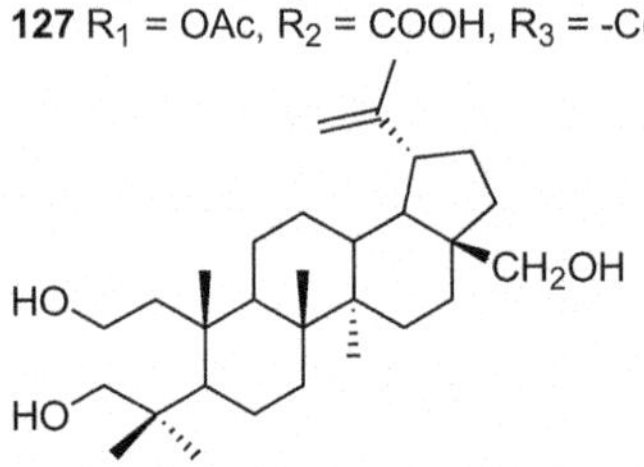

130

Fig. 3 Compounds with C_2 modification along with ring modification as mentioned along with the figure

Among them, a number of C-2-modified compounds are represented here (**121–127**) along with the cells in which they showed good cytotoxicity. Compounds **121–125**, **128,** and **129** showed good cytotoxicity against SK-MEL-2, A549, and B16F10 cells. Compounds **126** and **127** were found to be effective against CEM, HT-29, K562, K562 Tax, and PC3 cells, while compound **130** showed similar cytotoxicity as **126** and **127** except against PC3 cells (Table 2).

4 Nanoformulations of Betulinic Acid and Its Derivatives

Although BA has significant pharmacological activities as it shows wide-ranging biological and medicinal effects, it poses some disadvantages, such as poor solubility in water and tiny half-life in vivo, which limit its therapeutic use. To resolve these

Table 2 List of various cancer cells on which BA shows activity

Cells name	Tumor classification
CEM/ADR5000	Acute lymphoblastic leukemia
KB	Oral epidermoid carcinoma cells
Hep G2	Human liver cancer
A549	Adenocarcinomic human alveolar basal epithelial cells
CEM-SS	Leukemia
WEHI-3B	Mouse leukemia
HeLa	Cervical cancer
HCT116	Human colon cancer cells
HL-60	Leukemia
DBTRG05MG	Glioblastoma
Me-45	Melanoma
EPP85-181P	Pancreatic adenocarcinoma
EPP85-181RDB	Pancreatic adenocarcinoma
EPG85-257P	Gastric adenocarcinoma
MCF-7	Breast cancer
SW1736	Thyroid gland undifferentiated (anaplastic) carcinoma
A253	Salivary gland squamous cell carcinoma
518A2	Melanoma
NiH 3T3	Mouse embryonic fibroblast cells
DLD-1	Colon adenocarcinoma
8505C	Thyroid gland undifferentiated (anaplastic) carcinoma
MDA-MB-231	Breast adenocarcinoma
Daoy	Medulloblastoma
WS1	Skin fibroblast
PA-1	Ovarian teratocarcinoma
L132	Human papillomavirus-related endocervical adenocarcinoma
Colo-38	Amelanotic melanoma
A2780	Ovarian endometrioid adenocarcinoma
PC3	Prostate carcinoma
SF-295	Glioblastoma
T47D	Breast cancer cells
CCRF/CEM	Leukemia
SW707	Rectal adenocarcinoma
P388	Mouse lymphoma
G-361	Melanoma
MEL-2	Human melanoma
MGC-803	Gastric carcinoma
A375	Amelanotic melanoma
A431	Vulvar squamous cell carcinoma
Bcap-37	Human papillomavirus-related endocervical adenocarcinoma
A431	epidermoid carcinoma
SH-SY5Y	Neuroblastoma
Jurkat	Childhood T acute lymphoblastic leukemia
HT-29	Colon adenocarcinoma

(continued)

Table 2 (continued)

Cells name	Tumor classification
MOLT-4	Lymphoblastic leukemia
U-937	Adult acute monocytic leukemia
DU145	Prostate carcinoma

issues, lots of studies recently focused on increasing BA's aqueous solubility, half-life, and therapeutic efficiency by using nanoscale drug delivery methods. Several drug delivery techniques having a nano-sized range have been developed for the delivery of BA which includes polymeric nanoparticles, magnetic nanoparticles, nanoliposomes, nanofibers, polymeric conjugates, nanoemulsions, cyclodextrin-based nanocomplexes, and carbon nanotubes. In this segment, we have focused on the current proceedings of novel formulations having a nano-size that is used to deliver betulinic acid in order to enhance its efficacy [69, 70].

5 Nanoparticles

Nanoparticles (NPs) are the tiny colloidal particles of submicron size developed from biodegradable polymers, lipids, or inorganic materials [69]. A lot of biodegradable and biocompatible polymers have been investigated for delivering therapeutic molecules, as poly(D,L-lactic-coglycolic acid) (PLGA), chitosan (CS), cellulose, albumin, polycaprolactone, and gelatin. NPs provide numerous key advantages including enhanced aqueous solubility, better bioavailability and plasma half-life, changes in pharmacokinetics, and ease of surface modification using targeting ligand. Among the various biodegradable and biocompatible polymers used, PLGA has shown promising interests in researches which are approved by the US Food and Drug Administration (FDA) for systemic administration [70]. Hydrophobic drugs could be entrapped into the pores of PLGA and, hence, enhance solubility of the drug by forming stable and amorphous NPs [71, 72]. Furthermore, modification to the surface of PLGA nanocarriers with stealth polymers, such as polyethylene glycol (PEG) or chitosan (CS), could extend the systemic circulation time due to the reduction of protein opsonization, which reduces in nonspecific uptake toward normal tissues [73]. Considering these facts, BA encapsulated in monomethoxy polyethylene glycol-modified poly(D,L-lactide-co-glycolide) (mPEG–PLGA) nanoparticles was developed, which indicated that the half-life in vivo is increased by 7.21-fold resulting in better anticancer efficacy [74].

In 2016, Das et al. designed and developed PLGA-based NPs of BA derivatives and shown the comparison of therapeutic effect of NPs with native drug, BA against comutagenic sodium arsenite, and benzo(a)pyrene (BaP)-induced lung carcinogenesis in mice in vivo and A549 cells in vitro [75].

Cellulose-based nanoparticles have also gained significant importance due to its biocompatibility, nontoxicity, and mechanical strength [76]. In 2017, Dai and Si developed a copolymer, using atom transfer radical polymerization cellulose-grafted polymethyl methacrylate (CE-g-PMMA) and entrapped BA in the copolymer by using nanoprecipitation methods [77]. Superior in vivo efficacy was shown by the developed CE-g–PMMA–BA NPs in Lewis lung carcinoma (LLC) xenograft mice, showing 90.3% tumor growth inhibition (TGI) compared with native BA [78]. In a different study, Dai et al. also established BA-loaded cellulose-graft-poly(L-lactic acid) nanoparticles (CE-g-PLLA/BA NPs) and suggested better antitumor efficacy of the NPs than BA in a mouse tumor xenograft model [79].

Albumin, a resourceful protein carrier for drug delivery, is an ideal material to be used for nanoparticle preparation owing to its nontoxic, non-immunogenic, biocompatible, and biodegradable nature [80]. Owing to the advantages of albumin, Srivari and Chatterjee developed BA-loaded human serum albumin (HSA) NPs and shown the effects of solubilizer on the release of BA from HSA-binding sites [81]. In a different study, Dai et al. prepared a self-assembled bovine serum albumin (BSA) poly(L-lactic acid) NPs for the delivery of BA [82]. Further, this study demonstrated the greater efficacy of BSA (bovine serum albumin)–PLLA(poly(L-lactic acid))–BA NPs for in vitro cytotoxicity studies in A549 and LLC cells and in vivo antitumor efficacy in a mouse tumor xenograft model compared with native BA.

Recently, magnetic nanoparticles (MNPs) have also received considerable interest due to their super paramagnetism, biocompatibility, ability for targeted delivery, nontoxicity, low production cost, hyperthermia, magnetic resonance imaging (MRI), and controlled drug release [83, 84]. In a recent study, chitosan-coated magnetic nanoparticles (MNPs) loaded with BA were developed and BA–CS–MNPs showed enhanced in vitro cytotoxicity in comparison with BA in MCF-7 cells [85].

6 Nanoliposomes

Several researchers have reported the nanoliposomal formulation for the delivery of betulinic acid (BA). The liposomal formulation of BA was developed by Mullauer et al., and its efficacy was evaluated in nude mice xenografted with human colon and lung cancer tumors. It was observed that the administration of BA-loaded liposomes reduced the growth of tumors by more than 50% as compared to the control treatment. Moreover, no signs of systemic toxicity were observed after treatment with the liposomal formulation [86]. Polyethylene glycol-modified (PEGylated) BA liposomes were also synthesized by using an ethanol injection technique. The PEGylation of the liposomes resulted in more sustained drug release and also enhanced the antitumor activity of BA both in vitro and in vivo, as compared to BA alone or BA liposomes [87]. The same group of authors fabricated near-infrared (NIR) light-responsive gold nanoshell-coated BA liposomes (AuNS-BA-Lips). Glutathione was used as a linker molecule for combining gold nanoparticles with liposomes by the formation of Au–S bond. The irradiation with NIR resulted in higher uptake of AuNS-BA-Lips by tumor cells. In addition, it also exhibited an enhanced antitumor activity against tumor-bearing mice, thereby demonstrating a synergistic therapeutic effect of chemotherapy and thermotherapy [88]. Guo et al. developed folate–poly(ethylene glycol)3400–cholesterol conjugates (FA–PEG–Chol) by employing the enzyme lipase

B from *Candida antarctica* (Novozym 435). It was then incorporated into liposomes to prepare folate-decorated liposomes. The folate-functionalized liposomes exhibited an improved in vitro cytotoxicity in comparison to nontargeted liposomes containing BA in HepG2 cells [71, 72].

Many derivatives of betulinic acid are reported to exhibit anticancer activity. In another investigation, various betulin-derived compounds were synthesized and evaluated for anticancer activity. Among them, 3-O-acetylbetulinic aldehyde was used as an ideal starting material for the synthesis of 28-acetylenic derivatives. Further, these compounds were used for the synthesis of betulin-derived pyrazoles, 1,2,3-triazoles, and 3-substituted butenolides. The antitumor activity of these compounds was screened for 15 human cancer cells in a sulforhodamine B (SRB) assay and showed promising results. Moreover, apoptotic cell death was observed by incorporating these compounds into liposomal formulation [89].

7 Nanoemulsions

Nanoemulsion-based delivery systems are also one of the promising vehicles for the enhanced uptake of highly hydrophobic drugs [90, 91]. A comparative study of BA topical ointment and systemic nanoemulsion was performed by Ciurlea et al. The topical ointment of BA was used for the local treatment of skin pathologies by reducing the erythema and intense pigmentation. On the other hand, BA nanoemulsion delayed the metastatic process to the liver after intraperitoneal administration. Hence, both the topical ointment and systemic nanoemulsion of BA were found to possess significant activity [92]. The antiangiogenic activity against tumor cells was carried out by BA nanoemulsion by performing in vivo chick embryo chorioallantoic membrane (CAM) assay. The morphological alterations were observed in the CAM mesenchyme with a negative effect on the normal growth of the tumor capillaries [93]. The same group of authors also formulated betulin nanoemulsion to investigate its antiangiogenic as well as anticancer and anti-inflammatory effects and found promising results [94]. Cavazos-Garduño et al. employed response surface methodology (RSM) by a central composite design to evaluate the effects of various variables such as emulsification power, ultrasonication time, and emulsifier concentration in the characteristics of betulinic acid nanoemulsions. The optimal conditions for the preparation of nanoemulsions were 64 nm mean diameter by using 15% emulsifier (tween 60) and an ultrasonication power of 20 W during 220 s [95]. The betulinic acid nanoemulsion was administered topically for the treatment of skin carcinoma in mouse model. The nanoemulsion treated group showed a fewer number of skin lesions and tumors as compared to the non-treated group. Additionally, betulinic acid nanoemulsion delayed the development of skin tumors [96]. Similarly, betulin extracted from the bark of birch trees was also used for the preparation of nanoemulsion for the treatment of skin carcinoma [97]. BA nanoemulsion was also employed for improving its oral bioavailability and hepatoprotective activity [98].

8 Carbon Nanotubes

In modern drug delivery techniques and biomedical therapies, carbon nanotubes (CNTs) have proven its efficacy as a novel drug carrier with promising future applications owing to their diverse characteristics. Single-walled carbon nanotubes, functionalized with carboxylic acid (f-SWCNTs), were reacted with anticancer drug, BA, to produce f-SWCNTs-BA conjugate via π-π stacking interaction showing anticancer activity significantly higher in A549 than HepG2 cells [99, 100]. The same group formulated BA in oxidized multi-walled carbon nanotubes (MWCNT-COOH) for enhanced delivery efficiency into cancer cells with reduced cytotoxicity [99, 100]. In another study, biocompatibility was improved using some coating biomaterials [101].

9 Cyclodextrin-Based Nanocarriers

One more significant technique of delivering BA is cyclodextrin (CD)-based nanocarriers, which are macrocyclic oligosaccharides having hydrophilic outer surface and a hydrophobic inner cavity, capable of solubilizing hydrophobic molecules. Thermosensitive supramolecular hydrogels were developed using the addition complex between α-CD and the PEG chain of dual-drug-loaded NPs (CMC–BA/HCPT NPs) [78]. In vitro cytotoxicity and in vivo studies shown superior efficacy of the prepared BA hydrogels compared with native BA.

10 Nanofibers

One of the unique techniques for nanofabrication is self-assembly, which involves designing new structured molecules containing supramolecular entities. Various advantages are shown in self-assembly process in biological applications including modification of drug complexity and functional structure which accelerates the potency of drug compared to its native form [102]. Potential effects of self-assembled nano-sized betulinic acid (SA-BA) are observed by treating human leukemic cells (KG-1A and K562) and its normal counterpart showing better antileukemic efficacy of SA-BA over BA, and this modification will supplement the use of BA in cancer therapy [103, 104]. In another study, folate receptor-mediated self-assembled betulinic acid nanofibers (SA-BA) were developed and delivered to human leukemic cells, and they showed their specificity in induction of apoptosis in cancer cell [103, 104].

11 Conclusion and Future Perspectives

The diversity of BA derivatives and their efficacy toward multiple types of cancer cells reached this natural compound at the pinnacle of success in anticancer drug development. Although a huge number of potential cytotoxic agents have been discovered ever by modification of betulinic acid, very few of them have been tried to deliver as pharmaceutical dosage forms, especially nanoformulations. The number of BA derivatives successful by entered in clinical trials is also very low compared to those studied in the laboratory experiments. Structural modification along with the development of nanoformulated drug delivery may help to develop therapeutically effective novel dosage forms of BA derivatives. By observing structure activity relationships of BA derivatives, BA analogues highly selective to specific cancer types may be developed. It is also possible to develop broad-spectrum anticancer agents using combinatorial chemistry and molecular modeling approach to utilize the potential BA derivative.

References

1. Cragg GM, Pezzuto JM (2016) Natural products as a vital source for the discovery of cancer chemotherapeutic and chemopreventive agents. Med Princ Pract 25(Suppl 2):41–59
2. Fulda S, Jeremias I, Steiner HH, Pietsch T, Debatin KM (1999) Betulinic acid: a new cytotoxic agent against malignant brain-tumor cells. Int J Cancer 82:435–441
3. Zuco V, Supino R, Righetti SC, Cleris K, Marchesi E, GambacortiPasserini C, Formelli F (2002) Selective cytotoxicity of betulinic acid on tumor cells, but not normal cells. Cancer Lett 175:17–25
4. Pisha E, Chai H, Lee IS, Chagwedera TE, Farnsworth NR, Cordell AC, Beecher CWW, Fong HHS, Kinghorn AD, Brown DM, Wani MC, Wall ME, Hieken TJ, Das Gupta TK, Pezzuto JM (1995) Discovery of betulinic acid as a selective inhibitor of human melanoma that functions by induction of apoptosis. Nat Med 1:1046–1051
5. Zhang DM, Xu HG, Wang L, Li YJ, Sun PH, Wu XM, Ji WG, Chen WM, Ye WC (2015) Betulinic acid and its derivatives as potential antitumor agents. Med Res Rev 35:1127–1155
6. Liby KT, Yore MM, Sporn MB (2007) Triterpenoids and rexinoids as multifunctional agents for the prevention and treatment of cancer. Nat Rev Cancer 7(5):357–369
7. Petronelli A, Pannitteri G, Testa U (2009) Triterpenoids as new promising anticancer drugs. Anti-Cancer Drugs 20(10):880–892
8. Moghimi SM, Hunter AC, Murray JC (2001) Long-circulating and target-specific nanoparticles: theory to practice. Pharmacol Rev 53:283–318
9. Lessene G, Czabotar PE, Colman PM (2008) BCL-2 family antagonists for cancer therapy. Nat Rev Drug Discov 7:989–1000
10. Elmore S (2007) Apoptosis: a review of programmed cell death. Toxicol Pathol 35:495–516
11. Shen S, Kepp O, Michaud M (2011) Association and dissociation of autophagy, apoptosis and necrosis by systematic chemical study. Oncogene 30:4544–4556
12. Hanahan D, Weinberg RA (2011) Hallmarks of cancer: the next generation. Cell 144:646–674
13. Lin X, Liu M, Hu C, Liao DJ (2010) Targeting cellular proapoptotic molecules for developing anticancer agents from marine sources. Curr Drug Targets 11:708–715
14. Simon HU (2003) Targeting apoptosis in the control of inflammation. Eur Respir J 22(supl 44):20s–21s
15. Yuan J, Yankner BA (2000) Apoptosis in the nervous system. Nature 407:802–809
16. Schmidt ML, Kuzmanoff KL, Ling-Indeck L, Pezzuto JM (1997) Betulinic acid induces apoptosis in human neuroblastoma cells. Eur J Cancer 33:2007–2010
17. Wick W, Grimmel C, Wagenknecht B, Dichgans J, Weller M (1999) Betulinic acid-induced apoptosis in glioma cells: a sequential requirement for new protein synthesis, formation of reactive oxygen species, and caspase processing. J Pharmacol Exp Ther 289(3):1306–1312

18. Ji ZN, Ye WC, Liu GG, Hsiao WL (2002) 23-Hydroxybetulinic acid-mediated apoptosis is accompanied by decreases in bcl-2 expression and telomerase activity in HL-60 cells. Life Sciences 72(1):1–9
19. Thurnher D, Turhani D, Pelzmann M (2003) Betulinic acid: a new cytotoxic compound against malignant head and neck cancer cells. Head Neck 25(9):732–740
20. Tezuka Y, Stampoulis P, Banskota AH, Awale S, Tran KW, Saiki I, Kadota S (2000) Constituents of the Vietnamese medicinal plant *Orthosiphon stamineus*. Chem Pharm Bull 48(11):1711–1714
21. Kwon HJ, Shim JS, Kim JH, Cho HY, Yum YN, Kim SH, Yu J (2002) Betulinic acid inhibits growth factor-induced in vitro angiogenesis via the modulation of mitochondrial function in endothelial cells. Jpn J Cancer Res 4:93
22. Mukherjee R, Jaggi M, Rajendran P, Siddiqui MJ, Srivastava SK, Vardhan A, Burman AC (2004a) Betulinic acid and its derivatives as anti-angiogenic agents Bioorg. Med Chem Lett 14:2181–2184
23. Mukherjee R, Jaggi M, Siddiqui MJA, Srivastava SK, Rajendran P, Vardhan A, Burman AC (2004b) Synthesis and cytotoxic activity of 3-O-acyl/3-hydrazine/2-bromo/20,29-dibromo betulinic acid derivatives. Bioorg Med Chem Lett 14:4087–4091
24. Liu H, Wang S, Cai B, Yao X (2004) Anticancer activity of compounds isolated from *Engelhardtia serrata* stem bark. Arch Physiol Biochem 42:475–477
25. Wada SI, Tanaka R (2005) Betulinic acid and its derivatives, potent DNA topoisomerase II inhibitors, from the bark of Bischofia javanica. Chem Biodivers 2(5):689–694
26. Fu L, Zhang S, Li N, Wang J, Zhao M, Sakai J, Hasegawa T, Mitsui T, Kataoka T, Oka S, Kiuchi M, Hirose K, Ando M (2005) Three new triterpenes from Ner- ium oleander and biological activity of the isolated compounds. J Nat Prod 68(2):198–206
27. Abdille MH, Singh RP, Jayaprakasha GK, Jena BS (2005) Antioxidant activity of the extracts from *Dillenia indica* fruits. Food Chem 90:891–896
28. Ren W, Qin L, Xu Y, Cheng N (2010) Inhibition of betulinic acid to growth and angiogenesis of human colorectal cancer cell in nude mice. Chin J Clin Oncol 9(3):153–157
29. Kumar D, Mallick S, Vedasiromoni JR, Pal BC (2010) Anti-leukemic activity of Dilleniaindica L. Fruit Ex- tract and quantification of betulinic acid by HPLC. Phytomedicine 17(6):431–435
30. Gao Y, Jia Z, Kong X, Li Q, Chang DZ, Wei D, Le X, Huang S, Wang L, Xie K (2011) Combining betulinic acid and mithramycin a effectively suppresses pancreatic cancer by inhibiting proliferation, invasion, and angiogenesis. Cancer Res 71(15):5182
31. Chintharlapalli S, Papineni S, Lei P, Pathi S, Safe S (2011) Betulinic acid inhibits colon cancer cell and tumor growth and induces proteasome dependent and -independent downregulation of specificity proteins (Sp) transcription factors. BMC Cancer 11:371
32. Yang S, Liang N, Hu L, Hu WXD, Jin L, Yang QZS (2012a) Design, synthesis and biological evaluation of novel betulinic acid derivatives. Chem Cent J 6:14
33. Yang S, Liang N, Li H, Xue W, Hu D, Jin L, Zhao Q, Yang S (2012b) Design, synthesis and biological evaluation of novel betulinic acid derivatives. Chem Cent J 6:141
34. Reiner T, Parrondo R, de las Pozas A, Palenzuela D (2013) Betulinic acid selectively increases protein degradation and enhances prostate cancer-specific apoptosis: possible role for inhibition of deubiquitinase activity. PLoS One 8(2):e56234
35. Gomathi D, Ravikumar G, Kalaiselvi M, Devaki K, Uma C (2014) Antioxidant activity and functional group analysis of *Evolvulus alsinoides*. Chin J Nat Med 12(11):827–832
36. Chakraborty B, Dutta D, Mukherjee S, Das S, Maiti NC, Das P, Chowdhury C (2015) Synthesis and biological evaluation of a novel betulinic acid derivative as an inducer of apoptosis in human colon carcinoma cells (HT-29). Eur J Med Chem 102:93–105
37. Raisova M, Hossini AM, Eberle J (2001) The Bax/Bcl-2 ratio determines the susceptibility of human melanoma cells to CD95/Fas-mediated apoptosis. J Invest Dermatol 117(2):333–340
38. Selzer E, Pimentel E, Wacheck V (2000) Effects of betulinic acid alone and in combination with irradiation in human melanoma cells. J Invest Dermatol 114(5):935–940
39. Sjöblom T, Jones S, Wood LD, Parsons DW, Lin J, Barber TD, Mandelker D, Leary RJ, Ptak J, Silliman N, Szabo S, Buckhaults P, Farrell C, Meeh P, Markowitz SD, Willis J, Dawson

D, Willson JK, Gazdar AF, Hartigan J, Wu L, Liu C, Parmigiani G, Park BH, Bachman KE, Papadopoulos N, Vogelstein B, Kinzler KW, Velculescu VE (2006) The consensus coding sequences of human breast and colorectal cancers. Science 314:268–274
40. Connolly JD, Hill RA (2010) Triterpenoids. Nat Prod Rep 1:79–132
41. Ahmad FBH, Moghaddam MG, Basri M, Rahman MBA (2010) Anticancer activity of 3-Oacylated betulinic acid derivatives obtained by enzymatic synthesis. Biosci Biotechnol Biochem 7(5):1025–1029
42. Faujan NH, Alitheen NB, Yeap SK, Ali AM, Muhajir AH, Ahmad FBH (2010) Cytotoxic effect of betulinic acid and betulinic acid acetate isolated from Melaleuca cajuput on human myeloid leukemia (HL-60) cells. Afr J Biotechnol 9(38):6387–6396
43. Kommera H, Kaluderovic GN, Kalbitz J, Paschke R (2011) Lupane triterpenoids- botulin and betulinic acid derivatives induce apoptosis in tumor cells. Investig New Drugs 29:266–272
44. Kim DSHL, Pezzuto JM, Pisha E (1998) Synthesis of betulinic acid derivatives with activity against human melanoma. Bioorg Med Chem Lett 8:1707–1712
45. Mukherjee R, Kumar V, Srivastava SK, Agarwal SK, Burman AC (2006) Betulinic acid derivatives as anticancer agents: structure activity relationship. Anti Cancer Agents Med Chem 6:271–279
46. Rajendran P, Jaggi M, Singh MK, Mukherjee R, Burman AC (2008) Pharmacological evaluation of C-3 modified betulinic acid derivatives with potent anticancer activity. Investig New Drugs 26:25–34
47. Baratto LC, Porsani MV, Pimentel IC, Netto ABP, Paschke R, Oliviera BH (2013) Preparation of betulinic acid derivatives by chemical and biotransformation methods and determination of cytotoxicity against selected cancer cells. Eur J Med Chem 68:121–131
48. Thibeault D, Gauthier C, Legault J, Bouchard J, Dufour P, Pichette A (2007) Synthesis and structure–activity relationship study of cytotoxic germanicane- and lupane-type 3b-O monodesmosidic saponins starting from botulin. Bioorg Med Chem 15:6144–6157
49. Kvasnica M, Sarek J, Klinotova E, Dzubak P, Hajduch M (2005) Synthesis of phthalates of betulinic acid and betulin with cytotoxic activity. Bioorg Med Chem 13:3447–3454
50. Kommera H, Kaluderovic GN, Dittrich S, Kilbitz J, Drager B, Mueller T, Paschke R (2010) Carbamate derivatives of betulinic acid and betulin with selective cytotoxic activity. Bioorg Med Chem Lett 20:3409–3412
51. Bar FMA, Khanfar MA, Elnagar AY, Liu H, Zaghloul AM, Badria FA, Sylvester PW, Ahmad KF, Raisch KP, El Sayed KA (2009) Rational design and semisynthesis of betulinic acid analogues as potent topoisomerase inhibitors. J Nat Prod 72(9):1643–1650
52. Cui HW, He Y, Wang J, Gao W, Liu T, Qin M, Wang X, Gao C, Wang Y, Liu MY, Yi Z, Qiu WW (2015) Synthesis of heterocycle-modified betulinic acid derivatives as antitumor agents. Eur J Med Chem 95:240–248
53. Bildziukevich U, Rarova L, Saman D, Wimmer Z (2018) Picolyl amides of betulinic acid as antitumor agents causing tumor cell apoptosis. Eur J Med Chem 145:41–50
54. Csuk R, Nitsche C, Sczepek R, Schwarn S, Siewert B (2013a) Synthesis of antitumor-active betulinic acid-derived hydroxypropargylamines by copper-catalyzed Mannich reactions. Arch Pharm Chem Life Sci 346:232–246
55. Csuk R, Sczepek R, Siewert B, Nitsche C (2013b) Cytotoxic betulin-derived hydroxypropargylamines trigger apoptosis. Bioorg Med Chem 21:425–435
56. Drag-Zalesinska M, Drag M, Poreba M, Borska S, Kulbacka J, Saczko J (2017) Anticancer properties of ester derivatives of betulin in human metastatic melanoma cells (Me-45). Cancer Cell Int 17:4
57. Drag-Zalesinska M, Kulbacka J, Saczko J, Wysocka T, Zabel M, Surowiak P, Drag M (2009) Esters of betulin and betulinic acid with amino acids have improved water solubility and are selectively cytotoxic toward cancer cells. Bioorg Med Chem Lett 19:4814–4817
58. Jeong HJ, Chai HB, Park SY, Kim DSHL (1999) Preparation of amino acid conjugates of betulinic acid with activity against human melanoma. Bioorg Med Chem Lett 9:1201–1204

59. Bebenek W, Chrobak E, Wietrzyk J, Kadela M, Chrobak A, Kusz J, Ksiazek M, Jastrzebska M, Boryczka S (2016) Synthesis, structure and cytotoxic activity of acetylenic derivatives of betulonic and betulinic acids. J Mol Struct 1106:210–219
60. Orchel A, Kulczycka CE, Bebenek E, Borkoeska P, Boryczka KJ, Dzierzewicz Z (2014) Influence of betulin and 28-O-propynoylbetulin on proliferation and apoptosis of human melanoma cells (G-361). Postepy Hig Med Dosw (Online) 68:191–197
61. Santos RC, Salvador JAR, Marin S, Cascante M (2009) Novel semisynthetic derivatives of betulin and betulinic acid with cytotoxic activity. Bioorg Med Chem 17:6241–6250
62. Santos RC, Salvador JAR, Marin S, Cascante M, Moreira JN, Dinis TCP (2010) Synthesis and structure–activity relationship study of novel cytotoxic carbamate and N-acyl heterocyclic bearing derivatives of betulin and betulinic acid. Bioorg Med Chem 18:4385–4396
63. Majeed R, Sangwan PL, Chinthakindi PK, Khan I, Dangroo NA, Thota N, Hamid A, Sharma PR, Saxena AK, Koul S (2013) Synthesis of 3-O-propargylated betulinic acid and its 1,2,3-triazoles as potential apoptotic agents. Eur J Med Chem 63:782–792
64. Suresh C, Zhao H, Gumbs A, Chetty CS, Bose HS (2012) New ionic derivatives of betulinic acid as highly potent anti-cancer agents. Bioorg Med Chem Lett 22:1734–1738
65. Nedopekina DA, Gubaidullin RR, Odinokoz VN, Maximchik PV, Zhivotovsky B, Belskii YP, Khazanov VA, Manuylova AV, Gogvadze V, Spivak AY (2017) Mitochondria-targeted betulinic and ursolic acid derivatives: synthesis and anticancer activity. Med Chem Commun 8:1934
66. Dang TTA, Kim Tuyet NT, Pham TC, Thanh Nguyen H, Ba Thi C, Doan Duy T, Dhooghe M, Van Nguyen T (2014) Synthesis and cytotoxic evaluation of novel triterpenoid-AZT conjugates. Bioorg Med Chem Lett 24(22):5190–5194
67. Urban M, Sarek J, Klinot J, Korinkova G, Haduch M (2004) Synthesis of A-Seco derivatives of betulinic acid with cytotoxic activity. J Nat Prod 67:1100–1105
68. You YJ, Kim Y, Nam NH, Ahn BZ (2003) Synthesis and cytotoxic activity of A-ring modified betulinic acid derivatives. Biorg Med Chem Lett 13:3137–3140
69. Arora D, Jaglan SW (2017) Therapeutic applications of resveratrol Nano formulations. Environ Chem Lett. https://doi.org/10.1007/s10311-017-0660-0
70. Dubey RD, Saneja A, Qayum A et al (2016) PLGA nanoparticles augmented the anticancer potential of pentacyclic triterpenediol in vivo in mice. RSC Adv 6:74586–74597
71. Guo B, Xu D, Liu X, Yi J (2017a) Enzymatic synthesis and in vitro evaluation of folate-functionalized liposomes. Drug Des Devel Ther 11:1839
72. Guo YL, Yan SZ, Xu Q et al (2017b) Poly(lactic-coglycolic) acid nanoparticles improve oral bioavailability of hypocrellin A in rat. RSC Adv 7:42073–42082
73. Garinot M, Fiévez V, Pourcelle V et al (2007) PEGylated PLGA-based nanoparticles targeting M cells for oral vaccination. J Control Release 120:195–204
74. Saneja A, Kumar R, Singh A et al (2017) Development and evaluation of long-circulating nanoparticles loaded with betulinic acid for improved anti-tumor efficacy. Int J Pharm 531:153–166
75. Das J, Samadder A, Das S et al (2016) Nanopharmaceutical approach for enhanced anti-cancer activity of betulinic acid in lung-cancer treatment via activation of PARP: interaction with DNA as a target: anti-cancer potential of nano-betulinic acid in lung cancer. J Pharmacopuncture 19:37–44
76. Arora D, Sharma N, Sharma V et al (2016) An update on polysaccharide-based nanomaterials for antimicrobial applications. Appl Microbiol Biotechnol 100:2603–2615
77. Dai L, Si CL (2017) Cellulose-graft-poly (methyl methacrylate) nanoparticles with high biocompatibility for hydrophobic anti-cancer drug delivery. Mater Lett 207:213–216
78. Dai L, Liu R, Hu LQ et al (2017) Self-assembled PEG–carboxymethylcellulose nanoparticles/ α-cyclodextrin hydrogels for injectable and thermosensitive drug delivery. RSC Adv 7:2905–2912
79. Dai L, Yang T, He J et al (2014) Cellulose-graft-poly (l-lactic acid) nanoparticles for efficient delivery of anti-cancer drugs. J Mater Chem B 2:6749–6757

80. Elzoghby AO, Samy WM, Elgindy NA (2012) Albumin based nanoparticles as potential controlled release drug delivery systems. J Control Release 157:168–182
81. Srivari Y, Chatterjee P (2016) Factors influencing the fabrication of albumin bound drug nanoparticles (ABDns): Part I. Albumin-bound betulinic acid nanoparticles (ABBns). J Microencapsul 33:689–701
82. Dai L, Li CX, Liu KF et al (2015) Self-assembled serum albumin–poly (l-lactic acid) nanoparticles: a novel nanoparticle platform for drug delivery in cancer. RSC Adv 5:15612–15620
83. Tietze R, Zaloga J, Unterweger H et al (2015) Magnetic nanoparticle-based drug delivery for cancer therapy. Biochem Biophys Res Commun 468:463–470
84. Wong J, Prout J, Seifalian A (2017) Magnetic nanoparticles: new perspectives in drug delivery. Curr Pharm Des 23:2908–2917
85. Hussein-Al-Ali SH, Arulselvan P, Fakurazi S et al (2014) The in vitro therapeutic activity of betulinic acid nanocomposite on breast cancer cells (MCF-7) and normal fibroblast cell (3T3). J Mater Sci 49:8171–8182
86. Mullauer FB, van Bloois L, Daalhuisen JB, Ten Brink MS, Storm G, Medema JP, Schiffelers RM, Kessler JH (2011) Betulinic acid delivered in liposomes reduces growth of human lung and colon cancers in mice without causing systemic toxicity. Anti-Cancer Drugs 22(3):223–233
87. Liu Y, Gao D, Zhang X, Liu Z, Dai K, Ji B, Wang Q, Luo L (2016) Antitumor drug effect of betulinic acid mediated by polyethylene glycol modified liposomes. Mater Sci Eng C 64:124–132
88. Liu Y, Zhang X, Liu Z, Wang L, Luo L, Wang M, Wang Q, Gao D (2017) Gold nanoshell-based betulinic acid liposomes for synergistic chemo-photothermal therapy. Nanomedicine 13(6):1891–1900
89. Csuk R, Barthel A, Kluge R, Ströhl D (2010) Synthesis, cytotoxicity and liposome preparation of 28-acetylenic betulin derivatives. Bioorg Med Chem 18(20):7252–7259
90. Badruddoza AZ, Gupta A, Myerson AS, Trout BL, Doyle PS (2018) Low energy nano emulsions as templates for the formulation of hydrophobic drugs. Adv Ther 1(1):1–8
91. Harwansh RK, Mukherjee PK, Bahadur S, Biswas R (2015) Enhanced permeability of ferulic acid loaded nanoemulsion based gel through skin against UVA mediated oxidative stress. Life Sci 141:202–211
92. Ciurlea SA, Dehelean CA, Ionescu D, Berko S, Csanyi E, Hadaruga DI, Ganta S, Amiji MM (2010) A comparative study regarding melanoma activity of betulinic acid on topical ointment vs. systemic nanoemulsion delivery systems. J Agroaliment Process Technol 16(4):420–426
93. Dehelean CA, Feflea S, Ganta S, Amiji M (2011) Anti-angiogenic effects of betulinic acid administered in nanoemulsion formulation using chorioallantoic membrane assay. J Biomed Nanotechnol 7(2):317–324
94. Dehelean CA, Feflea S, Gheorgheosu D, Ganta S, Cimpean AM, Muntean D, Amiji M (2013) Anti-angiogenic and anti-cancer evaluation of betulin nanoemulsion in chicken chorioallantoic membrane and skin carcinoma in Balb/c mice. J Biomed Nanotechnol 9(4):577–589
95. Cavazos-Garduño A, Ochoa-Flores AA, Serrano-Niño JC, Beristain CI, García HS (2014) Operating and compositional variables for preparation of betulinic acid nanoemulsions. Rev Mex Ing Quím 13(3):689–703
96. Coricovac D, Danciu C, Ionescu D, Trandafirescu C, Simu G, Soica C, Kemeny L, Ardelean S, Dehelean C (2014) The influence of betulinic acid formulated as nanoemulsion on UVB activity as a tumor promoter in a mouse model of skin carcinoma. Toxicol Lett 229:S50–S51
97. Falamas A, Dehelean CA, Pinzaru SC (2018) Monitoring of betulin nanoemulsion treatment and molecular changes in mouse skin cancer using surface enhanced Raman spectroscopy. Vib Spectrosc 95:44–50
98. Harwansh RK, Mukherjee PK, Biswas S (2017) Nanoemulsion as a novel carrier system for improvement of betulinic acid oral bioavailability and hepatoprotective activity. J Mol Liq 237:361–371

99. Tan JM, Karthivashan G, Arulselvan P et al (2014a) Sustained release and cytotoxicity evaluation of carbon nanotube-mediated drug delivery system for betulinic acid. J Nanomater 2014:1
100. Tan JM, Karthivashan G, Arulselvan P et al (2014b) Characterization and in vitro studies of the anticancer effect of oxidized carbon nanotubes functionalized with betulinic acid. Drug Des Devel Ther 8:2333–2343
101. Tan JM, Karthivashan G, Gani SA et al (2016) Biocompatible polymers coated on carboxylated nanotubes functionalized with betulinic acid for effective drug delivery. J Mater Sci Mater Med 27:26
102. Pradeesh K et al (2013) Naturally self-assembled nanosystems and their templated structures for photonic applications. J Nanopart 2013:13. https://doi.org/10.1155/2013/531871
103. Dash SK, Chattopadhyay S, Dash SS et al (2015a) Self assembled nano fibers of betulinic acid: a selective inducer for ROS/TNF-alpha pathway mediated leukemic cell death. Bioorg Chem 63:85–100
104. Dash SK, Dash SS, Chattopadhyay S et al (2015b) Folate decorated delivery of self-assembled betulinic acid nano fibers: a biocompatible anti-leukemic therapy. RSC Adv 5:24144–24157

Quercetin-Loaded Nanomedicine as Oncotherapy

Sabya Sachi Das, P. R. P. Verma, Sweta Kar, and Sandeep Kumar Singh

1 Introduction

Cancer is a generic term used for an enormous group of diseases which can affect any body part and is the second leading cause of death globally and is liable for an estimated 9.6 million deaths, almost 1 in 6 deaths, as per [1] reports (https://www.who.int/news-room/fact-sheets/detail/cancer). Other terms used irrespective of cancer showing same significant results that of cancer are malignant tumors and neoplasms. Cancer is caused by frequent formation of malignant cells which grow abnormally, attack connecting body parts and extend to other organs, causing metastases which are known to be the major cause for cancer deaths. It is miserable that ~70% of deaths happen in low- and middle-income countries primarily due to cancer [2] which is attributed mainly to high body mass index (BMI), less fruit and vegetable consumption, absence of physical activity and consumption of tobacco and alcohol. Amongst various causes, tobacco consumption is the utmost vital threat aspect leading to 22% of cancer-related deaths. In low- and middle-income countries, some infections (hepatitis and human papilloma virus) are initiated due to cancer, and because of their inaccessible diagnosis and management, they are accountable for cancer cases [3]. As per the WHO [1] reports, the most common cancers include lung (2.09 million cases), breast (2.09 million cases), colorectal (1.80 million cases), prostate (1.28 million cases), skin cancer (non-melanoma) (1.04 million cases) and stomach (1.03 million cases). Furthermore, the most common causes of cancer deaths are cancers of the lung (1.76 million deaths), colon (862,000 deaths), stomach (783,000 deaths), liver (782,000 deaths) and breast (627,000 deaths) (https://www.who.int/news-room/fact-sheets/detail/cancer).

Cancer ascends from the conversion of normal cells into tumor cells in a multi-stage progression that usually evolves from a pre-cancerous lesion to form a

S. S. Das · P. R. P. Verma · S. Kar · S. K. Singh (✉)
Department of Pharmaceutical Sciences and Technology, Birla Institute of Technology, Ranchi, Jharkhand, India

© Springer Nature Singapore Pte Ltd. 2020

M. Rahman et al. (eds.), *Nanomedicine for Bioactives*,
https://doi.org/10.1007/978-981-15-1664-1_5

malignant tumor. Genetic factors of person to person vary, and interaction amongst these factors results in alterations in cells [2]. The three major classes of carcinogens include (1) physical (ultraviolet and ionizing radiation), (2) chemical (asbestos, constituents of tobacco and arsenic mostly present in water sources) and (3) biological (contagions from some significant viruses and infective agents). WHO reports have showed that ageing is one more vital factor for the evolution of cancer that occurs vividly due to loss of affinity for cellular healing mechanisms as a person grows older (https://www.who.int/news-room/fact-sheets/detail/cancer).

1.1 Quercetin: Anticancer Agent

Quercetin (3,3′,4′,5,7-pentahydroxyflavone) is a polyphenolic flavonoid which occurs commonly in various fruits and vegetables. The most important therapeutic activity of quercetin (QT) is anti-oxidative property which makes it useful in cancer management and is also known to have anti-inflammatory, antiviral and anti-allergic activities [5]. QT prominently hinders the proliferation of numerous kinds of cancers (lung, liver, breast, prostate, colon and cervical) and exhibits anticancer activity by different mechanisms involving cellular signaling and the capability to impede enzymes liable for the initiation of carcinogens [5, 6]. Also, QT exhibits anticancer properties centered on its affinity towards cellular receptors and proteins (Fig. 1). Methods for the treatment of cancer include surgery, radiotherapy and use of anticancer drugs (chemotherapy) in addition to other specified techniques. Moreover, chemotherapy has major limitations such as low solubility and bioavailability of chemotherapeutic agents, displeasing specificity which causes decrease of drug concentration at the tumor site(s), instigating numerous adverse effects and off-target toxicity [4].

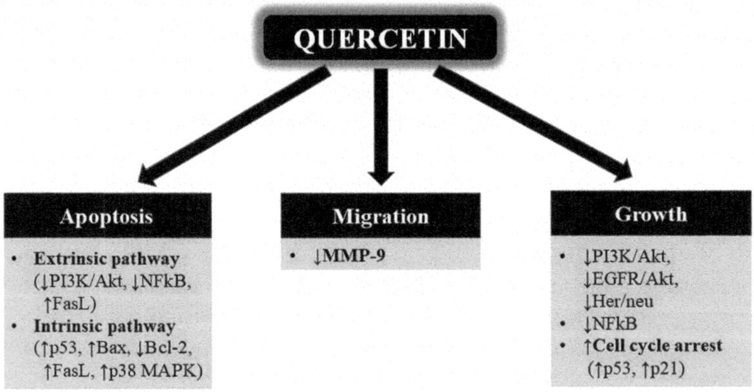

Fig. 1 Cellular processes modulated by QT (↓: downregulate and ↑: upregulate) (Reprinted from Nam, J.; Sharma, A.R.; Nguyen, L.T.; Chakraborty, C.; Sharma, G.; Lee, S. Application of Bioactive Quercetin in Oncotherapy: From Nutrition to Nanomedicine. *Molecules* [4], *21*, 108.© 2016 MDPI)

2 Nanotechnology-Based Approaches of Quercetin for Cancer Treatment

To overcome the issues related with chemotherapy, nanotechnology-based drug delivery systems are introduced which has displayed favourable effects for the management of cancer. QT-based nanoformulations (QTNFs) have been developed to overcome the bioavailability concerns with QT (Fig. 2). Targeted drug delivery by using QT nanoformulations (NFs) has improved drug delivery at tumor sites with improved therapeutic actions. Moreover, QT-loaded NFs such as liposomes [7, 8], polymeric nanoparticles [9–11], micelles [12], solid lipid nanoparticles [13], nanoemulsions [14], nanocapsules [15] and metal nanoparticles [16–18] have showed improvement in solubility, bioavailability, biodistribution, specificity and pharmacokinetics of anticancer agents/drugs targeted at the tumor site. The QTNFs have shown favourable results in its uptake by the epithelial system and also improved its targeted delivery at site [19, 20]. In this review, we have focused on the potential and significant activities of QTNFs for the management of cancer.

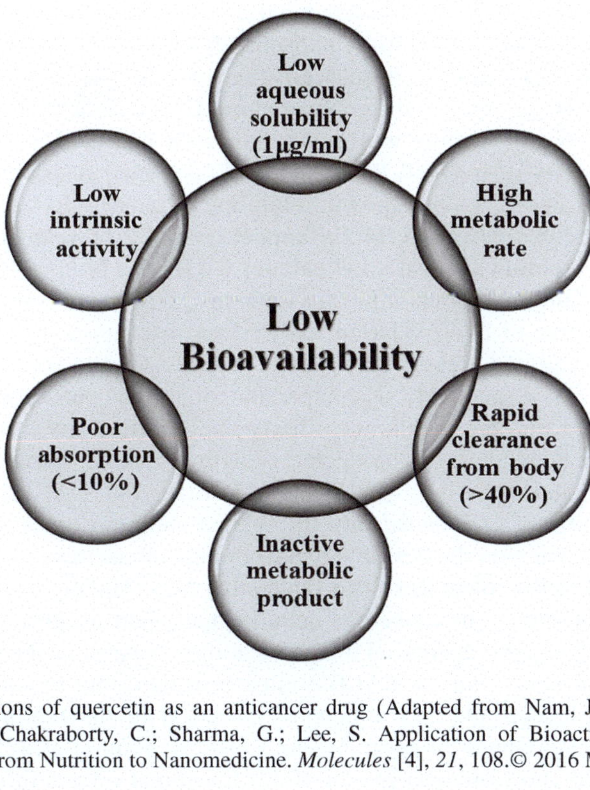

Fig. 2 Limitations of quercetin as an anticancer drug (Adapted from Nam, J.; Sharma, A.R.; Nguyen, L.T.; Chakraborty, C.; Sharma, G.; Lee, S. Application of Bioactive Quercetin in Oncotherapy: From Nutrition to Nanomedicine. *Molecules* [4], *21*, 108.© 2016 MDPI)

2.1 Liposome-Based NFs

Podhajcer et al. [21] developed QT-loaded liposomes (QTLs) which inhibited [3H] thymidine absorption and lactate release in Ehrlich ascites carcinoma (Ehrlich cells). The serum protein (albumin) exhibited no significant changes in the level of inhibition for QTLs having concentrations up to 40 mg/ml, and even the inhibitory effects of free QT on metabolic processes were eliminated by these serum proteins. The vesicle size of QTLs ranged from 50 to 100 nm. Moreover, the observations from negatively stained electron micrographs revealed the attachment of Ehrlich ascites tumor cells with QTLs. Also, the results showed that the protein eliminated the upsurge of cyclic adenosine 3′,5′-monophosphate level in the tumor cells produced by both free QT and QTLs. Spector et al. [22] solubilized the plasma membranes of Ehrlich ascites tumor cells, and the Na^+K^+-ATPase was purified and extracted. An ATP-dependent Na^+ flux with low efficiency was observed in the liposomes, reconstituted from these enzymes. It was observed that, with the addition of QT (appropriate amount), the Na^+ flux-ATP hydrolysis ratio was increased from 0.4 to 1.4. Goniotaki et al. [23] prepared liposomes entailing of egg phosphatidylcholine by a thin-film hydration method. These liposomes were used to explore the % encapsulation efficiency (% EE) of four flavonoids [QT, rutin (RT), isoscutellarein (IC) and isoscutellarein diglycoside (ICD)]. Moreover, results indicated that the % EE and flavonoid interaction with model membranes, dipalmitoylphosphatidylcholine (DPPC), was dependent on the molecular structure of respective flavonoids. The presence of a sugar moiety or hydroxyl group(s) at different locations on the molecular structure may play a crucial role in loading and interaction with DPPC. The results of the in vitro cytotoxicity of free QT, RT, IC and ICD and their respective liposomes showed only QT and QTLs had growth-inhibiting (GI_{50}) activity against the cancer cell lines SF268, MCF7 and H460. Tammela et al. [24] investigated the interactions of QT/luteolin (LT) and alkyl gallates (AG) with biological membranes in Caco-2 cells and membrane affinity assays in QT/LTLs. QT and LT seem to have greater degree of hydroxylation, showed no absorption across cell monolayer and also exhibited highest affinity with partition coefficients of 7.1 and 7.5 μM for LT and QT, respectively, as compared to other flavones. From the results, it was concluded that the membrane affinity and permeability through Caco-2 cells are highly dependent on the degree of hydroxylation, molecular arrangements and side chain length of QT/LT and AG. Yuan et al. [25, 26] investigated the biodistribution of QT loaded into PEGylated nanoliposomes (NLs) and its therapeutic efficiency on the development of malignant ascites of hepatocellular carcinoma in BALB/c mice. Results revealed that NLs (100 mg/kg) expressively improved the apoptosis in cancerous cells in malignant ascites, repressed the upsurge of body weight, decreased the permeability of peritoneal capillaries and sustained the residual time of BALB/c mice bearing tumor as compared with control group. Hence, QT-NLs efficiently amassed in the tumor tissues and inhibited the development of carcinomatous ascites. Yuan et al. [25, 26] developed PEGylated (polyethylene glycol 4000) liposomal quercetin (pL-QT) and investigated in vivo distribution of pL-QT and also antitumor efficiency (in vitro-in vivo). The antitumor activity of pL-QT was assessed in the immunocompetent C57BL/6N mice bearing LL/2 Lewis lung cancer and in BALB/c mice bearing CT26 colon adenocarcinoma and H22 hepatoma. In vitro results revealed that pL-QT induced apoptosis and in vivo results showed that pL-QT expressively reduced the tumor growth (dose-dependent manner). The liposomal formulation could efficiently accumulate in tumor tissues, extend the circulation time of QT in vivo, efficiently constrain numerous types of tumor growth and lengthen the residual time of tumor-bearing mice. Mandal et al. [27] formulated galactosylated liposomal QT (GLQT) and investigated its efficacy against hepato-carcinoma in rat model. GLQT suspension was given intravenously (IV) to rats (8.98 μM/kg) once in a week for 16 weeks. GLQT injections prevented diethylnitrosamine (DEN)-induced hepatocarcinoma and oxidative damage in rat liver. Ghosh et al. [28] evaluated the biological efficiency of QTLs and QT-polylactide nanocapsules (QT-p-NCs) against inhibiting arsenic (*As*)-mediated hepatic oxidative stress, fibrosis-related upregulation of gene expression and plasma TGF β (transforming growth factor β) in rats. Fibrosis caused by *As* caused upregulation of hepatic gene expression, value of plasma TGF β and cytochrome *c* (Cyto-*c*) release in cytoplasm. The results showed QT-p-NCs exhibited better results as compared to QTLs in combating *As*-mediated hepatic fibrosis in rat model, gene expression and plasma TGF β upregulation. Wong and Chiu [29] developed a liposomal formulation to solubilize QT and co-encapsulated and coordinated the release of QT and vincristine (VT) in their synergistic ratios for enhanced anticancer activity. In vitro 3-(4,5-dimethylthiahiazol-2-yl)-2,5-diphenyl tetrazolium (MTT) assays of the co-encapsulated liposomal formulation (QT:VT; 1:2) exhibited synergistic anticancer effect, with a combination index of 0.113 and a dose-reduction index value of 115 at ED_{50} for VT, and prolonged the exposure times for both drugs. Ghosh et al. [30, 31] evaluated the therapeutic efficacy of QTLs in eliminating *As*-induced oxidative injury in brain cells and hepatocytes in rats. Results showed that inorganic *As* most distinctly accumulated in neuronal (6.21 ± 0.090 μg/g tissue) and hepatic (9.32 ± 0.100 μg/g tissue) cells of rats, fed with 12 mg/kg body weight of arsenite. Wong and Chiu [32] formulated a novel system of QT-VT NLs for the management of breast cancer. Results showed that the co-loaded VT and QT (1:2) in NLs revealed enhanced antitumor efficiency and reduced toxicity in estrogen receptor (ER), progesterone receptor (PR) and trastuzumab-insensitive xenograft models synergistically. It was concluded that this drug delivery system could further develop into a conceivable clinical management option for HER-2 (human epidermal growth-factor receptor-2) overexpressing breast cancer. Yang et al. [33] explored that modulation of heat shock protein (HSP) has significant effects on tumor agglomeration due to combining radiofrequency (RF) ablation with adjuvant QTLs and/or doxorubicin (DOX) in a rat tumor model. Results showed that RF-QT combination expressively amplified the coagulation mass compared with either RF or liposomal QT alone. Triple therapy (QT-RF-DOX) exhibited greater coagulation diameter at 24 h than QT-RF or RF-DOX. Finally, it was concluded that suppression of HSP production using adjuvant liposomal QT can upsurge apoptosis and increase RF ablation-induced tumor destruction. Sinha et al. [34] modified the QT molecule which gave two substituents QT1 (4′-chloro-3-hydroxy-QT) and QT2 (4′-methoxy-3-hydroxy-QT). The reformed substituents QT1 and QT2 of QT interacted with head part of DPPC membrane model with altered levels and hence showed altered membrane permeability. The results indicated that the different substituents in the QT ring may have different effect on membrane permeability. It was observed that the energy barrier reduces and permeation enhances with the decrease in interface of molecule with polar head group. Additionally, the results of anticancer activity indicated that modification of QT maintained the anti-proliferative activity to a larger level while enhancing the permeability and thereby bioavailability. Wang et al. [35, 36] examined the cytotoxic effects of QTL and involvement of these effects on type III programmed cellular death (PCD) of C6 glioma cells (C6GCs). Results revealed that QTL induced the type III (necrotic) PCD in C6GC in time-dependent and dose-dependent manner. QTL (high concentration) induced cell necrosis, while free liposomes induced neither apoptosis nor significant necrosis in C6GC. Hence, non-apoptotic cell death induced by QTL (high concentration) for monitoring caspase-independent type III PCD may offer the base for novel healing ways to overcome evasion of apoptosis by cancerous cells. Yang et al. [37] investigated the effects of IV QTL individually and also with DOX-loaded liposomes (DOXL) on the growth of tumor in rat model and significance of end-point existence in the presence of RF. Results revealed that the variances in endpoint survival and tumor doubling time amongst the assemblies were extremely significant. Also, IV QTL in combination with RF ablation decreased the tumor growth rates and enhanced animal endpoint existence. Addition of DOX to QTL increased the endpoint survival, and hence targeting numerous post-ablation progressions with multidrug delivery systems can upsurge whole ablation efficiency. He et al. [38] investigated whether QTL could improve sensitivity to thermotherapy and thermo-chemotherapy. The effects of QTL were examined in vitro and in mouse colon cancer models bearing peritoneal carcinomatosis and subcutaneous tumor. The results revealed that QTL repressed the upregulation of heat shock proteins (especially Hsp70) and improved the apoptosis induced by hyperthermia and thermo-chemotherapy, and hence QTL might be used as a potential agent for treating cancer. Hu et al. [39] developed deformable QTLs to increase its distribution into human skin and to enhance its anti-UVB (ultraviolet B) effect. Results showed that the cell sustainability in UVB-induced HaCaT cells increased when treated with QTLs. Also, the ROS and malondialdehyde (MDA) levels were condensed. The permeation rate of QTLs was more (3.8 fold) as compared to QT suspensions. Furthermore, the edema and inflammation were improved by deformable QTL, and hence the potential of deformable QTL might be used as anti-UVB and anti-skin cancer agents. Liu et al. [40] tested the protective effect of QTL against radiation-induced acute pneumonitis (RIPI) in a murine C57BL/6J mice model. From the results, it was observed that QTL showed protective effect against RIPI and lung fibrosis, possibly by reducing oxidative damage. Moreover, QTL inhibited pulmonary oxidative stress, relieving RIPI and lung fibrosis. Wang et al. [9–11] investigated the molecular mechanisms of QTL-induced C6GC death and inspected whether QTL-induced PCD tangled (B-cell lymphoma 2) Bcl-2 family and mitochondrial pathway through STAT3 (signal transducers and activators of transcription 3) pathway. Results indicated that QTLs modulated the signaling cross point in mitochondrial pathway and the JAK2 (Janus kinase 2)/STAT3 signaling pathway in C6GC death; hence it was concluded that QTL combined with these signaling pathways may be an efficient way of treating chemical-resistant glioma. Zheng et al. [41] showed that QT or sodium butyrate (NaB) inhibited the human esophageal 9706 cancerous cellular growths with dose-dependent manner, and QT doped with NaB (QT-NaB) further inhibited Eca9706 cell proliferation in MTT assay. Zheng et al. [42] explored the expression of CD133/CD44 (stem cell biomarker) in esophageal carcinomas and Eca109/9706 cells. QTL combined with biomarkers downregulated NF-κBp65, histone deacetylase 1 (HDAC1), cyclin D1, upregulated caspase-3, reduced Histone deacetylase 1 (HDAC1) with nuclear translocation and stimulated E-cadherin expression. Hence, the QTLs doped with biomarkers induced better apoptotic effects than individual QTL, signifying an ideal anti-CSC management approach. Dabbagh-Bazarbachi et al. [43] established the potential of QT and epigallocatechin gallate (EG) as combined liposomic formulation rapidly increased labile zinc (Zn) in mouse hepatocarcinoma Hepa 1–6 cells. Results showed that only the co-encapsulated liposomal formulation activated a quick upsurge of FluoZin-3 (Zn-specific detector) fluorescence inside the nanoliposomes, thus signifying the ionophore action of QT, EG, and clioquinol on lipid membrane system. Moussa et al. [12] determined the effects of various drug-doped nanocarriers, liposomes and micelles (Mic), on the distribution and management efficiency for RF ablation (RFA) in the presence of nanocarriers. Results showed that beads with size of 20 nm had greater deposition and deeper tissue penetration in tumor as well as in the liver. Mic-DOX and Mic-QT suppressed periablational HIF-1a or HSP70 rim thickness more than liposomal preparations. Finally, it was concluded that with RFA, Mic (small sized) have greater permeation and more effective local molecular modulation. However, with RFA, DOXLs and QTLs resulted in better intratumoral accretion of drug for prolonged time and decreased tumor growth. Caddeo et al. [7] investigated anti-ROS activity of co-encapsulated QT and resveratrol (RES) liposomes. The results showed that co-encapsulated QT and RES effectively helped the entry of QT and RES in fibroblasts, and their amalgamation enhanced the anti-ROS activity and thus contributed to hinder the inflammatory effect triggered by 12-O-tetradecanoylphorbol 13-acetate (TPA) on the skin, which resulted in tissue repair and wound healing and may be used in the management of skin cancer. Caddeo et al. [8] prepared QT and RT liposomes (QT-RTLs) co-encapsulated in pH-sensitive succinyl-chitosan polymers for oral delivery. Results showed that the internal characteristics of the liposomal bilayers were not disturbed by alteration of liposomes with succinyl-chitosan, and hence this combination could be used for the management of intestinal circumstances related with inflammation and oxidative stress. Hu et al. [44] prepared and characterized a new type of 1,2-distearoyl-sn-glycero-3-phosphoethanolamine-N-[methoxy (polyethylene glycol)-2000] (DSPE-PEG2000) polymeric nanoliposomes (NLs) for the brain-targeted delivery of QT and temozolomide (TMZ). Results showed that the QT-TMZ nanoliposomes (QT-TMZ NLs) were evidenced to be more efficient at killing C6GC than free TMZ or QT. It was concluded that QT-TMZ NLs possessed characteristics, such as a greater drug encapsulation ratio, reduced in vitro release rate, slow drug clearance and extended circulation time in vivo. Ravichandiran et al. [45] prepared curcumin (CUR) NLs with/without QT by lipid hydration technique and the toxicity of the NLs were evaluated in-vitro and in-vivo efficiency were verified against Dalton's ascites lymphoma in mice model. Results revealed that QT-loaded NLs were more potent in enhancing the lifespan and mass of lymphoma-imposed mice associated to those deprived of QT. Likewise, the existence of QT in CUR NLs improved cytotoxicity of the NLs towards HT-29 cells and HCT-15 cells. Rodriguez et al. [46] prepared QTLs with phospholipids isolated from rice bran and investigated its effects on the distribution, control release, conformational stability protection and bioactivity. The results showed that the QTLs appeared to be as spherical NPs (157.33 ± 23.78 nm) with % entrapment efficiency (%EE; 84.92 ± 0.78) and also improved the radical-scavenging and anti-angiogenic actions of QT. Dos Santos et al. [47] studied the anti-/pro-oxidative activities of *Rapanea ferruginea* (*R. ferruginea*) leaf extracts (entailing QT flavonoids) by in vitro DPPH and TBARS studies. From MTT assay, it was observed that free form of extract decreased the C6GCs in vitro viability by 40%, whereas 70.8% cell viability was reduced with 500 μg/mL of liposomal extract. Moreover, phenolic content, membrane interface and variation showed significant role in the oxidative and antitumor actions of the *R. ferruginea* leaf extracts. Zhou et al. [48] prepared RGD (short peptide containing arginine-glycine-aspartic acid)-modified NLs containing QT-DSPE-PEG2000 forming DSPE-PEG2000-RGD/QT NLs for treating lung cancer. The encapsulation ratio and drug loading of NLs were 89.2% ± 7.4% and 9.2% ± 1.3%, respectively; mean diameter was 93.4 ± 7.2 nm. In vitro studies showed a relatively slow QT release profile from NLPs. Moreover, in vivo imaging results showed better tumor targeting capacity of RGD-modified LPs. Hence, DSPE-PEG2000-RGD/QT NLs exhibited a substantial antitumor activity in mice with A549 tumors. Kruszewski et al. [49] studied the effect of QT on the surface charge over PC-liposomes and human glioblastoma LN-229 and LN-18 cells by using micro-electrophoretic mobility calculations. The results indicated that QTLs altered the surface charge for LN-18 and LN-229 glioblastoma cells in alkaline pH medium only, while there were no major modifications in QT-induced vs. control cells in acidic pH medium. Concurrently, condensed feasibility approves that QT moves into the cells and consequences hindered propagation of LN-18 and LN-229 glioblastoma cells. Zuo et al. [50] investigated the antitumoral effects of 7-O-geranylquercetin (GQT) combined with survivin siRNA (siSuvi) or IL-10 siRNA (siIL-10) to treat breast cancer. It was observed that the liposomal formulation delivered siRNA to tumor efficiently. GQT- and siSuvi-loaded liposomes stimulated the anti-proliferation and pro-apoptosis activities of

GQT or siSuvi to MCF-7 cells and decreased the amount of survivin and increased the level of caspase-7 in cells. Hence, synergistic antitumor effect of GQT and siRNAs could be potentially used against breast cancer. Riaz et al. [51] prepared T7 surface-functionalized QT liposomes (T7-QTL) for lung cancer therapy by targeting transferrin receptors, which are overexpressed and restricted to tumor cells. In vitro studies showed that 2% of T7-QTL demonstrated improved cytotoxicity (~3-fold), enhanced apoptosis induction and S-phase cell-cycle seizure. T7-QTL expressively improved the anticancer activity of QT and lifespan of mice in orthotopic lung tumor-bearing mice via pulmonary administration. Soloviev and Kizub [52] revealed that the vascular dysfunction by infra-red rays (IR) could efficiently regenerate by QT-filled PC liposomes (QT-PCLs) and mesenchymal stem cell injections. Further, including RNA interference technique directed to various protein kinase C (PKC) isoforms. Yu et al. [53] prepared co-loaded NLs of QT and DOX (QT/DOX-NLs) which effectively suppressed the activity of P-gp (transmembrane protein) to inhibit multidrug resistance (MDR). Results showed QT/DOX NLs increased the blood concentration of DOX, extended the half-life of DOX in plasma and enhanced the cell toxicity to DOX-resistant cell strains (HL-6/ADR and MCF-7/ADR), and the inhibitory effects on the resistance of HL-6/ADR and MCF-7/ADR were amplified to 4.81-fold and 3.21-fold, respectively. Finally, in vivo studies showed that the QT/DOX-NLs treated tumors, reduced the size and growth and also increased the inhibition rate against tumor growth with maximum effects.

2.2 Cubosomes, Phytosomes and Niosome-Based NFs

Murgia et al. [54] prepared monoolein-based cubosomes doped with two fluorescent probes, dansyl and fluorescein, appropriately reformed with a hydrocarbon chain to upsurge their encapsulation efficiency inside the monoolein palisade. Later on, QT was loaded to the same nanocarriers. These nanocarriers showed significant anticancer activity against mouse 3T3 fibroblast cells; hence these novel fluorescent nanocarriers could be used for theranostic purposes. Minaei et al. [55] explored the activity of nano-QT (phytosome) in DOX-treated apoptosis. MTT assay showed the anti-proliferative effect of QT-based phytosomic formulations. Finally, results revealed that co-treatment with nanoQT-DOX phytosomes enhanced the percent apoptosis significantly against MCF-7 breast cancer cells. Karthivashan et al. [56] quantified QT, kaempferol (KP) and apigenin (AG) from the ethyl acetate fraction of *Moringa oleifera* leaves extract and assimilated into a single flavonosome by using sequential loading technique. The in vitro 1,1-diphenyl-2-picrylhydrazyl kinetics of QT/KP/AG ultimately showed the released kinetic activities of these flavonoids from the flavonosomes. The QT/KP/AG-loaded flavonosome showed no toxicity against human hepatoma cell line as shown by 3-(4,5-dimethylthiazol-2-yl)-2,5-diphenyltetrazolium bromide. Babazadeh et al. [57] reviewed and summarized the latest surveys concerning the promising application of phytosome complexes for chemotherapy, their formulation methods and mechanism of passage through phytosome. Tavano et al. [58] co-encapsulated niosomal delivery system of DOX-CUR and DOX-QT for the management of human breast cancer. Results showed that this formulation resulted in extended blood circulation and instinctively accumulated at the tumor site. Moreover, the in vitro studies showed enhanced anticancer activity of niosomes loaded with a DOX-CUR/QT as useful carriers for treatment of breast tumor. Hemati et al. [59] optimized the PEGylated niosomes comprising of DOX/QT (PEG-DOX/QT NiS) and small interfering/silencing RNA (siRNA) to improve the therapeutic activity against gastric, prostate and breast cancer and human foreskin fibroblast cells. Moreover, PEG-DOX/QT NiS showed greater toxicity against cancer cells when related with blank formulations. Moreover, it was observed that PEG-DOX/QT NiS showed enhanced anticancer activities against tumor cell death.

2.3 Micelle-Based NFs

Tan et al. [60] prepared QT-loaded nanomicelles (QTNM) embedded with PEG-derivatized phosphatidylethanolamine (PE). The %EE and solubilization (up to 110-fold) of QT were enhanced, and also the QTNM was nontoxic to the Caco-2 cells. QTNM showed better anticancer activity against A549 cancer cell line and murine xenograft model as compared to free QT. Wang et al. [35, 36] prepared QT-encapsulated monomethoxy-PEG-poly(ε-caprolactone) (mPEG-PCL) micelles, forming QT/mPEG-PCL micelles which further enhanced the in vitro release of QT for prolonged period of time. Moreover, in vivo studies showed that no toxicity was induced by QT/mPEG-PCL micelles to the bladder. Also, QT/mPEG-PCL micelles effectively decreased the inflammatory cell permeation and edema of the bladder in an *Escherichia coli*-induced acute cystitis model. Gao et al. [61] encapsulated QT into mPEG-PCL (QT/mPEG-PCL) for the treatment of ovarian cancer. In vitro studies showed that the QT/mPEG-PCL repressed A2780S ovarian cancer cells growth with a dose-dependent manner. The in vivo studies showed that the administration of QT/mPEG-PCL micelles (IV) expressively repressed the development of developed xenograft A2780S ovarian tumors by triggering apoptosis in cancerous cells and hindering angiogenesis. Xu et al. [62] improved the solubility of QT in water and/or enhanced the bioavailability by encapsulating QT into polymeric micellar (QTPM) formulations. Results showed improved effects of QTPM on inducing cell apoptosis, hindering tumor angiogenesis and restraining cell proliferation. Further, it was concluded that QTPM showed improved antitumor activity, which could be used as a promising anticancer agent against colon cancer. Wang et al. [63] encapsulated QT into freeze-dried NMs (QT-FD-NMs) which improved the bioavailability and solubility of QT and further could be used for glioma chemotherapy. Results revealed that QT-FD-NMs enhanced the intracellular upregulation into Caco-2 cells, effective permeation through blood brain barrier (BBB) and cytotoxicity on C6GCs. The in vivo studies showed that QT-FD-NMs amassed in tumor-bearing brain tissues and showed significant antitumor effects. Zhao et al. [64] evaluated the therapeutic efficacy of QTNMs surrounded by DSPE-PEG2000. The aqueous solubility of QT was increased (up to 450-fold). The in vitro studies indicated that the IC_{50} value of QTNMS was much lower than the free QT. Also, QTNMs were significantly more efficient than the free QT in proliferation suppression and apoptosis initiation of human androgen-free PC-3 cells; hence QTNMs exhibited greater anticancer activity and could be used for the management of prostate cancer. Patidar et al. [65] examined the effect of glucose over micellization. The solubilization of CUR and QT (anticancer drugs) showed enhanced control release kinetics and cytotoxicity. Finally, the variation in micellar behavior due to the presence of glucose opens fascinating probability of using Tetronic® 1307 (T1307) micelles as chemotherapeutic agents. Rich et al. [66] investigated the upregulation of QT into CaCo-2/TC7 cells in the existence and absence of blended micelles. Moreover, the NMs repressed the transport of QT into the CaCo-2/TC7 cells. It was observed that the segregation of QT into simple bile salt micelles was less and for mixed micelles was repressed by enhancing the bile salt concentration, which showed the possibilities of low bioavailability of QT. Sandhu et al. [67] developed surface-modified QT-salicylic acid (SA) NMs by the self-assembly method followed by quality by design (QbD) approach for effective management of skin carcinoma. The optimized formulation exhibited improved %permeation and enhanced in vitro cellular uptake, and annexin V apoptosis assay showed 60.86% cellular death. Also, 7,12-dimethylbenz(a) anthracene and histopathology studies showed the enhanced antineoplastic action of QT. Ahmad et al. [68] determined the significant role of hepatic RFA heating parameters combined with QTNMs and their initiation of HSPs in reducing distant tumor growth. It was observed that lower RF heating caused bigger distant tumors, while greater RF heating led to reduced distant tumor growth. Moreover, for lower dose, hepatic RFA showed more periablational HSP70 as compared to Anti-HSP70 QTNMs.

2.4 Nanocapsule-Based NFs

Ghosh et al. [69] evaluated the therapeutic efficacy of QT nanocapsules (QTNCs) against *As*-induced ROS-mediated oxidative impairment in brain cells and hepatocytes of a rat model. QTNCs prohibited the *As*-induced reduction in antioxidant levels in brain and hepatic sites. *As* induced a considerable reduction in brain and hepatic cell membrane microviscosities, and QTNC management ensued protection against such reductions. Ghosh et al. [28] evaluated the therapeutic efficiency of QTNCs against *As*-induced hepatic oxidative stress, fibrosis-related uptake of its gene expression and plasma TGF-β in rat model. It was observed that a single dose of 0.5 ml of QTNCs suspension, orally administered, protected the liver from *As*-induced decline of antioxidant levels and oxidative stress-related gene expression of the liver; hence QTNCs may be used for the treatment of hepatic fibrosis and cancer. Ghosh et al. [30, 31] synthesized QTNCs and dimercaptosuccinic acid nanocapsules (DMSA-NCs) either alone or co-encapsulated in polylactide-co-glycolide (PLGA) [QT/DMSA-PLGA NCs] to show their pharmacological effects in a rat model with chronic *As* toxicity. QT/DMSA-PLGA NCs showed significant synergistic effects, decreased brain and liver *As* levels. Also, QT/DMSA-PLGA NCs effectively stabilized modifications in mitochondrial activity, the development of ROS and liver damage. Ghosh et al. [70] evaluated the therapeutic efficiency of QTNCs against diethylnitrosamine (DEN)-induced hepatocarcinogenesis in rat models. QTNCs significantly reduced the antioxidant levels in hepatic cells, upregulated glutathione-S-transferase activity, secured the mitochondrial membrane of the liver from carcinoma cells and also prohibited the DEN-induced Cyto-*c* expression in the liver. Ding et al. [71] prepared folate-reformed QTNCs (QT-FNCs) by using phase inversion process and optimized it by using simplex lattice design method. The pharmacokinetic studies in rat model revealed that the QTNCs prolonged the release of QT. Moreover, in vitro anti-proliferative action and cellular uptake of QT-FNCs were assayed on HeLa and MCF-7/MDR cells, and the effects exhibited a robust cell-killing activity of QT-FNCs as compared with free QT. Finally, the in vivo antitumoral studies showed that both QTNCs and QT-FNCs inhibited tumor growth of H22 tumor-induced mice. El-Gogary et al. [72] formulated and characterized chemically modified polymeric QTNCs for passive and active targeting against tumors. Later on, folic acid (FA) was doped with PLGA to enable active targeting against cancer cells. MTT assay (against HeLa cells) evidenced the selective uptake and cytotoxicity of the FA-targeted NCs against the folate-induced cancerous cells in a folate-dependent approach. In conclusion, the in vivo studies showed that the established NC system provided a base for targeted delivery of QT (anticancer drug). Mandal et al. [73] investigated the regulatory activity and the efficiency of QTNCs on the inhibition of inflammatory mediators (MMP-13) in DEN-mediated hepatocarcinogenesis in Swiss albino rats. Results revealed that QTNCs limited all modifications, like generation of ROS, upregulation of TNF-α and IL-6, initiation of MMP-13, major oxidative damage and hyperplastic nodules with preneoplastic lesions, in DEN-induced hepatocarcinogenesis. Hatahet et al. [74] formulated lipid-based QTNCs (L-QTNCs) which improved the aqueous solubility of QT by more than 5000-fold. DPPH assay indicated that L-QTNCs did not modify the antioxidant activity of QT. Finally, it was observed that L-QTNCs showed protective actions of Tamm-Horsfall protein (THP-1, used to test leukemia cell lines) cells from oxidative stress by exogenous H_2O_2 (hydrogen peroxide) and hence could be targeted for the treatment of oxidative skin damages and skin carcinogenesis. Hatahet et al. [75] prepared and investigated three significant QT nanoformulations (QTLs, QTNCs and QT-smartCrystals®) to improve QT delivery through skin. No toxicity was witnessed in HaCaT cells with QT. The safety of QT was further confirmed on THP-1 cells with effective free radical scavenging ability. It was observed that QT-smartCrystals® enabled the apparent deposition of QT, and QTNCs allowed QT delivery to feasible epidermis and may be used for skin treatments. Omwenga et al. [15] examined the inhibitory effect of biofilm preparation, quorum sensing (QS) and the cytotoxicity effect of QT and baicalein (BC) on mammalian cells, in free state and conjugated with chitosan-doped NCs. QT and BC, in free form, showed minor inhibitory action on the QS response and biofilm establishment. Also, free QT and BC exhibited cytotoxicity against

MDCK-C7 cells at greater doses, though their nanoencapsulation showed cytoprotective effect.

2.5 Nanoemulsion-Based NFs

Schwingel et al. [76] compared the effects and activity of a group of antioxidant moieties on instantaneous progress of oxaliplatin (OXP)-induced hepato-/neurotoxicity in BALB/c mice. It was observed that animal groups, when treated with RES, RT, QT and QT nanoemulsion (QTNE), inhibited mechanical allodynia as compared to the OXP-treated group. These flavonoids were capable to lessen the immunopositivity for the apoptosis marker caspase-3, whereas only QT and QTNE treatments reduced neutrophil migration assessed by myeloperoxidase (MPO) activity, which could be useful for the treatment of OXP-induced hepato-/neurotoxicity. Pangeni et al. [77] prepared co-drug delivery, water-in-oil-in-water (W/O/W) nanoemulsion of pemetrexed (PT) and QT, and later on an ion-pairing complex of PT with N^{α}-deoxycholyl-L-lysyl-methylester (DCK) was designed and merged into QTNE to form PT/DCK-QT-NE. The results showed synergistic inhibitory effects of PT/DCK-QT-NE on A549 cell proliferation and migration. Oral bioavailability of PT and QT was enhanced by 4.51-fold and 23.9-fold, respectively. Finally, the oral administration of PT/DCK-QT-NE inhibited the tumor growth in A549 cell-bearing mice. Arbain et al. [78] prepared palm oil ester (POE)-based QTNE (POE-QTNE) and optimized it by using D-optimal mixture design (DOMD) for pulmonary delivery. In vitro studies showed that aerosol delivery of the optimized NE had an aerosol output, aerosol rate and median mass aerodynamic diameter of 99.31%, 0.19 g/min and 4.25 μm, respectively. Results of the physical characteristics and efficacy for aerosol delivery of POE-QTNE could be used for lung cancer treatment. Arbain et al. [79] prepared QTNE (oil-in-water; O/W) in the presence of POE/ricinoleic acid (oil phase) and optimized it by using DOMD for the delivery of QT to the lungs. The oral bioavailability was enhanced, and the release of QT was prolonged from the NE formulation when delivered from pulmonary route. The cytotoxicity study confirmed that the optimized QTNE could possibly induce cytotoxicity towards A549 lung cancer cells deprived of distressing the normal cells. Ojeda-Serna et al. [14] studied the physical characteristics of organogel-based NEs loaded with three different flavones (betulin (BL), CUR and QT), vegetable oil (oil phase) and its significant influence on their bioaccessibility. In vitro digestion (oral, gastric and intestinal phase), lipolysis assays, bioaccessibility and permeability trials were performed on Caco-2 cells. The results showed that emulsified organogels improved the bioaccessibility of QT, whereas no enhancement was observed for CUR and BL. Also, the permeability of BL was improved, while the permeability of CUR was not enhanced.

2.6 Solid Lipid Nanoparticle-Based NFs

Bose et al. [80] established a solvent-free solid lipid nanoparticle (SLN) and assessed for topical delivery of QT. In vitro release studies exhibited biphasic release of QT from the SLN system, showing early burst release trailed by prolonged release (up to 24 h). In vitro permeation studies showed localization of QT in the epidermal layers of skin, whereas it was not observed clearly in control sample. Thus, this accumulation tendency of QT in the skin is highly desired since the efficiency of QT in delaying UV radiation facilitated cell damage, and ultimately necrosis primarily arises in the epidermal cells/layers. Varshosaz et al. [81] formulated quercetin-loaded SLNs conjugated with three different sterols (cholesterol, stigmasterol and stigmastanol) and was optimized by D-optimal design for enhanced cellular penetration of QT against targeted hepatocellular carcinoma cells. MTT assay was performed to determine cytotoxicity of QT-SLNs on HepG-2 cells. The IC_{50} values of free QT and phytosterol SLNs were almost six and two times more than QT in cholesterol SLNs, respectively, and triggered more accumulation of QT in HepG2 cells. Chen et al. [13] prepared novel SLNs with redox-responsive properties based on disulfide (SS) bond, and then the QT-grafted glyceryl caprylate caprate (Gcc) was conjugated to form QT-SS-Gcc SLNs. It was observed that QT-SS-Gcc SLNs showed better entrapment of paclitaxel (PTX) due to π-π stacking amongst the aromatic rings of QT and PTX. Additionally, the in vitro studies revealed that QT-SS-Gcc SLNs specifically responded to high levels of reducing moieties by SS bond breakdown, consequently attained rapid and effective drug release and only disassociated frequently in tumor cells as compared to normal cells. Finally, it was concluded that such formulations showed redox-responsive drug release and drug efflux suppression in tumor cells through modified and improved SLNs.

2.7 Polymeric Nanoparticle-Based NFs

Khoee and Rahmatolahzadeh [82] prepared QT-loaded polymeric nanoparticles (PNPs) including synthesis of copolymer conjugated with three monomers [methacrylated PLGA (mPLGA)] as a lipophilic part, acrylated methoxy PEG (aMPEG) as hydrophilic domain and N-2-[(tert-butoxycarbonyl)amino]ethyl methacrylamide (Boc-AEMA)] as pH-responsive segment. It was observed that Boc-AEMA amalgated with QT PNPs showed effective anticancer properties and hence could be used as a novel carrier for the development of targeted anticancer drug delivery systems. Wang et al. [9–11] co-encapsulated QT and fluorescent silicon quantum dots (SiQDs) in PEG-block-polylactide (PEG-PLA) NPs (PEG-PLA-QTNPs) for concurrent in vitro imaging and to increase the biocompatibility of QT. Results showed that PEG-PLA-QTNPs suppressed HepG2 cell proliferation more efficiently rather than free QT. In addition, PEG-PLA-QT nanoparticles expressively inhibited the H_2O_2-induced DNA damage in HepG2 cells. Hence, it was concluded that the PEG-PLA NPs showed enhanced aqueous solubility and sustained release of the QT; hence this dual-encapsulation technique could be used in advanced cancer treatment. Sahu et al. [83] prepared QT-loaded ethylcellulose (biocompatible) by nanoprecipitation technique using ethylcellulose as polymer. Ex vivo results showed that the amount of QT released from PNPs was enhanced and the amount of QT could be retained in the skin for prolonged period of time, therefore reducing the dosing frequency of QT. Hence, QT-loaded ethylcellulose PNPs could be used for topical use and also for the management of skin cancer. Jain et al. [84] prepared and characterized the PLGA NPs co-encapsulated with QT and tamoxifen (TMX). DPPH free radical scavenging assay established that the antioxidant properties of QT were reserved in freeze-dried NPs. Greater cellular uptake, cytotoxicity and nuclear co-distribution of QT-TMX-NPs in MCF-7 cells showed greater efficacy of the polymeric nanoformulation with no assessable hepatotoxicity and oxidative stress. Also, the in vivo studies showed greater Caco-2 cell uptake by QT-TMX-NPs, which indicated its enhanced oral bioavailability as compared to the free TMX and QT. The multiple oral dosing of QT-TMX-NPs proficiently regulated the tumor angiogenesis as showed by standardized levels of relevant markers like MMP-2 and MMP-9. Wang et al. [9–11] prepared PNPs comprised of amine 1,2-distearoyl-sn-glycero-3-phosphoethanolamine-N-[amino(polyethylene glycol)-2000] (PEG2000-DPSE) coated with QTNPs (PEG2000-DPSE-QTNPs) and investigated NPs as potential carriers for anticancer agents/drugs induced PCD. Results indicated that PEG2000-DPSE-QTNPs exhibited dose-dependent cytotoxicity against C6GCs and improved ROS induced uptake of p53 protein and hence increased the level of Cyto-*c* and caspase-3 protein. Consequently, PEG2000-DPSE-QTNPs improved the anticancer effect against induced PCD on C6GCs and so could be used for cancer therapy. Zafar et al. [85] developed a Eudragit RL100-based co-loaded nanoparticulate system comprising of mitoxantrone dihydrochloride (MDC) along with QT (for Pgp) and hesperetin (HP, for BCRP) (both efflux transporter inhibitors), for targeted delivery of MDC/hyaluronic acid (HA) and concurrently to reverse the MDR in breast cancer. It was observed that IC_{50} values were reduced for inhibitor-doped PNPs in various cell lines. The PNPs also exhibited prolonged release with decreased macrophage uptake and enhanced residence time in blood. The enhanced cellular uptake of HA-coated PNPs in MCF7 cells showed their targeting potential. The HA-coated PNPs along with QT/HP exhibited greater tendency for targeted chemotherapy in CD44 overexpressing MDR breast cancer. Sharma et al. [86] synthesized mPEG-PLA-encapsulated QT nanoparticles (mPEG-PLA-QTNPs) for their anticancer efficiency. In vitro results exhibited that mPEG-PLA-QTNPs competently induced apoptosis in triple negative breast cancer cell line (MDA-MB-231) with greater quantity of QT in cell lysate treated with mPEG-PLA-QTNPs as compared to free QT. In vivo studies revealed that mPEG-PLA-QTNPs restricted the tumor growth of xenograft mice model. The terminal deoxynucleotidyl transferase (TdT) dUTP Nick-End Labeling (TUNEL) assay revealed more existence of apoptotic cells in mPEG-PLA-QTNP-treated tumors rather than free QT at the same dose. David et al. [87] developed QT and 5-fluorouracil (5FU) loaded in chitosan nanoparticles (QT-5FU-CNPs) for the treatment of pancreatic cancer. QT-5FU-CNPs showed enhanced %EE (quercetin 95% and 5-fluorouracil 75%) with the formulation ratio QT: 5-FU-C::1:2:3. The QT-5FU-CNPs showed significant toxicity against pancreatic cancer cells; hence this co-loaded drug delivery system could be possibly used for the management of pancreatic cancer. Du et al. [88] developed two different kinds of QT-loaded NPs based on O-carboxymethyl chitosan-cholic acid conjugates (QT-CMCA) and also investigated the effects on bioavailability of QT when altered with glycyrrhetinic acid (GA) (QT-GA-CMCA). QT-GA-CMCA exhibited more sensitivity to acidic pH condition and showed rapid drug release at pH 5.7 and pH 6.5 for QT-CMCA and QT-GA-CMCA, respectively. In vitro studies showed that QT-GA-CMCA improved the cellular cytotoxicity and apoptosis rate, whereas in vivo studies revealed that QT-GA-CMCA enhanced the residence time of QT, which would be helpful for the management of cancer. Pandey et al. [89] synthesized QT-loaded PLA nanoparticles (PLA-QTNPs) for the establishment of its anticancer efficacy against human breast cancer cells. Results showed that PLA-QTNPs retained sustained delivery of QT from the polymer matrix due to late diffusion and stronger interface. In vitro cytotoxicity study indicated that PLA-QTNPs killed almost ~50% of breast cancer cells. So, PLA-QTNPs could be efficiently used as carrier for the management of breast cancer. Zhu et al. [90] showed that QT-loaded PLGA-D-α-tocopheryl polyethylene glycol succinate (TPGS) nanoparticles (PLGA-TPGS-QTNPs) overcame low aqueous solubility of QT and improved its anti-UVB effect. PLGA-TPGS-QTNPs proficiently absorbed through epidermis, reached dermis and expressively blocked the UVB irradiation-induced COX-2 upregulation and NF-kB activation in HaCat cell line. It was concluded that PLGA-TPGS-QTNPs efficiently protected UVB-induced injuries and might be used for skin cancer therapy. Fatma et al. [91] designed PLGA-conjugated PNPs loaded with etoposide (ES) and QT, forming PLGA-ES/QT-NPs. MTT assay (cytotoxicity studies) showed a 9- and 11-fold reduction in the IC_{50} values for etoposide-loaded nanoparticles (ESNPs) and PLGA-ES/QT-NPs as compared to free ES, respectively. The pharmacokinetic studies on rats showed that PLGA-ES/QT-NPs increased the bioavailability of ES (anticancer agent) by 2.4-fold than ESNPs with no QT. Lv et al. [92] targeted co-delivery of DOX and QT to tumor cells in the presence of biotin (B)-conjugated PEG-PCL NPs for the management of MDR breast cancer. It was observed that the tumor targeting by DOX-QT-B-NPs showed significant benefits associated to management with the free DOX-QT-NPs, DOX-B-NPs and QT-B-NPs against MDR breast cancer cells (MCF-7/ADR cells). The DOX-QT-B- NPs significantly improved the in vivo anticancer efficacy of DOX-QT in MDR MCF-7/ADR xenograft model. Guan et al. [93] prepared QT-loaded PLGA-TPGS nanoparticles (QT-PLGA-TPGS-NPs) for the treatment of liver cancer. QT-PLGA-NPs and QT/coumarin-6 (C6, fluorescent marker)-loaded PLGA-TPGS NPs (QT-C6-PLGA-TPGS-NPs) showed significant activity for cellular uptake by HepG2 and HCa-F and also enhanced apoptosis of HepG2 cells. It was concluded that QT-PLGA-TPGS-NPs could be used as IV administration efficiently for the management of liver cancer with improved liver targeting. Saha et al. [94] co-encapsulated adriamycin (ADR) and MDC conjugated with QT-loaded PLGA NPs against MDR cancerous cells by surface modification using bovine serum albumin (BSA) or histones (His). In vitro studies showed that the release of QT from the nanoformulations persisted up to 20 days showing an initial burst release.

ADR-MDC-His-QT NPs showed better loading than ADR-MDC-BSA-QT NPs. Finally, it was concluded that these multidrug-loaded nanoformulations could be used to overcome MDR by specifically targeting the cancerous cells. Suksiriworapong et al. [95] investigated the effect of the different hydrophobic chain lengths of PCL-TPGS copolymers on the properties of NPs and efficacy of QT to get delivered to SKBR3 breast cancer cells. It was observed that longer PCL chains showed greater crystallinity and reduced the crystallinity of TPGS chain, which enhanced the particle size of PCL-TPGS NPs from 20 to 205 nm and improved the loading efficiency of QT but simultaneously retarded the release of QT. The PCL-TPGS NPs induced the toxicity of QT against SKBR3 cells almost by 2.9 times as compared to the QT solution; thus PCL-TPGS NPs loaded with QT might be used for the treatment of breast cancer. Xing et al. [96] developed the specific mitochondria-targeted triphenylphosphine (TPP)-conjugated QT nanoparticles (TPP-QTNPs), later on modified by adding PEG to form PEG-TPP-QTNPs. In vitro and in vivo studies revealed that PEG-TPP-QTNPs were a more efficient therapeutic compound as compared with free QT. The PEG-TPP-QTNPs induced mitochondria-facilitated apoptosis with ROS, depolarized potential of mitochondrial membrane and released Cyt *c* from mitochondria into cytosol, and caspase cascade was activated. Finally, these results showed that mitochondria-targeted delivery of hydrophobic anticancer drugs by conjugating with TPP-based NPs could be used in cancer therapy. Zhu et al. [97] developed lipid-polymeric nanocarriers (LPNs) loaded with dual- synergistic drugs VT and QT (VT/QT-LPNs) for treatment of lymphoma combination chemotherapy and their antitumor efficiency for in vitro cell culture models and a tumor xenograft mouse model. It was concluded that VT/QT-LPNs showed synergistic effect to overcome chemo-resistant lymphoma. Oliveira et al. [98] encapsulated QT into a novel pH-sensitive self-assembled amphiphilic chitosan NPs and formed QT-C-NPs. Results revealed that 83% of QT was entrapped in QT-C-NPs and the QT release was greater in acidic pH (5.0) as compared to physiological pH (7.4). The MTT assay results revealed that blank NPs showed no proliferative effect against the tested MCF-7 cells, whereas QT-C-NPs maintained its inhibitory effect against MCF-7 cells. Moreover, the haemocompatibility assays showed that the QT-C-NPs also exhibited blood compatibility and could act as effective agents for cancer treatment. Baksi et al. [99] developed QT-C-NPs with enhanced %EE and sustained drug release activity. In vitro cytotoxicity assay revealed that QT-C-NPs decreased the IC_{50} value as compared to free QT. IV treatment of QT-C-NPs in tumor xenograft mice with A549 and MDA MB 468 cells significantly reduced tumor growth as compared to control groups. Finally, it was concluded that QT-C-NPs with improved efficacy and antitumor effect could be used in cancer treatment. Yamina et al. [100] developed novel QT-loaded grafted halloysite nanotubes (HNT), conjugated with PEG-amine at external surface and formed HNT-g-PEG-QTNPs which were amalgated with carbon quantum dots (HNT-g- PEG-CD-QTNPs) for adjacent fluorescence activities and with biotin (B) for precise targeting of tumor tissue and higher cellular uptake, finally forming HNT-g-PEG-CD-B-QTNPs. The developed NPs showed a greater loading efficiency of QT with improved sustained release activity. The incubation of the HNT-g-PEG-CD-B-QTNPs with HeLa cells exhibited low toxicity and an appropriate biocompatibility. Also, MTT assay of the QTNPs possessed improved antitumor activity over the free QT.

2.8 Metallic Nanoparticle-Based NFs

2.8.1 Gold Nanoparticles

Bishayee et al. [101] prepared gold nanoparticles (GNPs) with QT and loaded QT-GNPs into PLGA NPs and tested the anticancer activity of the prepared NPs on HepG2 hepatocarcinoma cells. It was observed that PLGA-QT-GNPs efficiently triggered conformational alterations in DNA and several modified proteins associated to epigenetic alterations and cell cycle control. In vitro cytotoxicity assay showed that PLGA-QT-GNPs favourably killed cancerous cells as compared to the normal cells and also restrained actions of cell cycle-related proteins, such as p21WAF, cdk1 and pAkt. Also, PLGA-QT-GNPs prompted apoptosis in HepG2 cells by triggering p53-ROS and induced epigenetic changes that lead to cell cycle arrest. Lou et al. [102] showed that QT-GNPs that induced cell autophagy and apoptosis in human neuroglioma cell also actively inhibited tumor growth. Results of flow cytometry revealed that the quantity of the apoptosis cells was enhanced after QT-GNP treatment as compared to untreated group. Additionally, QT-GNPs down-regulated the existence of activated PI3K/AKT and Bcl-2 in human neuroglioma cells. Further, results showed that the existence level of LC3 and ERK proteins, cytoplasm p53 and inhibition of caspase-3-mediated poly(ADP ribose) polymerase (PARP) enzyme were certainly associated with the amount of QT-GNPs. Also, p-mTOR and G α interacting protein (GAIP) were inhibited by QT-GNPs in a dose-dependent manner. [103]) loaded QT-GNPs into PLGa NPs, formed PLGA-QT-GNPs which were targeted to cervical cancer cells. Also, the major mechanism behind the regulation of cervical cancer by QT was investigated. Results specified that JAK2, certainly, expressed greatly in cancer cell lines rather than the normal cervical cells; also apoptosis and autophagy were significantly observed in JAK2-repressed cancer cells by triggering caspase-3, and inhibiting Cyclin-D1 and mTOR controlled by (STAT) 3/5 and phosphatidylinositide 3-kinase/protein kinases (PI3K/AKT) signaling pathway; hence the cervical cancer cell proliferation was repressed. Remarkably, PLGA-QT-GNPs exhibited similar role with JAK2. Balakrishnan et al. [104] developed QT-GNPs which significantly reduced protein expression of vimentin (type III intermediate filament (IF) protein which is expressed in mesenchymal cells), N-cadherin (transmembrane protein), transcriptional factors (Snail, Slug, Twist), MMP-2 (matrix metalloproteinase-2), MMP-9 (matrix metalloproteinase-9), p-EGFR (phosphorylated-epidermal growth factor receptor), VEGFR-2 (vascular endothelial growth factor), p-PI3K (phosphoinositide 3-kinase), Akt (protein kinase B (PKB), also known as Akt) and p-GSK3β (glycogen synthase kinase 3 beta) and improved E-cadherin protein expression in response. Results showed that QT-GNPs repressed relocation and invasion of MCF-7 and MDA-MB-231 cells more expressively than free QT. Also, QT-GNPS-treated HUVECs showed decreased cell viability and inhibited new blood vessel development. It was concluded that QT-GNPs suppressed epithelial-mesenchymal transition (EMT), angiogenesis and metastasis of the breast cancer cells. Xu et al. [105] prepared PLGA-QT-GNPs and investigated their anti-oxidative stress and anti-inflammatory activities in high-fat diet-induced kidney failure. In vitro results showed that PLGA-QT-GNPs inhibited upregulation of TLR4/NF-κB and oxidative stress by endotoxin, eventually lessening podocyte injury. These conclusions directed that PLGA-QT-GNPs might reestablish the metabolic disorders initiated by high-fat diet, which suppressed insulin resistance, lipid metabolic disparity and pro-inflammatory cytokine release. Ren et al. [106] explored the activity and probable primary mechanism of QT-GNPs against antitumor activity in liver cancer cells. Results showed that QT-GNPs suppressed the liver cancer cell proliferation, cell migration and colony establishment, thus inhibiting liver cancer progression. QT-GNPs also upregulated apoptosis significantly. Results proposed that QT-GNPs showed an antitumor effect as it inactivated caspase/Cyto-c pathway, suppressed AP-2β/hTERT, suppressed NF-κB/COX-2 and impeded Akt/ERK1/2 signaling pathways; hence QT-GNPs could act as a potential agent for targeted liver cancer treatment. Balakrishnan et al. [16] prepared QT-GNPs and evaluated the effects of QT-GNPs in MCF-7 and MDA-MB-231 breast cancer cell lines. It was observed that QT-GNPs showed lower IC_{50} value compared to free QT. QT-GNPs increased apoptosis with amplified nuclear condensation. QT-GNPs upregulated caspase-3, down-regulated Bcl-2 and inhibited the epidermal growth factor receptor (EGFR) and its downstream signaling molecules PI3K/Akt/mTOR/GSK-3β. Thus, QT-GNPs induced apoptosis and suppressed EGFR signaling; hence QT-GNPs could be used for the treatment of breast cancer.

2.8.2 Iron Oxide Nanoparticles

Verma et al. [107] prepared PLGA-coated Fe_3O_4 magnetic nanoparticles (MNPs) loaded with QT (PLGA-QT MNPs) and investigated its targeting activity against lung cancer cells through intranasal delivery. Results showed that the developed PLGA-QT MNPs were biocompatible and showed no cytotoxicity against treated human lung epithelial cells. Moreover, in vivo studies indicated that intranasal treatment of mice with PLGA-QT MNPs showed greater endurance as measured by GLUT and IL-6 release assays. Also, PLGA-QT MNPs efficiently decreased the number of viable A549 cells; hence PLGA-QT MNPs could be used as a carrier for targeted delivery against lung cancer. Kumar et al. [108] prepared dextran (DT)-coated MNPs loaded with QT, formed DT-QT-MNPs and investigated its targeting activity against breast cancer cells. The results of MTT assay showed that no toxicity was inhibited by DT-MNPs against MCF7 cells, while DT-QT-MNPs enhanced the cytotoxicity considerably as compared with pure QT. The treatment of MCF-7 cells with DT-QT-MNPs showed noticeable alterations in cellular morphology which signified the applicability of DT-QT-MNPs as a noble carrier for the treatment of breast cancer. Mashhadi Malekzadeh et al. [109] designed a novel multifaceted drug delivery system comprised of QT-loaded MNPs coated with polycitric acid (PCA) and PEG and conjugated with FA (to achieve tumor cell-targeting property) and formed a complex of PCA-PEG-FA-QTMNPs. Results showed that the drug loading and release through PCA-PEG-FA-QTMNPs were enhanced. PCA-PEG-FA-QTMNPs showed greater cellular uptake by folate receptor-overexpressing HeLa cells as compared to PCA-PEG-QTMNPs. Furthermore, results showed that the cytotoxicity of NPs was significantly enhanced by folate conjugation and magnetic resonance images (MRI) of PCA-PEG-FA-QTMNPs exhibited concentration-dependent magnetic resonance signal. Henceforth, these nanocarriers could be used as chemotherapeutic agents as well as MRI contrast agents. Srisa-Nga et al. [17] co-encapsulated QT and superparamagnetic iron oxide nanoparticles (SPIONs) into mPEG-b-oligo(3-caprolactone) micelles (QT/SPIONs-mPEG-b-OCLM) for the suppression of hepatitis B virus-transmitted hepatocellular carcinoma (HepG2.2.15) cellular growth. It was observed that the SPION masses existing inside the centre of micellar formulation enhanced T2 relaxtivity value signifying decent MRI sensitivity of QT/SPIONs-mPEG-b-OCLM. The HepG2.2.15 cells incubated with QT/SPIONs-mPEG-b-OCLM exhibited greater cytotoxicity than QT. It was concluded that better anticancer activity of QT/SPIONs-mPEG-b-OCLM could be observed at superior magnetic field and mPEG-b-OCLM could be used as a potential co-delivery of QT/SPIONs for targeting hepatocellular carcinoma.

2.8.3 Silver Nanoparticles

Mittal et al. [110] synthetized biocompatible bimetallic silver-selenium (Ag/Se) NPs and doped with QT and GA (Ag/Se@QT-GA NPs). Results showed that Ag/Se@QT-GA NPs had good efficiency for in vitro antioxidant, antitumoral and antimicrobial (Gram-positive and Gram-negative bacteria) activities. It was concluded that Ag/Se enhanced the therapeutic efficacy of QT and GA synergistically. Also, Ag/Se@QT-GA NPs showed effective anticancer activity against Dalton lymphoma (DL) cells, and in vitro studies showed that cell viability was reduced by 80%. Martirosyan et al. [111] investigated the effects of QT on Ag-NPs against intestinal/colon cancer treatment. It was observed that Ag-NPs reduced cell viability and decreased thiol group sequences, tempted oxidative/nitrosative stress and lipid peroxidation and also altered the activity various antioxidant enzymes. QT-AgNPs showed significant anti-inflammatory effect exhibited in down-regulated IL-8 levels. Moreover, QT loaded AG-NPs with reduced cytotoxicity, oxidative stress and improved metabolic activity of Caco-2 cells, signifying the shielding activity of QT against the injurious activities of Ag-NPs. Yuan et al. [18] synthesized graphene oxide-silver nanoparticle (GO-AgNPs) nanocomposites in the presence of QT biomolecule (QT@GO-AgNPs) and also estimated the possible cytotoxicity and mechanism of QT@GO-AgNPs against human neuroblastoma cancer cells (SH-SY5Y). QT@GO-AgNPs showed considerably greater cytotoxicity against SH-SY5Y cells than GO due to decrease in cell viability, suppression of cell proliferation, enhancement in lactate dehydrogenase release, reduction in the level of mitochondrial membrane ability, decrease in mitochondria number, enhancement in in the level of ROS, increase in pro-apoptotic genes expression and decrease in anti-apoptotic genes expression. Also, QT@GO-AgNPs induced caspase-9/3-mediated apoptosis via DNA fragmentation; hence overall QT@GO-AgNPs showed enhanced cytotoxicity against neuroblastoma cancer cells.

2.9 Miscellaneous

Gismondi et al. [112] prepared novel nanodiamonds (NDs) loaded by ciproten (CT) and QT with different methods (oxidized, wet chemical reduced and plasma reduced) to check the effect of surface interaction on drug loading. It was observed that the anti-proliferative effects of CT and QT on HeLa and murine (B16F10) tumor cells were enhanced; in vitro pro-oxidant cytotoxicity and apoptosis activity were decreased after they were conjugated with QD. In conclusion, it was indicated that the chemical and structural alterations of ND surfaces affected the bioactivity of CT and QT. Gupta et al. [113] developed a single-phase reaction-precipitation technique to formulate QT-conjugated poly-β-amino ester (PβAE) nanogels (QT-PβAE-NG). Moreover, the formulation of QT-PβAE-NG enhanced the aqueous solubility of QT in the presence of PEG and controlled the size, and also the antioxidant activity was retained over prolonged period of time; hence QT-PβAE-NG might be used for the management of cancer. Lockhart et al. [114] developed co-encapsulated polyester nanosponge particles (NSPs) loaded with TMX and QT (TMX-QT-NSPs). Additionally, to estimate the effects on metabolism and drug release kinetics of TMX-QT-NSPs, nanocarrier system was divided into two different variations (4% and 8%; NSP-4% and NSP-8%). The stability of nanoformulations was performed in simulated GI fluids, and it was observed that the metabolism of TMX was reduced by twofold for NSP-4% and threefold for NSP-8%, respectively, while metabolism of QT was reduced by threefold for NSP-4% and fourfold for NSP-4%. Moreover, the consequences for enhanced bioavailability of the NSPs were backed by cytotoxicity results that exhibited a similar efficiency to TMX-QT-NSPs and also improved the anticancer effects. Cruz Dos Santos et al. [115] showed the probable interface between QT and asolectin (AL)-based magnetoliposomes (QT/AL-MLPs). Results showed that QT significantly reduced the efficacy of MLPs encapsulated with maghemite NPs by 11%, which signified possible interaction of QT with maghemite NPs in a definite region of MLPs. Also, QT reduced the magnetic capacity of MLPs, but did not affect the superparamagnetic activity of the nanoparticulate system. In vitro studies showed that the lipid peroxidation of the MLPs was prohibited (eightfold) in the presence of QT and QT also reduced the viability of C6GCs by 45%, hence QT loaded MLPs could efficiently be used as contrast and antitumoral agents. Alidadi et al. [116] investigated the nephron-protective effect of QT against renal injury insisted by titanium dioxide nanoparticles (TiO_2-NPs) in rat models. Results showed that TiO_2-NPs expressively enhanced the plasma level of the biomarkers, reduced catalase (CAT) and superoxide dismutase (SOD) activities and enhanced the malondialdehyde (MDA) levels. Also, TiO_2-NPs initiated proximal tubule injury, amassing of red blood cells and intrusion of inflammatory cells, decreased the diameter of glomerulus and persuaded apoptosis in the proximal tubules. The treatment of QT mitigated the histological alterations, stabilized the plasma biomarkers, inhibited oxidative stress, improved the activities of CAT and SOD and decreased apoptosis. Conclusively, QT reduced apoptosis and exhibited better protective effect against nephrotoxicity caused by TiO_2-NPs in rats. George et al. [117] extracted QT from onion peel waste, conjugated it with dialdehyde cellulose (DAC) modified from sugarcane bagasse (SCB) cellulose and efficiently cross-linked with chitosan hydrogel. Later on, the hydrogel complexes were inserted with zinc oxide (phyto-synthesized using musk melon seeds) nanoparticles (ZnO NPs). Results showed enhanced drug loading (39%) of QT in the hydrogel complexes in the presence of ZnO-NPs as compared to drug loading without ZnO NPs. The hydrogel complex inhibited the growth of *Staphylococcus aureus* (*S. aureus*) and *Trichophyton rubrum* (*T. rubrum*) strains. Also, the biocompatible hydrogel complex retained anticancer activity against L929 murine fibroblast cells and A431 human skin carcinoma cells.

3 Conclusion and Future Perspectives

Multi-faceted QT is a potential antioxidant drug and also serves as a potent anticancer agent, though its efficacy as a therapeutic drug is less due to its rapid metabolism in the body. Thus, the use of biodegradable and biocompatible nanocarrier-based delivery systems could improve the therapeutic ability of QT. The therapeutic activity of QT-loaded NPs is primarily influenced by the physico-chemical properties and carriers used for developing NFs. The stability of QT, its bioavailability and target accuracy are greatly influenced by these characteristics. Even though various in vitro and in vivo anticancer assays of QT-NFs have been studied, there are some restrictions for their clinical transformation such as formulation cost, safety and adverse effects. Mutually in vivo and in vitro studies have emphasized the anti-proliferative effect of QT-NPs on numerous types of cancer cells. However, the practice of such NPs at clinical level still desires more process optimization to improve their specificity and efficiency. Though numerous QT-NPs have been fabricated and also have showed their ability to improve bioavailability of QT in the body, still there is a necessity for more stable and target-specific NFs. This can be accomplished by the inclusion of cancer cell précised targeting compounds on the NPs, which would not only improve targeting ability of the delivery of QT-NFs but would also decrease their interface with the normal cells, avoiding any adverse effects of QT.

References

1. WHO Reports 2018. https://www.who.int/news-room/fact-sheets/detail/cancer
2. Ferlay J, Soerjomataram I, Dikshit R, Eser S, Mathers C, Rebelo M, Bray F (2015) Cancer incidence and mortality worldwide: sources, methods and major patterns in GLOBOCAN 2012. Int J Cancer 136(5):E359–E386
3. Plummer M, de Martel C, Vignat J, Ferlay J, Bray F, Franceschi S (2016) Global burden of cancers attributable to infections in 2012: a synthetic analysis. Lancet Glob Health 4(9):e609–e616
4. Nam JS, Sharma AR, Nguyen LT, Chakraborty C, Sharma G, Lee SS (2016) Application of bioactive quercetin in oncotherapy: from nutrition to nanomedicine. Molecules 21(1):E108

5. Rauf A, Imran M, Khan IA, Ur-Rehman M, Gilani SA, Mehmood Z, Mubarak MS (2018) Anticancer potential of quercetin: a comprehensive review. Phytother Res 32(11):2109–2130
6. Gibellini L, Pinti M, Nasi M, Montagna JP, De Biasi S, Roat E, Cossarizza A (2011) Quercetin and cancer chemoprevention. Evid Based Complement Alternat Med 2011:591356
7. Caddeo C, Diez-Sales O, Pons R, Carbone C, Ennas G, Puglisi G, Manconi M (2016) Cross-linked chitosan/liposome hybrid system for the intestinal delivery of quercetin. J Colloid Interface Sci 461:69–78
8. Caddeo C, Nacher A, Vassallo A, Armentano MF, Pons R, Fernandez-Busquets X, Manconi M (2016) Effect of quercetin and resveratrol co-incorporated in liposomes against inflammatory/oxidative response associated with skin cancer. Int J Pharm 513(1–2):153–163
9. Wang Q, Bao Y, Ahire J, Chao Y (2013) Co-encapsulation of biodegradable nanoparticles with silicon quantum dots and quercetin for monitored delivery. Adv Healthc Mater 2(3):459–466
10. Wang G, Wang JJ, Chen XL, Du SM, Li DS, Pei ZJ, Wu LB (2013) The JAK2/STAT3 and mitochondrial pathways are essential for quercetin nanoliposome-induced C6 glioma cell death. Cell Death Dis 4:e746
11. Wang G, Wang J, Luo J, Wang L, Chen X, Zhang L, Jiang S (2013) PEG2000-DPSE-coated quercetin nanoparticles remarkably enhanced anticancer effects through induced programed cell death on C6 glioma cells. J Biomed Mater Res A 101(11):3076–3085
12. Moussa M, Goldberg SN, Kumar G, Sawant RR, Levchenko T, Torchilin VP, Ahmed M (2014) Nanodrug-enhanced radiofrequency tumor ablation: effect of micellar or liposomal carrier on drug delivery and treatment efficacy. PLoS One 9(8):e102727
13. Chen SQ, Wang C, Tao S, Wang YX, Hu FQ, Yuan H (2018) Rational design of redox-responsive and P-gp-inhibitory lipid nanoparticles with high entrapment of paclitaxel for tumor therapy. Adv Healthc Mater 7(17):e1800485
14. Ojeda-Serna IE, Rocha-Guzman NE, Gallegos-Infante JA, Chairez-Ramirez MH, Rosas-Flores W, Perez-Martinez JD, Gonzalez-Laredo RF (2019) Water-in-oil organogel based emulsions as a tool for increasing bioaccessibility and cell permeability of poorly water-soluble nutraceuticals. Food Res Int 120:415–424
15. Omwenga EO, Hensel A, Shitandi A, Goycoolea FM (2018) Chitosan nanoencapsulation of flavonoids enhances their quorum sensing and biofilm formation inhibitory activities against an E.coli Top 10 biosensor. Colloids Surf B Biointerfaces 164:125–133
16. Balakrishnan S, Mukherjee S, Das S, Bhat FA, Raja Singh P, Patra CR, Arunakaran J (2017) Gold nanoparticles-conjugated quercetin induces apoptosis via inhibition of EGFR/PI3K/Akt-mediated pathway in breast cancer cell lines (MCF-7 and MDA-MB-231). Cell Biochem Funct 35(4):217–231
17. Srisa-Nga K, Mankhetkorn S, Okonogi S, Khonkarn R (2018) Delivery of superparamagnetic polymeric micelles loaded with quercetin to hepatocellular carcinoma cells. J Pharm Sci 108(2):1–11
18. Yuan YG, Wang YH, Xing HH, Gurunathan S (2017) Quercetin-mediated synthesis of graphene oxide-silver nanoparticle nanocomposites: a suitable alternative nanotherapy for neuroblastoma. Int J Nanomedicine 12:5819–5839
19. Davatgaran-Taghipour Y, Masoomzadeh S, Farzaei MH, Bahramsoltani R, Karimi-Soureh Z, Rahimi R, Abdollahi M (2017) Polyphenol nanoformulations for cancer therapy: experimental evidence and clinical perspective. Int J Nanomedicine 12:2689–2702
20. Kashyap D, Garg VK, Tuli HS, Yerer MB, Sak K, Sharma AK, Sandhu SS (2019) Fisetin and quercetin: promising flavonoids with chemopreventive potential. Biomolecules 9(5). https://doi.org/10.3390/biom9050174
21. Podhajcer OL, Friedlander M, Graziani Y (1980) Effect of liposome-encapsulated quercetin on DNA synthesis, lactate production, and cyclic adenosine 3′:5′ monophosphate level in Ehrlich ascites tumor cells. Cancer Res 40(4):1344–1350
22. Spector M, O'Neal S, Racker E (1980) Reconstitution of the $Na^{+}K^{+}$ pump of ehrlich ascites tumor and enhancement of efficiency by quercetin. J Biol Chem 255(12):5504–5507
23. Goniotaki M, Hatziantoniou S, Dimas K, Wagner M, Demetzos C (2004) Encapsulation of naturally occurring flavonoids into liposomes: physicochemical properties and biological activity against human cancer cell lines. J Pharm Pharmacol 56(10):1217–1224

24. Tammela P, Laitinen L, Galkin A, Wennberg T, Heczko R, Vuorela H, Vuorela P (2004) Permeability characteristics and membrane affinity of flavonoids and alkyl gallates in Caco-2 cells and in phospholipid vesicles. Arch Biochem Biophys 425(2):193–199
25. Yuan ZP, Chen LJ, Fan LY, Tang MH, Yang GL, Yang HS, Wei YQ (2006) Liposomal quercetin efficiently suppresses growth of solid tumors in murine models. Clin Cancer Res 12(10):3193–3199
26. Yuan ZP, Chen LJ, Wei YQ, Fan LY, Tang MH, Yang GL (2006) Nanoliposomal quercetin inhibits formation of malignant ascites of hepatocellular carcinoma. Ai Zheng 25(8):941–945
27. Mandal AK, Das S, Mitra M, Chakrabarti RN, Chatterjee M, Das N (2008) Vesicular flavonoid in combating diethylnitrosamine induced hepatocarcinoma in rat model. J Exp Ther Oncol 7(2):123–133
28. Ghosh D, Ghosh S, Sarkar S, Ghosh A, Das N, Das Saha K, Mandal AK (2010) Quercetin in vesicular delivery systems: evaluation in combating arsenic-induced acute liver toxicity associated gene expression in rat model. Chem Biol Interact 186(1):61–71
29. Wong MY, Chiu GN (2010) Simultaneous liposomal delivery of quercetin and vincristine for enhanced estrogen-receptor-negative breast cancer treatment. Anticancer Drugs 21(4):401–410
30. Ghosh S, Dungdung SR, Chowdhury ST, Mandal AK, Sarkar S, Ghosh D, Das N (2011) Encapsulation of the flavonoid quercetin with an arsenic chelator into nanocapsules enables the simultaneous delivery of hydrophobic and hydrophilic drugs with a synergistic effect against chronic arsenic accumulation and oxidative stress. Free Radic Biol Med 51(10):1893–1902
31. Ghosh A, Mandal AK, Sarkar S, Das N (2011) Hepatoprotective and neuroprotective activity of liposomal quercetin in combating chronic arsenic induced oxidative damage in liver and brain of rats. Drug Deliv 18(6):451–459
32. Wong MY, Chiu GN (2011) Liposome formulation of co-encapsulated vincristine and quercetin enhanced antitumor activity in a trastuzumab-insensitive breast tumor xenograft model. Nanomedicine (Lond) 7(6):834–840
33. Yang W, Ahmed M, Tasawwar B, Levchenko T, Sawant RR, Collins M, Goldberg SN (2011) Radiofrequency ablation combined with liposomal quercetin to increase tumour destruction by modulation of heat shock protein production in a small animal model. Int J Hyperthermia 27(6):527–538
34. Sinha R, Gadhwal MK, Joshi UJ, Srivastava S, Govil G (2012) Modifying effect of quercetin on model biomembranes: studied by molecular dynamic simulation, DSC and NMR. Int J Curr Pharm Res 4(1):70–79
35. Wang BL, Gao X, Men K, Qiu J, Yang B, Gou ML, Wei YQ (2012) Treating acute cystitis with biodegradable micelle-encapsulated quercetin. Int J Nanomedicine 7:2239–2247
36. Wang G, Wang JJ, Yang GY, Du SM, Zeng N, Li DS, Ye F (2012) Effects of quercetin nanoliposomes on C6 glioma cells through induction of type III programmed cell death. Int J Nanomedicine 7:271–280
37. Yang W, Ahmed M, Tasawwar B, Levchenko T, Sawant RR, Torchilin V, Goldberg SN (2012) Combination radiofrequency (RF) ablation and IV liposomal heat shock protein suppression: reduced tumor growth and increased animal endpoint survival in a small animal tumor model. J Control Release 160(2):239–244
38. He B, Wang X, Shi HS, Xiao WJ, Zhang J, Mu B, Wang YS (2013) Quercetin liposome sensitizes colon carcinoma to thermotherapy and thermochemotherapy in mice models. Integr Cancer Ther 12(3):264–270
39. Hu F, Bu YZ, Liang R, Duan RM, Wang S, Han RM, Skibsted LH (2013) Quercetin and daidzein beta-apo-14′-carotenoic acid esters as membrane antioxidants. Free Radic Res 47(5):413–421
40. Liu H, Xue JX, Li X, Ao R, Lu Y (2013) Quercetin liposomes protect against radiation-induced pulmonary injury in a murine model. Oncol Lett 6(2):453–459

41. Zheng N-G, Mo S-J, Li J-P, Wu J-L (2014) Anti-CSC effects in human esophageal squamous cell carcinomas and Eca109/9706 cells induced by nanoliposomal quercetin alone or combined with CD 133 Antiserum. Asian Pac J Cancer Prev 15(20):8679–8684
42. Zheng N-G, Wang J-L, Yang S-L, Wu J-L (2014) Aberrant epigenetic alteration in Eca9706 cells modulated by nanoliposomal quercetin combined with butyrate mediated via epigenetic-NF-κB signaling. Asian Pac J Cancer Prev 15(11):4539–4543
43. Dabbagh-Bazarbachi H, Clergeaud G, Quesada IM, Ortiz M, O'Sullivan CK, Fernandez-Larrea JB (2014) Zinc ionophore activity of quercetin and epigallocatechin-gallate: from Hepa 1-6 cells to a liposome model. J Agric Food Chem 62(32):8085–8093
44. Hu J, Wang J, Wang G, Yao Z, Dang X (2016) Pharmacokinetics and antitumor efficacy of DSPE-PEG2000 polymeric liposomes loaded with quercetin and temozolomide: Analysis of their effectiveness in enhancing the chemosensitization of drug-resistant glioma cells. Int J Mol Med 37(3):690–702
45. Ravichandiran V, Masilamani K, Senthilnathan B, Maheshwaran A, Wong TW, Roy P (2017) Quercetin-decorated curcumin liposome design for cancer therapy: in-vitro and in-vivo studies. Curr Drug Deliv 14(8):1053–1059
46. Rodriguez EB, Almeda RA, Vidallon MLP, Reyes CT (2018) Enhanced bioactivity and efficient delivery of quercetin through nanoliposomal encapsulation using rice bran phospholipids. J Sci Food Agric 99(4):1980–1989
47. Dos Santos DM, Rocha CVJ, da Silveira EF, Marinho MAG, Rodrigues MR, Silva NO, de Lima VR (2018) In vitro anti/pro-oxidant activities of R. ferruginea extract and its effect on glioma cell viability: correlation with phenolic compound content and effects on membrane dynamics. J Membr Biol 251(2):247–261
48. Zhou X, Liu HY, Zhao H, Wang T (2018) RGD-modified nanoliposomes containing quercetin for lung cancer targeted treatment. Onco Targets Ther 11:5397–5405
49. Kruszewski M, Kusaczuk M, Kotynska J, Gal M, Kretowski R, Cechowska-Pasko M, Naumowicz M (2018) The effect of quercetin on the electrical properties of model lipid membranes and human glioblastoma cells. Bioelectrochemistry 124:133–141
50. Zuo J, Jiang Y, Zhang E, Chen Y, Liang Z, Zhu J, Zhen Y (2019) Synergistic effects of 7-O-geranylquercetin and siRNAs on the treatment of human breast cancer. Life Sci 227:145–152
51. Riaz MK, Zhang X, Wong KH, Chen H, Liu Q, Chen X, Yang Z (2019) Pulmonary delivery of transferrin receptors targeting peptide surface-functionalized liposomes augments the chemotherapeutic effect of quercetin in lung cancer therapy. Int J Nanomedicine 14:2879–2902
52. Soloviev AI, Kizub IV (2019) Mechanisms of vascular dysfunction evoked by ionizing radiation and possible targets for its pharmacological correction. Biochem Pharmacol 159:121–139
53. Yu J, Chen H, Jiang L, Wang J, Dai J, Wang J (2019) Codelivery of adriamycin and P-gp inhibitor quercetin using PEGylated liposomes to overcome cancer drug resistance. J Pharm Sci 108(5):1788–1799
54. Murgia S, Bonacchi S, Falchi AM, Lampis S, Lippolis V, Meli V, Caltagirone C (2013) Drug-loaded fluorescent cubosomes: versatile nanoparticles for potential theranostic applications. Langmuir 29(22):6673–6679
55. Minaei A, Sabzichi M, Ramezani F, Hamishehkar H, Samadi N (2016) Co-delivery with nano-quercetin enhances doxorubicin-mediated cytotoxicity against MCF-7 cells. Mol Biol Rep 43(2):99–105
56. Karthivashan G, Masarudin MJ, Kura AU, Abas F, Fakurazi S (2016) Optimization, formulation, and characterization of multiflavonoids-loaded flavanosome by bulk or sequential technique. Int J Nanomedicine 11:3417–3434
57. Babazadeh A, Zeinali M, Hamishehkar H (2018) Nano-phytosome: a developing platform for herbal anti-cancer agents in cancer therapy. Curr Drug Targets 19(2):170–180
58. Tavano L, Mauro L, Naimo GD, Bruno L, Picci N, Ando S, Muzzalupo R (2016) Further evolution of multifunctional niosomes based on pluronic surfactant: dual active targeting and drug combination properties. Langmuir 32(35):8926–8933

59. Hemati M, Haghiralsadat F, Yazdian F, Jafari F, Moradi A, Malekpour-Dehkordi Z (2019) Development and characterization of a novel cationic PEGylated niosome-encapsulated forms of doxorubicin, quercetin and siRNA for the treatment of cancer by using combination therapy. Artif Cells Nanomed Biotechnol 47(1):1295–1311
60. Tan Q, Liu W, Guo C, Zhai G (2011) Preparation and evaluation of quercetin-loaded lecithin-chitosan nanoparticles for topical delivery. Int J Nanomedicine 6:1621–1630
61. Gao X, Wang B, Wei X, Men K, Zheng F, Zhou Y, Wei Y (2012) Anticancer effect and mechanism of polymer micelle-encapsulated quercetin on ovarian cancer. Nanoscale 4(22):7021–7030
62. Xu G, Shi H, Ren L, Gou H, Gong D, Gao X, Huang N (2015) Enhancing the anti-colon cancer activity of quercetin by self-assembled micelles. Int J Nanomedicine 10:2051–2063
63. Wang G, Wang JJ, Chen XL, Du L, Li F (2016) Quercetin-loaded freeze-dried nanomicelles: Improving absorption and anti-glioma efficiency in vitro and in vivo. J Control Release 235:276–290
64. Zhao J, Liu J, Wei T, Ma X, Cheng Q, Huo S, Liang XJ (2016) Quercetin-loaded nanomicelles to circumvent human castration-resistant prostate cancer in vitro and in vivo. Nanoscale 8(9):5126–5138
65. Patidar P, Pillai SA, Sheth U, Bahadur P, Bahadur A (2017) Glucose triggered enhanced solubilisation, release and cytotoxicity of poorly water soluble anti-cancer drugs fromT1307 micelles. J Biotechnol 254:43–50
66. Rich GT, Buchweitz M, Winterbone MS, Kroon PA, Wilde PJ (2017) Towards an understanding of the low bioavailability of quercetin: a study of its interaction with intestinal lipids. Nutrients 9(2). https://doi.org/10.3390/nu9020111
67. Sandhu PS, Kumar R, Katare OP, Singh B (2017) Surface-tailored nanomixed micelles containing quercetin-salicylic acid physical complex for enhanced cellular and in vivo activities: a quality by design perspective. Nanomedicine (Lond) 12(11):1281–1303
68. Ahmad N, Ahmad R, Naqvi AA, Alam MA, Ashafaq M, Abdur Rub R, Ahmad FJ (2018) Intranasal delivery of quercetin-loaded mucoadhesive nanoemulsion for treatment of cerebral ischaemia. Artif Cells Nanomed Biotechnol 46(4):717–729
69. Ghosh A, Mandal AK, Sarkar S, Panda S, Das N (2009) Nanoencapsulation of quercetin enhances its dietary efficacy in combating arsenic-induced oxidative damage in liver and brain of rats. Life Sci 84(3–4):75–80
70. Ghosh A, Ghosh D, Sarkar S, Mandal AK, Thakur Choudhury S, Das N (2012) Anticarcinogenic activity of nanoencapsulated quercetin in combating diethylnitrosamine-induced hepatocarcinoma in rats. Eur J Cancer Prev 21(1):32–41
71. Ding B, Chen P, Kong Y, Zhai Y, Pang X, Dou J, Zhai G (2014) Preparation and evaluation of folate-modified lipid nanocapsules for quercetin delivery. J Drug Target 22(1):67–75
72. El-Gogary RI, Rubio N, Wang JT, Al-Jamal WT, Bourgognon M, Kafa H, Al-Jamal KT (2014) Polyethylene glycol conjugated polymeric nanocapsules for targeted delivery of quercetin to folate-expressing cancer cells in vitro and in vivo. ACS Nano 8(2):1384–1401
73. Mandal AK, Ghosh D, Sarkar S, Ghosh A, Swarnakar S, Das N (2014) Nanocapsulated quercetin downregulates rat hepatic MMP-13 and controls diethylnitrosamine-induced carcinoma. Nanomedicine (Lond) 9(15):2323–2337
74. Hatahet T, Morille M, Shamseddin A, Aubert-Pouessel A, Devoisselle JM, Begu S (2017) Dermal quercetin lipid nanocapsules: Influence of the formulation on antioxidant activity and cellular protection against hydrogen peroxide. Int J Pharm 518(1–2):167–176
75. Hatahet T, Morille M, Hommoss A, Devoisselle JM, Muller RH, Begu S (2018) Liposomes, lipid nanocapsules and smartCrystals(R): a comparative study for an effective quercetin delivery to the skin. Int J Pharm 542(1–2):176–185
76. Schwingel TE, Klein CP, Nicoletti NF, Dora CL, Hadrich G, Bica CG, Morrone FB (2014) Effects of the compounds resveratrol, rutin, quercetin, and quercetin nanoemulsion on oxaliplatin-induced hepatotoxicity and neurotoxicity in mice. Naunyn Schmiedebergs Arch Pharmacol 387(9):837–848

77. Pangeni R, Panthi VK, Yoon IS, Park JW (2018) Preparation, characterization, and in vivo evaluation of an oral multiple nanoemulsive system for co-delivery of pemetrexed and quercetin. Pharmaceutics 10(3). https://doi.org/10.3390/pharmaceutics10030158
78. Arbain NH, Salim N, Wui WT, Basri M, Rahman MBA (2018) Optimization of quercetin loaded palm oil ester based nanoemulsion formulation for pulmonary delivery. J Oleo Sci 67(8):933–940. https://doi.org/10.5650/jos.ess17253
79. Arbain NH, Salim N, Masoumi HRF, Wong TW, Basri M, Abdul Rahman MB (2019) In vitro evaluation of the inhalable quercetin loaded nanoemulsion for pulmonary delivery. Drug Deliv Transl Res 9(2):497–507. https://doi.org/10.1007/s13346-018-0509-5
80. Bose S, Du Y, Takhistov P, Michniak-Kohn B (2013) Formulation optimization and topical delivery of quercetin from solid lipid based nanosystems. Int J Pharm 441(1–2):56–66
81. Varshosaz J, Jafarian A, Salehi G, Zolfaghari B (2014) Comparing different sterol containing solid lipid nanoparticles for targeted delivery of quercetin in hepatocellular carcinoma. J Liposome Res 24(3):191–203
82. Khoee S, Rahmatolahzadeh R (2012) Synthesis and characterization of pH-responsive and folated nanoparticles based on self-assembled brush-like PLGA/PEG/AEMA copolymer with targeted cancer therapy properties: A comprehensive kinetic study. Eur J Med Chem 50:416–427
83. Sahu S, Saraf S, Kaur CD, Saraf S (2013) Biocompatible nanoparticles for sustained topical delivery of anticancer phytoconstituent quercetin. Pak J Biol Sci 16(13):601–609
84. Jain AK, Thanki K, Jain S (2013) Co-encapsulation of tamoxifen and quercetin in polymeric nanoparticles: implications on oral bioavailability, antitumor efficacy, and drug-induced toxicity. Mol Pharm 10(9):3459–3474
85. Zafar S, Negi LM, Verma AK, Kumar V, Tyagi A, Singh P, Talegaonkar S (2014) Sterically stabilized polymeric nanoparticles with a combinatorial approach for multi drug resistant cancer: in vitro and in vivo investigations. Int J Pharm 477(1–2):454–468
86. Sharma G, Park J, Sharma AR, Jung JS, Kim H, Chakraborty C, Nam JS (2015) Methoxy poly(ethylene glycol)-poly(lactide) nanoparticles encapsulating quercetin act as an effective anticancer agent by inducing apoptosis in breast cancer. Pharm Res 32(2):723–735
87. David KI, Jaidev LR, Sethuraman S, Krishnan UM (2015) Dual drug loaded chitosan nanoparticles-sugar – coated arsenal against pancreatic cancer. Colloids Surf B Biointerfaces 135:689–698
88. Du H, Liu M, Yang X, Zhai G (2015) The role of glycyrrhetinic acid modification on preparation and evaluation of quercetin-loaded chitosan-based self-aggregates. J Colloid Interface Sci 460:87–96
89. Pandey SK, Patel DK, Thakur R, Mishra DP, Maiti P, Haldar C (2015) Anti-cancer evaluation of quercetin embedded PLA nanoparticles synthesized by emulsified nanoprecipitation. Int J Biol Macromol 75:521–529
90. Zhu X, Zeng X, Zhang X, Cao W, Wang Y, Chen H, Shi X (2016) The effects of quercetin-loaded PLGA-TPGS nanoparticles on ultraviolet B-induced skin damages in vivo. Nanomedicine (Lond) 12(3):623–632
91. Fatma S, Talegaonkar S, Iqbal Z, Panda AK, Negi LM, Goswami DG, Tariq M (2016) Novel flavonoid-based biodegradable nanoparticles for effective oral delivery of etoposide by P-glycoprotein modulation: an in vitro, ex vivo and in vivo investigations. Drug Deliv 23(2):500–511
92. Lv L, Liu C, Chen C, Yu X, Chen G, Shi Y, Li G (2016) Quercetin and doxorubicin co-encapsulated biotin receptor-targeting nanoparticles for minimizing drug resistance in breast cancer. Oncotarget 7(22):32184–32199
93. Guan X, Gao M, Xu H, Zhang C, Liu H, Lv L, Tian Y (2016) Quercetin-loaded poly (lactic-co-glycolic acid)-d-alpha-tocopheryl polyethylene glycol 1000 succinate nanoparticles for the targeted treatment of liver cancer. Drug Deliv 23(9):3307–3318
94. Saha C, Kaushik A, Das A, Pal S, Majumder D (2016) Anthracycline drugs on modified surface of quercetin-loaded polymer nanoparticles: a dual drug delivery model for cancer treatment. PLoS One 11(5):e0155710

95. Suksiriworapong J, Phoca K, Ngamsom S, Sripha K, Moongkarndi P, Junyaprasert VB (2016) Comparison of poly(epsilon-caprolactone) chain lengths of poly(epsilon-caprolactone)-co-d-alpha-tocopheryl-poly(ethylene glycol) 1000 succinate nanoparticles for enhancement of quercetin delivery to SKBR3 breast cancer cells. Eur J Pharm Biopharm 101:15–24
96. Xing L, Lyu JY, Yang Y, Cui PF, Gu LQ, Qiao JB, Jiang HL (2017) pH-Responsive de-PEGylated nanoparticles based on triphenylphosphine-quercetin self-assemblies for mitochondria-targeted cancer therapy. Chem Commun (Camb) 53(62):8790–8793
97. Zhu B, Yu L, Yue Q (2017) Co-delivery of vincristine and quercetin by nanocarriers for lymphoma combination chemotherapy. Biomed Pharmacother 91:287–294
98. Oliveira AI, Pinho C, Fonte P, Sarmento B, Dias ACP (2018) Development, characterization, antioxidant and hepatoprotective properties of poly(E-caprolactone) nanoparticles loaded with a neuroprotective fraction of Hypericum perforatum. Int J Biol Macromol 110:185–196
99. Baksi R, Singh DP, Borse SP, Rana R, Sharma V, Nivsarkar M (2018) In vitro and in vivo anticancer efficacy potential of Quercetin loaded polymeric nanoparticles. Biomed Pharmacother 106:1513–1526
100. Yamina AM, Fizir M, Itatahine A, He H, Dramou P (2018) Preparation of multifunctional PEG-graft-Halloysite Nanotubes for Controlled Drug Release, Tumor Cell Targeting, and Bio-imaging. Colloids Surf B Biointerfaces 170:322–329
101. Bishayee K, Khuda-Bukhsh AR, Huh SO (2015) PLGA-loaded gold-nanoparticles precipitated with Quercetin downregulate HDAC-Akt activities controlling proliferation and Activate p53-ROS crosstalk to induce apoptosis in hepatocarcinoma cells. Mol Cells 38(6):518–527
102. Lou M, Zhang LN, Ji PG, Feng FQ, Liu JH, Yang C, Wang L (2016) Quercetin nanoparticles induced autophagy and apoptosis through AKT/ERK/Caspase-3 signaling pathway in human neuroglioma cells: In vitro and in vivo. Biomed Pharmacother 84:1–9
103. Luo CL, Liu YQ, Wang P, Song CH, Wang KJ, Dai LP, Ye H (2016) The effect of quercetin nanoparticle on cervical cancer progression by inducing apoptosis, autophagy and antiproliferation via JAK2 suppression. Biomed Pharmacother 82:595–605
104. Balakrishnan S, Bhat FA, Raja Singh P, Mukherjee S, Elumalai P, Das S, Arunakaran J (2016) Gold nanoparticle-conjugated quercetin inhibits epithelial-mesenchymal transition, angiogenesis and invasiveness via EGFR/VEGFR-2-mediated pathway in breast cancer. Cell Prolif 49(6):678–697
105. Xu MX, Wang M, Yang WW (2017) Gold-quercetin nanoparticles prevent metabolic endotoxemia-induced kidney injury by regulating TLR4/NF-kappaB signaling and Nrf2 pathway in high fat diet fed mice. Int J Nanomedicine 12:327–345
106. Ren KW, Li YH, Wu G, Ren JZ, Lu HB, Li ZM, Han XW (2017) Quercetin nanoparticles display antitumor activity via proliferation inhibition and apoptosis induction in liver cancer cells. Int J Oncol 50(4):1299–1311
107. Verma NK, Crosbie-Staunton K, Satti A, Gallagher S, Ryan KB, Doody T, Gun'ko YK (2013) Magnetic core-shell nanoparticles for drug delivery by nebulization. J Nanobiotechnol 11:1
108. Kumar SR, Priyatharshni S, Babu VN, Mangalaraj D, Viswanathan C, Kannan S, Ponpandian N (2014) Quercetin conjugated superparamagnetic magnetite nanoparticles for in-vitro analysis of breast cancer cell lines for chemotherapy applications. J Colloid Interface Sci 436:234–242
109. Mashhadi Malekzadeh A, Ramazani A, Tabatabaei Rezaei SJ, Niknejad H (2017) Design and construction of multifunctional hyperbranched polymers coated magnetite nanoparticles for both targeting magnetic resonance imaging and cancer therapy. J Colloid Interface Sci 490:64–73
110. Mittal AK, Kumar S, Banerjee UC (2014) Quercetin and gallic acid mediated synthesis of bimetallic (silver and selenium) nanoparticles and their antitumor and antimicrobial potential. J Colloid Interface Sci 431:194–199
111. Martirosyan A, Grintzalis K, Polet M, Laloux L, Schneider YJ (2016) Tuning the inflammatory response to silver nanoparticles via quercetin in Caco-2 (co-)cultures as model of the human intestinal mucosa. Toxicol Lett 253:36–45

112. Gismondi A, Reina G, Orlanducci S, Mizzoni F, Gay S, Terranova ML, Canini A (2015) Nanodiamonds coupled with plant bioactive metabolites: a nanotech approach for cancer therapy. Biomaterials 38:22–35
113. Gupta P, Authimoolam SP, Hilt JZ, Dziubla TD (2015) Quercetin conjugated poly(beta-amino esters) nanogels for the treatment of cellular oxidative stress. Acta Biomater 27:194–204
114. Lockhart JN, Stevens DM, Beezer DB, Kravitz A, Harth E (2015) Dual drug delivery of tamoxifen and quercetin: regulated metabolism for anticancer treatment with nanosponges. J Control Release 220(Pt B):751–757
115. Cruz Dos Santos S, Osti Silva N, Dos Santos Espinelli JBJ, Germani Marinho MA, Vieira Borges Z, Bruzamarello Caon Branco N, Rodrigues de Lima V (2019) Molecular interactions and physico-chemical characterization of quercetin-loaded magnetoliposomes. Chem Phys Lipids 218:22–33
116. Alidadi H, Khorsandi L, Shirani M (2018) Effects of quercetin on tubular cell apoptosis and kidney damage in rats induced by titanium dioxide nanoparticles. Malays J Med Sci 25(2):72–81. https://doi.org/10.21315/mjms2018.25.2.8
117. George D, Maheswari PU, Begum K (2019) Synergic formulation of onion peel quercetin loaded chitosan-cellulose hydrogel with green zinc oxide nanoparticles towards controlled release, biocompatibility, antimicrobial and anticancer activity. Int J Biol Macromol 132:784–794

Polyunsaturated Fatty Acid-Loaded Nanomedicine for Solid Tumor

Suma Saad, Sarwar Beg, and Farhan J. Ahmad

1 Introduction

Solid tumor is one of the leading causes of deaths worldwide; each year, estimated 7.6 million individuals died accounting for 13% of all deaths [1]. According to the cancer-related mortality by 2030, the expected increment is 13.1 million [2]. Some estimates indicate that about 30% of cancer deaths are related with smoking or other lifestyle factors [3].

An abnormal mass of tissue in which there is no cyst or liquid is present is called solid tumor [4]. Solid tumors may be both cancerous (malignant) and not cancerous (benign) [5]. There are different types of solid tumors which are named according to the type of cells that form them.

Classification of Localized Solid Tumors

- Sarcomas—Cancers rising from supporting tissues, such as bone or muscle [5].
- Carcinomas—Cancers rising from the body's glandular cells and epithelial cells [6].
- Lymphomas—Cancers of the lymphoid organs which produce (like the lymph nodes, spleen, and thymus) and store infection-fighting cells. Almost all the tissues of the body have these cells and lymphomas, so these may grow in a wide variety of organs [7].

Cancer is the abnormal growth of the cell, and nowadays nanotechnology is used for the treatment of cancer because it has various advantages like small particle size and drug targeting, so first we talk about nanotechnology [8]. Nanotechnology includes the engineering of particle size (1–100 nm), structures, and materials [9]. They have unique properties like biological, optical, electrical, and magnetic that are in relationship with their size, shape, and surface chemistry [8]. Therefore the

S. Saad (✉) · S. Beg · F. J. Ahmad
Nanomedicine Research Lab, School of Pharmaceutical Education & Research, Jamia Hamdard, New Delhi, India

© Springer Nature Singapore Pte Ltd. 2020

M. Rahman et al. (eds.), *Nanomedicine for Bioactives*,
https://doi.org/10.1007/978-981-15-1664-1_6

nanotechnology is currently exploited in medicine to treating the disease. In the field of treating cancer, nanotechnology is a powerful tool [8]. There is nothing inherently dangerous about being nanosized. Nanoparticles are as old as the earth, but our capability to manipulate objects is recently discovered [10]. The safety of nanotechnology is continuously being tested. For the chemotherapeutics, the nanoparticles work as drug carriers which are less toxic than the drug they carry, and they are used for drug safety to tumor without damaging the organ [8].

Nanotechnology along with fatty acid has an increasingly important role in the treatment and diagnosis of cancer. Now we discuss about fats and fatty acids.

Fats are necessary for living organisms [11]. For the component of cells, they play a very important role they work as a fundamental building of the cell, tissues, and organs and also for the synthesis of some biologically active substances [12]. There are two main types of naturally occurring fatty acids: saturated and unsaturated [13]. According to the chemical structure, saturated fat has no double bonds, whereas an unsaturated fat has one (monounsaturated) or more double bonds (polyunsaturated) [14]. The polyunsaturated fats have two major classes, omega-3 and omega-6 acids, which are essential for the brain [15]. On the basis of the first double-bond location, PUFAs can be subdivided relative to the methyl terminus of the chain [16]. For example, *n*-3 and *n*-6 FAs are two of the most biologically significant PUFA classes, and on the third or sixth carbon, the first double bond is present [17]. The final carbon in the FA chain is also known as the omega carbon; hence the common reference to these FAs is omega-3 or omega-6 PUFAs [18]. Long-chain *n*-3 and *n*-6 PUFAs are synthesized from the essential FAs (EFAs) alpha-linolenic acid (ALA) and linoleic acid, respectively [19]. PUFAs are mostly considered to have favorable health effects [20]. On the metabolic function of *n*-3 and *n*-6, PUFAs have opposing effects in the body [20]. So there are various types of fatty acids which are described below.

1.1 Saturated Fatty Acids

One molecule of glycerol with three fatty acid chains is from dietary fats [21]. Saturated fatty acids have a double bond in the chain of fatty acids. These fatty acids are those which have a straight chain, and, at room temperature, they are in solid form [22]. They are inflexible and sluggish in the body. They are primarily found in animals edibles like fish, eggs, milk, meat and also found in coconuts fat and palm kernels fat [21].

1.2 Unsaturated Fatty Acids

One or more double bonds are found in unsaturated fatty acids. So, they are known as mono- or polyunsaturated fatty acids [22]. Due to their double bonds, it takes on a steric structure. They are more flexible than saturated fat and that is why they are more reactive [23]. A representative of the unsaturated fatty acids is omega-3, omega-6, or omega-9 fatty acids [11].

1.3 Monounsaturated Fatty Acids

Monounsaturated fatty acids have the capability to synthesize in human bodies. Thus, they are not necessary to be taken as food daily [24]. They can be present in various oils like olive oil, almond oil, and rapeseed oil and avocados or hazelnut [25].

1.4 Essential Fatty Acids

Those substances which are required for vital functions but in the body cannot be synthesized are known as essential fatty acids [26]. Examples are various vitamins, amino acids, and omega-3 fatty acids. The flexibility of cell membranes is effective due to the spatial structure of essential fatty acids [27].

1.5 Polyunsaturated Fatty Acids

The body can synthesize polyunsaturated fatty acids in small quantities. With the help of food, they must be ingested [28]. Due to their structure, there are two important groups, one is the omega-6 and the other omega-3 fatty acids [29].

1.6 Omega-6 Fatty Acids

Omega-6 fatty acid is known as linoleic acid. In the body, the GLA is produced in between in the metabolism of linoleic acid [30, 31]. And due to the deficiencies of vitamins and minerals (zinc, cobalt, etc.), the reaction is slow and further restricted during various reasons like nutritional and inflammatory conditions (arthritis, psoriasis) [30]. Some other diseases (hypertension, diabetes, etc.) also affect the activity of enzyme; due to this, the production of GLA is less in the body [16]. Stress, smoking, and alcohol are some factors which affect the lifestyle due to the inhabitation of saturated and trans fats and preformed arachidonic acid and eicosapentaenoic acid [32]. GLA is formed in the body or administered and extended to dihomo-γ-linolenic acid (DGLA) which is merged into the phospholipid cell membrane acyl transferases following acylation reaction [30, 31]. The GLA can have various roles in the metabolism, DGLA (dihomo-γ-linolenic acid)is the active form of GLA which mediates various physiological actions like a cell to cell dihomo-γ-linolenic acid, prostaglandins synthesis, vasodilation, and anti-inflammatory actions. With the help of delta-5-desaturase enzyme, a small quantity of DGLA can also be converted to arachidonic acid (AA) [17]. However, this reaction is slow that's why it is totally dependent on the dietary and environmental factors. By the action of enzyme, membrane-bound DGLA is released with the help of enzyme phospholipase A2 [33]. After release, for the production of cell-cell communications, it is participated with AA for enzymes cyclooxygenases (COX) and lipoxygenases (LOX) of these fatty acids [17, 31].

In the last four decades, GLA has played an important role for its anti-inflammatory [34–37] and anticancer [31, 38–40] actions. In diabetic patients, it also recovers nerve conduction velocity [41, 42] and improves blood flow. Due to these activities of GLA, numerous sources of GLA have been commercialized. The important commercial resources of GLA include borage (*Borago officinalis*, 20–26% GLA), black current (*Ribes nigrum*, 15–18%), and evening primrose (*Oenothera biennis* L., 8–12% GLA) [31].

With the help of lipid peroxidation and free radical generation, the cytotoxic action of polyunsaturated fatty acids (PUFAs) is thought to be mediated predominantly [43, 44]. In tumor cells, proliferation and metastasis play an important role in eicosanoid and leukotriene synthesis [45, 46]. For the biosynthesis of prostaglandins of the 2- and 1-series, arachidonic acid (AA, 20:4n6) and dihomo-γ-linolenic acid (DGLA, 20:3n6) are the main substrate [47]. To form DGLA and AA, GLA is metabolized. Various studies show that the GLA and other PUFAs increase prostanoid synthesis in cells with the help of PUFAs on eicosanoid synthesis [48–50].

1.7 Omega-3 Fatty Acids

Polyunsaturated fatty acids have different forms; one form is omega-3 fatty acids [18]. The most renowned omega-3 fatty acids are alpha-linolenic acid (ALA), eicosapentaenoic acid (EPA), and docosahexaenoic acid (DHA) [51]. ALA is an essential fatty acid that is why it is taken from the outside because the body is unable to synthesize this [51]. It is present in plants, and that is why it is referred to as vegetable omega-3. Omega-3 fatty acid is present in flax seed in the highest amount of all the vegetable oils [18].

Highly unsaturated EPA and DHA from ALA are synthesized in our body [52]. Main components of phospholipids are ALA, EPA, and DHA, which are an essential part of the structure of cell membranes [52]. In the phospholipids, the role of omega-3 fatty acids is double. Firstly, they increase the exchange of the substance between the blood and the cells, improve blood fluidity and thus improve the circulation of the blood in the body. Secondly, the phospholipids store the free fatty acids [15]. In the nerve cells, phospholipids are predominant, that's why omega-3 has the highest concentration in the brain [53, 54]. And this is the reason why a sufficient intake of ALA is important in a developing brain (pregnancy and while nursing) [53].

Since the late 1980s, polyunsaturated omega-3 fatty acids received strong attention from clinicians and epidemiologists for the treatment of cancer [55]. Omega-3 fatty acids are mainly derived from fish oil; a diet rich in omega-3 fatty acids may protect people from cancer [56].

2 Sources of *n*-3 and *n*-6 PUFAs

Main sources of *n*-6 FAs are found in the crop seeds and vegetable oils in the form of LA with low proportions of *n*-3 FAs. Because of their limited sources, the intake of *n*-3 FAs is usually inadequate [57]. There are various seeds like chia, perilla, and flax that have an ample amount of ALA [58]. Green leafy vegetables contain PUFA (ALA) in high concentration. Fish, fish oil, beef, etc. can also be the source of EPA, DPA, DHA, and ARA and can be directly used in the body for basic physiological functions [59]. For food which is based on plant and animal, the maximum amount of LC-PUFAs is found in various forms like fat-soluble vitamin esters, triacylglycerols (TAGs), diacylglycerols, phospholipids, and cholesterol esters [60].

3 Role of PUFAs in Infection, Inflammation, Cancer, and CVD

There are various major chronic diseases which are related to dietary and physical activity patterns like cancer, CVD, diabetes, obesity, osteoporosis, and bone fractures [61]. Protective effects of *n*-3 PUFAs are mediated by modifications in properties of cancer cells by reducing their proliferation, invasion, and metastasis though increasing their apoptosis [61]. With the help of regulating inflammation, immune response, and angiogenesis, they affect host cells. In maintaining membrane fluidity of the retina and the brain, DHA plays a very important role because neurological and cognitive functions are essential [62]. Eicosanoids are formed with the help of ARA, EPA, and DHA conversion which can control various sets of homeostatic and inflammatory processes with these several linked diseases, such as cancer, infection, inflammation [63]. The eicosanoids show different derivations with *n*-6 LC-PUFAs (pro-inflammatory activity) and *n*-3 LC-PUFAs (anti-inflammatory activities); in the person's diet, the amount of *n*-6 and *n*-3 FAs mostly affects the production of pro-inflammatory and anti-inflammatory eicosanoids [60].

4 PUFAs in Solid Tumor

There are various studies (in vitro and animal studies) which show that *n*-3 and *n*-6 PUFAs have contrasting effects on cancer. n−6 PUFA can boost cancer development although n−3 LC-FUFAs (like DHA and EPA) can suppress tumor carcinogens [64, 65]. There is no such propane mechanism yet. But there are few recommended mechanisms which are given below.

The mechanisms recommended are the following:

1. They might act with the help of cellular signaling mediators, including protein kinase C, mitogen-activated protein kinase (MAPK), and NF-κB [61].
2. For nuclear receptors, they might act directly as ligands like peroxisome proliferators-activated receptors (PPARs) and retinoid X receptor alpha [61].

3. They might regulate intracellular homeostasis by regulating Ca^{2+} channels on plasma membrane to start eIF2α kinase which can down-regulate oncogenes and G1 cyclins [61].
4. They might modify lipid composition of the plasma membrane which can affect membrane variability and interaction between T cells and antigen-presenting cells (APCs) [61].
5. They might control non-enzymatic lipid peroxidation in tumor cell which induces apoptosis [61].

The cariogenic effect of *n*-6 LC-PUFAs is the most widely studied due to their production of pro-inflammatory eicosanoids. Therefore, the ratio in dietary *n*-6/*n*-3 PUFAs is related with reduced risk of various types of carcinogenesis. But, this depends on several factors, including race/ethnicity, the source of *n*-3 PUFAs (fish oils, seed oils, purified PUFAs), and genetic differences [61]. Increasing dietary ratio of *n*-6/*n*-3 FAs is associated with lower prostate cancer risk among black men but higher risk in white men, in race/ethnicity-specific analyses [66], due to the genetic differences in a key enzyme which is involved in fatty acid metabolism, including COX and LOX [64].

Second, SNPs in modifier genes may influence the metabolism of *n*-6 and *n*-3 PUFAs. For instance, the enzyme of COX and LOX families can metabolize both *n*-6 and *n*-3 PUFAs to form eicosanoids with contrasting effects that could modulate cancer risk [67]. In addition to SNPs, there are other different parameters which may influence the outcome like gender, ethnicity, presence of other diseases, and medication [60]. Thus, to achieve consistent results, both genetic factors and individual parameters should be considered. Finally, the source of *n*-3 PUFAs and its nature also show significant roles in bioaccessibility and bioavailability [60].

In industrialized countries, the second leading cause of death is cancer. The anticancer drugs are non-selective to describing toxic effects on cancerous and normal cells that's why they show toxicity [38, 39]. GLA have the property to exert tumoricidal activity against a variety of cancers (cancers of breast, pancreas, colon, brain, etc.) [38–40, 68, 69]. GLA inhibits the metastasis of cancer cells [40, 70–72]. The combination of mechanism has been mediated in the inhibition of metastasis. GLA imposes its anticancer action through various mechanisms including angiogenesis inhibition in cancer cells, stimulation of cell death by apoptosis and gene activation. Various studies show that, when GLA is combined with anticancer agents, it improved the efficacy of anticancer drug and also reduced the side effects. GLA has the property to improve the efficacy of anticancer drugs such as paclitaxel, vinorelbine, doxorubicin, epirubicin, mitoxantrone, idarubicin, 5-fluorouracil, and gemcitabine; it is shown in various in vitro studies [73–76]. There are several mechanisms which are including down-regulation of estrogen receptors [73, 77], inhibition of 5-reductase enzyme, and inhibition of cell adhesion molecules like vCAM, ICAM, E-selectin, and urokinase-type plasminogen activator, etc. When GLA is given in combination with tamoxifen, GLA reduced the side effects and improved the efficacy of tamoxifen, along with time to response to therapy which is confirmed by the clinical trial in breast cancer patients [77]. According to the studies, when patients suffering from advanced glioma were injected with GLA into tumor cells, it produced significant reduction in tumor mass and enhanced survival [78].

5 Benefits of Nanotechnology for Solid Tumor

In comparison to human cells, the nanoscale devices are one hundred to ten thousand times smaller [79]. The size is similar as large biological molecules which are known as biomolecules, e.g., enzyme and receptor. Nanoscale devices can interact with biomolecules due to their small size [80]. That is why the nanocarriers have the access to so many areas of the body and have the potential to detect disease and treat the disorder [81, 82].

5.1 Passive Tumor Targeting

In the tumor, the cancer drug delivery should achieve high accumulation and spare the surrounding healthy tissues [83]. Due to extravasation through enhanced permeability and retention, the passive localization of many drugs and drug carriers works for tumors [84]. The mass of the tumor grows rapidly, that is why a network of blood vessels needs to expand quickly because tumor cells need oxygen and nutrient. This abnormal growth (i.e., angiogenesis) results in vessel walls with large pores (40 nm to 1 μm); these permeable vessels allow relatively large nanoparticles to extravasate into tumor masses. With the help of this EPR effect, by freely passing through large pores, nanoparticles can passively target tumors and achieve higher intratumoral accumulation. For solid tumor treatment, the majority of current nanomedicines depend on EPR effect to ensure high drug accumulation in this manner improved treatment efficacy; this drug delivery system is called passive targeting [85].

In solid tumors, the EPR effect is a unique phenomenon but the utilization of the effect of EPR for passive drug delivery for the targetting of the tumor is not without difficulty. The larger tumor mass and the central region of metastatic tumor do not exhibit EPR effect [82]. For this reason, in the clinics to artificially enhance EPR effect, other methods are used, such as slow infusion of angiotensin II to increase systolic blood pressure and therapy of photodynamic or hyperthermia-mediated vascular permeabilization in solid tumors [86].

For solid tumor treatment, the passive accumulation through EPR effect is the most conventional drug delivery system [87, 88]. Yet, nanoparticle size or molecular weight is not the individual factor of the EPR effect; other factors like surface charge, biocompatibility, and in vivo investigation system for macromolecules should not be unnoticed in scheming the nanomedicine for well-organized passive tumor growth [83].

5.2 Active Tumor Targeting

Passive tumor targeting system is responsible for accumulation of particles in the tumor region with the help of EPR. But, nanoparticle/drug cell internalization is required for some of the treatment because EPR does not promote uptake of nanoparticles into cells; however modalities depend on drug activation within the cell nucleus or cytosol [83]. EPR strength varies among different tumors and/or patients because it is heterogeneous. These reasons show that, in the field of nanoparticle therapeutics, active targeting is considered an essential feature [89]. To tumor cells, Active targeting of nanoparticles can be reached with the help of modification in the surface of nanoparticles with small molecules, peptides, and antibodies [83].

The process of extravasation of nanoparticles from the circulation to the tumor region is passive targeting (EPR effect) [90]. Nanoparticles with the drug molecules are released in the extracellular matrix and diffuse the tumor tissue. To facilitate active targeting of particles, they carry surface ligands to receptors which are present on target cell or tissue [90]. To enhance nanoparticle or drug accumulation in tumor, the active targeting is expected and also promotes their prospective cell uptake by receptor-mediated endocytosis [83, 90].

5.3 Transport Across Tissue Barriers

Before the drug can reach the tumor site, nano-drug delivery is prevented by tissue barriers. For efficient transporting of nano-drugs to tumor sites, tissue barriers include tumor stroma (e.g., biological barriers) and tumor endothelium barriers (e.g., functional barriers) [91]. Biological barriers are those physical constructs or cell formation that limits the movement of nanoparticles. Into the tumor mass, Functional barriers can have an impact on the transport of nanoparticles or nanomedicine [92]. To improve the treatment of cancer, it is important to design nanoparticles and strategies to overcome these barriers [93].

There is a dynamic system which is composed of abnormal vasculature, fibroblasts, and immune cells which are known as tumor microenvironment (TME), all embedded in an extracellular matrix (ECM). In the treatment of cancer to deliver the nano-drug, TME poses as both biological and functional [91]. Within a tumor mass, the interstitial fluid pressure is increased due to the escalation of cell density and abnormal vasculature. For free diffusion of the nanoparticles, such pressure gradient is critical and is often a restraining factor for the enhanced permeability and retention (EPR) effect [94].

Clinically EPR effect for nanocarriers' tumor penetration is a relevant phenomenon, to address the tumor endothelium barrier approaches which have been developed. There are various strategies which are given below:

- To reduce interstitial fluid pressure
- To improve tumor penetration containing ECM-targeting pharmacological interventions

- To normalize vasculature within TME
- Hypertonic solutions to shrink ECM cells
- Hyperthermia, radiofrequency (RF), or high-intensity focused ultrasound (HIFU)
- To enhance nano-drug transport and accumulation [95]

In larger tumor mass, these strategies can also reduce hypoxic conditions. In recent years, for nanocarrier stability, TME and tumor mass pose a harsh and acidic environment; pH-responsive nanocarrier designs leveraging this unique feature are gaining interest [95].

Blood-brain barrier is another formidable tissue barrier for drugs and nanotechnology [96]. In the central nervous system, there is a physical barrier which prevents harmful substances from entering the brain [97]. Around the capillaries, there are sealed in continuous tight junction which consists of endothelial cells. One layer of epithelial cell is present outside which is covered by astrocytes that further contribute to the selectivity of substance passage [96]. BBB protects the brain from harmful substances and restricts the therapeutic delivery for brain diseases (brain tumors and other neurological diseases) [98].

6 Significance of Nanoherbal Formulation

Herbal materials are intriguing restorative choices for the treatment of overall perceived maladies, for many diseases [99]. These medications generally have a low water solubility or low permeability and poor physicochemical steadiness; they are quickly metabolized in the body, gathered in tissues, or unstabilized on account of gastrointestinal pH. In the event that the novel medication delivery innovation is connected in herbal drugs, it might help in expanding the adequacy and lessening the symptoms of different herbal materials [100]. This is the fundamental thought behind fusing the novel technique for medication delivery in herbal drugs. There is an increasing interest in developing novel delivery strategies for the natural products [101]. Liposomes, micelles, nanoemulsions, solid lipid nanoparticles, nanostructured lipid carriers, etc. are biocompatible and biodegradable nanoparticles. The stability and solubility of phytochemicals are enhanced in those nanoparticles, show a sustained release property, protect them from premature enzymatic degradation or metabolism, and via passive or targeted delivery increase their absorption and bioavailability, prolong their circulation time, improve their target specificity to cancer cells or tumors, lower toxicity or side effects to normal cells or tissues through preventing them from prematurely interacting with the biological environment, and increase anticancer activities. For developing phytochemical-loaded nanoparticles, nanotechnology opens a door for prevention and treatment of cancer [102, 103]. In recent years, the use of nanotechnology to enhance delivery of phytochemicals to tumors or cancer cells for improving therapeutic efficiency has received considerable attention [104]. Many phytochemicals can be loaded into biocompatible and biodegradable nanoparticles, which can enhance, protect them from degradation by enzymes, enhance their stability, exhibit high differential uptake efficiency in cancer cells (or tumors) over other normal cells (or normal tissues), and lower toxicity through preventing them from prematurely interacting with the biological environment [105]. Several phytochemical nanoparticles have been commercialized. Compared with free phytochemicals, phytochemical nanoparticles not only show increased drug solubility but also other advantages [106]. The change in pharmacokinetics and biodistribution of phytochemicals as a result of forming nanoparticles can lead to improvement in the therapeutic index of these phytochemicals by lowering toxicity and increasing efficacy [106]. Potential antineoplastic phytochemicals are often extracted from edible plants, including medicinal plants, vegetables, and fruits. The abundance and easy availability of plants make phytochemicals diverse and economical to study. Moreover, phytochemicals derived from these edible sources are usually safe for human consumption and even beneficial for normal physiological functions. This means they exhibit low or no toxicity on normal human organs and cells when used at a physiologically relevant concentration. Further studies on pharmacological mechanisms of natural anticancer components showed that they usually not only show bioactivities such as anti-inflammation, antioxidation, and immuno-modulation but also target multiple cancer-related processes and signaling pathways [107]. Their anticancer bioactivities, including apoptosis-promoting, anti-proliferative, and anti-metastatic activities, work alone or in combination to kill cancer cells. Briefly, phytochemicals can act with multiple anticancer mechanisms [108].

7 GLA: Anticancer Drugs

It is interesting that GLA itself has no anticancer property but has the capability of potentiating the tumoricidal actions of other known anticancer medicines [109]. Numerous anticancer drugs and radiation supplement free radical generation and lipid peroxidation procedure and therefore bring about tumoricidal action [110–112].

Since PUFAs including in tumor cell GLA enhance free radical generation and lipid peroxidation process, it is expected that, when GLA and anticancer drugs are given together, they may exhibit synergistic tumoricidal action [68].

8 Summary

Polyunsaturated fatty acids have properties to kill cancer cells (breast, lung, and prostate cancer) with no adverse effect on normal cells which is present in the human. Fatty acids contain three, four, and five double bonds that is why the most consistent and selective effects were obtained. In the absence of polyunsaturated fatty acids when human cancer cells and normal human fibroblasts were co-cultured, the malignant cells overgrew the normal ones. The normal cells outgrew the malignant ones in the presence of fatty acid (eicosapentaenoic acid (EPA, 20:5*n*-3), gamma-linolenic acid (GLA, 18:3*n*-6), or arachidonic acid (AA, 20:4*n*-6). These observations suggest that with polyunsaturated fatty acids treatment of malignancy may have significant potential with a high level of safety.

One of the major aims of cancer therapy is without harming the normal cells to eliminate tumor cells effectively. But this is hardly achieved with the presently available modes of therapy; there are various therapies available but are not without side effects.

In vitro studies have shown that GLA shows lack of side effect and also does not harm normal cells in the present study suggesting that this may be so even in an in vivo situation. In almost all tissues of the body, GLA is present and is also antimutagenic in nature. So in the treatment of cancer cell, GLA may offer a new approach in the treatment.

Nowadays for cancer treatment, recent studies performed in vitro or in animals have directly investigated the biological effects of PUFA-containing nanomaterials. As yet, there are no clinical trials of these nanoformulations that have been tested. With the help of these fatty acids, different strategies have been used, because of their ability to induce healthy effects. Furthermore, sometimes these fatty acids are used as external part of some nanoparticles because these fatty acids have the tendency to target the cancer cells.

Remarkably, in nanoformulation, the addition of fatty acid may show both anti-inflammatory and anti-neoplastic activities which could be more bioavailable. PUFAs present various advantages including reducing adverse effects and toxicity improved targeting specificity towards cancerous cells thus preventing any on toward effect on normal cells.

References

1. Torre LA, Bray F, Siegel RL, Ferlay J, Lortet-Tieulent J, Jemal A (2015) Global cancer statistics, 2012. CA Cancer J Clin 65:87–108
2. Chen W, Zheng R, Baade PD, Zhang S, Zeng H, Bray F, Jemal A, Yu XQ, He J (2016) Cancer statistics in China, 2015. CA Cancer J Clin 66:115–132
3. Miller KD, Siegel RL, Lin CC, Mariotto AB, Kramer JL, Rowland JH, Stein KD, Alteri R, Jemal A (2016) Cancer treatment and survivorship statistics, 2016. CA Cancer J Clin 66:271–289
4. Coleman KM, Doherty MC, Bigler SA (2003) Solid-pseudopapillary tumor of the pancreas. Radiographics 23:1644–1648
5. Scully RE (1987) Classification of human ovarian tumors. Environ Health Perspect 73:15–24
6. Nicholson R, Gee J, Harper M (2001) EGFR and cancer prognosis. Eur J Cancer 37:9–15
7. Avlani D, Majee SB, Biswas GR (2016) Magnet-guided nanovectors as agents for magnetofection in therapeutic management of solid tumors. World J Pharm Res 5(7):469
8. Bertrand N, Wu J, Xu X, Kamaly N, Farokhzad OC (2014) Cancer nanotechnology: the impact of passive and active targeting in the era of modern cancer biology. Adv Drug Deliv Rev 66:2–25
9. Roco MC (1999) Nanoparticles and nanotechnology research. J Nanopart Res 1:1–6
10. Rogers B, Adams J, Pennathur S (2014) Nanotechnology: understanding small systems. CRC, New York
11. De Souza RJ, Mente A, Maroleanu A, Cozma AI, Ha V, Kishibe T, Uleryk E, Budylowski P, Schünemann H, Beyene J (2015) Intake of saturated and trans unsaturated fatty acids and risk of all cause mortality, cardiovascular disease, and type 2 diabetes: systematic review and meta-analysis of observational studies. BMJ 351:h3978

12. Rahman M, Ahmed MZ, Kazmi I et al (2012) Novel approach for the treatment of cancer: theranostic nanomedicines. Pharmacologia 3:371–376
13. Tattrie NH, Bennett J, Cyr R (1968) Maximum and minimum values for lecithin classes from various biological sources. Can J Biochem 46:819–824
14. Demel R, Geurts van Kessel W, van Deenen L (1972) The properties of polyunsaturated lecithins in monolayers and liposomes and the interactions of these lecithins with cholesterol. Biochim Biophys Acta Biomembr 266:26–40
15. Simopoulos AP (2002) The importance of the ratio of omega-6/omega-3 essential fatty acids. Biomed Pharmacother 56:365–379
16. Ander BP, Dupasquier CM, Prociuk MA, Pierce GN (2003) Polyunsaturated fatty acids and their effects on cardiovascular disease. Exp Clin Cardiol 8:164
17. Russo GL (2009) Dietary n− 6 and n− 3 polyunsaturated fatty acids: from biochemistry to clinical implications in cardiovascular prevention. Biochem Pharmacol 77:937–946
18. Simopoulos AP (1991) Omega-3 fatty acids in health and disease and in growth and development. Am J Clin Nutr 54:438–463
19. Gopalan C (2006) Essential fatty acids in maternal diets. Bull Nutr Found India 27(1)
20. Calder PC (2006) n-3 polyunsaturated fatty acids, inflammation, and inflammatory diseases. Am J Clin Nutr 83:1505S–1519S
21. Bell M, Henderson R, Sargent J (1986) The role of polyunsaturated fatty acids in fish. Comp Biochem Physiol B Comp Biochem 83:711–719
22. Lee JY, Sohn KH, Rhee SH, Hwang D (2001) Saturated fatty acids, but not unsaturated fatty acids, induce the expression of cyclooxygenase-2 mediated through Toll-like receptor 4. J Biol Chem 276:16683–16689
23. Leyton J, Drury P, Crawford M (1987) Differential oxidation of saturated and unsaturated fatty acids in vivo in the rat. Br J Nutr 57:383–393
24. Kris-Etherton PM (1999) Monounsaturated fatty acids and risk of cardiovascular disease. Circulation 100:1253–1258
25. Grundy SM (1987) Monounsaturated fatty acids, plasma cholesterol, and coronary heart disease. Am J Clin Nutr (USA). https://doi.org/10.1093/ajcn/45.5.1168
26. Simopoulos AP (1999) Essential fatty acids in health and chronic disease. Am J Clin Nutr 70:560s–569s
27. Das UN (2006a) Essential fatty acids: biochemistry, physiology and pathology. Biotechnol J Healthcare Nutr Technol 1:420–439
28. Mattson FH, Grundy SM (1985) Comparison of effects of dietary saturated, monounsaturated, and polyunsaturated fatty acids on plasma lipids and lipoproteins in man. J Lipid Res 26:194–202
29. Simopoulos AP (2008) The importance of the omega-6/omega-3 fatty acid ratio in cardiovascular disease and other chronic diseases. Exp Biol Med 233:674–688
30. Das U (1990) Gamma-linolenic acid, arachidonic acid, and eicosapentaenoic acid as potential anticancer drugs. Nutrition (Burbank, Los Angeles County, Calif.) 6:429–434
31. Kapoor R, Huang Y-S (2006) Gamma linolenic acid: an antiinflammatory omega-6 fatty acid. Curr Pharm Biotechnol 7:531–534
32. Robinson LE, Buchholz AC, Mazurak VC (2007) Inflammation, obesity, and fatty acid metabolism: influence of n-3 polyunsaturated fatty acids on factors contributing to metabolic syndrome. Appl Physiol Nutr Metab 32:1008–1024
33. Salem N (1999) Introduction to polyunsaturated fatty acids. Background 3:1–8
34. Brzeski M, Madhok R, Capell H (1991) Evening primrose oil in patients with rheumatoid arthritis and side-effects of non-steroidal anti-inflammatory drugs. Rheumatology 30:370–372
35. DeLuca P, Rossetti R, Alavian C, Karim P, Zurier R (1999) Effects of gammalinolenic acid on interleukin-1 beta and tumor necrosis factor-alpha secretion by stimulated human peripheral blood monocytes: studies in vitro and in vivo. J Invest Med 47:246–250

36. Gillis RC, Daley BJ, Enderson BL, Karlstad MD (2004) Inhibition of 5-lipoxygenase induces cell death in anti-inflammatory fatty acid-treated HL-60 cells. J Parenter Enter Nutr 28:308–314
37. Ziboh VA, Naguwa S, Vang K, Wineinger J, Morrissey BM, McIntyre J, Watnik M, Gershwin ME (2004) Suppression of leukotriene B 4 generation by ex-vivo neutrophils isolated from asthma patients on dietary supplementation with gammalinolenic acid-containing borage oil: possible implication in asthma. Clin Dev Immunol 11:13–21
38. Mainou-Fowler T, Proctor SJ, Dickinson AM (2001) γ-Linolenic acid induces apoptosis in B-chronic lymphocytic leukaemia cells in vitro. Leuk Lymphoma 40:393–403
39. Sagar PS, Das U, Koratkar R, Ramesh G, Padma M, Kumar GS (1992) Cytotoxic action of cis-unsaturated fatty acids on human cervical carcinoma (HeLa) cells: relationship to free radicals and lipid peroxidation and its modulation by calmodulin antagonists. Cancer Lett 63:189–198
40. Watkins G, Martin TA, Bryce R, Mansel RE, Jiang WG (2005) γ-Linolenic acid regulates the expression and secretion of SPARC in human cancer cells. Prostaglandins Leukot Essent Fat Acids 72:273–278
41. Coste T, Pierlovisi M, Leonardi J, Dufayet D, Gerbi A, Lafont H, Vague P, Raccah D (1999) Beneficial effects of gamma linolenic acid supplementation on nerve conduction velocity, Na+, K+ ATPase activity, and membrane fatty acid composition in sciatic nerve of diabetic rats. J Nutr Biochem 10:411–420
42. Jamal G (1994) The use of gamma linolenic acid in the prevention and treatment of diabetic neuropathy. Diabet Med 11:145–149
43. Begin ME, Ells G, Horrobin DF (1988) Polyunsaturated fatty acid-induced cytotoxicity against tumor cells and its relationship to lipid peroxidation. J Natl Cancer Inst 80:188–194
44. Ells GW, Chisholm KA, Simmons VA, Horrobin DF (1996) Vitamin E blocks the cytotoxic effect of γ-linolenic acid when administered as late as the time of onset of cell death—insight into the mechanism of fatty acid induced cytotoxicity. Cancer Lett 98:207–211
45. Damtew B, Spagnuolo P (1997) Tumor cell-endothelial cell interactions: evidence for roles for lipoxygenase products of arachidonic acid in metastasis. Prostaglandins Leukot Essent Fat Acids 56:295–300
46. Earashi M, Noguchi M, Kinoshita K, Tanaka M (1995) Effects of eicosanoid synthesis inhibitors on the in vitro growth and prostaglandin E and leukotriene B secretion of a human breast cancer cell line. Oncology 52:150–155
47. CRAWFORD MA (1983) Background to essential fatty acids and their prostanoid derivatives. Br Med Bull 39:210–213
48. Abou El-Ela SH, Prasse KW, Carroll R, Bunce OR (1987) Effects of dietary primrose oil on mammary tumorigenesis induced by 7, 12-dimethylbenz (a) anthracene. Lipids 22:1041–1044
49. Bell JG, Tocher DR, MacDonald FM, Sargent JR (1995) Diets rich in eicosapentaenoic acid and γ-linolenic acid affect phospholipid fatty acid composition and production of prostaglandins E1, E2 and E3 in turbot (Scophthalmus maximus), a species deficient in Δ5 fatty acid desaturase. Prostaglandins Leukot Essent Fat Acids 53:279–286
50. Rose DP, Connolly JM, Liu XH (1995) Effects of linoleic acid and γ-linolenic acid on the growth and metastasis of a human breast cancer cell line in nude mice and on its growth and invasive capacity in vitro. Nutr Cancer 24(1):33–45
51. Harnack K, Andersen G, Somoza V (2009) Quantitation of alpha-linolenic acid elongation to eicosapentaenoic and docosahexaenoic acid as affected by the ratio of n6/n3 fatty acids. Nutr Metab 6:8
52. Barre DE (2007) The role of consumption of alpha-linolenic, eicosapentaenoic and docosahexaenoic acids in human metabolic syndrome and type 2 diabetes—a mini-review. J Oleo Sci 56:319–325
53. Innis SM (2008) Dietary omega 3 fatty acids and the developing brain. Brain Res 1237:35–43
54. McNamara RK, Carlson SE (2006) Role of omega-3 fatty acids in brain development and function: potential implications for the pathogenesis and prevention of psychopathology. Prostaglandins Leukot Essent Fat Acids 75:329–349

55. Rose DP, Connolly JM (1999) Omega-3 fatty acids as cancer chemopreventive agents. Pharmacol Ther 83:217–244
56. Wall R, Ross RP, Fitzgerald GF, Stanton C (2010) Fatty acids from fish: the anti-inflammatory potential of long-chain omega-3 fatty acids. Nutr Rev 68:280–289
57. Meyer BJ, Mann NJ, Lewis JL, Milligan GC, Sinclair AJ, Howe PR (2003) Dietary intakes and food sources of omega-6 and omega-3 polyunsaturated fatty acids. Lipids 38:391–398
58. Tocher DR, Francis DS, Coupland K (2010) n-3 Polyunsaturated fatty acid-rich vegetable oils and blends. In: Fish oil replacement and alternative lipid sources in aquaculture feeds. CRC, New York, pp 209–244
59. Sargent J (1997) Fish oils and human diet. Br J Nutr 78:S5–S13
60. Saini RK, Keum Y-S (2018) Omega-3 and omega-6 polyunsaturated fatty acids: dietary sources, metabolism, and significance—a review. Life Sci 203:255–267
61. Berquin IM, Edwards IJ, Chen YQ (2008) Multi-targeted therapy of cancer by omega-3 fatty acids. Cancer Lett 269:363–377
62. Stillwell W, Wassall SR (2003) Docosahexaenoic acid: membrane properties of a unique fatty acid. Chem Phys Lipids 126:1–27
63. Dennis EA, Norris PC (2015) Eicosanoid storm in infection and inflammation. Nat Rev Immunol 15:511
64. Kimura Y, Kono S, Toyomura K, Nagano J, Mizoue T, Moore MA, Mibu R, Tanaka M, Kakeji Y, Maehara Y (2007) Meat, fish and fat intake in relation to subsite-specific risk of colorectal cancer: The Fukuoka Colorectal Cancer Study. Cancer Sci 98:590–597
65. Zheng J-S, Hu X-J, Zhao Y-M, Yang J, Li D (2013) Intake of fish and marine n-3 polyunsaturated fatty acids and risk of breast cancer: meta-analysis of data from 21 independent prospective cohort studies. BMJ 346:f3706
66. Williams CD, Whitley BM, Hoyo C, Grant DJ, Iraggi JD, Newman KA, Gerber L, Taylor LA, McKeever MG, Freedland SJ (2011) A high ratio of dietary n-6/n-3 polyunsaturated fatty acids is associated with increased risk of prostate cancer. Nutr Res 31:1–8
67. Siddiqui RA, Harvey KA, Zaloga GP (2008) Modulation of enzymatic activities by n-3 polyunsaturated fatty acids to support cardiovascular health. J Nutr Biochem 19:417–437
68. Das UN (2004) From bench to the clinic: γ-linolenic acid therapy of human gliomas. Prostaglandins Leukot Essent Fat Acids 70:539–552
69. Menendez JA, Colomer R, Lupu R (2004) ω-6 polyunsaturated fatty acid γ-linolenic acid (18: 3n-6) is a selective estrogen-response modulator in human breast cancer cells: γ-linolenic acid antagonizes estrogen receptor-dependent transcriptional activity, transcriptionally represses estrogen receptor expression and synergistically enhances tamoxifen and ICI 182,780 (Faslodex) efficacy in human breast cancer cells [2]. Int J Cancer 109:949–954
70. Jiang WG, Bryce R, Mansel R (1997a) Gamma linolenic acid regulates gap junction communication in endothelial cells and their interaction with tumour cells. Prostaglandins Leukot Essent Fat Acids 56:307–316
71. Jiang WG, Hiscox S, Horrobin DF, Bryce RP, Mansel RE (1997b) Gamma linolenic acid regulates expression of maspin and the motility of cancer cells. Biochem Biophys Res Commun 237:639–644
72. Jiang WG, Hiscox S, Bryce R, Horrobin D, Mansel RE (1998) The effects of n-6 polyunsaturated fatty acids on the expression of nm-23 in human cancer cells. Br J Cancer 77:731
73. Davies C, Loizidou M, Cooper A, Taylor I (1999) Effect of γ-linolenic acid on cellular uptake of structurally related anthracyclines in human drug sensitive and multidrug resistant bladder and breast cancer cell lines. Eur J Cancer 35:1534–1540
74. Menéendez JA, Ropero S, del Mar Barbacid M, Montero S, Solanas M, Escrich E, Cortés-Funes H, Colomer R (2002) Synergistic interaction between vinorelbine and gamma-linolenic acid in breast cancer cells. Breast Cancer Res Treat 72:203–219
75. Menendez J, del Mar Barbacid M, Montero S, Sevilla E, Escrich E, Solanas M, Cortes-Funes H, Colomer R (2001) Effects of gamma-linolenic acid and oleic acid on paclitaxel cytotoxicity in human breast cancer cells. Eur J Cancer 37:402–413

76. Whitehouse P, Cooper A, Johnson C, Adrian TE (2003) Synergistic activity of gamma-linolenic acid and cytotoxic drugs against pancreatic adenocarcinoma cell lines. Pancreatology 3:367–374
77. Kenny FS, Pinder SE, Ellis IO, Gee JM, Nicholson RI, Bryce RP, Robertson JF (2000) Gamma linolenic acid with tamoxifen as primary therapy in breast cancer. Int J Cancer 85:643–648
78. Das U, Prasad V, Reddy DR (1995) Local application of γ-linolenic acid in the treatment of human gliomas. Cancer Lett 94:147–155
79. Cattaneo AG, Gornati R, Sabbioni E, Chiriva-Internati M, Cobos E, Jenkins MR, Bernardini G (2010) Nanotechnology and human health: risks and benefits. J Appl Toxicol 30:730–744
80. Chow EK-H, Ho D (2013) Cancer nanomedicine: from drug delivery to imaging. Sci Transl Med 5:216rv214
81. Fang J, Nakamura H, Maeda H (2011) The EPR effect: unique features of tumor blood vessels for drug delivery, factors involved, and limitations and augmentation of the effect. Adv Drug Deliv Rev 63:136–151
82. Farokhzad OC, Langer R (2009) Impact of nanotechnology on drug delivery. ACS Nano 3:16–20
83. Torchilin VP (2010) Passive and active drug targeting: drug delivery to tumors as an example, drug delivery. Springer, New York, pp 3–53
84. Kamaly N, Xiao Z, Valencia PM, Radovic-Moreno AF, Farokhzad OC (2012) Targeted polymeric therapeutic nanoparticles: design, development and clinical translation. Chem Soc Rev 41:2971–3010
85. Zhu L, Staley C, Kooby D, El-Rays B, Mao H, Yang L (2017) Current status of biomarker and targeted nanoparticle development: the precision oncology approach for pancreatic cancer therapy. Cancer Lett 388:139–148
86. Anchordoquy TJ, Barenholz Y, Boraschi D, Chorny M, Decuzzi P, Dobrovolskaia MA, Farhangrazi ZS, Farrell D, Gabizon A, Ghandehari H (2017) Mechanisms and barriers in cancer nanomedicine: addressing challenges, looking for solutions. ACS, Washington, DC
87. Rahman M, Akhter S, Ahmad MZ, Ahmad J, Addo RT, Ahmad FJ, Pichon C (2015) Emerging advances in cancer nanotheranostics with graphene nanocomposites: opportunities and challenges. Nanomedicine. 10(15):2405–2422
88. Rahman M, Zaki Ahmad M, Kazmi I, Akhter S, Afzal M, Gupta G, Ranjan SV (2012) Emergence of nanomedicine as cancer targeted magic bullets: recent development and need to address the toxicity apprehension. Curr Drug Discov Technol 9(4):319–329
89. Shi J, Kantoff PW, Wooster R, Farokhzad OC (2017) Cancer nanomedicine: progress, challenges and opportunities. Nat Rev Cancer 17:20
90. Danhier F, Feron O, Préat V (2010) To exploit the tumor microenvironment: passive and active tumor targeting of nanocarriers for anti-cancer drug delivery. J Control Release 148:135–146
91. Klepper J, Wang D, Fischbarg J, Vera JC, Jarjour IT, O'Driscoll KR, Darryl C (1999) Defective glucose transport across brain tissue barriers: a newly recognized neurological syndrome. Neurochem Res 24:587–594
92. Juillerat-Jeanneret L (2008) The targeted delivery of cancer drugs across the blood–brain barrier: chemical modifications of drugs or drug-nanoparticles? Drug Discov Today 13:1099–1106
93. Petros RA, DeSimone JM (2010) Strategies in the design of nanoparticles for therapeutic applications. Nat Rev Drug Discov 9:615
94. Kobayashi H, Watanabe R, Choyke PL (2014) Improving conventional enhanced permeability and retention (EPR) effects; what is the appropriate target? Theranostics 4:81
95. McNeil SE (2011) Unique benefits of nanotechnology to drug delivery and diagnostics. In: Characterization of nanoparticles intended for drug delivery. Springer, New York, pp 3–8
96. Lockman P, Mumper R, Khan M, Allen D (2002) Nanoparticle technology for drug delivery across the blood-brain barrier. Drug Dev Ind Pharm 28:1–13

97. Begley DJ (1996) The blood-brain barrier: principles for targeting peptides and drugs to the central nervous system. J Pharm Pharmacol 48:136–146
98. Gabathuler R (2010) Approaches to transport therapeutic drugs across the blood–brain barrier to treat brain diseases. Neurobiol Dis 37:48–57
99. Zeng Y, Song JX, Shen XC (2012) Herbal remedies supply a novel prospect for the treatment of atherosclerosis: a review of current mechanism studies. Phytother Res 26:159–167
100. Saraf S (2010) Applications of novel drug delivery system for herbal formulations. Fitoterapia 81:680–689
101. Devi VK, Jain N, Valli KS (2010) Importance of novel drug delivery systems in herbal medicines. Pharmacogn Rev 4:27
102. Chuan L, ZHANG J, Yu-Jiao Z, Shu-Fang N, Jun C, Qian W, Shao-Ping N, Ze-Yuan D, Ming-Yong X, Shu W (2015) Biocompatible and biodegradable nanoparticles for enhancement of anti-cancer activities of phytochemicals. Chin J Nat Med 13:641–652
103. Rahman M, Ahmad MZ, Kazmi I, Akhter S, Afzal M, Gupta G, Jalees Ahmed F, Anwar F (2012) Advancement in multifunctional nanoparticles for the effective treatment of cancer. Expert Opin Drug Deliv 9(4):367–381
104. Kievit FM, Zhang M (2011) Cancer nanotheranostics: improving imaging and therapy by targeted delivery across biological barriers. Adv Mater 23:H217–H247
105. Rahman M, Ahmad MZ, Kazmi I, Akhter S, Kumar Y, Ahmad FJ, Anwar F (2012) Novel approach for the treatment of cancer: theranostic nanomedicine. Pharmacologia 3(9):371–376
106. Xie J, Yang Z, Zhou C, Zhu J, Lee RJ, Teng L (2016) Nanotechnology for the delivery of phytochemicals in cancer therapy. Biotechnol Adv 34:343–353
107. Wang H, Oo Khor T, Shu L, Su Z-Y, Fuentes F, Lee J-H, Tony Kong A-N (2012) Plants vs. cancer: a review on natural phytochemicals in preventing and treating cancers and their druggability. Anticancer Agents Med Chem 12:1281–1305
108. Wu W, Luo L, Wang Y, Wu Q, Dai H-B, Li J-S, Durkan C, Wang N, Wang G-X (2018) Endogenous pH-responsive nanoparticles with programmable size changes for targeted tumor therapy and imaging applications. Theranostics 8:3038
109. Xu Y, Qian SY (2014) Anti-cancer activities of ω-6 polyunsaturated fatty acids. Biom J 37:112
110. Das UN (2006b) Tumoricidal and anti-angiogenic actions of gamma-linolenic acid and its derivatives. Curr Pharm Biotechnol 7:457–466
111. Backes J, Anzalone D, Hilleman D, Catini J (2016) The clinical relevance of omega-3 fatty acids in the management of hypertriglyceridemia. Lipids Health Dis 15:118
112. Menendez JA, Vellon L, Colomer R, Lupu R (2005) Effect of γ-linolenic acid on the transcriptional activity of the Her-2/neu (erbB-2) oncogene. J Natl Cancer Inst 97:1611–1615

Nanomedicine Approaches for the Delivery of Herbal Anticancer Drugs

Sobiya Zafar, Gaurav Kumar Jain, and Farhan Jalees Ahmad

1 Introduction

Cancer is the most prevalent and resistant disease that indiscriminately affects people of any age, gender or social status accounting for about 18 million new cases and approximately 10 million deaths worldwide in 2018 and thus poses a huge challenge for the scientific community [1]. Molecular changes in the cellular behaviour that cause the development and progression of cancer are usually linked to genetic alterations which may be inherited or induced by endogenous and exogenous carcinogenic factors. The six hallmarks of cancer that dictate the malignant growth include self-sufficiency in growth signals, insensitivity to the anti-growth signals, tissue invasion and metastasis, unlimited replicative potential, sustained angiogenesis and evasion of apoptosis. Chemotherapy, the mainstay of cancer treatment, though reduced the mortality/morbidity, still the efficacy, safety and convenience for the patient remain a great challenge. The anticancer agents have a narrow therapeutic index, develop multidrug resistance (MDR) that can be innate and/or adaptive and non-specifically distribute to tissues upon intravenous administration resulting in severe side effects to healthy tissues [2].

Natural products derived from plants, animals and minerals have been in use since ancient times for the treatment and prevention of several diseases and still considered as a major source for the modern drugs. The earlier knowledge on the medicinal application of plants in the Greco-Roman culture followed the empirical basis, without mechanistic information on their pharmacological properties or the active phytoconstituents. It was only with the work of Anton von Störck on the poisonous herbs (aconite and colchicum) and William Withering on the use of foxglove for edema, in the eighteenth century, the rational clinical investigation of the

S. Zafar · G. K. Jain · F. J. Ahmad (✉)
Nanomedicine Research Lab, School of Pharmaceutical Education & Research, Jamia Hamdard, New Delhi, India
e-mail: fjahmad@jamiahamdard.ac.in

© Springer Nature Singapore Pte Ltd. 2020

M. Rahman et al. (eds.), *Nanomedicine for Bioactives*,
https://doi.org/10.1007/978-981-15-1664-1_7

medicinal plants came into existence. From the advent of nineteenth century and until now, the rational drug discovery from the natural sources, such as plants, has laid the scientific and financial foundation of the modern pharmaceutical industry [3]. A substantial proportion of new chemical entities discovered were derived or inspired from the nature. The prominent examples of plant-derived compounds which have become indispensable in cancer treatment regimens include anti-cancer agents, such as paclitaxel and its derivatives docetaxel and cabazitaxel obtained from yew (*Taxus*) species (sp.), alkaloids such as vincristine and vinblastine discovered from Madagascar periwinkle (*Catharanthus roseus* sp.), artemisinin originally derived from Chinese herb *Artemisia annua* L. and camptothecin obtained from Chinese tree *Camptotheca acuminata* Decne [4].

2 Phytochemicals in Cancer

Phytochemicals have gained immense interest from the scientific community for their potential in cancer chemoprevention and therapy. The word phyto refers to "plants" in Greek, and phytochemicals are the naturally occurring plant-derived compounds which may be the secondary metabolites or the bioactive constituents of the plants. A growing body of evidence for the chemotherapeutics-associated toxicity (even at therapeutic doses) has shifted the interest to the investigation of natural products for cancer treatment. Diverse herbal drugs or phytochemicals have been reported possessing anticancer activity at various stages of cancer development and progression [5]. Being pleiotropic in nature, they down-regulate multiple signaling pathways and cause the modulation of miRNAs, cellular signaling and epigenome preventing progression and proliferation of cancer cells, causing apoptosis/cell cycle arrest and suppressing the invasion, metastasis and angiogenesis of cancerous cells [6]. Phytomolecules produce their action by affecting the major signaling targets/pathways, some of which may include PI3K/Akt/mTOR pathway, Ras/Raf signaling, MAPK/ERK pathway, insulin-like growth factor receptor (IGFR), epidermal growth factor receptor (EGFR) family receptors, sonic hedgehog signaling, Wnt/β--catenin signaling pathway, etc. [7–10]. Resveratrol, a phytoalexin found in many plants including grapes, peanuts and berries, has been shown to affect the nuclear factor κB (NF-κB) signaling pathway and IGF-1R/Akt/Wnt pathways and activate p53, influencing cell development and tumourigenesis. The inhibition of PI3K/Akt pathway by resveratrol regulates the cell differentiation, growth and proliferation [7]. Curcumin, a major yellow pigment and spice in turmeric, has been proven to be a very potent anticancer agent. The common molecular targets of curcumin include transcription factors, inflammatory mediators, protein kinases and enzymes like protein reductases and histone acetyltransferase, including TNF-α, IL-1,6,8,12, NF-κB, p38 MAPK, cyclin D1, VEGF, glutathione and others [8]. Curcumin has shown effectiveness in every stage of cancer progression including induction of tumour apoptosis, inhibition of tumour proliferation, angiogenesis, invasion and metastasis, regulation of cell cycle and cancer stem cell and may also increase chemotherapy sensitivity to treatment [11]. One of the most promising, intensively studied herbs for tumour suppression is *Nigella sativa* (*N. sativa*). Thymoquinone, one of the major bioactive phytochemical constituents of the oil and other extracts of *N. sativa* seeds, has garnered much attention for the anticancer activity in several in vitro and xenograft models of solid tumour. Thymoquinone has been shown to induce the mitochondrial pro-apoptotic pathway, leading to loss of mitochondrial membrane potential, enhanced cleavage of Poly (ADP-ribose) polymerase (PARP), activation of caspase-3 and caspase-9 and up-regulated expression of Bax, p53 and cytochrome c, and also mediates suppression of the expression of Bcl-2, Bcl-xL and survivin [9, 12]. Epigallocatechin gallate (EGCG), the most abundant polyphenol in green tea, induces autophagy and cell apoptosis [10]. Genistein, an isoflavone compound of soy products, produces potent effects on inflammation, cellular proliferation and modulation of epigenetic changes. It also regulates the cyclo-oxygenases-2 (COX-2) pathway, inhibits matrix metalloproteinase (MMP) and activates the G2/M cell cycle arrest [13]. Luteolin is another plant flavonoid with antimetastatic, anti-angiogenic and cell proliferation inhibitory activity mediated through suppression of cell survival pathways such PI3K/Akt, NF-κB and X-linked inhibitor of apoptosis protein (XIAP) and stimulates the apoptotic pathways by inducing tumour suppressor gene, p53 [14]. Furthermore, the combined effects of three nutraceuticals have been observed in breast cancer patients undergoing tamoxifen therapy. A daily supplement of CoQ10 (100 mg), riboflavin (10 mg) and niacin (50 mg) (CoRN), one dosage per day along with tamoxifen (TAM) 10 mg twice a day for 45 days and 90 days, resulted in a significant reduction in pro-angiogenic marker levels and increased the levels of anti-angiogenic factors which indicated good prognosis and efficacy of the treatment [15].

Alterations in drug efflux mechanisms are a common cause of multidrug resistance. Overexpression of ATP-binding cassette (ABC) transporter proteins, including P-glycoprotein (P-gp), MDR-associated protein 1 (MRP1) and breast cancer resistance protein (BCRP), pumps the therapeutic agents out of the intracellular milieu into the extracellular matrix, reducing the effective intracellular drug concentrations [16]. Common substrates for P-gp include anthracyclines, taxanes, vinca alkaloids and antimetabolites. Natural compounds have shown substantial effectiveness in overcoming the MDR associated with commonly used chemotherapeutic agents. Tea catechins, epicatechin gallate (ECG) and EGCG augment doxorubicin (DOX) induced cell killing of resistant hepatocellular carcinoma (HCC) cells by reversal of MDR, involving the suppression of MDR1 expression and enhancing the intracellular DOX accumulation sensitizing the chemoresistant HCC cells to DOX [17]. Chrysin and biochanin A at concentrations of 0.5 or 1 μM were found to be the most potent BCRP inhibitors, producing significant increase in mitoxantrone accumulation in BCRP-positive MCF-7 MX100 breast cancer cells [18]. Epithelial to mesenchymal transition (EMT) process is the initiating point for tumour metastasis. The epithelial cells could be transformed into the cells which have mesenchymal morphology and lose polarization and tight connection via EMT. The down-regulation of cell adhesion factor, E-cadherin, and the up-regulation of mesenchymal markers, N-cadherin, Snail, Slug, Vimentin, ZEB, Twist, etc. have been observed [19]. Kaempferol, a phytoestrogen abundantly present in tea, grapes, berries, Brussels sprouts and broccoli, has shown inhibitory effect on EMT by causing a reduction in EMT related markers in non-small cell lung cancer cells. It can also suppress TGF-α-induced migration and EMT by abrogating the expression of matrix metalloproteinase (MMP) and increasing E-cadherin expression [20]. The combination of paclitaxel (PTX) and phenethyl isothiocyanate (PEITC), a member of Brassicaceae family, was observed to have a profound synergistic inhibitory effect on the growth of breast cancer cells evident with a decreased IC50 value of the combination with increased apoptosis of over more than twofold and significant increase in the cells at the G2/M phase arrest [21]. Herbal drugs can also overcome the toxicity issues with synthetic chemotherapeutics. Combination of resveratrol (10 mg/kg) with DOX (15 mg/kg) increased the mean survival time of Ehrlich ascites carcinoma bearing mice to 30 days and protected the xenograft models against the DOX-induced cardiotoxicity [22]. Thymoquinone has shown effectiveness in preventing DOX-induced cardiotoxicity, carbon tetrachloride-evoked hepatotoxicity and cisplatin-induced nephropathy [23].

Till date, a plethora of nutraceuticals have been investigated for cancer prevention and management worldwide, but the poor aqueous solubility and permeability of the nutraceuticals limit their therapeutic effectiveness. Nanotechnology has emerged as a solution to circumvent these problems. Nanocarriers have demonstrated immense potential to enhance the bioavailability of nutraceuticals.

3 Nanomedicine Approaches for Herbal Drug Delivery

The extraordinary chemical diversity, the favourable chemical and biological properties and less toxicity with the natural products make them suitable leads in the discovery of novel drugs for cancer treatment. The tangible importance of plant-based therapeutics in cancer treatment is unquestionable. However, their efficacy is frequently scarce attributed to their poor water solubility and an intrinsic dissolution rate, low absorption, high first-pass metabolism, scarce biodistribution and poor penetration leading to low bioavailability, larger dosage requirement with increased frequency of dose administration, physical or chemical instability and low targeting efficacy [24]. The development of appropriate drug delivery systems, such as nanocarriers, can offer a more advanced approach for overcoming the potential limitations of phytochemicals.

Nanotechnology in therapeutic delivery was first described in the 1960s with the advent of liposomal carriers for the delivery of drugs and proteins to various diseases. Major breakthrough has been observed in the nanocarrier drug delivery systems and clinical therapeutics for cancer therapy with over two dozen nanotechnology-based therapeutic products approved by the Food and Drug Administration (FDA) for clinical use and more being evaluated under clinical trials. Paclitaxel (PTX)-bound albumin nanocarrier Abraxane and DOX-encapsulated liposomal nanocarrier Doxil (Johnson & Johnson, Piscataway, NJ) are the significant contribution of nanotechnology for the treatment of various cancers [25, 26]. Nanoparticle formulations have an edge over the conventional formulation or the combinations of free drugs. Nanoparticles offer controlled drug release that can normalize the pharmacokinetics, biodistribution and stability of chemically dissimilar drugs that may independently have disparate pharmacological behaviours. They can be endowed with long-circulating property that causes the continuous drug release at controlled ratios. Stimuli-responsive nanocarriers respond to the external or the internal stimuli that can co-release drugs resulting in increased therapeutic efficacy with reduced toxicity preventing off-target exposure. The nanoparticles can be targeted to the cancer cells by two approaches, namely, passive and active targeting. Passive targeting facilitates the effective localization of nanoparticles in the tumour interstitium but is limited in promoting their uptake by the cancer cells. This can be achieved by actively targeting nanoparticles to receptors or other surface membrane proteins that are overexpressed on the target cancer cells. Multifunctional nanoparticles can target specific tissues allowing the transport of the therapeutic agents across the biological barriers. The nano-size and the targeting ligands facilitate their cellular uptake by the target cells leading to efficient drug delivery and hence therapeutic action at the targeted location. Target molecules can be proteins, sugars or lipids present in the diseased organs or on the surface of cells. The promising potential of many isolated natural products in cancer treatment has encouraged the use of nanotechnologists for the design of useful nanoformulations for improvising the solubility, stability, specificity, tolerability and therapeutic index of the phytoconstituents [24, 27]. Some of the important contributions in the field of phytoconstituent-loaded cancer nanotherapeutics is highlighted in Table 1.

3.1 Lipid-Based Nanoparticles

Liposomes represent the first generation of lipid-based nanocarriers and one of the first nanosized drug delivery systems ever to be produced. The constituents of liposomes primarily include natural and/or synthetic phospho- and sphingolipids and other membrane bilayer constituents like cholesterol and hydrophilic polymer-conjugated lipids positioned randomly around each liposomal vesicle. Cholesterol reduces liposomal membrane fluidity decreasing the permeability of water-soluble molecules through the liposomal membrane, improving its stability in biologic fluids. The conjugation of poly(ethylene glycol) (PEG) to the liposomal surface reduces its interaction with plasma proteins such as opsonins through steric hindrance, which reduces opsonization and uptake of liposomes by the reticuloendothelial system (RES) extending circulation time of liposomes within the body, consequently increasing their accumulation within tumours [73]. Pegylated nanoliposomes encapsulating silibinin (obtained from dried extracts of milk thistle seeds) and glycyrrhizic acid (obtained from *Glycyrrhiza glabra* (L.)) produced synergistic effects in HepG2 cells with lower IC_{50} value of combination compared to single agent alone [28]. Curcumin, being a natural photosensitizer, was investigated for photodynamic therapy of cancer using nanoliposomes and PVP-capped gold nanoparticles. The light irradiation augmented the cytotoxicity of both curcumin liposomes and gold nanoparticles (GNPs), with former being superior over the

Table 1 In vitro and in vivo studies evaluating the anticancer effects of phytoconstituent-loaded nanoparticles

Phytoconstituent	Type of NPs	Cancer	Cell lines	In vivo model	Dose	Key outcomes	Ref.
Silibinin and glycyrrhizic acid	Liposome	Hepatocellular carcinoma	HepG2	—	—	Enhanced cytotoxicity and synergistic effects	Ochi et al. [28]
Curcumin	Liposome	Hepatocellular carcinoma	HepG2	Ehrlich solid tumour model	20 mg/kg	Light irradiation significantly augmented the cytotoxicity Superior regression in tumour volume	Fadel et al. [29]
Aloe emodin	Cubosome	Breast cancer	MCF-7	—	7.5 mg/kg	Good serum stability and hemocompatibility Higher cellular uptake Significant increase in the bioavailability	Freag et al. [30]
Epigallocatechin-3-gallate	Ethosome	Skin cancer	A431	7,12-Dimethylbenz[a]anthracene (DMBA)-induced skin cancer	1 mg/cm^2	Reasonable skin deposition and preservation of the inherent antioxidant properties of EGCG Strong chemotherapeutic and chemopreventive potential	El-Kayal et al. [31]
Wogonin	SLNs	Breast cancer	MCF-7	—	—	Exhibited enhanced cytotoxicity and inhibited poly(ADP-ribose) polymerase	Baek et al. [32]

Linalool	SLNs	Liver and lung cancer	HepG2 and A549	—	—	Controlled drug release profiles for 72 h Enhanced anti-proliferative effects	Rodenak-Kladniew et al. [33]
Resveratrol	SLNs	Glioma	C6	—	2 mg/kg	Higher area under the curve, plasma half-life and distribution in brain tissue	Vijayakumar et al. [34]
Linalool	SNEDDs	Epithelial ovarian cancer (EOC)	A2780, HeyA8, and SKOV3ip1	HeyA8 and SKOV3ip1 xenografts and PDX	100 mg/kg 50 mg/kg + PTX (6 mg/kg)	Increased apoptosis in EOC cells mediated via ROS generation, decrease in mitochondrial membrane potential and increase in caspase-3 levels Remarkable decrease in tumour weight in both SKOV3ip1 xenografts and PDX at reduced dose of 50 mg/kg in combination with PTX	Han et al. [35]

(continued)

Table 1 (continued)

Phytoconstituent	Type of NPs	Cancer	Cell lines	In vivo model	Dose	Key outcomes	Ref.
Curcumin	LNCs	Glioma	9L	—	10 mg/kg	Produced enhanced anti-proliferative and apoptotic effect Improved pharmacokinetic properties	Lollo et al. [36]
Diferuloylmethane	Nanoemulsion	Lung cancer	A549	—	50 mg/kg	Improved the absorption and bioavailability	Sun et al. [37]
Mangiferin	SPNMS	Breast cancer	MCF-7 and MDA-MB-231	—	30 mg/Kg	Higher lymphatic uptake 6.6- and 3.0-folds augmentation in permeability and bioavailability	Khurana et al. [38]
Honokiol	mPEG-PCL-g-PEI PMs	Nasopharyngeal carcinoma	HNE-1	Mouse HNE-1 xenograft model	3 mg/kg	Remarkable suppression of both CD 31 and Ki 67 in the tumour tissues Effective suppression of tumour metabolism and proliferation, caused cell cycle arrest and induced apoptosis Increased the median survival time to 57.5 days	Yang et al. [39]

Ginsenoside-Rg3	PCL-PEG PMs	Glioma	C6	—	—	Targeted NPs could cross the blood-brain barrier more easily with enhanced cellular uptake.	Su et al. [40]
Emodin	PLGA-TPGS NPs	Liver cancer	HepG2 and HCa-F	HCa-F xenograft model	10 mg/kg	More effective cellular uptake and enhanced anti-tumour effects due to NPs small size and presence of TPGS which prevented drug efflux	Liu et al. [41]
Bufalin	VES-CSO/ TPGS mixed PMs	Colon	LoVo/ ADR HCT116/LOHP	LoVo/ADR-xenografts	2 mg/kg	Overcome MDR due to enhanced apoptosis rate and P-gp efflux inhibition Improved pharmacokinetic and tumour targeting	Yuan et al. [42]
Oleanolic acid	mPEG-PLGA/ PLA NPs	Lung and liver cancer	A549 cells and HepG2	—	—	Induced apoptosis and enhanced anti-proliferative effects	Man et al. [43]
Curcumin and Chrysin	PLGA NPs	Colorectal cancer	Caco-2	—	—	Inhibited hTERT gene expression	Lotfi-Attari et al. [44]

(continued)

Table 1 (continued)

Phytoconstituent	Type of NPs	Cancer	Cell lines	In vivo model	Dose	Key outcomes	Ref.
Quercetin	PLGA NPs	Cervical cancer	Caski, Hela and Siha	—	—	Inhibited multiple signalling pathways promoting anti-proliferative, anti-migratory and apoptosis	Luo et al. [45]
Piperine	PEG-PLGA NPs	Multiple cancers	MDA-MB-231, KB, A549, HepG2, HeLa, MCF-7	—	—	Suppressed P-gp activity in paclitaxel resistant MCF-7 cells Enhanced the therapeutic potential of PTX	Pachauri et al. [46]
Apigenin	PLGA NPs	Colorectal cancer	HT-29 and HCT-116	1,2-Dimethylhydrazine (DMH)-induced colon cancer	20 and 2.5 mg/kg	Decreased glutathione (GSH) level Enhanced cellular uptake due to aptamer binding to overexpressed EpCAM receptor More colonic accumulation and less hepatic clearance	Dutta et al. [47]

Garcinol	PLGA NPs	Melanoma	B16F10, HepG2 and KB	B16F10 melanoma tumour-bearing mice	40 mg/kg	Inhibition of cancer cells proliferation through an apoptotic pathway Gradually increased tumour to muscle and tumour to blood ratios	Gaonkar et al. [48]
Quercetin	Chitosan NPs	Breast and lung cancer	A549 and MDA MB 468	A549 and MDA MB 468 xenografts	25 mg/kg	Significant reduction of tumour volume Markedly increased Superoxide dismutase (SOD) level	Baksi et al. [49]
Silibinin	IONs	Lung cancer	A549	—	—	Dose- and time-dependent cytotoxicity Reduced hTERT expression	Amirsaadat et al. [50]
Curcumin	IONs	Sarcoma	-	Sarcoma-180 solid tumour-bearing mice	20 mg/kg	Greater accumulation within the tumour tissue due to folate targeting and magnetically induced heating ability of IONs lead to more effective killing of the tumour cell	Thu Huong et al. [51]

(continued)

Table 1 (continued)

Phytoconstituent	Type of NPs	Cancer	Cell lines	In vivo model	Dose	Key outcomes	Ref.
Curcumin	MSNs	Hepatocellular carcinoma	HepG2	—	—	Sustained drug release Enhanced cytotoxicity, cellular uptake and apoptosis following receptor mediated endocytosis	Lv et al. [52]
Thymoquinone	MSNs	Breast and cervical cancer	MCF-7 and HeLa	—	—	Enhanced cytotoxicity and anti-invasion ability due to increased ROS generation	Goel and Mishra [53]
Curcumin	MSNs	Breast cancer	MDA-MB-231	MDA-MB-231 xenografts	8 mg/kg	More precise targeting and higher accumulation in tumours with FA targeted NPs compared to HA modified NPs	Li et al. [54]
Ursolic acid	MSNs	Cervical carcinoma	HeLa and HepG2	HeLa-bearing tumour model	—	pH-sensitive drug release Induced cell cycle arrest in G0/G1 stage Suppressed the tumour progression and lung metastasis	Jiang et al. [55]

Baicalin	GNPs	Breast cancer	MCF-7	—	—	Improved cytotoxic and apoptotic effects	Lee et al. [56]
Hesperetin	GNPs	Liver cancer	-	Diethylnitrosamine (DEN) induced liver cancer in Wistar rats	20 mg/kg	Better efficacy against lipid peroxidation, hepatic cell damage and protects the antioxidant system in liver carcinogenesis model	Gokuladhas et al. [57]
Linalool	GNPs	Breast cancer	MCF-7	—	—	Exerted cell growth arrest and induced apoptosis	Jabir et al. [58]
Curcumin	GO nanocomposites	Breast and lung cancer	MCF-7 and A549	—	10 mg/kg	Improved hemocompatibility Exhibited significantly stronger photothermal effect Prolonged retention time	Shi et al. [59]
Artemisinin	MWCNTs	Sarcoma and breast cancer	MCF-7	S180 tumour-bearing mice	37.5 mg/kg	With laser irradiation of 808 nm, demonstrated highest inhibitory effects on tumours	Zhang et al. [60]

(continued)

Table 1 (continued)

Phytoconstituent	Type of NPs	Cancer	Cell lines	In vivo model	Dose	Key outcomes	Ref.
Genkwanin	Nanosuspension	Multiple cancers	4T1, MCF-7, MDA-MB-453, HeLa, HepG2, BT474, and A549	MCF-7 xenografts	60 mg/kg	Suitable for both oral and intravenous administration Similar therapeutic efficacy as PTX	Li et al. [61]
Curcumin	Nanosuspension	Hepatoma	Hela, 4T1, HCT-8, and HepG2	H22 tumour-bearing mice	2.5, 5.0 or 10.0 mg/kg	Better therapeutic efficacy	Hong et al. [62]
Bufalin	Albumin NPs	Hepatocellular carcinoma	SMMC-7721	SMMC-7721 xenografts	1 mg/kg	Higher liver uptake and stronger antitumour activity	Zhang et al. [63]
Resveratrol	Albumin NPs	Pancreatic tumour	PANC-1	PNAC-1 xenografts	10 mg/kg	Prolonged blood circulation and increased half-life by 5.43-fold Significant tumour growth suppression and no systemic toxicity	Geng et al. [64]
Carvacrol	Albumin NPs	Gastric cancer	AGS	—	—	Controlled and sustained drug release	Maryam et al. [65]
Saporin	Nanogel	Breast cancer	-	4T1-luc metastatic breast cancer model	3.33 and 13.3 nmol/kg	Significant inhibition of tumour metastasis to lung	Chen et al. [66]
Curcumin + doxorubicin	LipoNiosome	Osteosarcoma	Saos-2, MG-63 and KG-1	—	—	pH-sensitive sustained release	Naderinezhad et al. [67]

Malaridine + Doxorubicin	Nanodiamond	Breast cancer	MCF-7 and MCF-7/ADR	—	—	Greater drug release at low pH of tumour microenvironment Improved cytotoxic performance against resistant cancer	Zhu et al. [68]
Curcumin + camptothecin	PNPs	Colon cancer	Colon-26	—	—	Suppressed Bcl-2 mRNA level, inducing apoptosis	Xiao et al. [69]
Ursolic acid + sorafenib	MSNs	Hepatocellular carcinoma	SMMC-7721	H22 tumour-bearing mice model	20.5 mg/kg	pH-responsive drug release Enhanced bioavailability Down-regulated the expression of EGFR and VEGFR2 proteins related with cell proliferation and tumour angiogenesis.	Zhao et al. [70]
Naringin + celecoxib	PNPs	Lung cancer	A549	—	—	Favourable nebulization properties	Said-Elbahr et al. [71]
Rapamycin + berberine	Lipid core nanocomposites	Lung cancer	A549	Urethane induced lung cancer	1.5 mg/kg	Remarkable decrease in lung weight, and reduction in both number and diameters of lung adenomatous foci and angiogenic markers	Kabary et al. [72]

latter, and produced sufficient tumour regression and comparable survival post 1 month photodynamic treatment [29]. Cubosomes, the liquid crystalline nanoparticles (LCNPs), have also found suitable applications in the delivery of phytoconstituents for cancer treatment. Cubosomes are the self-assembled structures formed upon mixing polar lipids with polar solvent in the presence of stabilizer producing non-lamellar structures such as bicontinuous cubic (Q2) and reversed hexagonal (H2) phases and discrete micellar cubic phase (I2). Freag et al. developed aloe emodin (AE) (an anthraquinone obtained from *Aloe barbadensis* Miller or *Rheum palmatum* L.) loaded cubosome to enhance its water solubility and enable its anticancer potential. The pegylated LCNPs reduced the monoolein hemolytic tendency and increased the serum stability of the NPs and exhibited enhanced cytotoxicity and a 5.4-fold increase in $t_{1/2}$ compared to free AE [30]. Ubidecarenone-loaded cubosomes prevented doxorubicin (DOX)-induced cardiotoxicity [74]. Green tea catechins, EGCG when encapsulated in ethosomes and transethosomes, an ultradeformable colloidal vesicular systems, preserved the antioxidant activity and photostability of EGCG. The topical application of EGCG-loaded vesicular systems displayed strong chemotherapeutic and chemopreventive properties against skin cancer [31].

Solid lipid nanoparticles (SLNs) are the submicron particles in the size range of 50–100 nm prepared from lipids which are solid at room and body temperature, wherein the lipid material is selected from a variety of lipids, including mono-, di- or triglycerides, glyceride mixtures and lipid acids. The lipid matrix is then stabilized by biocompatible surfactants. The production of SLNs can be facilitated by a number of techniques, such as high-pressure homogenization, microemulsion and precipitation of lipid particles by solvent evaporation [75]. A flavonoid, wogonin isolated from the radix of *Scutellaria baicalensis*, when encapsulated in SLNs, produced enhanced cytotoxicity and inhibited PARP in breast cancer cells [32]. Linalool, a terpenoid alcohol present in several herbs (mint, laurel, cinnamon, and citrus fruits, herbal essential oils and in both green and black teas), exerted multiple effects on the mevalonate pathway and anti-proliferative activity in cancer cells. SLNs with mean diameters of 90–130 nm and with different compositions (Myristyl Myristate (MM), Cetyl Esters (SS) and cetyl palmitate (CP)) stabilized with Pluronic®F68 were developed for the encapsulation of Linalool. All the SLNS, particularly SLN (MM), produced enhanced cellular uptake and anti-proliferative effects on hepatocarcinoma (HepG2) and lung adenocarcinoma (A549) cell lines in a dose-dependent manner [33]. Similarly, resveratrol-loaded TPGS-coated SLNs showed sustained release up to 48 h and significantly higher cytotoxicity against C6 glioma cells. The brain distribution of resveratrol from SLNs was found to be 9.23 times higher compared to resveratrol solution [34].

Nanostructured lipid carriers (NLCs), an advanced form of SLNs, have been developed for encapsulating anticancer phytochemicals. NLCs are composed of solid and liquid lipids creating imperfect matrix which can accommodate more drug molecules increasing drug encapsulation efficiency. The larger surface area-to-volume ratio and the biodegradable and biocompatible nature of the excipients provide an added advantage to the NLCs. Thymoquinone is known to act on several hallmarks of cancer including inhibition of tumour cell proliferation, carcinogen activation and inflammation, induction of apoptosis, suppressed tumour angiogenesis, invasion and metastasis. Encapsulation of thymoquinone in the NLCs improved the survival rate of 4T1 tumour-bearing mice and inhibited the lung metastasis causing alterations in the expression of MMP-2 and inducing the intrinsic apoptotic pathway in the tumour cells [76]. Curcumin NLCs of 100 nm size were found to be effective treatment modality for brain cancer due to time-dependent cellular uptake, reactive oxygen species (ROS) production, markedly increased bioavailability and tumour-targeting effect [77].

Several other lipid-based systems such as self-nanoemulsifying drug delivery systems (SNEDDS), nanoemulsion and lipid nanocapsules have also been investigated for the delivery of phytoconstituents as anticancer therapeutics. Linalool was found to be effective against resistant epithelial ovarian cancer (EOC) when developed as SNEDDS [35]. SNEDDS are thermodynamically stable preconcentrate of drugs in oils, surfactants and cosurfactant, which on gentle agitation form oil-in-water (o/w) nanoemulsion in the gastrointestinal fluids. SNEDDs improve drug bioavailability by enhancing the dissolution rate of lipophilic drugs and absorb through selective lymphatic transport system avoiding hepatic first pass metabolism [78]. Linalool SNEDDS induced apoptosis in ovarian cancer cells by reducing the mitochondrial membrane potential and increasing the caspase-3 levels after ROS generation. The in vivo study in patient-derived xenografts (PDX) and ovarian cancer xenografts demonstrated a significant reduction in tumour growth inhibiting microvessel density (MVD) with SNEDDS formulation when combined with PTX, an established chemotherapeutic agent [35]. Lipid nanocapsules (LNCs), a hybrid between the polymeric nanocapsules and liposomes, are produced by a solvent-free process. LNCs provide several benefits, such as possible protection of the phytoconstituent from degradation by entrapping it inside the liquid core, improved efficacy and bioavailability and ability to overcome MDR. LNCs for curcumin were prepared using the phase inversion technique for the treatment of glioma. A twofold decrease in IC_{50}, induction of apoptosis and a strong release of ATP were observed in 9L glioma cells upon LNC treatment. Increase in C_{max} and area under the curve (AUC) and decrease in clearance (CL) were observed with CUR-LNCs indicating improvement in curcumin bioavailability [36]. Similarly, a nanoemulsion lipid hybrid nanosystem, termed as DNHLN, was developed for the effective delivery of diferuloylmethane (DIF) to lung cancer cells. The better hydrophilicity, higher dispersibility and the greater permeability of the DNHLNs facilitated the improved bioavailability and the uptake of DIF by the tumour cells inhibiting the cancer cell growth in a time- and dose-dependent manner. Additionally, the surfactants and oil phase of the DNHLNs have the ability to inhibit P-gp/CYP450 preventing efflux of DIF from the cancer cells [37]. Another type of delivery system, the self-assembled phospholipidic nano-mixed miceller system (SPNMS) was developed encapsulating mangiferin, a xanthone bioactive. The TPGS-functionalized SPNMs demonstrated greater killing and faster cellular uptake in MCF-7 and MDA-MB-231 breast cancer cells. The higher lymphatic uptake and augmentation in permeability and bioavailability of mangiferin by 6.6- and 3.0-fold indicated improved performance of mangiferin by SPNMS [38].

3.2 Polymer-Based Nanoparticles

The polymeric system offers broad scope for chemical modification conferring them wide utility for targeted drug delivery for effective cancer treatment. Polymeric NPs are defined by their morphology and composition in the core and periphery. The drug molecule can be either conjugated to the nanoparticles surface or encapsulated inside the polymeric core. They can be of different types, including polymer conjugate, polymeric micelle, polymersome, dendrimer and polyplexes. Polymeric NPs are the solid colloidal systems wherein the therapeutic agent is either dissolved, entrapped, encapsulated or adsorbed onto the constituent polymer matrix. Several polymers, including poly(lactide-co-glycolide) (PLGA), polyglycolide (PGA), polylactide (PLA), polycaprolactone (PCL), chitosan and PLGA-polyethylene glycol (PEG), have been investigated for passive and active-targeted delivery of therapeutic molecules [79].

Honokiol, a major phenolic constituent of root and stem bark of *Magnolia officinalis*, incorporated in folic acid-targeted poly(ε-caprolactone) (PCL) NPs targeted the folate receptors expressing nasopharyngeal carcinoma cells and caused significant inhibition of tumour growth, metabolism, proliferation and micro-vessel generation in mouse HNE-1 xenograft model [39]. Angiopep-2 (ANG)-functionalized ginsenoside-Rg3 (Rg3)-loaded PCL-PEG polymeric micelles (PMs) were developed by Su et al., for targeted delivery of Rg3 to glioma, one of the most common and aggressive intracranial malignancies. Rg3 is the protopanaxadiol type of dammarane ginsenoside which is the main active component of ginseng. Angiopep-2, a Kunitz domain-derived peptide from protease enzyme, can target the low-density lipoprotein receptor-related protein-1 (LRP-1) overexpressed in malignant astrocytomas, especially in glioma cells. ANG-Rg3-NP NPs crossed the blood-brain barrier more easily and exhibited stronger inhibitory effect against C6 glioma cells [40]. PLGA-TPGS NPs containing emodin produced better anti-tumour effects mainly ascribed to the smaller particle size of the NPs and the presence of TPGS that could reduce the P-gp-mediated MDR and improve the permeability of emodin across the cellular membrane producing greater effects against the liver cancer [41]. Mixed micelles composed of vitamin E succinate-chitosan oligosaccharide (VES-CSO) targeted with RGD peptide were developed for the targeted delivery of bufalin to colon cancer. The bufalin-loaded micelles enhanced the levels of pro-apoptotic Bax, cleaved the caspase-3 and caspase-9, decreased the expressions of anti-apoptotic molecule Bcl-2 and inhibited P-gp expression in LoVo/ADR cells. It also caused the inhibition of tumour growth by 65% in LoVo/ADR xenografts [42]. mPEG-PLGA/PLA NPs were developed for the delivery of oleanolic acid (OA), a natural triterpenoid with anticancer properties. The OA-loaded NPs induced apoptosis and produced significant cytotoxicity in lung (A549) and liver (HepG2) cancer cells [43]. Another flavone, chrysin, when incorporated in PLGA NPs, produced dose-dependent cytotoxic effects against Caco-2 cells, inhibited hTERT gene expression and significantly arrested the growth of cancer cells [44]. Quercetin, a dietary flavonoid found in red onions, apples, berries, citrus fruits, tea and red wine, was investigated after encapsulation in PLGA NPs for its anti-cancer effects in cervical cancer cells. The quercetin

PLGA NPs were found to act on multiple signaling pathways such as JAK2, signal transducer and activator of transcription (STAT) 3/5 and PI3K/AKT signaling pathway, inhibiting them resulting in anti-proliferative, anti-migratory and anti-invasive effects. Moreover, reduction in tumour size and weight was also observed [45]. The limitations of piperine including poor aqueous solubility, reproductive toxicity and immunotoxicity were overcome by encapsulating it in PLGA NPs. The aptamer-targeted nanoparticles could target Mucin 1 (MUC1), a transmembrane glycoprotein overexpressed on many epithelial cancers and served as adjuvant cancer chemotherapy against PTX-resistant breast cancers [46]. Another aptamer-functionalized apigenin-loaded NPs were developed for targeting EpCAM receptor-overexpressing colorectal cancer cells. Increased cellular uptake of NPs, depolarization of mitochondrial transmembrane potential, enhanced level of NADH activity and decreased level of succinate dehydrogenase (SDH) activity were observed which ultimately resulted in apoptosis in HCT-116 cells [47]. Polyisoprenylated phenolic compound, garcinol (GAR)-loaded PLGA NPs, produced a sustained release of GAR over 36 h and increased the AUC value compared to free GAR suspension indicating its improved bioavailability. Furthermore, GAR accumulation was observed in the B16F10 melanoma tumour-bearing mice at 2 h post injection which gradually increased up to 8 h and also increased the tumour to muscle and tumour to blood ratios with time throughout the study period [48]. Quercetin-loaded chitosan nanoparticles showed significant reduction in IC_{50} value of QCT-CS NPs compared to free quercetin and exerted remarkable decrease in tumour volume in the A549 and MDA MB 468 cells bearing mouse xenograft models [49].

3.3 Inorganic Nanoparticles

Increasing attention has been directed towards the use of superparamagnetic iron oxide nanoparticles (SPIONs) as drug delivery vehicles. SPIONs are the intrinsic magnetic resonance imaging (MRI) contrast agent and biocompatible nano-device with controllable targeting properties. The instability and less contrast enhancement (compared to the gadolinium-based T1-contrast agent) limit the clinical application of SPIONs for imaging and drug therapy [80]. Silibinin-loaded PLGA-PEG-Fe_3O_4 NPs produced inhibitory effects on the growth of A549 lung cancer cells and hTERT gene expression in the cancer cells [50]. A combination of stimulus-responsive and actively targeted NPs was investigated to increase the targeting efficacy to the cancer cell. Folic acid-targeted curcumin-loaded iron oxide-O-carboxymethyl chitosan NPs (Fe_3O_4/OCMCs/Cur NPs) adhered to the folate receptor-overexpressing S-180 tumours, where the higher Fe_3O_4 concentration resulted in higher temperature and initial heating rate producing cancer hyperthermia which effectively kills the cancer cells. Thus, a combined effect of curcumin and cancer hyperthermia is obtained which effectively kills the cancer cells [51].

Mesoporus silica nanoparticles (MSNs) are one of the most valuable cargos for the delivery of anticancer agents, due to several advantages they offer: uniform and tunable particle size, pore size, and morphology, high surface area and pore volume, facile surface functionalization, etc. [81]. Glycyrrhetinic acid (GA)-functionalized curcumin-loaded MSNs targeted the asialoglycoprotein receptors on the surface of HCC cells exhibiting satisfactory loading capacity and increased curcumin uptake by GA receptor-positive cells [52]. Thymoquinone-loaded MSNs of particle size 188 ± 3 nm were able to overcome the hydrophobicity, low aqueous and photostability concerns of THQ and induced apoptosis and enhanced anticancer activity against MCF-7 and HeLa cells [53]. Another active targeted MSN system was developed for selective delivery of curcumin to CD44 and folate receptor-overexpressing breast cancer cells. MSNs conjugated to hyaluronic acid (HA) or polyethyleneimine-folic acid (PEI-FA) allowed their cellular uptake by the tumour cells where high GSH concentrations induced curcumin release. The FA-modified MSNs displayed slightly superior targeting effects compared to HA-modified MSNs which can be attributed to the low molecular weight of FA which facilitates its efficient binding to its receptors. The MSNs also inhibited the tumour growth in MDA-MB-231 xenografts [54]. Folate-functionalized pH-responsive multifunctional MSNs were developed for the delivery of ursolic acid to cervical cancer cells. The MSNs exhibited pH-sensitive drug release whereby greater release of ursolic acid was observed in the acidic pH of the tumour microenvironment, inhibited growth, invasion and migration of cancer cells through cell cycle arrest in G0/G1 stage and regulated the PARP/Bcl-2/MMP-9/CD44/PTEN/P53. Suppression in tumour progression and lung metastasis was observed in HeLa tumour-bearing mice [55].

Recent reports have studied the use of GNPs for combination chemotherapy. Gold nanocages (AuNCs) that have hollow interiors and regulatable localized surface plasmon resonance peaks (LSPR) in the near infrared (NIR) region have received much attention in the biomedical applications such as controlled drug delivery and photoacoustic imaging. Baicalin delivered through GNPs conjugated with thiolated beta-cyclodextrin (AuNP-S-β-CD) produced anti-proliferative and apoptotic effects in breast cancer cells [56]. Hesperetin-loaded GNPs (AumPEG(5000)-S-HP) provided effective anti-cancer effects in diethylnitrosamine (DEN)-induced liver cancer in Wistar rats. The Au-mPEG(5000)-S-HP NPs increased the solubility of hesperetin and provided sustained drug delivery to the cancer cells. Further, they possess more potent free radical scavenging and antioxidant activities, can markedly modulate the oxidative stress by activating antioxidant defence system and exhibited potent chemopreventive properties [57].

Graphene, the two-dimensional honeycomb lattice can be effectively utilized to impart solubility and as a drug delivery agent. Graphene oxide (GO) is an atomic sheet of graphite having several oxygenated functional groups, exhibiting good biocompatibility, physiological stability and large specific surface area providing enhanced drug loading and other functionalities. The photothermal characteristics, excellent electrical conductivity and fast mobility of the charge carriers have evoked enormous interest in using GO as drug delivery carriers [82]. Graphene oxide is synthesized using strong oxidizing agent, which are ultimately carried forward to the end product. GO is highly acidic and can cause damage to the normal cells. Therefore, reduced graphene oxide was synthesized with the simple functionalization of GO with an amphiphilic triblock co-polymer such as pluronic (PF-127) which would enhance the solubility and stability of the therapeutic agent. The hydrophobic drugs, curcumin and PTX, could be loaded onto the empty spaces on GO through π–π interactions between the drugs and the aromatic structures of GO. The developed nanocarrier system further demonstrated enhanced cytotoxicity against lung cancer cells and induced early-stage apoptosis [83]. The role of electric and electrochemical properties of GO in modulating and tuning drug delivery in tumour environment was studied. The electrode system was designed where the tumour environment was simulated at pH 6.2 with 0.1 V pulse voltage, while the physiological environment was at pH 7.4 with 0.2 V. Berberine-bearing GO-based nanocomposite exhibited a burst release on its special "skin effect", whereas the PEGylated particles released slowly owing to the electroviscous effect of the polymer. GO exhibited a higher sensitivity both in neutral and acidic environment with about 46% of berberine being released in 4 h at pH 6.2 under 0.1 V, while PEG-GO presented poor electric-triggered release behaviour with almost no change in drug release at the same pH and under same voltage pulse. Such electrically modulated drug release may be the result of both electromigration and diffusion actions [82]. Heparin-grafted curcumin-loaded GO nanosheets (rGO-UFH/Cur) of size ~85 nm developed using modified Hummers' method exhibited greatly improved hemocompatibility and stronger photothermal effect compared to random large-sized GO nanosheets. Enhanced anticancer efficacy of rGO-UFH/Cur was observed against MCF-7 and A549 cells which further increased under 808 nm laser irradiation indicating potential combinational chemotherapy and photothermal therapy [59].

Besides these, multi-walled carbon nanotubes (MWCNTs) have also been developed as efficient carriers for anticancer drug molecule. MWCNTs have several applications as DNA and protein biosensors and ion channel blockers and can shuttle the drugs and biomolecules across the cellular membrane through endocytosis. They also exhibit a potential to be used for cancer photothermal therapy owing to the excellent photothermal conversion property in NIR region, where biological systems exhibit low absorption and high transparency. Zhang et al. developed a multi-functional tumour-targeting nanocarrier system employing hyaluronic acid-derivatized MWCNTs (HA-MWCNTs) as drug carriers, transferrin (Tf) as targeting ligand and artemisinin (ART) (a sesquiterpene lactone extracted from the Chinese herb *Artemisia annua* L.) as anticancer molecule. The nanocarrier system exhibited synergistic anti-tumour effects in S-180 tumour-bearing murine model and simultaneously delivered both the heat and drug to the tumourigenic region facilitating combined chemotherapy and photothermal therapy [60].

3.4 Others

Nanosuspensions have attracted significant interest in drug delivery applications due to their ability to increase the solubility of hydrophobic compounds and their suitability for various administration routes. Li et al. developed genkwanin nanosuspensions (GKA-NSps) using TPGS as the stabilizer and precipitation homogenization method and examined its effects in MCF-7 tumour-bearing mice. A sustained drug release and stronger cytotoxicity were observed against several cancer cells and resulted in 62% inhibition of tumours having comparable efficacy to PTX [61]. Similarly, curcumin nanosuspensions (CUR-NSps) prepared by precipitation ultrasonication method using mPEG2000-DSPE and soybean lecithin as the stabilizer exhibited a high drug payload of 67.07% and a mean particle size of 186.33 ± 2.73 nm. CURNSps exhibited greatly improved AUC and prolonged mean residence time (MRT) and improved antitumour therapeutic efficacy [62].

Albumin NPs are another versatile drug carrier system exhibiting favourable properties such as nontoxicity, non-immunogenicity, biocompatibility and biodegradability. Bufalin-loaded bovine serum albumin (BSA)-NPs showed higher drug concentration and longer residence in the liver that could target the HCC cells [63]. Resveratrol-loaded RGD-conjugated HSA-NPs (HRP-RGD NPs) were developed for highly effective targeted pancreatic tumour therapy. Improved cellular uptake, cytotoxicity and apoptosis were observed in PANC-1 cells. HRP-RGD NPs also prolonged the blood circulation of RV, increasing $t_{1/2}$ by approximately 5.43-fold and a higher accumulation and no significant systemic toxicity over 35 days of treatment in the tumour tissue of PNAC-1 xenografts [64]. A monoterpene phenol, carvacrol-loaded HSA-NP prepared using desolvation and emulsion/desolvation methods produced enhanced cytotoxic effects on gastric cancer cells line (AGS) [65].

Targeting metastasis is a potential treatment modality for many cancers. Chen et al. prepared epidermal growth factor receptor (EGFR) and CD44 dual-targeted hyaluronic acid nanogels (EGFR/CD44-NGs) affording enhanced targetability of saporin to metastatic 4T1 breast cancer-bearing mice. The hyaluronic acid and GE11 peptide (YHWYGYTPQNVI) can target the CD44 and EGFR receptors of cancer cells. Furthermore, the nanogels also produced significant inhibition of tumour metastasis to the lung in 4T1-luc metastatic breast cancer model in mice [66].

Recently combination chemotherapy of phytochemicals and chemotherapeutic agents has attracted huge interest from the scientific community for the treatment of several cancers. Co-delivery of DOX with curcumin in LipoNiosomes could overcome the MDR associated with cancer chemotherapy. The LipoNiosome with different lipid and surfactant composition could provide pH-sensitive drug release and synergistic toxic effect against Saos-2, MG-63 and KG-1 cell lines [67]. Similarly, the combination of DOX and malaridine encapsulated in nanodiamond produced synergistic effects on resistant breast cancer cells (MCF-7/Adr) [68]. Chitosan-functionalized CPT/curcumin-loaded polymeric NPs developed with different weight ratios of CPT and curcumin exhibited enhanced therapeutic effects against colon cancer cells [69]. Lactobionic acid-targeted pH-sensitive MSNs delivered sorafenib and ursolic acid (UA) at a controlled rate to ASGPR-overexpressing HCC cells. The nanocomplex provided sustained release of the two drugs, could attenuate the adhesion and migration of ASGPR overexpressing SMMC-7721 liver cancer cells, down-regulated the expression of EGFR and VEGFR2 proteins resulting in reduced cellular proliferation and tumour angiogenesis, enhanced the bioavailability and provided efficient tumour targeting effects by reducing the tumour burden in H22 xenografts and inhibited the lung metastasis in the H22 lung metastasis models [70]. Metastasis targeting was also achieved using celecoxib- and naringin-loaded polymeric NPs. Biodistribution studies suggested greater deposition of the NPs within the lungs and also to the bones, brain and liver which are common metastatic sites of lung cancer [71]. Inhalable multi-compartmental nanocomposites for localized and modulated release of rapamycin and berberine were developed for lung cancer therapy. Layer-by-layer-coated LNCs were fabricated by enveloping berberine hydrophobic ion pair-lipid nanocore within the shell of rapamycin-phospholipid complex bilayer which was further coated with cationic lactoferrin and anionic hyaluronate exhibited favourable aerosolization efficiency and enhanced internalization and cytotoxicity against lung cancer cells. The inhalable nanocomposites produced remarkable decrease in lung weight and reduction in both number and diameters of lung adenomatous foci and angiogenic markers in lung cancer-bearing mice [72]. Natural polyphenols can also counter the chemotherapeutic agent-induced toxicity. The hepato- and nephrotoxicity associated with bicalutamide (BCT), a non-steroidal anti-androgen for the treatment of locally advanced non-metastatic prostate cancer, were alleviated with co-administration of hesperetin (HSP) in the form of SNEDDS [84]. Improved bioavailability was evident with 1.23- and 3.42-fold increase in AUC and 1.24- and 8.81-fold increase in C_{max} for BCT and HSP, with BCT-HSP SNEDDS-treated group. The 14-day sub-acute toxicity study with the SNEDDS-administered group indicated no significant increase in the hepatotoxicity markers and significantly decreased levels of nephrotoxicity markers [84]. Naringenin provides Pgp inhibitory action improving treatment outcome with tamoxifen therapy in breast cancer when co-loaded in SNEDD formulation [85]. Similar effects were observed in tamoxifen- and quercetin-loaded solid SNEDDS [86].

4 Conclusion and Future Perspectives

Medicinal plants have a long history of use as a rich source for development of successful pharmacological leads with antibiotic, antioxidant, antiviral, antifungal and immunostimulating properties. Amongst these, the most prominent medicinal property of these natural compounds is their anticancer effect for management and treatment of cancer which is claimed to be the second leading cause of death worldwide. These phytochemicals have been found to exert chemotherapeutic and chemopreventive effects by interfering with almost all the hallmarks of cancer, suppressing the growth of cancer cells, inhibiting their metastasis and invasion and inducing apoptosis. Suppression in tumour growth has been observed in several preclinical cancer xenograft models. We have also enlisted combinations of phytoconstituents and standard chemotherapeutic agents which on co-delivery can synergize the overall anticancer activity. The future investigations should therefore be more focused on combination chemotherapy rather than monotherapy with isolated (single) phytoconstituent. However, a large body of evidence suggesting the potential effectiveness of medicinal plants against cancer lack the scientific evidence for their safety in humans.

Despite their great utility for the treatment of several cancers, these bioactive compounds have various limitations such as poor water solubility, poor stability, susceptibility to photodegradation, high drug dose, lack of selectivity and targetability which restrict their full exploitation as successful anticancer therapeutics. The science of nanomedicine has essentially paved a way to counter the challenging aspects of phytomedicines. This is primarily attributed to the reliable and targeted delivery with less toxicity to the surrounding environment, increased drug efficiency and improved cellular interaction with the nanoparticle-based therapeutics for cancer treatment. The successful translation of the phytochemical-based nanotherapeutics from preclinical proof of concept to clinical stage is nevertheless very challenging due to several reasons. The lack of reproducibility of the synthetic procedure, formulation complexity, insufficient information on the relationship between the critical quality attributes of the nanoformulation and the efficacy, regulatory perspectives and scale-up issues are some of the challenging aspects of nanomedicine products. Furthermore, the effect of tumour pathophysiology on the pharmacokinetics and pharmacodynamic aspects of nanoparticles, the use of more clinically relevant tumour model should also be given consideration while developing a nanomedicine product. Additionally, the economic considerations should not be undervalued. Nanomedicine has revolutionized the way to discover and administer the phytoconstituents in the biological systems.

References

1. Ferlay J, Colombet M, Soerjomataram I, Mathers C, Parkin D, Piñeros M, Znaor A, Bray F (2019) Estimating the global cancer incidence and mortality in 2018: GLOBOCAN sources and methods. Int J Cancer 144:1941–1953
2. Hanahan D, Weinberg RA (2000) The hallmarks of cancer. Cell 100:57–70. https://doi.org/10.1016/S0092-8674(00)81683-9
3. Atanasov AG, Waltenberger B, Pferschy-Wenzig E-M, Linder T, Wawrosch C, Uhrin P, Temml V, Wang L, Schwaiger S, Heiss EH, Rollinger JM, Schuster D, Breuss JM, Bochkov V, Mihovilovic MD, Kopp B, Bauer R, Dirsch VM, Stuppner H (2015) Discovery and resupply of pharmacologically active plant-derived natural products: a review. Biotechnol Adv 33:1582–1614. https://doi.org/10.1016/j.biotechadv.2015.08.001
4. Cragg GM, Newman DJ (2013) Natural products: a continuing source of novel drug leads. Biochim Biophys Acta BBA Gen Subj 1830:3670–3695
5. Bonam SR, Wu YS, Tunki L, Chellian R, Halmuthur MSK, Muller S, Pandy V (2018) What has come out from phytomedicines and herbal edibles for the treatment of cancer? ChemMedChem 13:1854–1872
6. Li Y, Go VLW, Sarkar FH (2015) The role of nutraceuticals in pancreatic cancer prevention and therapy: targeting cellular signaling, miRNAs and epigenome. Pancreas 44:1
7. Berman AY, Motechin RA, Wiesenfeld MY, Holz MK (2017) The therapeutic potential of resveratrol: a review of clinical trials. NPJ Precis Oncol 1:35
8. Kunnumakkara AB, Bordoloi D, Padmavathi G, Monisha J, Roy NK, Prasad S, Aggarwal BB (2017) Curcumin, the golden nutraceutical: multitargeting for multiple chronic diseases. Br J Pharmacol 174:1325–1348
9. Majdalawieh AF, Fayyad MW, Nasrallah GK (2017) Anti-cancer properties and mechanisms of action of thymoquinone, the major active ingredient of Nigella sativa. Crit Rev Food Sci Nutr 57:3911–3928

10. Yuan C, Horng C, Lee C, Chiang N, Tsai F, Lu C, Chiang J, Hsu Y, Yang J, Chen F (2017) Epigallocatechin gallate sensitizes cisplatin-resistant oral cancer CAR cell apoptosis and autophagy through stimulating AKT/STAT3 pathway and suppressing multidrug resistance 1 signaling. Environ Toxicol 32:845–855
11. Zhou S, Zhang S, Shen H, Chen W, Xu H, Chen X, Sun D, Zhong S, Zhao J, Tang J (2017) Curcumin inhibits cancer progression through regulating expression of microRNAs. Tumor Biol 39:1010428317691680
12. Gali-Muhtasib H, Roessner A, Schneider-Stock R (2006) Thymoquinone: a promising anti-cancer drug from natural sources. Int J Biochem Cell Biol 38:1249–1253
13. Mukund V, Mukund D, Sharma V, Mannarapu M, Alam A (2017) Genistein: its role in metabolic diseases and cancer. Crit Rev Oncol Hematol 119:13–22
14. Lin Y, Shi R, Wang X, Shen H-M (2008) Luteolin, a flavonoid with potential for cancer prevention and therapy. Curr Cancer Drug Targets 8:634–646
15. Premkumar VG, Yuvaraj S, Sathish S, Shanthi P, Sachdanandam P (2008) Anti-angiogenic potential of CoenzymeQ10, riboflavin and niacin in breast cancer patients undergoing tamoxifen therapy. Vasc Pharmacol 48:191–201
16. Fletcher JI, Haber M, Henderson MJ, Norris MD (2010) ABC transporters in cancer: more than just drug efflux pumps. Nat Rev Cancer 10:147
17. Liang G, Tang A, Lin X, Li L, Zhang S, Huang Z, Tang H, Li QQ (2010) Green tea catechins augment the antitumor activity of doxorubicin in an in vivo mouse model for chemoresistant liver cancer. Int J Oncol 37:111–123
18. Zhang S, Yang X, Morris ME (2004) Flavonoids are inhibitors of breast cancer resistance protein (ABCG2)-mediated transport. Mol Pharmacol 65:1208–1216
19. Tiwari N, Gheldof A, Tatari M, Christofori G (2012) EMT as the ultimate survival mechanism of cancer cells. Elsevier, Amsterdam, pp 194–207
20. Lee G-A, Choi K-C, Hwang K-A (2017) Kaempferol, a phytoestrogen, suppressed triclosan-induced epithelial-mesenchymal transition and metastatic-related behaviors of MCF-7 breast cancer cells. Environ Toxicol Pharmacol 49:48–57
21. Liu K, Cang S, Ma Y, Chiao JW (2013) Synergistic effect of paclitaxel and epigenetic agent phenethyl isothiocyanate on growth inhibition, cell cycle arrest and apoptosis in breast cancer cells. Cancer Cell Int 13:10
22. Osman A-MM, Al-Harthi SE, AlArabi OM, Elshal MF, Ramadan WS, Alaama MN, Al-Kreathy HM, Damanhouri ZA, Osman OH (2013) Chemosensetizing and cardioprotective effects of resveratrol in doxorubicin-treated animals. Cancer Cell Int 13:52
23. Ince S, Kucukkurt I, Demirel HH, Turkmen R, Zemheri F, Akbel E (2013) The role of thymoquinone as antioxidant protection on oxidative stress induced by imidacloprid in male and female Swiss albino mice. Toxicol Environ Chem 95:318–329
24. Bilia AR, Piazzini V, Guccione C, Risaliti L, Asprea M, Capecchi G, Bergonzi MC (2017) Improving on nature: the role of nanomedicine in the development of clinical natural drugs. Planta Med 83:366–381
25. Hare JI, Lammers T, Ashford MB, Puri S, Storm G, Barry ST (2017) Challenges and strategies in anti-cancer nanomedicine development: an industry perspective. Adv Drug Deliv Rev 108:25–38
26. Senapati S, Mahanta AK, Kumar S, Maiti P (2018) Controlled drug delivery vehicles for cancer treatment and their performance. Signal Transduct Target Ther 3:7
27. Patra JK, Das G, Fraceto LF, Campos EVR, del Pilar Rodriguez-Torres M, Acosta-Torres LS, Diaz-Torres LA, Grillo R, Swamy MK, Sharma S (2018) Nano based drug delivery systems: recent developments and future prospects. J Nanobiotechnol 16:71
28. Ochi MM, Amoabediny G, Rezayat SM, Akbarzadeh A, Ebrahimi B (2016) In vitro co-delivery evaluation of novel pegylated nano-liposomal herbal drugs of silibinin and glycyrrhizic acid (nano-phytosome) to hepatocellular carcinoma cells. Cell J Yakhteh 18:135
29. Fadel M, Kassab K, Abd El Fadeel DA, Nasr M, El Ghoubary NM (2018) Comparative enhancement of curcumin cytotoxic photodynamic activity by nanoliposomes and gold

nanoparticles with pharmacological appraisal in HepG2 cancer cells and Erlich solid tumor model. Drug Dev Ind Pharm 44:1809–1816
30. Freag MS, Elnaggar YS, Abdelmonsif DA, Abdallah OY (2016) Stealth, biocompatible monoolein-based lyotropic liquid crystalline nanoparticles for enhanced aloe-emodin delivery to breast cancer cells: in vitro and in vivo studies. Int J Nanomed 11:4799
31. El-Kayal M, Nasr M, Elkheshen S, Mortada N (2019) Colloidal (-)-epigallocatechin-3-gallate vesicular systems for prevention and treatment of skin cancer: a comprehensive experimental study with preclinical investigation. Eur J Pharm Sci 137:104972
32. Baek J-S, Na Y-G, Cho C-W (2018) Sustained cytotoxicity of wogonin on breast cancer cells by encapsulation in solid lipid nanoparticles. Nano 8:159
33. Rodenak-Kladniew B, Islan GA, de Bravo MG, Durán N, Castro GR (2017) Design, characterization and in vitro evaluation of linalool-loaded solid lipid nanoparticles as potent tool in cancer therapy. Colloids Surf B Biointerfaces 154:123–132
34. Vijayakumar MR, Kumari L, Patel KK, Vuddanda PR, Vajanthri KY, Mahto SK, Singh S (2016) Intravenous administration of trans-resveratrol-loaded TPGS-coated solid lipid nanoparticles for prolonged systemic circulation, passive brain targeting and improved in vitro cytotoxicity against C6 glioma cell lines. RSC Adv 6:50336–50348
35. Han HD, Cho Y-J, Cho SK, Byeon Y, Jeon HN, Kim H-S, Kim B-G, Bae D-S, Lopez-Berestein G, Sood AK, Shin BC, Park Y-M, Lee J-W (2016) Linalool-incorporated nanoparticles as a novel anticancer agent for epithelial ovarian carcinoma. Mol Cancer Ther 15:618–627. https://doi.org/10.1158/1535-7163.MCT-15-0733-T
36. Lollo G, Ullio-Gamboa G, Fuentes E, Matha K, Lautram N, Benoit J-P (2018) In vitro anti-cancer activity and pharmacokinetic evaluation of curcumin-loaded lipid nanocapsules. Mater Sci Eng C 91:859–867
37. Sun L, Wan K, Hu X, Zhang Y, Yan Z, Feng J, Zhang J (2016) Functional nanoemulsion-hybrid lipid nanocarriers enhance the bioavailability and anti-cancer activity of lipophilic diferuloylmethane. Nanotechnology 27:085102
38. Khurana RK, Gaspar BL, Welsby G, Katare O, Singh KK, Singh B (2018) Improving the biopharmaceutical attributes of mangiferin using vitamin E-TPGS co-loaded self-assembled phosholipidic nano-mixed micellar systems. Drug Deliv Transl Res 8:617–632
39. Yang B, Ni X, Chen L, Zhang H, Ren P, Feng Y, Chen Y, Fu S, Wu J (2017) Honokiol-loaded polymeric nanoparticles: an active targeting drug delivery system for the treatment of nasopharyngeal carcinoma. Drug Deliv 24:660–669
40. Su X, Zhang D, Zhang H, Zhao K, Hou W (2018) Preparation and characterization of angiopep-2 functionalized ginsenoside-Rg3 loaded nanoparticles and the effect on C6 glioma cells. Pharm Dev Technol:1–37. https://doi.org/10.1080/10837450.2018.1551901
41. Liu H, Gao M, Xu H, Guan X, Lv L, Deng S, Zhang C, Tian Y (2016) A promising emodin-loaded poly (lactic-Co-glycolic acid)-d-α-tocopheryl polyethylene glycol 1000 succinate nanoparticles for liver cancer therapy. Pharm Res 33:217–236. https://doi.org/10.1007/s11095-015-1781-4
42. Yuan Z, Yuan Y, Han L, Qiu Y, Huang X, Gao F, Fan G, Zhang Y, Tang X, He X (2018) Bufalin-loaded vitamin E succinate-grafted-chitosan oligosaccharide/RGD conjugated TPGS mixed micelles demonstrated improved antitumor activity against drug-resistant colon cancer. Int J Nanomedicine 13:7533
43. Man DK, Casettari L, Cespi M, Bonacucina G, Palmieri GF, Sze SC, Leung GP, Lam JK, Kwok PC (2015) Oleanolic acid loaded PEGylated PLA and PLGA nanoparticles with enhanced cytotoxic activity against cancer cells. Mol Pharm 12:2112–2125
44. Lotfi-Attari J, Pilehvar-Soltanahmadi Y, Dadashpour M, Alipour S, Farajzadeh R, Javidfar S, Zarghami N (2017) Co-delivery of curcumin and chrysin by polymeric nanoparticles inhibit synergistically growth and hTERT gene expression in human colorectal cancer cells. Nutr Cancer 69:1290–1299
45. Luo C, Liu Y, Wang P, Song C, Wang K, Dai L, Zhang J, Ye H (2016) The effect of quercetin nanoparticle on cervical cancer progression by inducing apoptosis, autophagy and anti-proliferation via JAK2 suppression. Biomed Pharmacother 82:595–605

46. Pachauri M, Gupta ED, Ghosh PC (2015) Piperine loaded PEG-PLGA nanoparticles: preparation, characterization and targeted delivery for adjuvant breast cancer chemotherapy. J Drug Deliv Sci Technol 29:269–282
47. Dutta D, Chakraborty A, Mukherjee B, Gupta S (2018) Aptamer-conjugated apigenin nanoparticles to target colorectal carcinoma: a promising safe alternative of colorectal cancer chemotherapy. ACS Appl Bio Mater 1:1538–1556
48. Gaonkar RH, Ganguly S, Dewanjee S, Sinha S, Gupta A, Ganguly S, Chattopadhyay D, Debnath MC (2017) Garcinol loaded vitamin E TPGS emulsified PLGA nanoparticles: preparation, physicochemical characterization, in vitro and in vivo studies. Sci Rep 7:530
49. Baksi R, Singh DP, Borse SP, Rana R, Sharma V, Nivsarkar M (2018) In vitro and in vivo anticancer efficacy potential of quercetin loaded polymeric nanoparticles. Biomed Pharmacother 106:1513–1526
50. Amirsaadat S, Pilehvar-Soltanahmadi Y, Zarghami F, Alipour S, Ebrahimnezhad Z, Zarghami N (2017) Silibinin-loaded magnetic nanoparticles inhibit hTERT gene expression and proliferation of lung cancer cells. Artif Cells Nanomed Biotechnol 45:1649–1656
51. Thu Huong LT, Nam NH, Doan DH, My Nhung HT, Quang BT, Nam PH, Thong PQ, Phuc NX, Thu HP (2016) Folate attached, curcumin loaded Fe3O4 nanoparticles: a novel multifunctional drug delivery system for cancer treatment. Mater Chem Phys 172:98–104. https://doi.org/10.1016/j.matchemphys.2015.12.065
52. Lv Y, Li J, Chen H, Bai Y, Zhang L (2017) Glycyrrhetinic acid-functionalized mesoporous silica nanoparticles as hepatocellular carcinoma-targeted drug carrier. Int J Nanomedicine 12:4361
53. Goel S, Mishra P (2019) Thymoquinone loaded mesoporous silica nanoparticles retard cell invasion and enhance in vitro cytotoxicity due to ROS mediated apoptosis in HeLa and MCF-7 cell lines. Mater Sci Eng C 104:109881
54. Li N, Wang Z, Zhang Y, Zhang K, Xie J, Liu Y, Li W, Feng N (2018) Curcumin-loaded redox-responsive mesoporous silica nanoparticles for targeted breast cancer therapy. Artif Cells Nanomed Biotechnol 46:921–935
55. Jiang K, Chi T, Li T, Zheng G, Fan L, Liu Y, Chen X, Chen S, Jia L, Shao J (2017) A smart pH-responsive nano-carrier as a drug delivery system for the targeted delivery of ursolic acid: suppresses cancer growth and metastasis by modulating P53/MMP-9/PTEN/CD44 mediated multiple signaling pathways. Nanoscale 9:9428–9439
56. Lee D, Ko W-K, Hwang D-S, Heo DN, Lee SJ, Heo M, Lee K-S, Ahn J-Y, Jo J, Kwon IK (2016) Use of baicalin-conjugated gold nanoparticles for apoptotic induction of breast cancer cells. Nanoscale Res Lett 11:381
57. Gokuladhas K, Jayakumar S, Rajan B, Elamaran R, Pramila CS, Gopikrishnan M, Tamilarasi S, Devaki T (2016) Exploring the potential role of chemopreventive agent, hesperetin conjugated pegylated gold nanoparticles in diethylnitrosamine-induced hepatocellular carcinoma in male Wistar albino rats. Indian J Clin Biochem 31:171–184. https://doi.org/10.1007/s12291-015-0520-2
58. Jabir MS, Taha AA, Sahib UI, Taqi ZJ, Al-Shammari AM, Salman AS (2019) Novel of nano delivery system for Linalool loaded on gold nanoparticles conjugated with CALNN peptide for application in drug uptake and induction of cell death on breast cancer cell line. Mater Sci Eng C 94:949–964
59. Shi X, Wang Y, Sun H, Chen Y, Zhang X, Xu J, Zhai G (2019) Heparin-reduced graphene oxide nanocomposites for curcumin delivery: in vitro, in vivo and molecular dynamics simulation study. Biomater Sci 7:1011–1027
60. Zhang H, Ji Y, Chen Q, Jiao X, Hou L, Zhu X, Zhang Z (2015) Enhancement of cytotoxicity of artemisinin toward cancer cells by transferrin-mediated carbon nanotubes nanoparticles. J Drug Target 23:552–567
61. Li Y, Hong J, Li H, Qi X, Guo Y, Han M, Wang X (2017) Genkwanin nanosuspensions: a novel and potential antitumor drug in breast carcinoma therapy. Drug Deliv 24:1491–1500. https://doi.org/10.1080/10717544.2017.1384519

62. Hong J, Liu Y, Xiao Y, Yang X, Su W, Zhang M, Liao Y, Kuang H, Wang X (2017) High drug payload curcumin nanosuspensions stabilized by mPEG-DSPE and SPC: in vitro and in vivo evaluation. Drug Deliv 24:109–120
63. Zhang H, Huang N, Yang G, Lin Q, Su Y (2017) Bufalin-loaded bovine serum albumin nanoparticles demonstrated improved anti-tumor activity against hepatocellular carcinoma: preparation, characterization, pharmacokinetics and tissue distribution. Oncotarget 8:63311
64. Geng T, Zhao X, Ma M, Zhu G, Yin L (2017) Resveratrol-loaded albumin nanoparticles with prolonged blood circulation and improved biocompatibility for highly effective targeted pancreatic tumor therapy. Nanoscale Res Lett 12:437
65. Maryam K, Shakeri S, Kiani K (2015) Preparation and in vitro investigation of antigastric cancer activities of carvacrol-loaded human serum albumin nanoparticles. IET Nanobiotechnol 9:294–299
66. Chen J, He H, Deng C, Yin L, Zhong Z (2019) Saporin-loaded CD44 and EGFR dual-targeted nanogels for potent inhibition of metastatic breast cancer in vivo. Int J Pharm 560:57–64
67. Naderinezhad S, Amoabediny G, Haghiralsadat F (2017) Co-delivery of hydrophilic and hydrophobic anticancer drugs using biocompatible pH-sensitive lipid-based nano-carriers for multidrug-resistant cancers. RSC Adv 7:30008–30019. https://doi.org/10.1039/C7RA01736G
68. Zhu H, Wang Y, Hussain A, Zhang Z, Shen Y, Guo S (2017) Nanodiamond mediated co-delivery of doxorubicin and malaridine to maximize synergistic anti-tumor effects on multi-drug resistant MCF-7/ADR cells. J Mater Chem B 5:3531–3540
69. Xiao B, Si X, Han MK, Viennois E, Zhang M, Merlin D (2015) Co-delivery of camptothecin and curcumin by cationic polymeric nanoparticles for synergistic colon cancer combination chemotherapy. J Mater Chem B 3:7724–7733
70. Zhao R, Li T, Zheng G, Jiang K, Fan L, Shao J (2017) Simultaneous inhibition of growth and metastasis of hepatocellular carcinoma by co-delivery of ursolic acid and sorafenib using lactobionic acid modified and pH-sensitive chitosan-conjugated mesoporous silica nanocomplex. Biomaterials 143:1–16
71. Said-Elbahr R, Nasr M, Alhnan MA, Taha I, Sammour O (2016) Nebulizable colloidal nanoparticles co-encapsulating a COX-2 inhibitor and a herbal compound for treatment of lung cancer. Eur J Pharm Biopharm 103:1–12
72. Kabary DM, Helmy MW, Abdelfattah E-ZA, Fang J-Y, Elkhodairy KA, Elzoghby AO (2018) Inhalable multi-compartmental phospholipid enveloped lipid core nanocomposites for localized mTOR inhibitor/herbal combined therapy of lung carcinoma. Eur J Pharm Biopharm 130:152–164
73. Maruyama K (2011) Intracellular targeting delivery of liposomal drugs to solid tumors based on EPR effects. Adv Drug Deliv Rev 63:161–169
74. Muheem A, Shakeel F, Warsi MH, Jain GK, Ahmad FJ (2017) A combinatorial statistical design approach to optimize the nanostructured cubosomal carrier system for oral delivery of ubidecarenone for management of doxorubicin-induced cardiotoxicity: in vitro–in vivo investigations. J Pharm Sci 106:3050–3065. https://doi.org/10.1016/j.xphs.2017.05.026
75. Yingchoncharoen P, Kalinowski DS, Richardson DR (2016) Lipid-based drug delivery systems in cancer therapy: what is available and what is yet to come. Pharmacol Rev 68:701–787
76. Ong YS, Saiful Yazan L, Ng WK, Abdullah R, Mustapha NM, Sapuan S, Foo JB, Tor YS, How CW, Abd Rahman N (2018) Thymoquinone loaded in nanostructured lipid carrier showed enhanced anticancer activity in 4T1 tumor-bearing mice. Nanomedicine 13:1567–1582
77. Chen Y, Pan L, Jiang M, Li D, Jin L (2016) Nanostructured lipid carriers enhance the bioavailability and brain cancer inhibitory efficacy of curcumin both in vitro and in vivo. Drug Deliv 23:1383–1392
78. Tripathi S, Kushwah V, Thanki K, Jain S (2016) Triple antioxidant SNEDDS formulation with enhanced oral bioavailability: Implication of chemoprevention of breast cancer. Nanomed Nanotechnol Biol Med 12:1431–1443
79. Brigger I, Dubernet C, Couvreur P (2012) Nanoparticles in cancer therapy and diagnosis. Adv Drug Deliv Rev 64:24–36

80. Mahmoudi M, Sant S, Wang B, Laurent S, Sen T (2011) Superparamagnetic iron oxide nanoparticles (SPIONs): development, surface modification and applications in chemotherapy. Adv Drug Deliv Rev 63:24–46
81. Li X, Wu M, Pan L, Shi J (2016) Tumor vascular-targeted co-delivery of anti-angiogenesis and chemotherapeutic agents by mesoporous silica nanoparticle-based drug delivery system for synergetic therapy of tumour. Int J Nanomedicine 11:93
82. Yu D, Ruan P, Meng Z, Zhou J (2015) The structure-dependent electric release and enhanced oxidation of drug in graphene oxide-based nanocarrier loaded with anticancer herbal drug berberine. J Pharm Sci 104:2489–2500
83. Muthoosamy K, Abubakar IB, Bai RG, Loh H-S, Manickam S (2016) Exceedingly higher co-loading of curcumin and paclitaxel onto polymer-functionalized reduced graphene oxide for highly potent synergistic anticancer treatment. Sci Rep 6:32808
84. Arya A, Ahmad H, Tulsankar S, Agrawal S, Mittapelly N, Boda R, Bhatta RS, Mitra K, Dwivedi AK (2017) Bioflavonoid hesperetin overcome bicalutamide induced toxicity by co-delivery in novel SNEDDS formulations: Optimization, in vivo evaluation and uptake mechanism. Mater Sci Eng C 71:954–964
85. Sandhu PS, Kumar R, Beg S, Jain S, Kushwah V, Katare O, Singh B (2017) Natural lipids enriched self-nano-emulsifying systems for effective co-delivery of tamoxifen and naringenin: Systematic approach for improved breast cancer therapeutics. Nanomed Nanotechnol Biol Med 13:1703–1713
86. Jain AK, Thanki K, Jain S (2014) Solidified self-nanoemulsifying formulation for oral delivery of combinatorial therapeutic regimen: Part I. Formulation development, statistical optimization, and in vitro characterization. Pharm Res 31:923–945

Nutraceutical-Loaded Chitosan Nanoparticles for Healthcare Applications

Entesar Hanan and Farhan J. Ahmad

1 Introduction

Nutraceuticals refer to a class of compounds found naturally in foods that are said to have beneficial effects on human health and well-being [1]. It includes compounds such as bioactive peptides, carotenoids, phenolic compounds, lipids, vitamins, etc., and the list continues to grow as more and more compounds are being isolated and identified from numerous sources [2]. Nutraceutical, a combination term of "nutrition" and "pharmaceutical," was devised by Stephen DeFelice in 1989 [3]. Nutraceuticals find wide application in the cure of various chronic diseases and treatment of various disorders. They are said to be effective in the treatment of malignancy and tumors, problems associated with high cholesterol, high blood glucose, hypertension, aging, infectious disease, inflammation, obesity, bone disorders, etc. [4]. Most of the nutraceuticals are said to have a potent antioxidant activity that aids in the prevention of diseases by alterations in redox state [5]. A majority class of the nutraceuticals are found to have manifold therapeutic properties. Nutraceuticals that are more prevalent include green tea, omega-3, lycopene, lutein, curcumin, and glucosamine. Currently nutraceuticals are attaining important attention due to their pronounced nutritive benefits and therapeutic applications. According to the latest market survey, the nutraceutical worldwide market is expanding and was valued to be around US $ 266 billion in 2018, and prospective marketability is expected to touch around US$ 403 billion by 2024, with an anticipated CAGR (compound annual growth rate) of 7% from 2019 to 2024 [4].

E. Hanan (✉)
Nano Formulation Lab, Department of Food Technology, SIST, Jamia Hamdard, New Delhi, India

F. J. Ahmad
Nano Formulation Lab, Department of Pharmaceutics, SPER, Jamia Hamdard, New Delhi, India

© Springer Nature Singapore Pte Ltd. 2020

M. Rahman et al. (eds.), *Nanomedicine for Bioactives*,
https://doi.org/10.1007/978-981-15-1664-1_8

Though the overall nutraceutical market data is interesting, the potent benefits of many nutraceuticals are not recognized because of their poor bioavailability [6] and degradation of bioactive compounds which is a challenging task. For example, polyphenols and antioxidants are subjected to degradation resulting in reduced efficacy of the nutraceuticals [7]. Moreover, the chemically unstable nature [1] and long-term stability issues [8] of nutraceuticals like ω-3 fatty acid, carotenoids, lycopene, and curcumin are equally challenging. The increase in the nutraceutical dose due to poor aqueous solubility of bioactives is an additional hindrance in the effective delivery of nutraceutical [9].

In order to combat the issues of nutraceutical delivery, nanotechnology has emerged as a suitable vehicle. The potential of nanotechnology to encapsulate bioactives and improve their stability, quality, safety, and nutrition has revolutionized the field of nutritional medicine and healthcare [10]. Nutraceutical-laden nanocarriers show an increased SA:volume fraction and enhanced absorption and bioavailability [7]. Different technologies like sub-micron emulsion, lipid nanoparticles, liposomes, inorganic nanoparticles and nanocomposites, polymeric micelles, and polymeric nanoparticles have been exploited to overcome the limitations offered by nutraceuticals [4].

In order to increase the bioavailability, solubility, and half-life of bioactives and functionality of hydrophobic nutraceutical compounds, various nanosystems are used such as sub-micron emulsion system, SNEDS and SMEDS, liposomal delivery system, dendrimer and inorganic nanoparticles, and polymeric micelles. To mend the bioavailability issue of poorly soluble bioactives, polymeric nanoparticles are suitable candidates [11]. Polymeric nanoparticles obtained from naturally occurring sources are effective vehicles for the controlled delivery of bioactives. They also offer shield from severe acidic environment and aid in the reduction of toxic effects.

Biopolymeric nanoparticles such as chitosan have diverse advantages for the efficacious delivery of nutraceutical. It is a widely available, biodegradable, and biocompatible polymer. It is said to possess antimicrobial property and show non-toxicity. Chitosan, a polymeric compound of N-acetylglucosamine, is formed via partial alkaline deacetylation of chitin. It entails β-(1,4)-linked-D-glucosamine units along with the randomly acetylated amine moieties. Due to the special properties of amine and –OH groups, chitosan finds its applicability in diverse fields. Owing to its versatile properties, chitosan has been explored as a capable material for the nanoparticle distribution of nutraceuticals. Diverse approaches have been employed for the development of chitosan nanoparticles. The most commonly used ones include ionotropic gelation method, cross-linking method, coacervation method, spray-drying, emulsion coalescence technique, reverse micellar technique, sieving method, etc. Dhiman and Bhalla [12] encapsulated lycopene in chitosan nanoparticle via ionic gelation method for the effective delivery of lycopene. Nallamuthu et al. [13] preserved the antioxidant activity of chlorogenic acid by encapsulating it in chitosan nanoparticle. Thamaket and Raviyan [14] reported the successful encapsulation of carotenoids in chitosan nanoparticles. The developed nanoparticles depicted improved solubility.

2 Nutraceuticals

The word nutraceutical is an amalgam of two words nutrition and pharmaceutical and was claimed by Dr. Stephen L. DeFelice in 1989 [15]. Nutraceuticals are described as food products that are said to deliver healthiness and medicinal benefits, including the deterrence of diseases and management of disorders. Nutraceutical is a broad category which includes many classes and subclasses. Its contribution to the healthcare system is well established. The two most specific categories of nutraceuticals include phytochemicals and antioxidants. Phytochemicals are said to provide protection against the peril of certain long-lasting diseases such as cancer, diabetes, heart disease, and hypertension. However, antioxidants by means of prevention to oxidative damage help in alleviating the risk of chronic diseases. Nutraceuticals are said to have many advantages like lesser side effects and elevated health benefits [16, 17]. Bioactive compounds, phenolic compounds, carotenoids, flavonoids, curcuminoids, vitamins, etc. are all listed under the nutraceutical umbrella. Nutraceuticals help in the regulation of blood glucose and cholesterol level and aid in the promotion of cognitive health. They tend to reduce inflammation, preclude cardiovascular disease, and impede the development of cancer, etc. [2].

Nutraceuticals due to their nutrition potential, less toxicity, and beneficial effects have received a noteworthy attention. The global nutraceutical market reached a value of US$ 266 billion in 2018, recording a CAGR of nearly 6% during 2011–2018. Now, the market is expected to approach a value of US$ 403 billion by 2024, growing at a CAGR of around 7% during 2019–2024 (www.imarcresearch.com). Researchers have revealed the involvement of nutraceuticals in dealing of many ailments such as insomnia and melancholy, digestive problems, cancer, hypertension, coronary heart disease, delayed gastrointestinal emptying, and many more conditions which need utmost attention. The application of nutraceutical compounds in healthcare is illustrated in Table 1.

2.1 Applications of Nutraceuticals in Healthcare

2.2 Challenges in the Delivery of Nutraceuticals

Nutraceuticals cater to the manifold health benefits. They accord to multitude healthcare applications. They have an effective role in the prevention and reduction of diseases and disorders. However, their effective delivery is a challenging task. Nutraceuticals encounter a number of technological challenges which include solubility of the bioactive compound in aqueous or lipid phase, sensitivity of the compound to heat and pH [38], reduced absorption, little solubility, reduced bioavailability, chemical instability [7], and fast metabolic rate [39].

Table 1 Nutraceutical compunds their sources and health benefits

Nutraceutical compounds	Sources	Health benefits	References
B-Carotene	Carrots, pumpkin, maize, and tomato	Cardioprotective and anticancer	Mayne [18]
Lycopene	Tomato, grapefruit, and apricot	Anticancer (prostate cancer)	Holzapfel et al. [19]
Astaxanthin	*Phaffia* yeast, shrimp, crab, and salmon	Antioxidant, anticancer, and prevents diabetes, CVD, and neurodegenerative disorders	Ambati et al. [20]
Lutein	Spinach, lettuce, and eggs	Antioxidant and prevents cataracts, diabetic retinal vascular disease, and eye detachment	O'Brienon [21]
Saponin	Soyabeans, legumes, and maize	Anti-inflammatory, hypocholesterolemic, and antifungal	Marrelli et al. [22]
Quercetin	Onion, apple, and grapes	Antioxidant	David et al. [23]
Kaempferol	Tea, strawberries, and tomatoes	Anticancer	Chen and Chen [24]
Myricetin	Grapes, walnut, and berries	Anti-diabetic	Li and Ding [25]
Rutin	Buckwheat and citrus fruit	Antibacterial, antitumor, anti-inflammatory, antiallergic, vasoactive, hypolipidemic, antiplatelet, and antihypertensive	Patel and Patel [26]
Naringenin	Citrus fruit	Antioxidant and effective in lowering visceral adiposity and plasma lipids	Rimm et al. [27]
Chrysin	Honey and propolis	Antioxidant, anti-inflammatory, and anticancer. Lowers neurotoxicity and oxidative stress	Mani and Natesan [28]
Apigenin	Celery and parsley	Anti-inflammatory, anti-toxicant, and anticancer	Ali et al. [29]
Catechin and EGCG	Tea, apples, chocolate, and green tea	Antioxidative and effective in the reduction of hypertension, inflammation, and hyperlipidemia	Gadkari and Balaraman [30]
Genistein	Soy-based foods	Anticancer	Spagnuolo et al. [31]
Resveratrol	Grape skin and seeds	Anti-inflammatory, anti-carcinogenic, cardioprotective, vasorelaxant, and neuroprotective	Salehi et al. [32]
Curcumin	Turmeric	Antioxidant, anti-inflammatory, anti-mutagenic, antimicrobial, and anticancer	Hewlings and Kalman [33]
Chlorogenic acid	Green coffee and apples	Effective against cancer, diabetes, inflammation, and obesity issues	Tajik et al. [34]
Cinnamic acid	Cinnamon	Anticancer and anti-diabetic	Johansson [35]

(continued)

Table 1 (continued)

Nutraceutical compounds	Sources	Health benefits	References
Capsaicin	Pepper	Effective against atherosclerosis, metabolic syndrome, diabetes, and obesity	McCarty et al. [36]
Gingerol	Ginger	Antioxidant and anti-inflammatory. Reduces gastric problems and diabetes	Pratap et al. [37]

The bioavailability of nutraceutical is limited mainly due to three reasons.

1. Its inability to be released from the food matrix, e.g., carotenoids are partially released into the GIT as they are trapped within the cellular structures [40].
2. Its undesirable interaction with other components of the food matrix, i.e., the environment and components. Polyphenols when pass through the GIT, actively interact with the food matrix and microflora of gut which affects their bioavailability. Rimbach et al. [41] stated that the nutraceuticals tend to form aggregates of particles with other compounds in the GIT resulting in the insoluble complex formations that impede their uptake.
3. Limited absorption of the nutraceutical in the GIT due to the mucus and phospholipid bilayer covering the epithelial lining which obstructs the passage of nutraceutical through it. The bioavailability of nutraceutical is the most challenging issue. Moelants et al. [42] suggested that the constrained release of the bioactive compound corresponds to poor bioavailability of nutraceuticals. The pharmacokinetic studies reveal that the rate and level of absorption of active metabolites are affected by the chemical structure of a compound which in turn affects the amount of nutraceutical available at the site of action. Actis-Goretta et al. [43] stated that the reduced permeability across the mucus layer or epithelium cells also contributes to reduced bioavailability.

The chemical degradation and transformation and long-term stability issue of the nutraceutical also hamper its efficacy. Some nutraceuticals are subjected to alteration in the biological activity. They undergo redox reactions or hydrolysis reactions that alter their activity. For example, polyphenols and antioxidants are subjected to degradation resulting in the reduced efficacy of the nutraceuticals [7]. Carail et al. [44] reported that ω-3 fatty acids or carotenoids oxidize in GIT due to the presence of pro-oxidants. Wickham et al. 2009 stated that the proteins and peptides may undergo hydrolization in the enzymatic and acidic milieu of the stomach. Besides one of the critical concerns in the delivery of nutraceutical is that it should confer to the intended health benefit after ingestion as some of the nutraceuticals are metabolized by the enzymes present in the GIT [45]. Koga et al. [46] reported that nobiletin and genistein are prone to oxidation by CYP450 enzyme system. Metzler et al. [47] stated that under the influence of a series of reaction, a variety of metabolites are formed by curcumin in the GIT. Metabolism of the nutraceuticals may be either deleterious or advantageous for the overall performance of nutraceuticals.

3 Chitosan

Chitosan is a naturally taking place biocompatible, mucoadhesive polymer which is white, hard, inelastic, cationic, and highly basic in nature. It is a chitin derivative, produced by the removal of acetate moiety from chitin [48]. It is essentially composed of β-(1→4)-linked glucosamine units coupled with some proportion of N-acetylglucosamine units [49]. Chitosan is present only in the cell wall of some fungi and hence is less abundantly found in nature. On commercial scale it is manufactured by deacetylation of chitin which is a ubiquitous biopolymer next to cellulose. Chitin has remarkable properties such as biocompatibility, bioactivity, biodegradability, or high mechanical strength, but due to its poor solubility, it finds limited utility which limits its usability and shifts attention toward chitosan. The conversion of chitin to chitosan is an enzymatic or chemical-mediated process; however, due to lower cost and its suitability for mass production, the chemical conversion is preferred [50]. The chitosan polymer is composed of 44.1, 6.8, and 7.9% of carbon, hydrogen, and nitrogen with a viscosity mean molecular weight of ~5.3×10^5 Da [51]. Chitosan due to its nontoxic, biodegradable, and antimicrobial properties finds manifold applications in biomed industries, agricultural field, genetic engineering, food industry, environmental pollution control, water treatment, paper manufacturing industry, photography, and so on [52]. The USFDA confirms chitosan as a safe and biocompatible polymer. Baldrick [53] reported the chitosan LD50 value of 16 g/kg in mice upon oral administration. In humans, 4.5 g/day of chitosan on oral administration taken consistently for 12 weeks indicated no ill symptoms. Kean and Thanou [54] compared the toxicity of chitosan with toxic compounds such as sulfide and reported that at LD_{50} less than 20 μg/mL against cancer and kidney cell lines, chitosan showed no symptoms of toxicity.

3.1 Chitosan Nanoparticles and Approaches for the Preparation of Chitosan Nanoparticles for Nutraceutical Delivery

In the recent decade, chitosan has been studied as the most promising biopolymeric material for the development of nanoparticles. The small particle size and greater surface-to-volume ratio are one of the characters unique to the nanoparticle. Chitosan nanoparticles are mucoadhesive and hydrophilic in nature. They have a widespread application as drug carriers as they have good biocompatibility and biodegradability. They tend to enhance the drug solubility and stability by their controlled drug release behavior and hence increase the overall efficiency and efficacy [55]. The various methods for the preparation of chitosan nanoparticles for the delivery of nutraceuticals include the following.

3.1.1 Ionotropic Gelation Method

Calvo et al. [56] first reported the development of chitosan nanoparticle via ionic gelation method. The mechanism underlying the development of chitosan nanoparticles via ionotropic gelation technique is founded on the concept of electrostatic interaction among the amine group and negatively charged polyanion moiety of chitosan and tripolyphosphate (TPP), respectively. The process involves the preparation in aqueous method. Chitosan is first liquefied in acetic acid in the presence or absence of stabilizing agent. These agents can be supplemented in the chitosan solution afore or before the addition of polyanion. Polyanion is then mixed, and nanoparticles are formed instantaneously under automated stirring at room temperature [57]. Sinha et al. [58] classified polyanions for ionic gelation into three categories, viz., (1) low-molecular-weight polyanions which include sodium pyrophosphate [SPP], tripolyphosphate, tetrapolyphosphate, octapolyphosphate, and hexametaphosphate; (2) hydrophobic polyanions such as sodium alginate and k-carrageenan; and (3) high-molecular-weight ions, e.g., octyl, dodecyl, hexadecyl, and cetylstearyl sulfate. STPP is the most commonly used polyanion for ionic gelation method. Alishahi et al. [59] encapsulated Vit C in chitosan reacted with STPP in order to increase the efficiency of Vit C.

3.1.2 Cross-Linking Method

This technique involves numerous variants, the most common being emulsion cross-linking method. The method involves the mixing of a chitosan in acetic acid with a liquid oil or paraffin to produce an oil-in-water (o/w) emulsion. The emulsion is later supplemented with a cross-linking agent to form the microspheres which is then consequently removed (by filtration) and later washed and dried. The use of formaldehyde (FA) and glutaraldehyde (GA) as covalent cross-linking agents is widely recognized. This method finds use in the encapsulation of water-insoluble active ingredients, such as fat-soluble vitamins or phenolic compounds, which are dispersed together with chitosan. Thamaket and Raviyan [14] successfully encapsulated carotenoids in the chitosan cross-linked tripolyphosphate (TPP) which was made with different ratios of chitosan:TPP. The nanoparticles developed had more aqueous solubility and proficient loading and release.

3.1.3 Coacervation Method

The coacervation process involves the dissolution of chitosan in acetic acid and adding it into the mixture of sodium hydroxide and methanol by means of a nozzle to generate the coacervate droplets. In some complex cases of coacervation, an additional polymer with an opposite charge is added to the system. The method best fits for the delivery of immobilized enzymes, lipoproteins, and cell cultures [60]. Mansoori et al. [61] prepared Chitosan DNA and Follic acid Chitosan DNA nanoparticles for gene therapy using complex coacervation process. Qv et al. [62] encapsulated lutein in chitosan nanoparticles by coacervation method that resulted in the improvement of its stability and physiochemical properties. Alencastre et al. [63] reported the encapsulation of Vit E via complex coacervation method using a carboxymethylcellulose-chitosan system.

3.1.4 Reverse Micellar Technique

Reverse micelles are thermodynamically steady liquid mixtures consisting of aqueous phase, an oil, and a surfactant. This method is advantageous as it allows the creation of nanoparticles of diameters <200 nm and low polydispersity indices. Brunel et al. [64]. prepared chitosan nanoparticles via reverse micellar method and illustrated that the lesser the molar mass of chitosan, the more improved the particle size and distribution. Mansouri et al. [65] successfully encapsulated bovine serum albumin into the chitosan nanoparticles via the reverse micellar technique.

3.1.5 Sieving Technique

Agnihotri et al. [66] reported this novel technique for the development of chitosan nanoparticles. In this method chitosan gel mass loaded with the bioactive compound is first prepared and then cross-linked with glutaraldehyde. This mixture is passed through sieves having an appropriate pore size to get nanoparticles. The nanoparticles are then finally washed with 0.1 N NaOH solution to remove the un-reacted excess glutaraldehyde and desiccated in an oven overnight at 400 °C.

3.1.6 Emulsion Coalescence Technique

This technique is a simple two-step process based on coalescence and precipitation. Two separate emulsions are prepared which are later mixed and agitated so as the droplets collide with each other effectively resulting in coalescence so that small-sized particles are precipitated.

4 Encapsulation of Different Nutraceuticals in Chitosan Nanoparticles

4.1 Lycopene

Lycopene, a naturally present red-colored carotenoid pigment, is found mostly in high amount in tomatoes but is also present in watermelons, pink grapefruits, apricots, and guavas. It is a non-provitamin A carotenoid and an isomer of β-carotene (acyclic). It is said to be defensive against prostate cancer [19], age-related macular degeneration [67], gastrointestinal diseases [68], diabetes [69], heart disease, asthma, and diabetic neuropathy [70]. However, lycopene has a very low aqueous solubility and restricted absorption, following oral administration [71]. Besides its bioavailability is affected by the presence and dosage of other carotenoids also, e.g., β-carotene.

Dhiman and Bhalla [12] formulated lycopene-loaded chitosan nanoparticle by ionic gelation method by means of several concentrations of chitosan (100–200 mg) and STPP (50–100 mg) and by changing stirring speed in the range of 10–20 min. The optimized lycopene-loaded chitosan nanoparticle was found to have 150 mg of chitosan, 75 mg of STPP, and 20 mg of drug lycopene. The lycopene-loaded chitosan nanoparticle showed the entrapment efficiency of 89.4% with 15 min of stirring time, and the percent release of the same at the end of 6 h was found to be 83.5%, against percent release of pure lycopene which was observed to be 79.6%, thus depicting that the lycopene-loaded chitosan nanoparticles enhance the accumulation of lycopene in the cellular system and are thus promising candidates for the effective delivery of lycopene. Al-Eisa [72] reported the development of lycopene-loaded chitosan nanoparticles to study its effect against oxidative stress of aging and hepatotoxicity. The results depicted that both native chitosan nanoparticles and chitosan nanoparticles loaded with lycopene showed an improvement in the lipid profile with a significant decrease in tumor necrosis and lipid peroxidation, thus offering shield against oxidative stress and hepatotoxicity as a consequence of aging.

4.2 Catechin

Catechin is a polyphenol compound belonging to flavonoid class. Tea leaves are the most abundant source of catechins [73]. Catechin is a non-esterified having a strong biological potential against many degenerative diseases. Catechins are said to possess anti-carcinogenic, anti-cataract, anti-hypercholesterolemic, anti-mutagenic, antiproliferative, anti-viral, and antibacterial properties. Besides it is also said to have a good anti-inflammatory effect and antioxidative effect [30]. However, the critical issue that environs catechin is its instability that leads to its decreased bioavailability. Catechins are extremely unsteady against pH, temperature, and light shifts [74]. They undergo phenomena of epimerization which causes loss of activity [75].

Manikkam and Pitchai [76] developed catechin-loaded chitosan-alginate and chitosan-carrageenan nanoparticles. The nanoparticles were formed via ionotropic pre-gelation of an alginate/carrageenan core followed by chitosan polyelectrolyte complexation. The investigation by TEM and FTIR revealed nanoparticles with a diameter of 20–50 nm and an encapsulation efficiency of (65–97%) and stability of (99.5–94.5%). Further, chitosan-alginate polymeric nanoparticles were found to be better than the chitosan-carrageenan nanoparticles. These nanoparticles were reported to be nontoxic and were found to have anticancer activity. Dudhani and Kosaraju [77] used ionic gelation method between the chitosan and sodium tripolyphosphate for the formulation of catechin-loaded chitosan nanoparticles. The developed nanoparticles were smooth and spherical with a particle size of 130 nm. The in vitro release studies revealed that catechin sustained control release in the GI tract. Hu et al. [78] also fabricated the chitosan tripolyphosphate (CS-TPP) nanoparticles via polyanion-initiated gelation process, projected to be used as effective carriers for conveying tea catechins. The nanoparticles developed had an encapsulation efficiency of 24–53%. Tang et al. [79] reported the delivery of catechins via self-assembled nanoparticles, comprised of chitosan (CS) and an edible polypeptide, poly(γ-glutamic acid) (γ-PGA). The results validated that the developed nanoparticles retained the antioxidant activity of tea catechins and thus can be suggested as a good carrier for oral delivery of tea catechins. Zhang et al. [80] used β-chitosan nanoparticles (β-CS NPs) and ionic gelation technology for the encapsulation of catechins (CAT) or CAT-Zn complex to evaluate their antibacterial potential and suggested that encapsulation of CAT-Zn complex in β-CS NPs improved the antibacterial activity of CAT and CAT-Zn complex.

4.3 Epigallocatechin-3-gallate (EGCG)

EGCG is a strong natural antioxidant and is made up of epigallocatechin and gallic acid. It is the most abundant tea catechin. EGCG is normally present in high concentration in green tea [81]. Epigallocatechin-3-gallate is said to have potent antioxidant and anti-carcinogenic properties [82]. It is effective against some types of cancers like melanoma, tumors, and squamous cell carcinoma. It is said to have a positive effect in the treatment of chronic lymphocytic leukemia, multiple sclerosis [83], liver fibrosis and cirrhosis, and Alzheimer's disease [30]. However, the oral bioavailability of EGCG is quite low. Only a small amount of EGCG finds way to blood circulation and active sites accredited mainly to its poor stability and low absorption in intestines.

Hu et al. [84] encapsulated EGCG in monodispersed nanoparticles coupled together from bioactive caseinophosphopeptide (CPP) and chitosan (CS). The developed nanoparticles had an encapsulation efficiency of 70.5–81.7% with a controlled release rate. The antioxidant assays depicted that EGCG nanoparticles exhibited a much strong activity against free radical scavenging than the free EGCG and hence can thus effectively enhance the antioxidant activity of biological systems. Hong et al. [85] reported the development of EGCG-loaded nanoparticle in self-assembled nanoparticles of chitosan (CS) and aspartic acid (PAA). The developed EGCG-CS-PAA nanoparticles had an average diameter of 102 nm. Oral administration of EGCG-CS-PAA nanoparticles to rabbits showed an average ratio of lipid deposition of (16.9 ± 5.8%) that was found close to simvastatin (15.6 ± 4.1%) depicting effectiveness of EGCG against rabbit atherosclerosis. Dube et al. [86] encapsulated EGCG in chitosan-tripolyphosphate nanoparticles for enhancing its oral delivery. The oral absorption of EGCG evaluated in Swiss outbred mice revealed a 2.3-fold increase in the exposure of EGCG to the jejunum following CS NP encapsulation which in turn increased plasma concentrations of EGCG.

4.4 Resveratrol

Resveratrol, a naturally occurring polyphenolic compound (phytoalexin), belongs to family stilbenes. It is found mostly in grape skins and peanuts. Resveratrol is found to possess a strong antiplatelet activity and vasodilating properties [87]. It is found to aid in the prevention against kidney diseases [88] and is found to possess anticancer activity by inhibition of angiogenesis [89]. It is also known to have anti-inflammatory [90] and cardioprotective effects [91]. It is found beneficial for neuroprotection and a curative for Alzheimer's disease [92]. However, resveratrol has poor water solubility [93] and lower chemical stability [94] which leads to its decreased bioavailability after oral administration.

Wu et al. [95] prepared resveratrol-loaded CS-TPP nanoparticles via ionic cross-linking method. The nanoparticle formulated had a size of 172–217 nm along with extensive storage and UV light stability. Compared to the free resveratrol, the resveratrol-loaded nanoparticle showed lesser cytotoxicity on normal hepatocyte cells L02 and antiproliferative activity on liver carcinoma cells SMMC 7721 indicative of its effective use in chemotherapy. Min et al. [96] developed resveratrol-loaded trimethyl chitosan (TMC) nanoparticles. Resveratrol was encapsulated in TMC-TPP (TP NPs), TMC-alginate (TA NPs), and TMC-alginate-TPP (TAP NPs). A lower particle size and more sustained release were shown by TA NPs and TAP NPs compared to TP NPs. However, higher encapsulation efficiency was found to be in TP NPs. The result illustrated that TMC is a good carrier to increase the cellular uptake of resveratrol. Iglesias et al. [97] demonstrated the preparation of resveratrol-loaded chitosan nanoparticle by optimizing RES/polymer ratio and temperature using Box-Behnken experimental design, in order to evaluate their effect on encapsulation efficiency. The nanoparticles developed had better encapsulation efficiency achieved in 24 h at 39 °C and at RES/polymer ratio of 0.75:1 *w/w*. However, the final cumulative drug release showed a significant difference being 79–81% in NP and 29–34% CTS NP composites.

4.5 Curcumin

Curcumin is a polyphenol also called diferuloylmethane. It is mainly found in the rhizome of *Curcuma longa*, i.e., turmeric [98]. Curcumin is recognized to have immense health benefits. It is the most widely explored nutraceutical augmented to have a vast therapeutic potential. Curcumin is found to have antioxidant activity attributing to its property to scavenge free radicals, ROS, and nitrogen species [99, 100]. It has an anti-inflammatory property [101] and is said to be a potent suppressant of inflammation which is a root cause for the development of many diseases which include Alzheimer's disease (AD), Parkinson's disease, multiple sclerosis, cerebral injury, CVD, metabolic syndrome, cancer, psoriasis, diabetes, obesity, etc. [102]. The anti-arthritic, osteoarthritic, and rheumatoid arthritic effect of curcumin is well established in various studies [103]. Curcumin is also reported to be a potent anticancer [104] and antibacterial agent [105]. Curcumin is said to attenuate various metabolic syndromes such as insulin resistance, hyperglycemia, hypertension, and obesity [106–109]. Despite numerous health benefits of curcumin, the major issue encountered includes its poor bioavailability.

Vijayakurup et al. [110] developed curcumin-loaded chitosan nanoparticle in order to study carcinogenesis. The formulated nanoparticle had a particle size of 170–200 nm with a sustained release over a period of 180 h. The results further certified that the encapsulated curcumin was more effective in reducing tumor incidence and multiplicity than free curcumin. Samrot et al. [111] developed curcumin-loaded chitosan nanoparticle using sodium tripolyphosphate and barium chloride as chelators. The nanoparticles developed had a size below 500 nm as confirmed via AFM and a time-dependent encapsulation efficiency. The results further illustrated that the drug release was found to be best when a higher concentration of chelating agents was used. Mirnejad et al. [112] developed curcumin-loaded chitosan-TPP nanoparticles via ionotropic gelation methods to elucidate the antibacterial effect of curcumin. The developed nanoparticles were spiral shaped and revealed a size of

160 ± 10 nm with an encapsulation efficiency of 75 ± 2%. The studies revealed curcumin nanoparticle as a potent agent in the cure of various infections like *Staphylococcus aureus* and *Pseudomonas aeruginosa* infections. Nair et al. [113] synthesized curcumin-loaded nanoparticles for transdermal delivery. The formulation revealed a spherical shape with a particle size of 167.3 ± 3.8 nm and an entrapment efficiency of 80%. The cytotoxicity assay on human keratinocyte (HaCat) cells revealed enhanced percentage cell viability, superior drug release, and enhanced transdermal permeation of curcumin. Khan et al. [114] reported the development of curcumin-loaded chitosan nanoparticles via ionic gelation of chitosan with sodium tripolyphosphate (TPP). The cervical cancer cell line SiHa studies illustrated it to be effective in the treatment of cervical cancer. The cell line studies demonstrated curcumin-loaded chitosan nanoparticles showed increased uptake and released more lactate and lower ATP as compared to native curcumin. Duse et al. [115] developed the curcumin-loaded chitosan nanoparticle via ionic gelation technique for photodynamic therapy (PDT), i.e., a technique in which tumor tissues are selectively destroyed by minimally invasive tumor targeting strategy. With the use of novel LED device ($\lambda_{ex} = 457$ nm), the curcumin nanoparticles were able to destroy tumor cells by generation of ROS. The results conferred the efficacy of curcumin against cancer.

4.6 Quercetin

Quercetin is a type of flavonol (flavonoid) which is found most commonly in various foods including fruits, vegetables, grains, and leaves [116]. It is naturally found in plants as glycone or carbohydrate conjugates [117]. Structurally it consists of five hydroxyl groups. The highest concentration of quercetin is found in red onions and tomatoes [118]. Quercetin is said to have many health-promoting benefits. It is said to be a potent cardioprotectant having vasodilatory effect [119]. It is an effective anticancer agent which is found to have antiproliferative effects in cancer cell types, like breast, leukemia, squamous cell, colon, ovary, endometrial, and gastric cell [120]. It is a powerful antioxidant effective against ischemia-reperfusion injury, toxic compounds, and oxidative stress [121]. Quercetin is also effective against inflammation and also is proven to have well-established effects on diabetes [122]. It is reported to have antibacterial effects as well [123]. However, the bioavailability of quercetin is poor and highly variable [124].

Zhang et al. [125] described the development of quercetin-loaded chitosan nanoparticle via ionic gelation of chitosan with tripolyphosphate anions for enhancing its bioavailability. The results confirmed the antioxidant activity of quercetin-loaded chitosan nanoparticles evaluated by free radical scavenging activity test and reducing power test. Tzankova et al. [126] explored the hepatoprotective and antioxidant effects of quercetin-loaded chitosan nanoparticle via gelation method. Quercetin was encapsulated in chitosan-alginate nanoparticle. The hepatoma HepG2 cell induced with H2O2 oxidative stress depicted a decrease in cell viability on pre-treatment with encapsulated quercetin. The vivo results suggested a significant diminution in the amplified levels of serum transaminases ALT and AST, reduced the lipid peroxidation, and restored the levels of glutathione. The results were found to be better in encapsulated quercetin compared to free quercetin suggesting it to be effective for hepatoprotection. Nan et al. [127] studied the protective effect of quercetin against UVB radiation by entrapment of quercetin in TPP-chitosan nanoparticles (TCs). The results displayed that the encapsulated quercetin has better permeation through the epidermis with better stability and low cytotoxicity. The developed nanoparticle showed enhanced effect in improving the skin edema caused by UVB radiation and hence confers to its therapeutic topical use against UVB radiation. Baksi et al. [128] formulated the quercetin-loaded chitosan nanoparticles (QCT-CS NPs) having a particle size of <200 nm with an encapsulation efficiency of 79.78% and drug release of 67.28% over a period of 12 h. A significant reduction was found to be in the IC50 value and an increase in the level of SOD of QCT-CS NPs as compared to free QCT suggesting improved antitumor effect by the quercetin nanoparticles. The A549 and MDA MB 468 cells showed a significant reduction of tumor volume upon the intravenous treatment of QCT-CS NPs in comparison to disease control groups, thus making it efficacious in cancer therapy.

4.7 Naringenin

Naringenin is a natural flavonoid aglycone of naringin, a naturally occurring bioactive flavanone (flavonoid) known to have a constructive effect on human well-being [129]. It is predominantly found in citrus fruits such as grapefruits and oranges [130]. Naringenin is said to have immense biological activities. It is said to exhibit a potent antioxidant activity and protection against oxidative stress by decreasing lipid peroxide and nitric oxide and increasing glutathione level [131]. Naringenin is also found to be antifibrinogenic and hepatoprotective in nature suggesting it to be suitable in the treatment of hepatic fibrosis [132]. It also has an anti-inflammatory effect owing to its ability to scavenge ROS [133]. Naringenin is also said to be cardioprotective and anti-carcinogenic in nature. It is found to inhibit carcinogenesis and tumor growth [134]. Naringenin is said to decrease the visceral adiposity and lower plasma lipids, hence reducing obesity [27]. However, the poor bioavailability of naringenin is a challenging task in its efficacious delivery.

Winarti et al. [135] developed naringenin-loaded chitosan nanoparticles via ionic gelation method to evaluate its potential against cancer. The result illustrated an increase in the cellular uptake and cytotoxicity of naringenin against T47D breast cancer cell line. Furthermore, it was found that both free naringenin and naringenin nanoparticles were able to induce apoptosis, hence suggesting naringenin as an effective anticancer nutraceutical. Malathy and Iyer [136] synthesized naringin-loaded chitosan nanoparticle for bone redevelopment. The results confirmed that both naringin and naringin-loaded chitosan nanoparticle promote the secretion of bone formation proteins along with the enhancement of the proliferation and osteogenic differentiation of bone marrow stromal cells (BMSCs). The developed naringin nanoparticles were also evaluated for other biological activities like anti-inflammation, anti-coagulation, antioxidant, cytotoxicity, and anticancerous activity. Naringin's potential for prolonged and sustained release makes it suitable for bone tissue engineering. Kumar et al. [137] explored the potential of naringenin-loaded chitosan nanoparticles for lung cancer cells. Naringenin was loaded into the chitosan nanoparticle by ionic gelation method. The encapsulated nanoparticles showed an increased free radical scavenging activity and a particle size of 407.47 nm with an encapsulation efficiency of approximately 70% and 80% as determined by FC and HPLC methods, respectively. Cell line studies suggested it to be nontoxic toward normal fibroblast 3T3 cells; however, cytotoxic effects were remarked against lung cancer cells (A549) proposing it to be an effective agent against anti-cancer activities.

4.8 Rutin

Rutin, a glycoside flavonoid, is also known as rutoside, quercetin-3-rutinoside, and sophorin [138]. Chemically it is composed of quercetin, a flavonolic aglycone, and rutinose, a disaccharide. It is a found in abundance in plants and food stuff, such as buckwheat, tea, and apple [139]. Rutin has been perceived for its health-promoting and nutraceutical effects. It is demonstrated to have multiple effects in the reduction of various ailments. Rutin is anticancerous and is proven to be effective against various cancers like leukemia, colon cancer, neuroblastoma, colorectal cancer, pulmonary metastasis, melanotic melanoma, and hepatic carcinoma [140]. Rutin is also reported to have a neuroprotective effect. It is said to prevent neuroinflammation and hence is effective in the treatment of Alzheimer's [141]. Besides it is said to have anti-convulsant and sedative activity [142]. Rutin is said to show potential benefits on the endocrine system owing to its anti-diabetic [143] and anti-hypercholesterolemic effect [144]. It is also a well-established cardioprotectant with antihypertensive ([145], anti-coagulant [146], and antiplatelet effect [147]. Administration of rutin is said to prevent cataractogenesis, thus exhibiting anti-cataract effect [148]. Besides rutin is also reported to be anti-fungal [149], anti-asthmatic [150], anti-ulcer [151], antiosteoporotic, and antiosteopenic [152]. Rutin is said to possess an oral bioavailability of around 20%. The reason of its poor bioavailability is attributed to its poor aqueous solubility.

Patil and Jobanputra [153] prepared rutin-loaded chitosan nanoparticles and evaluated it for antimicrobial potential. The formulated nanoparticles showed enhanced antibacterial potential and effectiveness against dental caries pathogen. Significant differences were observed in the MIC and MBC of rutin-encapsulated chitosan nanoparticle and free rutin or chitosan against *Bacillus pumilus* and *Enterococcus faecalis*. In order to improve the aqueous solubility of rutin, Mujtaba et al. [154] developed chitosan-alginate nanoparticles (CANPs) of rutin via ionotropic pre-gelation of an alginate core trailed by chitosan polyelectrolyte complexation. The rutin-loaded CANP had an average particle size of 141.29 ± 5.11 nm. The release profile showed an initial burst effect in the media at the first 2 h, followed by a continuous controlled release phase within 24 h. The study suggested that solubility of rutin can be improved by CANP system which can help in its effective delivery. Ahmad et al. [155] prepared rutin-encapsulated chitosan nanoparticles (RUT-CS NPs) by ionic gelation method for the treatment of cerebral ischemia. The results depicted a particle size of <100 nm with an entrapment efficiency of 84.98% ± 4.18%. The bioavailability for RUT in the brain as compared to intravenous administration was found to be higher. The i.n. administration of RUT-CS NPs in MCAO-induced cerebral ischemic rat model showed enhanced neurobehavioral activity and histopathology with lesser infarction volume effects. Thus, it can be concluded that RUT-CS NPs are effective in brain targeting.

4.9 Kaempferol

Kaempferol is a naturally occurring flavonoid. It is present in several edible plants (e.g., tea, broccoli, cabbage, tomato, strawberries, and grapes) and in plants of traditional medicine (e.g., *Ginkgo biloba*, *Moringa oleifera*, and propolis) [156]. Epidemiological studies suggested kaempferol to be effective for reducing the risk of several disorders. Kaempferol is said to possess ability to act as an antioxidant by decreasing the superoxide anion, hydroxyl radical, and peroxynitrite levels to considerably low concentrations. It is said to be a good superoxide scavenger, with a 0.5 M of IC50 [157]. The anti-inflammatory effect of kaempferol is attributed to its property to reduce the cellular levels of reactive oxygen species [158]. Besides kaempferol is said to be a potent compound for the reduction of coronary diseases [99, 100] and atherosclerosis [159] facilitated by its capability to reduce LDLs. It is also said to lower the prevalence of myocardial infarction [120] and brain disease [160]. It also deliberates to vasodilator effects and antiplatelet and antithrombotic effect [161]. Kaempferol has shown effective results in chemoprotection. It is reported to be effective in the suppression of proliferation of breast cancer [162]. Due to its antiestrogenic activity, it is also said to inhibit breast and prostate cancer [163]. Kaempferol is said to possess low oral bioavailability. In order to elevate its therapeutic properties, the high metabolism of kaempferol should be overcome.

Luo et al. [164] studied cancer cell viability by the development of kaempferol-loaded chitosan nanoparticles. The nanoparticles developed were found to have a particle size of 230 ± 70 nm; however, the results illustrated no significant reduction in cancer cell viability. Ilk et al. [165] studied a very unique process, i.e., quorum sensing (QS) which is measured to be the new objective for antimicrobial chemotherapy. It is a cell density-dependent expression of species in bacteria mediated by compounds called autoinducers (AI). Kaempferol-loaded chitosan nanoparticles were formulated and investigated for modulating QS. QS inhibition by nanoparticles was measured in a time-dependent manner by violacein pigment produced using the biosensor strain *Chromobacterium violaceum* CV026 using acylated homoserine lactone (AHL) as AI. The results indicated that kaempferol-loaded chitosan/TPP had a particle size of 192.27 ± 13.6 nm with loading and encapsulation efficiency of 78 and 93%, respectively. The formulated nanoparticles on a 30-day storage significantly inhibited production of violacein pigment in *Chromobacterium violaceum* CV026 signifying QS to be an exciting new strategy for antimicrobial chemotherapy. Ilk et al. [166] studied the anti-fungal activity of kaempferol loaded into lecithin/chitosan nanoparticles (LC NPs) against fungus *Fusarium oxysporum*. The nanoparticles formed had an encapsulation efficiency of 93.8 ± 4.28% with an inhibition efficacy of 67% at the end of the 60-day storage period and a good physicochemical stability. Thus, it could be inferred that kaempferol nanoparticle formulation could tackle the problems of solubility and loss of kaempferol and can be used as an effective control of fungal pathogens.

4.10 Piperine

Piperine is an amide alkaloid derived from *P. nigrum*. The pungency in black pepper is due to the presence of piperine. It finds place in the vanilloid family. It is said to possess multiple health benefits. Piperine exhibits a potent anti-inflammatory effect by inhibiting the production of inflammatory factors [167]. It is said to possess an antihypertensive and antiplatelet effect [168]. It is a good antioxidant and helps in the effective reduction of various tumors [169]. Besides it is also an antiasthmatic [170], antipyretic, and analgesic compound. It is also found to be an effective immunomodulant and hepatoprotective alkaloid [171]. Piperine initially is tasteless but has a strong burning aftertaste. It is essentially insoluble in water and its low solubility limits its use.

Anissian et al. [172] elaborated the neuroprotective effect of piperine. Piperine-loaded chitosan-sodium tripolyphosphate nanoparticles (CS-STPP NPs) were developed, and its effect on seizure behavior was evaluated using PTZ-induced kindling model. Nissl staining and immunostaining against NeuN were used to study its neuroprotective effects. The results suggested that piperine NPs tend to inhibit the progress of seizure parameters compared to the free piperine groups. Additionally, the levels of cell damage and astrocyte activation were reduced in piperine NP groups. Thus, it could be concluded that piperine nanoparticles can enhance neuroprotection. Elnaggar et al. [173] explored the neuroprotectant potential of Alzheimer's disease (AD). The piperine-loaded chitosan nanoparticles (CS NPs) were developed and studied for brain targeting. The results revealed spherical-shaped nanoparticles with a particle size of 248.50 nm and an entrapment efficiency of 81.70%. The biochemical and behavioral testing of AD-induced Wistar rats illustrated the effective cognitive function of PIP NPs with additional advantages of twin mechanism, Ach esterase inhibition and antioxidant effect, compared to a standard drug donepezil injection. The nanoparticles developed resulted in effective, non-toxic, and non-invasive PIP delivery with 20-fold decrease in oral dose. Thus, it can be concluded that these NPs can be explored to lower the rates of AD morbidity.

4.11 Chrysin

Chrysin is a flavonoid found to occur naturally in honey and propolis with potential pharmacological properties. Chrysin is an excellent antioxidant, anti-inflammatory, anticancer, anti-diabetic, and antihypertensive compound [28]. However, its vast potential is undermined due to its poor solubility and low bioavailability. Sangeetha et al. [174] stated the formation of chrysin-loaded chitosan nanoparticle via ionic gelation method. The developed nanoparticles where spherical in shape and showed the lowest mean particle size, maximum entrapment efficiency and controlled release, thus suggesting an improved solubility and prolonged delivery.

4.12 Phloretin

Phloretin is a type of natural phenol (dihydrochalcone). It is found to be present in apples Picinelli et al. [175]. Phloretin is said to exhibit diverse activities such as antioxidant, anticancer, and anti-inflammatory activities Crespy et al. [176]. Ma et al. [177] and Schaefer et al. [178] claimed the anticancer activities of phloretin by inducing apoptosis in a variety of human cancer cells. Mariadoss et al. [179] studied the effect of phloretin-loaded chitosan nanoparticles on oral squamous cell carcinoma (OSCC) induced by DMBA in hamsters. The administration of PhC NPs via oral route reduced the mitochondrial-mediated apoptosis suggesting phloretin to be a rich antioxidant and anti-carcinogen.

4.13 Nobiletin

Nobiletin, a dietary polymethoxy flavonoid, is found in citrus fruits and is said to possess multiple health benefits. It is said to be good for neurological protection, against CVD, anti-metabolic disorder, anticancer and tumor growth, anti-inflammation, and antioxidation [180]. Luque-Alcaraz et al. [181] explored nobiletin-loaded chitosan nanoparticles for antiproliferative activity. The results depicted effective inhibition of cancerous cell with an IC 50 = 8 μg/mL, thus making it a potential candidate for cancer chemotherapy.

4.14 Ferulic Acid

Ferulic acid is hydroxycinnamic acid. It is found widely distributed in cereals and fruits like tomato and sweet corn. It is found effective against diseases like cancer, diabetes, and cardiovascular and neurodegenerative diseases [182]. However, its beneficial effects remain nascent because of its bioavailability and absorption issues. Panwar et al. [183] evaluated the anticancer activity against human cervical cancer cell lines (ME-180) by the development of ferulic acid-loaded chitosan nanoparticle via ionic gelation method. The results revealed development of smooth and spherical nanoparticles with average diameter of 125 nm at a concentration of 40 μM FA. The developed nanoparticles showed higher cytotoxicity against cervical cancer cell lines compared to the native FA, thus stating the exceptional potential of ferulic acid's loaded chitosan nanoparticles against cancer cell proliferation.

4.15 Chlorogenic Acid (CGA)

CGA, a phenolic compound, is an ester of certain *trans*-cinnamic acids (caffeic, ferulic, and coumaric acids) and quinic acid [184]. It is found in abundance in fruits and vegetables. Green coffee is one of the richest sources of CGA (4–14%) [185]. CGA is reported to be anticancerous [186] and an effective candidate for treating Alzheimer's disease [187]. Nallamuthu et al. [13] reported that the encapsulation of chlorogenic acid in chitosan nanoparticle was carried by ionic gelation method. The particle size, zeta potential, and shape of the particle were analyzed. The particles showed a size of 210 nm and zeta potential of 33 mV. Scanning electron microscopy (SEM) revealed particles with a regular spherical shape. FTIR confirmed the entrapment of CGA. The CGA-loaded nanoparticles with loading efficiency of 5.2% had an encapsulation efficiency of about 59%, percent release of 69% at the end of 100th hours, and well-preserved antioxidant property that can be targeted for the effective distribution of CGA in the body to confer to manifold health benefits. Pharmacokinetic analysis revealed lower level of C_{max}, longer T_{max}, longer MRT, and larger AUC_{0-t} and $AUC_{0-\infty}$ for the CGA nanoparticles compared to free CGA depicting that the encapsulated CGA had the slow and sustained release of CGA over long duration and therefore the bioavailability is more upon oral administration.

5 Conclusion

In view of their multitude health benefits, nutraceuticals have, of late, evoked much research interest particularly their pivotal role in preclusion and disease control. However, their solubility, bioavailability, and effective delivery concerns encumber their use. The delivery of nutraceutical via chitosan-based nanoparticles is a promising approach to overcome these limitations. Chitosan nanoparticles due to their supreme advantages of safety, availability, and easy biodegradation are regarded as a suitable vehicle for the loading of nutraceuticals. The nutraceutical-loaded chitosan nanoparticles conferred with immense therapeutic uses find a colossal applicability in the healthcare system.

References

1. McClements DJ, Li F, Xiao H (2015) The nutraceutical bioavailability classification scheme: classifying nutraceuticals according to factors limiting their oral bioavailability. Annu Rev Food Sci Technol 6:299–327
2. Ting Y, Jiang Y, Ho CT, Huang Q (2014) Common delivery systems for enhancing in vivo bioavailability and biological efficacy of nutraceuticals. J Funct Foods 7:112–128
3. Kumar K, Kumar S (2015) Role of nutraceuticals in health and disease prevention: a review. South Asian J Food Technol Environ 1:116–121
4. Surve DH, Paul AT, Jindal AB (2019) Nanotechnology based delivery of nutraceuticals. In: Environmental Nanotechnology. Springer, Cham, pp 63–107
5. Nasri H, Baradaran A, Shirzad H, Rafieian-Kopaei M (2014) New concepts in nutraceuticals as alternative for pharmaceuticals. Int J Prev Med 5(12):1487
6. Rein MJ, Renouf M, Cruz-Hernandez C, Actis-Goretta L, Thakkar SK, da Silva Pinto M (2013) Bioavailability of bioactive food compounds: a challenging journey to bioefficacy. Br J Clin Pharmacol 75(3):588–602
7. Acosta E (2009) Bioavailability of nanoparticles in nutrient and nutraceutical delivery. Curr Opin Colloid Interface Sci 14(1):3–15
8. McClements DJ, Decker EA, Park Y, Weiss J (2009) Structural design principles for delivery of bioactive components in nutraceuticals and functional foods. Crit Rev Food Sci Nutr 49:577–606
9. Aklakur M, Asharf Rather M, Kumar N (2016) Nanodelivery: an emerging avenue for nutraceuticals and drug delivery. Crit Rev Food Sci Nutr 56(14):2352–2361
10. Huang Q, Yu H, Ru Q (2010) Bioavailability and delivery of nutraceuticals using nanotechnology. J Food Sci 75(1):R50–R57
11. Mora-Huertas CE, Fessi H, Elaissari A (2010) Polymer-based nanocapsules for drug delivery. Int J Pharm 385(1–2):113–142
12. Dhiman A, Bhalla D (2019) Development and evaluation of lycopene loaded chitosan nanoparticles. Curr Nanomed (Formerly: Recent Patents on Nanomedicine) 9(1):61–75
13. Nallamuthu I, Devi A, Khanum F (2015) Chlorogenic acid loaded chitosan nanoparticles with sustained release property, retained antioxidant activity and enhanced bioavailability. Asian J Pharm Sci 10(3):203–211
14. Thamaket P, Raviyan P (2015) Preparation and physical properties of carotenoids encapsulated in chitosan cross-linked tripolyphosphate nanoparticles. Food Appl Biosci J 3(1):69–84
15. Brower V (1998) Nutraceuticals: poised for a healthy slice of the healthcare market? Nat Biotechnol 16(8):728
16. Sapkale Anita P, Thorat Mangesh S, Vir Prasad R, Singh Meera C (2012) Nutraceuticals-global status and applications: a review. Int J Pharm Chem Sci 1(3):1166–1181
17. Tank Dharti S, Gandhi S, Shah M (2010) Nutraceuticals-portmanteau of science and nature. Int J Pharm Sci Rev Res 5(3):33–38
18. Mayne ST (1996) Beta-carotene, carotenoids, and disease prevention in humans. FASEB J 10(7):690–701
19. Holzapfel, N., Holzapfel, B., Champ, S., Feldthusen, J., Clements, J., & Hutmacher, D. (2013). The potential role of lycopene for the prevention and therapy of prostate cancer: from molecular mechanisms to clinical evidence. International journal of molecular sciences, 14(7), 14620–14646
20. Ambati R, Phang SM, Ravi S, Aswathanarayana R (2014) Astaxanthin: sources, extraction, stability, biological activities and its commercial applications—a review. Mar Drugs 12(1):128–152
21. Sharon O'Brienon. Lutein and zeaxanthin: benefits, dosage and food sources. July 11, 2018
22. Marrelli M, Conforti F, Araniti F, Statti G (2016) Effects of saponins on lipid metabolism: a review of potential health benefits in the treatment of obesity. Molecules 21(10):1404
23. David AVA, Arulmoli R, Parasuraman S (2016) Overviews of biological importance of quercetin: a bioactive flavonoid. Pharmacogn Rev 10(20):84

24. Chen AY, Chen YC (2013) A review of the dietary flavonoid, kaempferol on human health and cancer chemoprevention. Food Chem 138(4):2099–2107
25. Li Y, Ding Y (2012) Minireview: therapeutic potential of myricetin in diabetes mellitus. Food Sci Human Wellness 1(1):19–25
26. Patel K, Patel DK (2019) The beneficial role of rutin, a naturally occurring flavonoid in health promotion and disease prevention: a systematic review and update. In: Bioactive food as dietary interventions for arthritis and related inflammatory diseases. Academic Press, Cambridge, pp 457–479
27. Rimm EB, Katan MB, Ascherio A, Stampfer MJ, Willett WC (1996) Relation between intake of flavonoids and risk for coronary heart disease in male health professionals. Ann Intern Med 125(5):384–389
28. Mani R, Natesan V (2018) Chrysin: Sources, beneficial pharmacological activities, and molecular mechanism of action. Phytochemistry 145:187–196
29. Ali F, Rahul, Naz F, Jyoti S, Siddique YH (2017) Health functionality of apigenin: a review. Int J Food Prop 20(6):1197–1238
30. Gadkari PV, Balaraman M (2015) Catechins: sources, extraction and encapsulation: a review. Food Bioprod Process 93:122–138
31. Spagnuolo C, Russo GL, Orhan IE, Habtemariam S, Daglia M, Sureda A, Nabavi SF, Devi KP, Loizzo MR, Tundis R, Nabavi SM (2015) Genistein and cancer: current status, challenges, and future directions. Adv Nutr 6(4):408–419
32. Salehi B, Mishra A, Nigam M, Sener B, Kilic M, Sharifi-Rad M, PVT F, Martins N, Sharifi-Rad J (2018) Resveratrol: a double-edged sword in health benefits. Biomedicine 6(3):91
33. Hewlings S, Kalman D (2017) Curcumin: a review of its' effects on human health. Foods 6(10):92
34. Tajik N, Tajik M, Mack I, Enck P (2017) The potential effects of chlorogenic acid, the main phenolic components in coffee, on health: a comprehensive review of the literature. Eur J Nutr 56(7):2215–2244
35. Rhonda Johansson. Cinnamic acid sources, health benefits and uses. September 12, 2017
36. McCarty MF, DiNicolantonio JJ, O'keefe JH (2015) Capsaicin may have important potential for promoting vascular and metabolic health. Open Heart 2(1):e000262
37. Pratap SR, Gangadharappa HV, Mruthunjaya K (2017) Ginger: a potential nutraceutical, an updated review. Int J Pharmacogn Phytochem Res 9:1227–1238
38. Augustin MA, Sanguansri L (2012) Challenges in developing delivery systems for food additives, nutraceuticals and dietary supplements. In: Encapsulation technologies and delivery systems for food ingredients and nutraceuticals. Woodhead Publishing, Cambridge, pp 19–48
39. Espín JC, García-Conesa MT, Tomás-Barberán FA (2007) Nutraceuticals: facts and fiction. Phytochemistry 68(22–24):2986–3008
40. Panozzo A, Lemmens L, Van Loey A, Manzocco L, Nicoli MC, Hendrickx M (2013) Microstructure and bioaccessibility of different carotenoid species as affected by high pressure homogenisation: a case study on differently coloured tomatoes. Food Chem 141(4):4094–4100
41. Rimbach G, Pallauf J, Moehring J, Kraemer K, Minihane AM (2008) Effect of dietary phytate and microbial phytase on mineral and trace element bioavailability: a literature review. Curr Top Nutraceut Res 6(3):131–144
42. Moelants KR, Lemmens L, Vandebroeck M, Van Buggenhout S, Van Loey AM, Hendrickx ME (2012) Relation between particle size and carotenoid bioaccessibility in carrot-and tomato-derived suspensions. J Agric Food Chem 60(48):11995–12003
43. Actis-Goretta L, Lévèques A, Rein M, Teml A, Schäfer C, Hofmann U, Li H, Schwab M, Eichelbaum M, Williamson G (2013) Intestinal absorption, metabolism, and excretion of (−)-epicatechin in healthy humans assessed by using an intestinal perfusion technique. Am J Clin Nutr 98(4):924–933
44. Carail, M., Goupy, P., Reynaud, E., Dangles, O., & Caris-Veyrat, C. (2013). Oxidative cleavage products of lycopene: production and reactivity in a biomimetic experimental model

of oxidative stress. In Carotenoid Cleavage Products (pp. 191–205). American Chemical Society.
45. Chen Z, Zheng S, Li L, Jiang H (2014) Metabolism of flavonoids in human: a comprehensive review. Curr Drug Metab 15(1):48–61
46. Koga N, Ohta C, Kato Y, Haraguchi K, Endo T, Ogawa K et al (2011) In vitro metabolism of nobiletin, a polymethoxy-flavonoid, by human liver microsomes and cytochrome P450. Xenobiotica 41(11):927–933
47. Metzler M, Pfeiffer E, Schulz SI, Dempe JS (2013) Curcumin uptake and metabolism. Biofactors 39(1):14–20
48. Mohammed M, Syeda J, Wasan K, Wasan E (2017) An overview of chitosan nanoparticles and its application in non-parenteral drug delivery. Pharmaceutics 9(4):53
49. Peniche H, Peniche C (2011) Chitosan nanoparticles: a contribution to nanomedicine. Polym Int 60(6):883–889
50. Muxika A, Etxabide A, Uranga J, Guerrero P, De La Caba K (2017) Chitosan as a bioactive polymer: processing, properties and applications. Int J Biol Macromol 105:1358–1368
51. Soutter W. Chitosan Nanoparticles - Properties and Applications. Available from: http://www.azonano.com/ article.aspx. [Last Accessed on 2015 Sep 18]
52. Divya K, Jisha MS (2018) Chitosan nanoparticles preparation and applications. Environ Chem Lett 16(1):101–112
53. Baldrick P (2010) The safety of chitosan as a pharmaceutical excipient. Regul Toxicol Pharmacol 56(3):290–299
54. Kean T, Thanou M (2010) Biodegradation, biodistribution and toxicity of chitosan. Adv Drug Deliv Rev 62(1):3–11
55. Aljebory AM, Alsalman TM (2017) Chitosan nanoparticles. Imp J Interdiscip Res 3(7):10
56. Calvo P, Remunan-Lopez C, Vila-Jato JL, Alonso MJ (1997) Novel hydrophilic chitosan-polyethylene oxide nanoparticles as protein carriers. J Appl Polym Sci 63(1):125–132
57. Rajan, A., Gopukumar, S. T., & Praseetha, P. K. Journal of Global Trends in Pharmaceutical Sciences
58. Sinha VR, Singla AK, Wadhawan S, Kaushik R, Kumria R, Bansal K, Dhawan S (2004) Chitosan microspheres as a potential carrier for drugs. Int J Pharm 274(1–2):1–33
59. Alishahi A, Mirvaghefi A, Tehrani MR, Farahmand H, Shojaosadati SA, Dorkoosh FA, Elsabee MZ (2011) Shelf life and delivery enhancement of vitamin C using chitosan nanoparticles. Food Chem 126(3):935–940
60. Kumar MNVR (2000) Nano and microparticles as controlled drug delivery devices. J Pharm Pharm Sci 3(2):234–258
61. Mansouri S, Cuie Y, Winnik F, Shi Q, Lavigne P, Benderdour M, Beaumont E, Fernandes JC (2006) Characterization of folate-chitosan-DNA nanoparticles for gene therapy. Biomaterials 27(9):2060–2065
62. Qv XY, Zeng ZP, Jiang JG (2011) Preparation of lutein microencapsulation by complex coacervation method and its physicochemical properties and stability. Food Hydrocoll 25(6):1596–1603
63. Alencastre JB, Bentley MVLB, Garcia FS, Moragas MD, Viladot JL, Marchetti JM (2006) A study of the characteristics and in vitro permeation properties of CMC/chitosan microparticles as a skin delivery system for vitamin E. Rev Bras Ciênc Farm 42(1):69–76
64. Brunel F, Véron L, David L, Domard A, Delair T (2008) A novel synthesis of chitosan nanoparticles in reverse emulsion. Langmuir 24(20):11370–11377
65. Mansouri M, Khorram M, Samimi A, Osfouri S (2012) Preparation of bovine serum albumin loaded chitosan nanoparticles using reverse micelle method. Int J Polymer Mater 61:1079–1090
66. Agnihotri SA, Mallikarjuna NN, Aminabhavi TM (2004) Recent advances on chitosan-based micro-and nanoparticles in drug delivery. J Control Release 100(1):5–28
67. Chan CM, Hung CF (2014) Lycopene and retinal pigment epithelial cells: molecular aspects. In: Handbook of nutrition, diet and the eye. Academic Press, Cambridge, pp 587–598

68. Başaran N, Bacanlı M, Başaran AA (2017) Lycopenes as antioxidants in gastrointestinal diseases. In: Gastrointestinal tissue. Academic Press, Cambridge, pp 355–362
69. Shidfar F, Arjomand GN (2015) Glucose intake and utilization in pre-diabetes and diabetes: tomato and diabetes. In: Glucose intake and utilization in pre-diabetes and diabetes. Academic Press, Cambridge, pp 301–313
70. da Silva GF, Rocha LW, Quintão NLM (2015) Nutraceuticals, dietary supplements, and functional foods as alternatives for the relief of neuropathic pain. In: Bioactive nutraceuticals and dietary supplements in neurological and brain disease. Academic Press, Cambridge, pp 87–93
71. Nazemiyeh E, Eskandani M, Sheikhloie H, Nazemiyeh H (2016) Formulation and physicochemical characterization of lycopene-loaded solid lipid nanoparticles. Adv Pharm Bull 6(2):235
72. Al-Eisa RA (2018) Synergistic antioxidant capacity of chitosan nanoparticles and lycopene against aging hepatotoxicity induced by D-galactose in male rats. Int J Pharmacol 14(6):811–825
73. Khalaf NA, Shakya AK, Al-Othman A, El-Agbar Z, Farah H (2008) Antioxidant activity of some common plants. Turk J Biol 32(1):51–55
74. Komatsu Y, Suematsu S, Hisanobu Y, Saigo H, Matsuda R, Hara K (1993) Effects of pH and temperature on reaction kinetics of catechins in green tea infusion. Biosci Biotechnol Biochem 57(6):907–910
75. Yoshida Y, Kiso M, Goto T (1999) Efficiency of the extraction of catechins from green tea. Food Chem 67(4):429–433
76. Manikkam R, Pitchai D (2013) Catechin loaded chitosan nanoparticles as a novel drug delivery system for cancer–synthesis and in vitro and in vivo characterization. World J Pharm Pharm Sci 3:1553–1577
77. Dudhani AR, Kosaraju SL (2010) Bioadhesive chitosan nanoparticles: preparation and characterization. Carbohydr Polym 81(2):243–251
78. Hu B, Pan C, Sun Y, Hou Z, Ye H, Hu B, Zeng X (2008) Optimization of fabrication parameters to produce chitosan– tripolyphosphate nanoparticles for delivery of tea catechins. J Agric Food Chem 56(16):7451–7458
79. Tang DW, Yu SH, Ho YC, Huang BQ, Tsai GJ, Hsieh HY, Sung H-W, Mi FL (2013) Characterization of tea catechins-loaded nanoparticles prepared from chitosan and an edible polypeptide. Food Hydrocoll 30(1):33–41
80. Zhang H, Jung J, Zhao Y (2016) Preparation, characterization and evaluation of antibacterial activity of catechins and catechins–Zn complex loaded β-chitosan nanoparticles of different particle sizes. Carbohydr Polym 137:82–91
81. Yang J, Liu RH (2013) The phenolic profiles and antioxidant activity in different types of tea. Int J Food Sci Technol 48(1):163–171
82. Wang H, Bian S, Yang CS (2011) Green tea polyphenol EGCG suppresses lung cancer cell growth through upregulating miR-210 expression caused by stabilizing HIF-1α. Carcinogenesis 32(12):1881–1889
83. Wu D, Guo Z, Ren Z, Guo W, Meydani SN (2009) Green tea EGCG suppresses T cell proliferation through impairment of IL-2/IL-2 receptor signaling. Free Radic Biol Med 47(5):636–643
84. Hu B, Ting Y, Zeng X, Huang Q (2013) Bioactive peptides/chitosan nanoparticles enhance cellular antioxidant activity of (−)-epigallocatechin-3-gallate. J Agric Food Chem 61(4):875–881
85. Hong Z, Xu Y, Yin JF, Jin J, Jiang Y, Du Q (2014) Improving the effectiveness of (−)-epigallocatechin gallate (EGCG) against rabbit atherosclerosis by EGCG-loaded nanoparticles prepared from chitosan and polyaspartic acid. J Agric Food Chem 62(52):12603–12609
86. Dube A, Nicolazzo JA, Larson I (2011) Chitosan nanoparticles enhance the plasma exposure of (−)-epigallocatechin gallate in mice through an enhancement in intestinal stability. Eur J Pharm Sci 44(3):422–426

87. Bertelli AA, Migliori M, Panichi V, Origlia N, Filippi C, Das DK, Giovannini L (2002) Resveratrol, a component of wine and grapes, in the prevention of kidney disease. Ann N Y Acad Sci 957(1):230–238
88. Falchi M, Bertelli A, Galazzo R, Viganò P, Dib B (2010) Central antalgic activity of resveratrol. Arch Ital Biol 148(4):389–396
89. Olas B, Wachowicz B (2005) Resveratrol, a phenolic antioxidant with effects on blood platelet functions. Platelets 16(5):251–260
90. Torres-López JE, Ortiz MI, Castañeda-Hernández G, Alonso-López R, Asomoza-Espinosa R, Granados-Soto V (2002) Comparison of the antinociceptive effect of celecoxib, diclofenac and resveratrol in the formalin test. Life Sci 70(14):1669–1676
91. Das DK, Maulik N (2006) Resveratrol in cardioprotection: a therapeutic promise of alternative medicine. Mol Interv 6(1):36
92. Marambaud P, Zhao H, Davies P (2005) Resveratrol promotes clearance of Alzheimer's disease amyloid-β peptides. J Biol Chem 280(45):37377–37382
93. Sessa M, Tsao R, Liu R, Ferrari G, Donsì F (2011) Evaluation of the stability and antioxidant activity of nanoencapsulated resveratrol during in vitro digestion. J Agric Food Chem 59(23):12352–12360
94. Zupančič Š, Lavrič Z, Kristl J (2015) Stability and solubility of trans-resveratrol are strongly influenced by pH and temperature. Eur J Pharm Biopharm 93:196–204
95. Wu J, Wang Y, Yang H, Liu X, Lu Z (2017) Preparation and biological activity studies of resveratrol loaded ionically cross-linked chitosan-TPP nanoparticles. Carbohydr Polym 175:170–177
96. Min JB, Kim ES, Lee JS, Lee HG (2018) Preparation, characterization, and cellular uptake of resveratrol-loaded trimethyl chitosan nanoparticles. Food Sci Biotechnol 27(2):441–450
97. Iglesias N, Galbis E, Díaz-Blanco MJ, Lucas R, Benito E, de-Paz M (2019) Nanostructured chitosan-based biomaterials for sustained and colon-specific resveratrol release. Int J Mol Sci 20(2):398
98. Aggarwal BB, Kumar A, Bharti AC (2003) Anticancer potential of curcumin: preclinical and clinical studies. Anticancer Res 23(1/A):363–398
99. Lin J, Rexrode KM, Hu F, Albert CM, Chae CU, Rimm EB, Stampfer MJ, Manson JE (2007a) Dietary intakes of flavonols and flavones and coronary heart disease in US women. Am J Epidemiol 165(11):1305–1313
100. Lin YG, Kunnumakkara AB, Nair A, Merritt WM, Han LY, Armaiz-Pena GN, Kamat AA, Spannuth WA, Gershenson DM, Lutgendorf SK, Aggarwal BB (2007b) Curcumin inhibits tumor growth and angiogenesis in ovarian carcinoma by targeting the nuclear factor-κB pathway. Clin Cancer Res 13(11):3423–3430
101. Marchiani A, Rozzo C, Fadda A, Delogu G, Ruzza P (2014) Curcumin and curcumin-like molecules: from spice to drugs. Curr Med Chem 21(2):204–222
102. Panahi Y, Hosseini MS, Khalili N, Naimi E, Simental-Mendía LE, Majeed M, Sahebkar A (2016) Effects of curcumin on serum cytokine concentrations in subjects with metabolic syndrome: a post-hoc analysis of a randomized controlled trial. Biomed Pharmacother 82:578–582
103. Henrotin Y, Priem F, Mobasheri A (2013) Curcumin: a new paradigm and therapeutic opportunity for the treatment of osteoarthritis: curcumin for osteoarthritis management. Springerplus 2(1):56
104. Yallapu MM, Jaggi M, Chauhan SC (2012) Curcumin nanoformulations: a future nanomedicine for cancer. Drug Discov Today 17(1–2):71–80
105. Gunes H, Gulen D, Mutlu R, Gumus A, Tas T, Topkaya AE (2016) Antibacterial effects of curcumin: an in vitro minimum inhibitory concentration study. Toxicol Ind Health 32(2):246–250
106. Hlavačková L, Janegová A, Uličná O, Janega P, Černá A, Babál P (2011) Spice up the hypertension diet-curcumin and piperine prevent remodeling of aorta in experimental L-NAME induced hypertension. Nutr Metab 8(1):72

107. Mohammadi A, Sahebkar A, Iranshahi M, Amini M, Khojasteh R, Ghayour-Mobarhan M, Ferns GA (2013) Effects of supplementation with curcuminoids on dyslipidemia in obese patients: a randomized crossover trial. Phytother Res 27(3):374–379
108. Na LX, Li Y, Pan HZ, Zhou XL, Sun DJ, Meng M, Li XX, Sun CH (2013) Curcuminoids exert glucose-lowering effect in type 2 diabetes by decreasing serum free fatty acids: a double-blind, placebo-controlled trial. Mol Nutr Food Res 57(9):1569–1577
109. Sahebkar A (2014) Curcuminoids for the management of hypertriglyceridaemia. Nat Rev Cardiol 11(2):123
110. Vijayakurup V, Thulasidasan AT, Retnakumari AP, Nandan CD, Somaraj J, Antony J, Alex VV, Vinod BS, Liju VB, Sundaram S, Kumar GV (2019) Chitosan encapsulation enhances the bioavailability and tissue retention of curcumin and improves its efficacy in preventing B [a] P-induced lung carcinogenesis. Cancer Prev Res 12(4):225–236
111. Samrot AV, Burman U, Philip SA, Shobana N, Chandrasekaran K (2018) Synthesis of curcumin loaded polymeric nanoparticles from crab shell derived chitosan for drug delivery. Informatics in Medicine Unlocked 10:159–182
112. Mirnejad R, Jahromi M, Ali M, Al-Musawi S, Pirestani M, Fasihi Ramandi M, Ahmadi K, Rajayi H, Hassan ZM, Kamali M (2014) Curcumin-loaded chitosan tripolyphosphate nanoparticles as a safe, natural and effective antibiotic inhibits the infection of Staphylococcus aureus and Pseudomonas aeruginosa in vivo. Iran J Biotechnol 12(3):1–8
113. Nair RS, Morris A, Billa N, Leong CO (2019) An evaluation of curcumin-encapsulated chitosan nanoparticles for transdermal delivery. AAPS PharmSciTech 20(2):69
114. Khan MA, Zafaryab M, Mehdi SH, Ahmad I, Rizvi M, Moshahid A (2018) Physicochemical characterization of curcumin loaded chitosan nanoparticles: implications in cervical cancer. Anticancer Agents Med Chem (Formerly Current Medicinal Chemistry-Anti-Cancer Agents) 18(8):1131–1137
115. Duse L, Baghdan E, Pinnapireddy SR, Engelhardt KH, Jedelská J, Schaefer J, Quendt P, Bakowsky U (2018) Preparation and characterization of curcumin loaded chitosan nanoparticles for photodynamic therapy. Phys Status Solidi A 215(15):1700709
116. Baghel SS, Shrivastava N, Baghel RS, Agrawal P, Rajput S (2012) A review of quercetin: antioxidant and anticancer properties. World J Pharm Pharm Sci 1(1):146–160
117. Lakhanpal P, Rai DK (2007) Quercetin: a versatile flavonoid. Internet J Med Update 2(2):22–37
118. Mitchell AE, Hong YJ, Koh E, Barrett DM, Bryant DE, Denison RF, Kaffka S (2007) Ten-year comparison of the influence of organic and conventional crop management practices on the content of flavonoids in tomatoes. J Agric Food Chem 55(15):6154–6159
119. Russo M, Spagnuolo C, Tedesco I, Bilotto S, Russo GL (2012) The flavonoid quercetin in disease prevention and therapy: facts and fancies. Biochem Pharmacol 83(1):6–15
120. Hertog MG, Feskens EJ, Kromhout D, Hollman PCH, Katan MB (1993) Dietary antioxidant flavonoids and risk of coronary heart disease: the Zutphen Elderly Study. Lancet 342(8878):1007–1011
121. Alrawaiq NS, Abdullah A (2014) A review of flavonoid quercetin: metabolism, bioactivity and antioxidant properties. Int J Pharm Tech Res 6(3):933–941
122. Adewole SO, Caxton-Martins EA, Ojewole JA (2007) Protective effect of quercetin on the morphology of pancreatic β-cells of streptozotocin-treated diabetic rats. Afr J Tradit Complement Altern Med 4(1):64–74
123. Li M, Xu Z (2008) Quercetin in a lotus leaves extract may be responsible for antibacterial activity. Arch Pharm Res 31(5):640–644
124. Kaşıkcı MB, Bağdatlıoğlu N (2016) Bioavailability of quercetin. Curr Res Nutr Food Sci J 4(Special Issue Nutrition in Conference October 2016):146–151
125. Zhang Y, Yang Y, Tang K, Hu X, Zou G (2008) Physicochemical characterization and antioxidant activity of quercetin-loaded chitosan nanoparticles. J Appl Polym Sci 107(2):891–897
126. Tzankova V, Aluani D, Kondeva-Burdina M, Yordanov Y, Odzhakov F, Apostolov A, Yoncheva K (2017) Hepatoprotective and antioxidant activity of quercetin loaded chitosan/

alginate particles in vitro and in vivo in a model of paracetamol-induced toxicity. Biomed Pharmacother 92:569–579
127. Nan W, Ding L, Shi X, Sui XB (2018) Topical use of quercetin-loaded chitosan nanoparticles against ultraviolet B radiation. Front Pharmacol 9:826
128. Baksi R, Singh DP, Borse SP, Rana R, Sharma V, Nivsarkar M (2018) In vitro and in vivo anti-cancer efficacy potential of Quercetin loaded polymeric nanoparticles. Biomed Pharmacother 106:1513–1526
129. Rahigude A, Bhutada P, Kaulaskar S, Aswar M, Otari K (2012) Participation of antioxidant and cholinergic system in protective effect of naringenin against type-2 diabetes-induced memory dysfunction in rats. Neuroscience 226:62–72
130. Felgines C, Texier O, Morand C, Manach C, Scalbert A, Régerat F, Rémésy C (2000) Bioavailability of the flavanone naringenin and its glycosides in rats. Am J Physiol Gastrointest Liver Physiol 279(6):G1148–G1154
131. Kim JH, Lee JK (2015) Naringenin enhances NK cell lysis activity by increasing the expression of NKG2D ligands on Burkitt's lymphoma cells. Arch Pharm Res 38(11):2042–2048
132. Lee YS, Reidenberg MM (1998) A method for measuring naringenin in biological fluids and its disposition from grapefruit juice by man. Pharmacology 56(6):314–317
133. Middleton E Jr, Kandaswami C (1992) Effects of flavonoids on immune and inflammatory cell functions. Biochem Pharmacol 43(6):1167–1179
134. Verbeek R, Plomp AC, van Tol EA, van Noort JM (2004) The flavones luteolin and apigenin inhibit in vitro antigen-specific proliferation and interferon-gamma production by murine and human autoimmune T cells. Biochem Pharmacol 68(4):621–629
135. Winarti, L., Sari, K., Ruma, L. O., & Nugroho, A. E. (2015). NARINGENIN-LOADED CHITOSAN NANOPARTICLES FORMULATION, AND ITS IN VITRO EVALUATION AGAINST T47D BREAST CANCER CELL LINE. Indonesian Journal of Pharmacy/ Majalah Farmasi Indonesia, 26(3)
136. Malathy S, Iyer PR (2018) Naringin loaded chitosan nanoparticle for bone regeneration: a preliminary in vitro study. J Nanomed Nanotechnol 9:507
137. Kumar SP, Birundha K, Kaveri K, Devi KR (2015) Antioxidant studies of chitosan nanoparticles containing naringenin and their cytotoxicity effects in lung cancer cells. Int J Biol Macromol 78:87–95
138. Kreft S, Knapp M, Kreft I (1999) Extraction of rutin from buckwheat (Fagopyrum esculentum Moench) seeds and determination by capillary electrophoresis. Journal of agricultural and food chemistry 47(11):4649–4652
139. Harborne JB (1986) Nature, distribution and function of plant flavonoids. Prog Clin Biol Res 213:15–24
140. Ganeshpurkar A, Saluja AK (2017) The pharmacological potential of rutin. Saudi Pharm J 25(2):149–164
141. Javed H, Khan MM, Ahmad A, Vaibhav K, Ahmad ME, Khan A, Ashafaq M, Islam F, Siddiqui MS, Safhi MM (2012) Rutin prevents cognitive impairments by ameliorating oxidative stress and neuroinflammation in rat model of sporadic dementia of Alzheimer type. Neuroscience 210:340–352
142. Fernández SP, Wasowski C, Loscalzo LM, Granger RE, Johnston GA, Paladini AC, Marder M (2006) Central nervous system depressant action of flavonoid glycosides. Eur J Pharmacol 539(3):168–176
143. Hsu CY, Shih HY, Chia YC, Lee CH, Ashida H, Lai YK, Weng CF (2014) Rutin potentiates insulin receptor kinase to enhance insulin-dependent glucose transporter 4 translocation. Mol Nutr Food Res 58(6):1168–1176
144. Nagem TJ, Pinto AS, Albino LF, Pinto JG (2001) Hypocholesterolemic effect of naringin and rutin flavonoids. Arch Latinoam Nutr 51(3):258–264
145. Kim DW, Hwang IK, Lim SS, Yoo KY, Li H, Kim YS, Kwon DY, Moon WK, Kim DW, Won MH (2009) Germinated buckwheat extract decreases blood pressure and nitrotyrosine immunoreactivity in aortic endothelial cells in spontaneously hypertensive rats. Phytother Res 23(7):993–998

146. Chan APE, Hegde A, Chen X (2009) Effect of rutin on warfarin anticoagulation and pharmacokinetics of warfarin enantiomers in rats. J Pharm Pharmacol 61(4):451–458
147. Chen WM, Jin M, Wu W (2002) Experimental study on inhibitory effect of rutin against platelet activation induced by platelet activating factor in rabbits. Zhongguo Zhong Xi Yi Jie He Za Zhi 22(4):283–285
148. Isai M, Sakthivel M, Ramesh E, Thomas PA, Geraldine P (2009) Prevention of selenite-induced cataractogenesis by rutin in Wistar rats. Mol Vis 15:2570
149. Johann S, Mendes BG, Missau FC, Resende MAD, Pizzolatti MG (2011) Antifungal activity of five species of Polygala. Braz J Microbiol 42(3):1065–1075
150. Jung CH, Lee JY, Cho CH, Kim CJ (2007) Anti-asthmatic action of quercetin and rutin in conscious guinea-pigs challenged with aerosolized ovalbumin. Arch Pharm Res 30(12):1599
151. Dubey S, Ganeshpurkar A, Shrivastava A, Bansal D, Dubey N (2013) Rutin exerts antiulcer effect by inhibiting the gastric proton pump. Indian J Pharmacol 45(4):415
152. Horcajada-Molteni MN, Crespy V, Coxam V, Davicco MJ, Rémésy C, Barlet JP (2000) Rutin inhibits ovariectomy-induced osteopenia in rats. J Bone Miner Res 15(11):2251–2258
153. Patil AG, Jobanputra AH (2015) Rutin-chitosan nanoparticles: fabrication, characterization and application in dental disorders. Polym-Plast Technol Eng 54(2):202–208
154. Mujtaba MA, Hassan KA, Imran M (2018) Chitosan-alginate nanoparticles as a novel drug delivery system for rutin. Int J Adv Biotechnol Res 9(1):1895–1903
155. Ahmad N, Ahmad R, Naqvi AA, Alam MA, Ashafaq M, Samim M, Iqbal Z, Ahmad FJ (2016) Rutin-encapsulated chitosan nanoparticles targeted to the brain in the treatment of Cerebral Ischemia. Int J Biol Macromol 91:640–655
156. Calderon-Montano JM, Burgos-Morón E, Pérez-Guerrero C, López-Lázaro M (2011) A review on the dietary flavonoid kaempferol. Mini Rev Med Chem 11(4):298–344
157. Wang L, Tu YC, Lian TW, Hung JT, Yen JH, Wu MJ (2006) Distinctive antioxidant and anti-inflammatory effects of flavonols. J Agric Food Chem 54(26):9798–9804
158. Willerson JT, Ridker PM (2004) Inflammation as a cardiovascular risk factor. Circulation 109(21_suppl_1):II–I2
159. Tu YC, Lian TW, Yen JH, Chen ZT, Wu MJ (2007) Antiatherogenic effects of kaempferol and rhamnocitrin. J Agric Food Chem 55(24):9969–9976
160. Knekt P, Kumpulainen J, Järvinen R, Rissanen H, Heliövaara M, Reunanen A, Hakulinen T, Aromaa A (2002) Flavonoid intake and risk of chronic diseases. Am J Clin Nutr 76(3):560–568
161. Hannum SM (2004) Potential impact of strawberries on human health: a review of the science. Crit Rev Food Sci Nutr 44(1):1–17
162. Smith JA, Poteet-Smith CE, Xu Y, Errington TM, Hecht SM, Lannigan DA (2005) Identification of the first specific inhibitor of p90 ribosomal S6 kinase (RSK) reveals an unexpected role for RSK in cancer cell proliferation. Cancer Res 65(3):1027–1034
163. Tang X, Zhu X, Liu S, Nicholson RC, Ni X (2008) Phytoestrogens induce differential estrogen receptor β-mediated responses in transfected MG-63 cells. Endocrine 34(1–3):29–35
164. Luo H, Jiang B, Li B, Li Z, Jiang BH, Chen YC (2012) Kaempferol nanoparticles achieve strong and selective inhibition of ovarian cancer cell viability. Int J Nanomedicine 7:3951
165. Ilk S, Saglam N, Özgen M (2017a) Kaempferol loaded lecithin/chitosan nanoparticles: Preparation, characterization, and their potential applications as a sustainable antifungal agent. Artificial Cells Nanomed Biotechnol 45(5):907–916
166. Ilk S, Sağlam N, Özgen M, Korkusuz F (2017b) Chitosan nanoparticles enhances the anti-quorum sensing activity of kaempferol. Int J Biol Macromol 94:653–662
167. Dzoyem JP, McGaw LJ, Kuete V, Bakowsky U (2017) Anti-inflammatory and anti-nociceptive activities of African medicinal spices and vegetables. In: Medicinal spices and vegetables from Africa. Academic Press, Cambridge, pp 239–270
168. Taqvi SIH, Shah AJ, Gilani AH (2008) Blood pressure lowering and vasomodulator effects of piperine. J Cardiovasc Pharmacol 52(5):452–458
169. Manoharan S, Balakrishnan S, Menon VP, Alias LM, Reena AR (2009) Chemopreventive efficacy of curcumin and piperine during 7, 12-dimethylbenz (a) anthracene-induced hamster buccal pouch carcinogenesis. Singap Med J 50(2):139

170. Parganiha R, Verma S, Chandrakar S, Pal S, Sawarkar HA, Kashyap P (2011) In vitro anti-asthmatic activity of fruit extract of Piper nigrum (Piperaceae). Inter J Herbal Drug Res 1:15–18
171. Matsuda H, Ninomiya K, Morikawa T, Yasuda D, Yamaguchi I, Yoshikawa M (2008) Protective effects of amide constituents from the fruit of Piper chaba on d-galactosamine/TNF-α-induced cell death in mouse hepatocytes. Bioorg Med Chem Lett 18(6):2038–2042
172. Anissian D, Ghasemi-Kasman M, Khalili-Fomeshi M, Akbari A, Hashemian M, Kazemi S, Moghadamnia AA (2018) Piperine-loaded chitosan-STPP nanoparticles reduce neuronal loss and astrocytes activation in chemical kindling model of epilepsy. Int J Biol Macromol 107:973–983
173. Elnaggar YS, Etman SM, Abdelmonsif DA, Abdallah OY (2015) Intranasal piperine-loaded chitosan nanoparticles as brain-targeted therapy in Alzheimer's disease: optimization, biological efficacy, and potential toxicity. J Pharm Sci 104(10):3544–3556
174. Sangeetha KS, Umamaheswari S, Reddy CUM, Kalkura SN (2017) Chrysin loaded chitosan nanoparticle: formulation and in-vitro characterization. Int J Pharm Sci Res 8(3):1102–1109
175. Picinelli A, Dapena E, Mangas JJ (1995) Polyphenolic pattern in apple tree leaves in relation to scab resistance. A preliminary study. Journal of Agricultural and Food Chemistry 43(8):2273–2278
176. Crespy V, Aprikian O, Morand C, Besson C, Manach C, Demigné C, Rémésy C (2001) Bioavailability of phloretin and phloridzin in rats. The Journal of nutrition 131(12):3227–3230
177. Ma L, Wang R, Nan Y, Li W, Wang Q, Jin F (2016) Phloretin exhibits an anticancer effect and enhances the anticancer ability of cisplatin on non-small cell lung cancer cell lines by regulating expression of apoptotic pathways and matrix metalloproteinases. Int J Oncol 48(2):843–853
178. Schaefer S, Baum M, Eisenbrand G, Dietrich H, Will F, Janzowski C (2006) Polyphenolic apple juice extracts and their major constituents reduce oxidative damage in human colon cell lines. Mol Nutr Food Res 50(1):24–33
179. Mariadoss AVA, Vinayagam R, Xu B, Venkatachalam K, Sankaran V, Vijayakumar S, Bakthavatsalam SR, Mohamed AS, David E (2019) Phloretin loaded chitosan nanoparticles enhance the antioxidants and apoptotic mechanisms in DMBA induced experimental carcinogenesis. Chem Biol Interact 308:11–19
180. Huang H, Li L, Shi W, Liu H, Yang J, Yuan X, Wu L (2016) The multifunctional effects of nobiletin and its metabolites in vivo and in vitro. Evid Based Complement Alternat Med 2016:2918796
181. Luque-Alcaraz AG, Lizardi J, Goycoolea FM, Valdez MA, Acosta AL, Iloki-Assanga SB, Higuera-Ciapara I, Argüelles-Monal W (2012) Characterization and antiproliferative activity of nobiletin-loaded chitosan nanoparticles. J Nanomater 2012:100
182. Srinivasan M, Sudheer AR, Menon VP (2007) Ferulic acid: therapeutic potential through its antioxidant property. J Clin Biochem Nutr 40(2):92–100
183. Panwar R, Sharma AK, Kaloti M, Dutt D, Pruthi V (2016) Characterization and anticancer potential of ferulic acid-loaded chitosan nanoparticles against ME-180 human cervical cancer cell lines. Appl Nanosci 6(6):803–813
184. Clifford MN (2000) Chlorogenic acids and other cinnamates–nature, occurrence, dietary burden, absorption and metabolism. J Sci Food Agric 80(7):1033–1043
185. Perrone D, Donangelo R, Donangelo CM, Farah A (2010) Modeling weight loss and chlorogenic acids content in coffee during roasting. J Agric Food Chem 58(23):12238–12243
186. Kasai H (1997) Analysis of a form of oxidative DNA damage, 8-hydroxy-2′-deoxyguanosine, as a marker of cellular oxidative stress during carcinogenesis. Mutation Research/Reviews in Mutation Research 387(3):147–163
187. Lee K, Lee JS, Jang HJ, Kim SM, Chang MS, Park SH, Kim KS, Bae J, Park JW, Lee B, Choi HY (2012) Chlorogenic acid ameliorates brain damage and edema by inhibiting matrix metalloproteinase-2 and 9 in a rat model of focal cerebral ischemia. Eur J Pharmacol 689(1–3):89–95

Quercetin-Loaded Nanomedicine as Nutritional Application

Sabya Sachi Das, P. R. P. Verma, and Sandeep Kumar Singh

1 Introduction

Flavonoids are plant secondary metabolites and have been reported for their wide range of therapeutic and pharmacological activities. Among all the reported flavonoids, quercetin (3,3′,4′,5,7-pentahydroxyflavone), being abundantly found in onion, apple, berries, tea, and brassica vegetables, as well as in many nuts, seeds, barks, flowers, and leaves, has gained superior considerations for its impending therapeutic activities [1]. Quercetin (QT) is found to be one of the major flavonoids which also tends to be the backbone of several other flavonoids [2]. Furthermore, QT is an essential constituent of countless food supplements and also has been reported for its numerous pharmacological activities including anticancer [3], antioxidant [4], anti-inflammatory [5], anti-infective [6, 7], antidiabetic [8], and many more (Fig. 1). Though QT has been proven to have numerous pharmacological activities, it has several issues including poor aqueous solubility (0.17–7.7 mg/mL) and low bioavailability (16.2%), so its clinical application is limited [10]. QT suffers with autoxidation in organic medium and in aqueous medium at pH above 8.0. Autoxidation relies on the buffer type and pH of medium and is also associated with the dissociation of the hydroxyl (-OH) groups in the molecule. In addition, QT, like other flavonoids, is photosensitive [11–13]. To overcome these issues, nano-based drug delivery systems have been developed for efficient delivery of QT.

Nanotechnology deals with particle size in the range of 10–200 nm and is a novel and prolific method to deliver lipophilic drugs into the biological system [14]. Furthermore, because of their exceptionally small size, nanocarriers have established many advantages which include improved aqueous solubility, enhanced gastrointestinal tract (GIT) time, enhanced physicochemical stability in GIT, improved intestinal permeation and controlled release in GIT, and

S. S. Das · P. R. P. Verma · S. K. Singh (✉)
Department of Pharmaceutical Sciences and Technology, Birla Institute of Technology, Ranchi, Jharkhand, India

© Springer Nature Singapore Pte Ltd. 2020

M. Rahman et al. (eds.), *Nanomedicine for Bioactives*,
https://doi.org/10.1007/978-981-15-1664-1_9

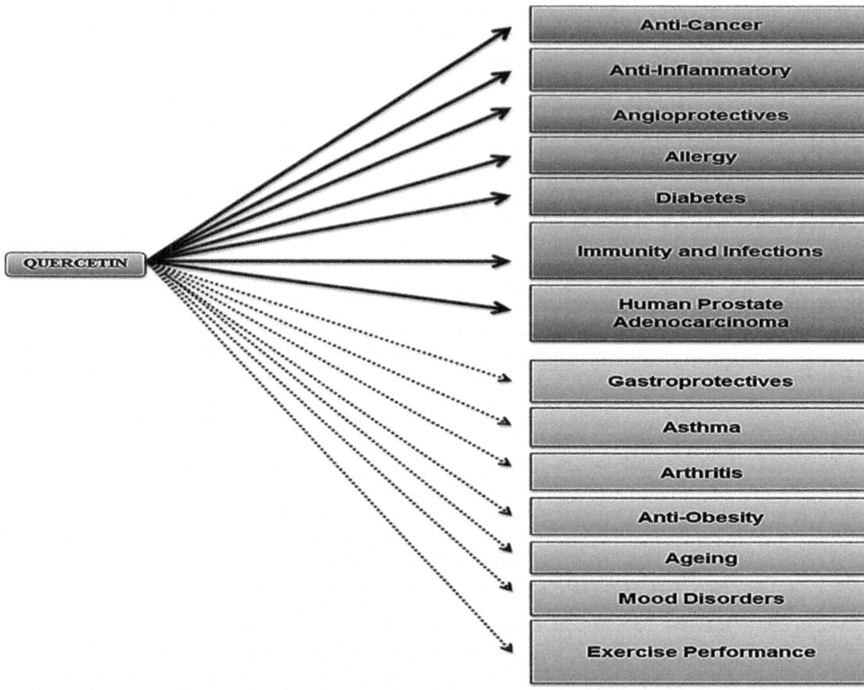

Fig. 1 Therapeutic applications of quercetin (*thick arrows*, most suggested uses in human; *thin arrows*, uses mainly derived from animal studies) [reprinted and modified from D'Andrea, G., Quercetin: A flavonol with multifaceted therapeutic applications? *Fitoterapia [9]*, *106*, 256–71. © 2015, Elsevier]

intracellular and transcellular delivery [15, 16]. Basically, based on the types of composition, nanoformulations (NFs) have been classified into three distinct categories: (1) lipid- and surfactant-based delivery systems (nanoliposome, nanoemulsion, solid lipid nanoparticles, nanostructured lipid carriers); (2) nanocarriers based on polysaccharides (polymer nanoparticle, polymeric micelles, and inclusion complex); and (3) protein-based nano-transporters (Fig. 2) [17, 18]. Moreover. QT-loaded NFs such as liposomes [19, 20], polymeric nanoparticles [21], micelles [22], solid lipid nanoparticles [23–24], nanoemulsions [25], nanocapsules [26], and metal nanoparticles [27–29] have shown improvement in solubility, bioavailability, biodistribution, specificity, and pharmacokinetics of QT. In this review, we have tried to describe the key potential and therapeutic activities of QT as well as its limitations and its relation to nanoformulations, including their advantages, disadvantages, and applications, in increasing the solubility and bioavailability of this bioactive compound.

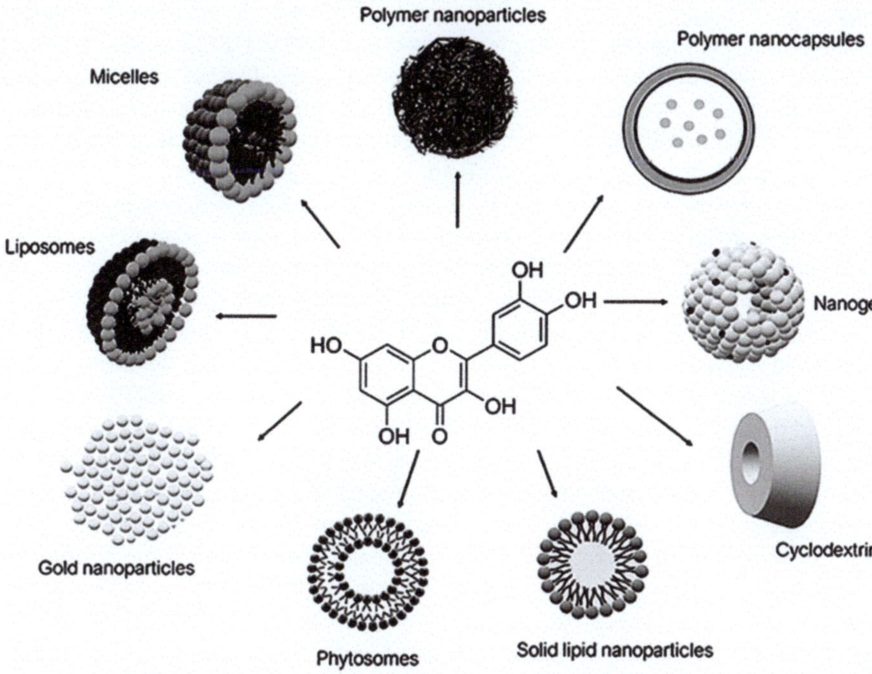

Fig. 2 Quercetin nanoformulations used for various therapeutic applications. [reprinted and modified from Wong, K. E.; Ngai, S. C.; Chan, K.-G.; Lee, L.-H.; Goh, B.-H.; Chuah, L.-H., Curcumin nanoformulations for colorectal cancer: A review. *Frontiers in Pharmacology* 2019, *10*. © 2019, Frontiers media]

2 Therapeutic Effects of Quercetin

2.1 Anti-Infective Effect

2.1.1 Antibacterial and Antifungal

Mittal et al. [30] synthesized QT and gallic acid (GA)-conjugated silver (Ag)-selenium (Se) NPs at room temperature. Results showed NPs possessed greater therapeutic efficiency in terms of antioxidant, antimicrobial (Gram negative and Gram positive), and anticancer activity against Dalton lymphoma (DL) cells; also the in vitro studies revealed that the conjugated NPs at concentration of 50 μg/mL reduced the cell viability by 80%. Yun et al. [31] examined the fungicidal activity and the basic mechanism of isoquercetin (IQT) against *C. albicans*. Moreover, propidium iodide and potassium release studies established that IQT initiated membrane impairment, thus increased permeability. The giant and large unilamellar vesicles loaded with IQT disturbed the membrane, and the degree of disruption was valued to be in a range of 2.3 nm to 3.3 nm, and hence, IQT exerted its fungicidal activity by hindering the cell membrane. Bidone et al. [32] explored the inhibitory effects of *Achyrocline satureioides* extract (ASE;

comprising of major flavonoids QT, luteolin (LT), and 3-O-methylQT)-loaded nanoemulsion (NE) on herpes simplex virus type 1 (HSV-1/KOS strain) replication. The topical delivery of ASE-NE exhibited more distinct effects against HSV-1 replication as compared to ASE or pure QT; hence, it was concluded that topical delivery of ASE-NE has significant actions against herpes infections. Sun et al. [33–34] fabricated AgNPs combined with QT (QT-AgNPs) which showed that QT improved the antibacterial activity of AgNPs synergistically. The QT-AgNPs were stabilized with a layer of siRNA molecules (siRNa-QT-AgNPs) which enhanced the antibacterial effect against *Pseudomonas aeruginosa* (*P. aeruginosa*) and *B. subtilis* (most prominent drug-resistant species) and consequently damaged the cellular wall, increased cellular uptake, and inhibited proliferation of *B. subtilis*. Moreover, the in vivo results showed that the intravenous (i.v.) injection of QT-AgNPs reduced the level of bacterial accumulation in the blood and organs and also in the inflamed cells in the tissues. Jain and Mehata [6] synthesized AGNPs (leaves of *Ocimum sanctum*) and also isolated QT (*Ocimum sanctum*) as precursors and investigated its bactericidal activity against *E. coli* (Gram-negative bacteria) strain. Results showed that QT as a precursor was primarily responsible for the reduction of AgNPs and also enhanced the antibacterial activity of AgNPs against *E. coli* strain. Sun et al. [35] synthesized QT-AgNPs and showed that QT synergistically improved the bactericidal effects of AgNPs against the drug-resistant bacteria *E. coli* and *Staphylococcus aureus* (*S. aureus*). Results of RNA sequencing and gene ontology (GO) assays revealed the molecular mechanism of bactericidal effects of QT-AgNPs. In was concluded that the method of transcriptome analysis offered novel perceptions into molecular mechanism of antibacterial assays. Yuan et al. [29] synthesized QT-AgNPs and investigated its antibacterial activity against two multidrug-resistant (MDR) bacterial species, *P. aeruginosa* and *S. aureus* (both isolated from milk samples of goats diseased with mastitis). Results showed that AgNPs exerted antibacterial activity in a time- and dose-dependent manner. Also, the antibacterial activity of QT-AgNPs was due to secretion of reactive oxygen species (ROS), malondialdehyde (MDA), and release of proteins and sugars in bacterial cells. The QT-AgNPs lowered the lactate dehydrogenase activity (LDH) and levels of adenosine triphosphate (ATP) in bacterial strains as compared to control. Additionally, bacterial strains treated with QT-AgNP showed downregulation of glutathione (GSH), superoxide dismutase (SOD), and catalase (CAT) and upregulation of glutathione S-transferase (GST). Nguyen et al. [36] developed QT nanoparticles (NPs) loaded with of CS, sulfobutyl-ether-β-cyclodextrin (Captisol®), and/or pentasodium tripolyphosphate by using ionotropic gelation method and evaluated their impending activity to inhibit QS in *E. coli*. The release of QT in QTNPs was prolonged due to Captisol®. Bioassay results showed that QT-CS-NPs inhibited QS and antibacterial effects. Hence, CS-based NPs could be used efficiently as nanocarriers to deliver hydrophobic phytocompounds locally at the bacterial site and also could increase anti-QS and antibacterial activities significantly. Omwenga et al. [26] prepared QT- and baicalein (BL)-loaded chitosan (CS) nanocapsules (QT-BL-CSNCs) and evaluated its inhibitory, biofilm formation and quorum sensing (QS) activity against strains of *Escherichia coli* (*E. coli*) and also their cytotoxicity effects on MDCK-C7 cells. Results revealed that free form of QT and BL showed minor inhibitory effects on the QS response and biofilm formation. Moreover, free form of QT and BL appeared to be cytotoxic to MDCK-C7 cells at greater doses, though, upon nanoencapsulation, a cytoprotective effect was noticed. Li et al. [37] developed QT and catechins (CC) loaded on genipin (GP)-modified CS-NPs (GP-CC-QT NPs) by using ionic gelation method. Results showed that GP-CC-QT NPs enhanced the %entrapment efficiency (%EE) of CC and QT. In vitro release study revealed that CC and QT followed sustained release from GP-CC-QT NPs. Also, GP-CC-QT NPs showed greater potency in scavenging 1, 1-diphenyl-2-picrylhydrazyl (DPPH), 2,2′-azinobis (3-ethylbenzothiazoline-6-sulfonic acid ($ABTS^{\bullet+}$), OH, and $O_2^{\bullet-}$ radicals and in suppressed growth of *E. coli*, *Bacillus subtilis* (*B. subtilis*), and *Staphylococcus aureus* (*S. aureus*) as compared to CC, QT and, GP (alone or blended). Yu et al. [38] performed antibacterial and anti-biofilm assays of QT, Ag NPs, and QT-AgNPs against *E. coli* strain ECDCM1. Results showed that QT-AgNPs exhibited greater antibacterial and anti-biofilm properties in a MDR *E. coli* strain isolated from a dairy cow infected with mastitis as compared with free QT and AgNPs. It was also observed that the anti-biofilm effect of QT-AgNPs was due to reduced transcription of biofilm-related genes (bcsA, csgA, fliC, fimA, motA, and wcaF). Lotha et al. [39] isolated afzelin (AZ) and QT from *Crotalaria tetragona*, which were further used to fabricate AZ/QT-AgNPs by the process of green synthesis, and investigated their antibacterial and anti-biofilm activity against *P. aeruginosa* and *Salmonella typhimurium* (*S. typhimurium*). AZ/QT-AgNPs showed potential anti-biofilm activity against *S. typhimurium* and showed that the anti-biofilm activity is independent of antibacterial effect. It was observed that AZ/QT-AgNPs exhibited anti-biofilm activity due to decreased cell surface hydrophobicity, reduced surface adherence, and loss of membrane potential. AZ/QT-AgNPs showed comparatively less toxicity and were more efficient in restricting the bioburden of *S. typhimurium* in infected zebrafish by >3 log fold. Sowmya et al. [40] synthesized AgNPs from the aqueous extracts of *Phyllanthus acidus* (*P. acidus*), having QT as the major flavonoid, and formed PA-AgNPs. Results showed that PA-AgNPs inhibited the cell growth proliferation of *E. coli* and hence have potent antibacterial effect on *E. coli* cells. Additionally, the bactericidal activity PA-AgNPs *E. coli* strains were confirmed by the scanning electron microscopy (SEM), which showed damaged cell membranes of *E. coli* in the presence of PA-AgNPs leading to arrest of cell growth. Zhou and Tang [41] fabricated AgNP-coated silk materials using QT and rutin (RT). The results exhibited that QT-synthesized AgNPs were produced more rapidly but showed lesser size consistency as compared to RT-synthesized AgNPs, whereas the RT-synthesized AgNPs displayed greater uniform distribution on silk as compared to QT-synthesized AgNPs. Moreover, the antibacterial activity of AgNP-treated silk persisted over 90% against *E. coli* and *S. aureus* strains, but the antioxidant activity of the AgNP-treated silk progressively reduced during washing.

2.1.2 Antileishmanial

Mittra et al. [42] prepared QT- and LT-loaded formulation individually and investigated their ALA against *L. donovani* promastigotes. Results showed that LT (IC_{50}-12.5 μM) and QT(IC_{50}-45.5 μM) suppressed the growth of *L. donovani* promastigotes and amastigotes in vitro, inhibited DNA synthesis in promastigotes, and promoted topoisomerase II-mediated linearization of kDNA minicircles. Also, these compounds arrested cell cycle progression in *L. donovani* promastigotes, leading to apoptosis. Sarkar et al. [43] prepared various types of QT-loaded vesicular delivery systems including liposomes, niosomes, microspheres, and nanocapsules (all loaded with approximately the same amount of QT). These QT-loaded nanovesicular systems were studies for ALA of QT simultaneously against *L. donovani* promastigotes. Results showed that judging from the efficacy in lowering of spleen parasite load and capacity to reduce overall toxicity, QT could very well be tried in the clinics and best result could be expected when it is delivered in nonencapsulated form. Obviously, these biocompatible and biodegradable polymeric nanocapsules have longer circulation time in blood, and they remain unaffected by circulating lipases, protecting the drug from bio-environment. Gupta et al. [44] prepared QT-loaded o/w microemulsion system and investigated the phase behavior of composition (clove oil/Tween 20/water: 5:30:65). A particular was chosen as the drug delivery system from the clear oil-in-water zone of the pseudoternary system. Results showed that QT-loaded microemulsion system exhibited hepatotoxicity and nephrotoxicity, hence enhancing the ALA against experimental leishmaniasis in hamster model than the free QT. Tasdemir et al. [2] surveyed a large set of flavonoid aglycones and glycosides, as well as a panel of other related compounds of phenolic and phenylpropanoid nature, for their in vitro activities against *Trypanosoma brucei rhodesiense* (*T. brucei rhodesiense*), *Trypanosoma cruzi* (*T. cruzi*), and *L. donovani*. Overall, the best in vitro trypanocidal activity for *T. brucei rhodesiense* was exerted by 7,8-dihydroxyflavone, followed by 3-hydroxyflavone, rhamnetin, and 7,8,3′,4′-tetrahydroxyflavone and catechol. The activity against *T. cruzi* was moderate, and only chrysin dimethylether and 3-hydroxydaidzein showed activity. Fisetin, 3-hydroxyflavone, LT, and QT showed the most potent ALA. 7,8-Dihydroxyflavone and QT appeared to ameliorate parasitic infections in mouse models. Sen et al. [45] revealed the efficiency of QT moiety which inhibited the development of anemia during the leishmanial infection (postinfection period). QT inhibited the oxidation of proteins and lipids on the red cell membranes and lowered the parasite load in the spleen of infected animals. Combination therapy of QT with the antileishmanial drug stibanate (SN) produced a better decay of SOH in the erythrocytes of the infected animals compared to that induced by QT or drug treatment alone. Sen et al. [46] described the mechanism underlying the shortened lifespan of erythrocytes in visceral leishmaniasis and proposed a combination therapy with QT and hamster serum albumin (HSA) toward its rectification. Moreover, QT-HSA combination increased the bioavailability of QT. The combination targeted hemoglobin oxidation and produced an effective attenuation of heme degradation. This led to decreased iron decompartmentalization, thereby increasing the life span of erythrocytes during leishmanial infection. Sen et al. [47] studied the effect of combination treatment of QT-SA on the proliferation of amastigotes of *L. donovani*. Results showed that QT-SA enhanced the bioavailability of QT and affected the repression of splenic parasite load in *L. donovani*. Results have shown that QT-SA improved ALA targeting ribonucleotide reductase (RR) and hence could be used for future drug design based on interference with the parasite's iron metabolism under in vivo conditions. Marin et al. [48] investigated the ALA of various flavonoids, kaempferol (KF) (1), QT (2), trifolin(TF) (3), octa-*O*-acetyl hyperoside(O-o-A) (4), and acetyl hyperoside(AH) (5) and their *O*-acetyl derivatives (1a, 2a, 3a) isolated from *Consolida oliveriana* against promastigote as well as amastigote forms of *Leishmania* spp. Results of the cellular proliferation and metabolic and ultrastructural studies showed that the acetylated compounds 2a, 3a, and 4 were highly active against *Leishmania peruviana* (*L. peruviana*), while compounds 2a as well as 4 were effective against *Leishmania braziliensis* (*L. braziliensis*), and no cytotoxicity was observed in any compound. Fonseca-Silva et al. [49] studied the mechanism for the ALA of QT against *Leishmania amazonensis* (*L. amazonensis*) promastigotes. QT inhibited *L. amazonensis* promastigote growth in a dose- and time-dependent manner. QT increased ROS generation in a dose-dependent manner. The antioxidant compounds, glutathione (GSH) and *N*-acetyl-L-cysteine (NAC), each significantly reduced QT-induced cell death. Taken together, results showed that QT exerted its ALA on *L. amazonensis* promastigotes due to the generation of ROS and interrupted the mitochondrial function of *L. amazonensis*. Das et al. [50] developed QT-loaded gold nanoparticles (QTGNPs) by one pot synthesis method and evaluated its ALA against *L. donovani* promastigotes. Results showed that the QTGNPs exhibited ALA against sodium stibogluconate-resistant strain and the paromomycin-resistant strains. Also, TEM results exhibited that the macrophage uptake of QTGNPs was completed within an hour. Fonseca-Silva et al. [51] reported the mechanism of the ALA of QT against the intracellular amastigote form of *L. amazonensis*. QT reduced the infection index in *L. amazonensis*-infected macrophages in a dose-dependent manner and also increased ROS generation in a dose-dependent manner. Taken together, these results suggested that ROS production plays a crucial role in the mechanism of action of QT in the control of intracellular amastigotes of *L. amazonensis*. Sousa-Batista et al. [52] prepared QT and its penta-acetylated derivative (PQT)-loaded lipid-core nanocapsules (LNCs) of poly (ε-caprolactone) (QT/PQT-PCL-LNCs) and evaluated their ALA against BALB/c mice infected with *L. amazonensis*. The ALA of PQT-PCL-LNCs was improved but was not as good as QT-PCL-LNCs. From the results, it was concluded that the oral efficiency of QT or PQT was increased after encapsulation into LNCs. Also, QT/PQT-PCL-LNCs showed enhanced ALA and hence could significantly be used for the treatment of cutaneous leishmaniasis.

2.2 Antidiabetic Effect

Mukhopadhyay et al. [53] developed an QT-alginate/succinyl chitosan-conjugated polymeric nanoparticles (QT-A/SC PNPs) for the treatment of diabetes mellitus (DM). In vitro results showed that the QT-A/SC PNPs appeared to be spherical with particle size of ~91.58 nm and %entrapment efficiency of ~95% and exhibited controlled release with non-Fickian kinetics trend. In vivo studies revealed that, as compared to free QT, QT-A/SC PNPs showed a marked and enhanced hypoglycemic effect against diabetic rat model with no cytotoxicity. Ebrahimpour et al. [54] synthesized QT-conjugated superparamagnetic iron oxide nanoparticles (QT-SPIONs) and treated it against streptozotocin-induced diabetic rats for the treatment of diabetes-associated memory impairment. Results showed that QT-SPIONs exhibited better efficiency than free QT in memory performance assessment. QT-SPIONs showed greater potency in improving diabetes-associated memory impairment. Also, SPIONs showed minimal toxicity on body weight, blood glucose level, and body tissues and hence could be a favorable carrier for oral delivery of QT. Maity et al. [55] measured the amplitude and frequency of the equilibrium fluctuations of fluorescently labeled small oligomers of hIAPP (intrinsically disordered proteins (IDPs) associated with type 2 diabetes) in a physiological solution, using nanosecond fluorescence cross-correlation spectroscopy. All the fluctuations were affected in a dose-dependent manner with a degree of interaction: resveratrol (RES) < epicatechin (ECC) ~ QT < Congo red < epigallocatechin-3-gallate (EGCG). It was observed that the ability of a molecule to restrain intramolecular fluctuations was related with its ability to reduce membrane affinity. Rathee and Kamboj [8] developed QT-loaded phytosomes (QTPs), isolated from the fruits of *Citrullus colocynthis* (L.), *Momordica balsamina*, and *Momordica dioica*, and optimized by using three-factor, three-level Box-Behnken design. Results revealed that optimized QTPs exhibited spherical shape with particle size of 450 nm, total flavonoid content of 10.0 ± 0.002 μg/g, and %EE of 92.1 ± 5.1%. Also, QTPs exhibited more antidiabetic activity as compared to metformin in low dose.

2.3 Anti-inflammatory Effect

Landi-Librandi et al. [56] prepared both soya phosphatidylcholine (SP) and cholesterol (CL) or cholesteryl ethyl ether (CLEE)-loaded QT liposomes (SP-CL/CLEE-QTLs), and also blank liposomes (SP-CL/CLEE-Ls), and evaluated their role in the activation of human complement system proteins. The results indicated that blank liposomes consumed more components than the others for both classical (CP) and alternative (AP) pathways. Furthermore, replacement of CHOL with CHOL-OET reduced complement consumption via both CP and AP. SP-CL/CLEE-QTLs activated hCS by CP and AP and hence could be used as an alternative for drug therapy of various diseases in which oxidative stress and inflammation are involved, such as cancer and autoimmune disorders. Kumari et al. [57] prepared QT-loaded poly-D,L-lactide (PLA) nanoparticles (QT-P-NPs) by solvent evaporation method to improve the solubility, permeability, and stability of QT. The in vitro release kinetics of QT under physiological condition revealed initial burst release followed by sustained release. Also, QT-P-NPs enhanced the therapeutic activity of QT for intestinal anti-inflammatory effect. Wang et al. [21] studied the anti-inflammatory activity of QT-conjugated monomethoxy poly (ethylene glycol)-poly(ε-caprolactone) micelles (QT-MPEG-PCLMs). The in vivo results specified that intravesical (IV) application of QT-MPEG-PCLMs did not induce any toxicity to the bladder; moreover, the therapeutic efficiency of IV administration of QT-MPEG-PCLMs reduced the edema and inflammatory cell infiltration of the bladder in an *E. coli*-induced acute cystitis model. Hence, from the results, it was concluded that QT-MPEG-PCLMs could be used in acute cystitis therapy. Landi-Librandi et al. [58, 59] prepared liposomes loaded with QT, myricetin (MT), KF, and galangin (GG) s were evaluated as potential antioxidant carriers, in attempt to overcome their poor solubility and stability. Results showed that the inhibitory effect of liposomal flavonoids on CLlum and CLluc was inversely related to the number of hydroxyl groups in the flavonoid B ring. Moreover, phagocytosis of liposomes by neutrophils does not seem to necessarily promote such activity, as the liposomal flavonoids are also able to reduce CL when the cells are pretreated with cytochalasin B. Under assessed conditions, the antioxidant liposomes are not toxic to the human neutrophils and do not interfere with IC-induced phagocytosis. Azofeifa et al. [60] evaluated the antioxidant and anti-inflammatory activities of QT and ellagic acid (EA), extracted from blackberries. The antioxidant activity measured via oxygen radical absorbance capacity (ORAC) was higher for the blackberry extract than for QT and EA. Results showed that the extract inhibited superoxide production by NADPH oxidase in THP-1 cells and nitrite production in J774A.1 cells stimulated with LPS + IFNγ, with decreased nitrite production. Caddeo et al. [19, 20] prepared QT-chitosan-conjugated liposomes (QT-CLs) which enhanced the bioavailability of QT and also enhanced its release in the intestine. Results showed that QT-CLs provided resistance to acidic conditions and promoted the release of QT in alkaline pH. The proposed hybrid system represented a promising combination of nanovesicles and chitosan for the delivery of QT to the intestine in the therapy of oxidative stress/inflammation-related disorders. Lee et al. [61] synthesized QT-loaded silica nanoparticles (QT-S-NPs) using an oil-in-water (O/W) microemulsion method. The resulting QT-loaded silica nanoparticles were spherical, highly monodispersed, and stable in an aqueous system. Superoxide radical scavenging effects were found for the QT-loaded silica nanoparticles as well as free QT. The QT-loaded silica nanoparticles showed cell viability comparable to that of the controls. The amounts of pro-inflammatory cytokines produced by macrophages, such as interleukin 1 beta, interleukin 6, and tumor necrosis factor alpha, were reduced significantly for QT-S-NPs. These results suggest that the antioxidative and anti-inflammatory activities of QT are maintained after encapsulation in silica. Strugala et al. [62] synthesized QT-7-O-β-D-(4″-O-methyl) by glycosylation and determined the anti-oxidative and anti-inflammatory properties of QT 7-MeGlu as compared to free QT. The results showed that QT 7-MeGlu and free QT both increased the packing order, not only in the hydrophilic but also in hydrophobic regions of the membrane (QT > QT 7-MeGlu). Also, QT-7-MeGlu and free QT both inhibited the human serum albumin (HSA) intrinsic fluorescence through a static quenching mechanism. The studies have also showed that QT-7-MeGlu inhibited the activity of the anti-inflammatory enzymes COX-1 and COX-2 more as compared to free QT. Abd El-Fattah et al. [63] prepared QT-loaded phytosomes (QTPs) using the thin-film hydration method. Results showed that QTPs exhibited very high % EE. Also, QTPs showed significant increase in serum calcium, inorganic phosphorus, and glutathione content, whereas, it significantly decreased serum alkaline phosphatase, acid phosphatase, malondialdehyde level, tumor necrosis factor-alpha, and glucose level and improved lipid profile. Consequently, the results obtained confirm the superiority of QTPs over free QT at the same doses as a promising hormone replacement therapy. Aithal et al. [64] fabricated QT-loaded nanoemulsion gel (QTNG) and showed its activity as a potent antimicrobial and anti-inflammatory agent for the treatment of periodontitis that could improve its solubility and bioavailability. The release of QT from QTNG was significantly more as compared to that of pure QT-loaded gel. Hence, QTNG could be used as a potent carrier for clinical testing in periodontitis. Ozdal et al. [65] synthesized QT-loaded gold (Au) nanoparticles (Au-QTNPs) and investigated its effect on LPS-induced neuro-inflammation in BV-2 microglial cells. Results showed that Au-QTNPs inhibited the inflammatory stimulation of the BV-2 cells, prostaglandin, E2, nitric oxide (NO), upregulated cyclooxygenase-2, inducible NO synthase mRNA, and protein levels, without causing any cytotoxic effects. Hence, Au-QTNPs showed protective effects against inflammatory neurodegenerative disease via suppression of acute microglial activation.

2.4 Antioxidative Effect

Ratty et al. [66] studied antioxidant activities (AOA) of various flavonoids (QT, quercetrin, RT, MT, phloretin, phloridzin, CC, morin (MR), and taxifolin (TAX)) with the DPPH free radical. Also, flavonoid-DPPH interaction was investigated for the absence and presence of liposomes so as to reveal some information on bilayers. It was observed that the interaction of the flavonoids on the lipid bilayer occurred in the polar zone of the lipid bilayers. The flavonoids were able to scavenge free radicals and could do so in biomembranes. Among all flavonoids, QT was the most potent molecule, and it inhibited the lipoxygenase in the liposomal suspension by about 42%, while the other flavonoids inhibited the enzyme by about 14–23%. Terao et al. [67] examined the AOA of CCs [(−)-ECC, (−)-ECGC] and flavonoid (QT) by measuring the inhibition of lipid peroxidation in large unilamellar liposomes composed of egg yolk phosphatidylcholine (PC). Their inhibitory effects lasted longer than that of α-tocopherol. It was concluded that CCs and QT served as powerful antioxidants against lipid peroxidation when phospholipid bilayers are exposed to aqueous oxygen radicals. Saija et al. [68] studied the AOA of four flavonoids (QT, hesperetin (HPT), naringenin (NG), RT), using two different in vitro experimental models: (1) Fe^{2+}-induced linoleate peroxidation (Fe^{2+}-ILP), by detection of conjugated dienes, and (2) autoxidation of rat cerebral membranes (ARCM), by using thiobarbituric acid for assay of free malondialdehyde production. The antilipoperoxidant activity followed the order RT > HPT > QT > > NG in the Fe^{2+}-ILP test and QT > RT > > HPT > NG in the ARCM test. Gordon and Roedig-Penman [69] investigated the AOA of QT, MT, and α-tocopherol in small unilamellar liposomes. Liposomes loaded with MT were more effective than α-tocopherol. Cupric chloride and ferric chloride strongly reduced the antioxidant activity of MT and QT with cupric chloride causing a stronger reduction in activity than ferric chloride. Also, at a pH of 7.4, QT was less effective than α-tocopherol at lower concentration of phospholipids, but its activity increased more strongly with concentration, and it was very effective at higher concentration of phospholipid. Liao and Yin [70] compared the AOA of seven phenolic agents against Fe^{2+}-induced lipid oxidation with α-tocopherol, β-carotene, and vitamin C in human erythrocyte membrane ghosts and liposome systems. The AOA of five test flavonoids followed the order CC > ECC > RT > QT > MT in both systems. It was observed that the partition coefficients of phenolic agents play an important role in determining their individual and combined antioxidant activities with other antioxidants. Shirai et al. [71] studied the AOA of QT 3-O-â-D-glucuronide (QT-3-G), one of the QT metabolites in the blood, after intake of QT-rich food. Results showed that QT-3-G inhibited lipid peroxidation significantly, although its inhibitory effect was lower than that of free QT. Hence, QT-3-G inhibited membranous phospholipid peroxidation induced by ROS and peroxidative enzymes. Beninger and Hosfield [4] isolated various flavonoids from seeds of *Phaseolus vulgaris* L. Antioxidant results showed that the pure anthocyanins delphinidin 3-O-glucoside, petunidin 3-O-glucoside, and malvidin 3-O-glucoside and QT 3-O-glucoside exhibited significantly higher antioxidant activity than the Fe^{2+} control. Filip and Ferraro [72] isolated and quantified various compounds (chlorogenic acid, caffeic acid, 3,4-dicaffeoylquinic acid, 3,5- dicaffeoylquinic acid, 4,5-dicaffeoylquinic acid, RT, QT, and KF) from *Ilex paraguariensis* St. Hilaire (Aquifoliaceae). Results showed that extracts exhibited high antioxidant and choleretic activities. It also induced an increase in intestinal transit in the experimental conditions employed and did not produce any sign of toxicity at the analyzed doses. Zou et al. [73] quantitatively extracted a series of flavonoids (RT, hyperoside, Iso-QT, avicularin, quercitrin, and QT) from *Hypericum perforatum* L. (*FEHP*). Results of free radical scavenging activity and antioxidant activity of FEHP showed that FEHP was most potent against quenching DPPH and superoxide radical. Also, FEHP decreased the peroxidation of linoleic acid in dose-dependent manner. Murata et al. [74] investigated the AOA of derivatives of ethyl docosahexaenoate (DHA-Et) dissolved in an ethanol solution or embedded in liposomes. The degradation rate of DHAEt embedded in liposomes was higher than that of DHAEt dissolved in ethanol. Antioxidants suppressed the degradation of DHA-Et embedded in liposomes, the order of activity of the antioxidants being LT > fisetin>KF > QT > RT. These results suggest that the hydrophobicity (log P) of an antioxidant is one of the determinants for antioxidative activity but that a vicinal diol structure in the B ring is not favorable for the antioxidative activity. Ly et al. [75] isolated a series of phenolic compounds (QT derivatives) from methanolic extract of dry outer scales of onion (*Allium cepa* L.). It was observed that the flavonoid compounds having o-dihydroxy substituent in the B ring were shown to be effective antioxidants against nonenzymic lipid peroxidation. Also, the compounds suppressed the autoxidation of methyl linoleate in bulk phase and the radical-initiated peroxidation of soybean phosphatidylcholine in liposomes. Therefore, the outer scale extract of onion is expected to be a resource for food ingredients. Gabrielska et al. [76] investigated the protective action of QT on phosphatidylcholine liposome membranes exposed to the prooxidative action of diphenyltin dichloride (DPhT) and triphenyltin chloride (TPhT) induced by UV radiation. Results revealed that the antioxidative activity followed the sequence QT:TPhT > QT:DPhT > QT. It was concluded that QT's AOA was due to formation of complexes with phenyltins, though it protects against the peroxidation caused by the free radical forms of phenyltins. Ghosh et al. [77] evaluated the therapeutic efficacy of QTLs and QTNCs in combating *As*-induced hepatic oxidative stress and fibrosis-associated upregulation of its gene expression and plasma TGF ß (transforming growth factor ß) in rat model. Among the two vesicular delivery systems formulated with QT, polylactide-QTNCs showed a promising result compared to QTLs in controlling *As*-induced hepatic alteration. Liang et al. [78] reported that antioxidant synergism (or antagonism) between plant (iso)flavonoids (daidzein, baicalein, and QT) or their glycosides (puerarin, baicalin, and RT) and β-carotene in phosphatidylcholine liposomes (pH 7.4) with oxidation was initiated thermally by the lipophilic free radical initiator 2,20-azobis(2,4-dimethylvaleronitrile) (AMVN) followed by the formation of conjugated. Bernardy et al. [79] prepared QT-loaded nanocapsules and nanospheres, comprising of lecithin (surfactant) and miglyol 812 (cosurfactant), and also investigated the effects of the monomer/cosurfactant ratio. QT lowered the polymerization rates of the reaction, which showed that QT inhibited the redox pair(H_2O_2). Also, greater QT recovery was attained for nanocapsules as compared with the QT nanospheres. Sahoo et al. [80] enhanced the solubility of QT, by loading it into nanocrystal systems (QTNs) using high-pressure homogenization. Results of X-ray diffraction (XRD) and DSC studies showed that the QTNs exhibited crystalline behavior after high-pressure homogenization. It was found that the solubility of QTNs was much higher than that of the free QT at pH 6.8 and 1.2. The AOA and reducing ability of the QTNs were more efficient than the free QT. Parmar et al. [81] examined the solubilization efficacy of four phenolic antioxidants including p-hydroxybenzoic acid (PHBAA), syringic acid, sinapic acid, and QT loaded into micelles at different temperatures and pH and in the presence of NaCl. The critical micellar concentration (CMC), critical micellar temperature (CMT), and cloud point (CP) of micelles reduced due to solubilization and interfaces of antioxidants. The solubilization was preferred at higher temperature and pH and in the existence of NaCl and followed the order PHBA > syringic acid > sinapic acid > QT. Furthermore, the structure and hydrophobicity of antioxidants were found to be principal factors for their interaction and positioning in the micelles. Landi-Librandi et al. [58, 59] prepared blank and flavonoid-loaded liposomes and evaluated their antioxidant activity. Results showed that AOA of flavonoids was dependent on concentration and chemical structure of active compound. QT and MT were the most active flavonols, followed by KF and galangin. Further, it was concluded that antioxidant-loaded liposomes may be promising tools for therapy of diseases where oxidative stress is involved. Bennet and Kim [1] synthesized QT nanoemulsion (QTNE) comprising of poly (lactic-co-glycolic acid) (PLGA), hyaluronic acid (HA), and emulsifier (Tween 20) for the treatment of oxidative stress-related diseases. The FT-IR result showed no interaction between QT and PLGA that proves excellent compatibility. QTNE showed a high %EE, less toxicity, improved controlled delivery with improved transdermal drug permeation, and effective scavenging of free radicals. Weiss-Angeli et al. [82] explored the AOA of QT-loaded lipid nanocapsules (QT-LNCs) comprising of octyl methoxycinnamate and sorbitan monostearate (lipophilic core) and PCL (polymer). QT-LNCs showed higher AOA as compared to QT solution against the damage caused by H_2O_2 using the model of yeast cells (*Saccharomyces cerevisiae*). Additionally, the aqueous suspension of QT-LNCs was nontoxic and sustained AOA. Park et al. [83] developed a two-step delivery system to enhance transdermal permeation of QT and its glycoside RT. Further, QT/RT-loaded liposome-in-hydrogel complex systems were prepared by incorporating ceramide liposomes into cellulose hydrogel. RT showed greater %EE and better in vitro release properties than QT. However, QT demonstrated greater skin permeability than RT. Also, liposome-in-hydrogel complex systems improved skin permeability of QT and RT compared to control (phosphate buffer, pH 7.4) or single systems of hydrogel or liposome. Bose et al. [84] prepared solid lipid (glyceryl dibehenate)-based nanosystem (SLN) using a probe ultrasonication method for topical delivery of QT. In vitro release studies exhibited biphasic release of QT from the SLN formulation, with an initial burst release followed by prolonged release for up to 24 h. In vitro permeation studies using full-thickness human skin showed higher concentration of QT localized within the skin as compared to a control formulation with particles in the micrometer range. Such accumulation of QT in the skin is highly desirable since the efficacy of QT in delaying UV radiation-mediated cell damage and eventual necrosis mainly occur in the epidermis. Yagolnik et al. [85] investigated the AOA of flavonoids (QT, CC, and TAX) and prepared their liposomes conjugated with iron (III) forming complex. It was observed that the lipophilicity of complexes depends on an iron:flavonoid ratio and increases at a ratio of 1:1, while complexes with ratio of 2:1 were the most effective to slow down the lipid peroxidation and restore radical DPPH. Also, in a phospholipid bilayer, the process of complex formation of flavonoids with iron cations may differ from those in water solution. Tavano et al. [86] developed QT-, GA-, ascorbic acid (AA)-, and curcumin (CUR)-loaded niosomes as single agents and in combination. Results suggested that the co-encapsulations of GA/CUR and AA/QT mixture not only influenced their physicochemical properties and their %EE with respect to the formulations containing the single antioxidant but also enhanced the release of the molecules, and their mixture promoted the capability of reducing free radicals, due to a synergistic AOA. Jeon et al. [87] developed multilayered QTLs based on layer-by-layer technique with deposition of hyaluronate-chitosan for transdermal delivery of QT. Membrane stability in the presence of the surfactant Triton X-100 increased with an increase in bilayers as compared to uncoated liposomes. Multilayered QTLs enhanced the release and AOA of QT. In an in vitro skin permeation study, negatively charged (HA-CH)-Ls and positively charged CH-Ls were observed to have similar skin permeability, which were superior to uncoated liposomes. Linkeviciute et al. [88] examined the efficiency of mixtures of soy phosphatidylcholine (SPC) and glycerol dioleate (GDO) as encapsulation matrices for encapsulating QT. Results showed that SPC/GDO-based formulations incorporated QT in high amounts and served as liquid crystalline delivery vehicles in the form of bulk phases or colloidal dispersions. Also, the nanoencapsulation of QT into these matrix systems enhanced its AOA. Gupta et al. [89] prepared QT-conjugated poly (β-amino esters) (PβAE) nanogels (QT-P-NGs). Further, these spherical QT-P-NGs were coated with PEG derivates, which enhanced their aqueous stability and stealth characteristics. Moreover, uniform release of QT was observed over 45–48 h due to PβAE ester hydrolysis under physiological conditions; also QT-P-NGs retained AOA of QT over prolonged period. Zorzi et al. [90] prepared NE of the ethanolic extracts of *Achyrocline satureioides* (*A. satureioides*) (ASNE) and isolated QT (QTNE). ASNE showed lower lipoperoxidation as compared to QTNE, though no significant difference between ASNE and QTNE for lipoperoxidation was observed. Conclusively, this study showed the formulation effect on the physicochemical properties of NEs as well as on the skin retention and AOA of QT. Balestrin et al. [11] developed topically applied antioxidant hydrogels composed of ASNE. ASNE-doped hydrogels (HASNE) exhibited a non-Newtonian pseudoplastic behavior. A higher release of 3-o-methylquercetin (3-MQT) from ASNE was observed when compared with HASNE. From the Franz diffusion assay, it was observed that the amount of formulation provided to donor compartment played a crucial role. Also, higher retention of 3-MQT was observed as compared with the other flavonoids (QT and LT). The overall results showed that ASNE and HASNE prevented oxidative stress on the skin due to UVA/UVB light. Kaur et al. [91] prepared QT-loaded o/w nanoemulsion (QTNE) using food-grade surfactants (saponin and Tween 80). Results showed that rate of degradation decreased from 9 ± 1%, 11 ± 1% at pH 7.4, 8.0, respectively, as compared to 42 ± 2% in water/ethanol system. The free radical scavenging activity of DPPH was compared between QTNE and CURNE. QT showed lowest IC_{50} values as compared to CUR and AA and hence showed strong AOA. Also, QT showed strong interaction with sodium cholate (bile salt), which may be due to more electron density on oxygen atom of carboxylate ion. Rezaei-Sadabady et al. [92] prepared QTLs to increase the hydrophilicity and drug delivery capability of QT. Results showed that QTLs improved the solubility and bioavailability of QT and was used as an effective antioxidant for ROS protection within the polar cytoplasm, and the nano-sized QTLs enhanced the cellular uptake (cancer cell human MCF-7). Cell viability and ROS assays established that QT penetrated into the polar medium inside the cells and protected them against the highly toxic and deadly belongings of cumene hydroperoxide. The activity of QTLs was better than that of free QT at equimolar concentrations. Altunkaya et al. [93] investigated the effect of pH on the AOA of combinations of *lettuce extract* (LE) with QT, green tea extract (GTE), or grape seed extract (GSE) using direct electron spin resonance (ESR) spectroscopy and in L-a-phosphatidylcholine liposome peroxidation assay. The phenolic antioxidants increased the radical scavenging effect with increase in pH values. QT, GTE, and GSE showed synergistic effects with LE against oxidation of peroxidating liposomes and QT showed the most. It was concluded that the pH-dependent increase of the AOA of QT, GTE, and GSE is due to an increase of their electron-donating ability upon deprotonation and to their stabilization in alkaline solutions leading to polymerization reaction. Moretti et al. [94] assessed the in vitro effects of QT and QTLs on motility, viability, and chromatin integrity of swim-up selected human sperm. Antioxidant power was assayed against tert-butyl hydroperoxide-induced lipid peroxidation (LPO) using C11-BODIPY581/591 fluorescent probe and transmission electron microscopy. QTLs decreased the toxicity for sperm motility and viability and increased DNA damage compared to free QT. QT alone appeared to be more effective than QTLs. Anwer et al. [95] developed PLGA-coated QT polymeric nanoparticles (PLGA-QT-PNPs) by single emulsion-solvent evaporation technique. SEM images of PLGA-QT-PNPs established that prepared PNPs were spherical in shape with a smooth surface. In vitro results revealed that PLGA-QT-PNPs showed better AOA as compared to that of free QT. Furthermore, in vivo diuretic activity of PLGA-QT-PNPs was much superior as compared to pure QT. Aluani et al. [96] developed and evaluated the safety profile of chitosan/alginate-coated QT nanoparticles (C/A-QTNPs). In vitro results revealed that the C/A-QTNPs exhibited no cytotoxicity and caused no cell injury and membrane damage in Hep G2 cells and isolated hepatocytes. Also, in vivo results showed that C/A-QTNPs did not alter the body weight, relative weight of rat liver, liver histology, hematology, and biochemical parameters. The AOA of C/A-QTNPs, formulated with higher amount of chitosan, was greater than the C/A-QTNPs, formulated with higher amount of sodium alginate. Hence, C/A-NPs could be used as safe drug delivery carrier for QT against oxidative stress-related damages. Huang et al. [97] prepared liposomes encapsulated with three flavonoids (QT, LT, and KF) and investigated their AOA as well as their interaction mechanisms. It was observed that the loading of flavonoids into liposomal membrane was strongly dependent on their structure and loading concentration. Liposomes loading QT and LT exhibited relatively small size and homogeneous suspension compared to those loading KF. After encapsulation, QT displayed the strongest DPPH scavenging and lipid peroxidation inhibition capacity, followed by KF and LT. Singh et al. [98] studied the interactions between flavonoids (QT and KF) and surfactants, sodium bis(2-ethylhexyl) sulfosuccinate (AOT), and sodium bis (2-ethylhexyl) phosphate (NaDEHP). The results obtained from UV-visible absorption, fluorescence emission, and differential pulse voltammetry (DPV) studies revealed that B and C rings of QT/KF get solubilized in AOT micelles, while opposite results demonstrate that A and C rings are solubilized in NaDEHP micelles. Dynamic light scattering (DLS) measurements showed that the hydrodynamic diameter (Dh) of AOT and NaDEHP micelles in the presence of QT/KF was increased. Furthermore, radical scavenging activity (RSA) revealed that the AOA of QT/KF was enhanced in NaDEHP micelles while it decreased in AOT micelles. Rodriguez et al. [99] prepared QT nanoliposomes (QTNLs) coated with rice bran phospholipids for its efficient delivery and controlled release, protection of its structural stability, and enhancement of its bioactivity. QTNLs appeared to be spherical in shape with high %EE. Also, QTNLs exhibited a slow, restricted release pattern in simulated gastric fluid (SGF) and an initial burst release followed by a slow constant releasing pattern in simulated intestinal fluid (SIF). A 1004-fold increase in the DPPH radical-scavenging activity was observed in QTNLs, compared to free QT. Similarly, QTNLs enhanced the antiangiogenic activity of QT by two- to fivefold. Singh et al. [100] incorporated insulin (IL) and QT into liquid crystalline nanoparticles (IL-QT-LCNPs). IL-QT-LCNPs showed excellent stability in SGF. IL-QT-LCNPs exhibited higher uptake of fluorescein isothiocyanate in Caco-2 cells. Significant decrease in oxidative stress was confirmed by reduction in malondialdehyde level. Overall,

IL-QT-LCNPs not only improved the poor oral bioavailability of IL and QT but also prevented the generation of ROS, responsible for diabetes-mediated complications. Riva et al. [15] prepared QT-loaded Phytosome® (QTPs) and compared its significance with free QT in terms of solubility in SGF and oral absorption in a randomized crossover pharmacokinetic study of healthy volunteers. In vitro and in vivo results suggested that QTPs improved solubility and oral absorption as compared to free QT in healthy volunteers in a human clinical study. So, QTPs could be used as a ROS scavenging agent, hence enhancing the AOA. Farrag et al. [101] prepared QT-loaded starch (originated from various botanical sources) NPs (QT-S-NPs) by nanoprecipitation technique, using 0.1 M HCl as non-solvent. The QT-S-NPs were spherical in shape, and their sizes varied depending on the origin and the concentration of the starch solution. In vitro results showed that starch origin affected the %EE of QT, release kinetics, and the AOA of the QT-S-NPs. The QT-S-NPs from cereal origin exhibited the lowest %EE as compared with QT-S-NPs from tuber and legume origin. Further, the release kinetics of the QT from the QT-S-NPs was controlled primarily by Fickian diffusion, revealed from Peppas-Sahlin model. Najafabadi et al. [102, 103] studied the effect of QT (antioxidant molecule)-conjugated superparamagnetic iron oxide nanoparticles (QT-SPION) against the prevention of weight loss. Results showed that rats fed with 50 and 100 mg/kg SPION indicated a major weight loss while those that fed with 50 mg/kg QT-SPION did not, although weight loss was observed in rats treated with 100 mg/kg of QT-SPION. It was concluded that QT prevented weight loss due to the presence of SPION.

2.5 Brain Targeted

Sarkar and Das [104] investigated the AOA of QT-loaded in mannosylated (QC 7.2 mmol/kg b.wt.) liposomes (QT-MS-Ls) using an in vivo model of cerebral ischemia and reperfusion on Sprague-Dawley rats (old and young). The activity of various antioxidant enzymes including superoxide dismutase, catalase, glucose-6-phosphate dehydrogenase, glutathione reductase, and glutathione S-transferase activities was lowered in the brain of normal-aged rat. Further, these antioxidant enzymes were reduced in the brain of aged rat by the induction of cerebral ischemia and reperfusion. Further, treatment with QT-MS-Ls resulted in a significant protection of the activities of antioxidant enzymes and a marked inhibition of cellular edema formation in neuronal cells of young and old rats. Matsuzaki et al. [105] studied the effects of nordihydroguaiaretic acid (NDGA), rifampicin (RIF), tannic acid (TA), and QT on amyloid β-protein (Aβ) aggregation, since aggregation and deposition of Aβ are considered to be a crucial step in the etiology of Alzheimer's disease (AD) and fibril destabilization in the presence of GM1-containing raft-like liposomes. Results showed that NDGA and RIF inhibited the binding of Aβ to GM1 liposomes by competitively binding to the membranes and/or direct interaction with Aβ in solution. Conjugation of Aβ with NDGA, RIF, and QT in the presence of GM1 liposomes resulted in elongate particles, whereas the presence of TA yielded protofibrillar structures. TA and RIF also destabilized fibrils. Priprem et al. [106] evaluated the anxiolytic and cognitive activities of QT (antioxidant molecule) in male Wistar rats. Oral QT (300 mg/kg body weight/day) was compared with oral and intranasal QTLs (20 μg/day). Results showed that both conventional and QTLs exhibited anxiolytic and cognitive-enhancing effects. A lower dose and a faster rate were observed with intranasal QTLs when compared with oral QT, conventional and liposomal. The intranasal QTLs were potent in the delivery of QT to the central nervous system and hence could be used for the treatment of various neuropharmacological disorders. Rivera et al. [107] studied the protective effects of QTLs in the rat model after 1 and 4 h of arterial occlusion. Results showed that after administration of QTLs (30 min), significant protective effect against ischemic lesion was observed with an increase in numbers of cells in the striatum and cortex, together with a partial reversal of motor deficits. Also, glutathione-stimulating hormone (GSH) levels reduced after ischemia in the ipsilateral striatum and cortex, and the QTLs reversed these effects after the occlusion. It was concluded that endogenous brain GSH was a critical factor for defense mechanisms after ischemia, as an important mediator of the defensive effects of the QTLs. Dhawan et al. [108] formulated QT-SLNs comprising of compritol (lipid) and Tween 80 (surfactant) using micro-emulsification technique followed by optimization using 3^2 central composite design. QT-SLNs enhanced the permeation of QT across the BBB to get into CNS and ultimately improved the therapeutic efficacy of QT for the management of AD. In vivo results showed that the rats treated with QT-SLNs exhibited better memory retention as compared to pure QT-treated rats. Ghosh et al. [109] evaluated the therapeutic efficiency of QTNCs in combating ischemia-reperfusion-induced neuronal damage in young and aged Swiss albino rats. Results showed that QTNCs reduced the mitochondrial membrane microviscosity and amplified ROS activity in different brain regions of young and aged rats. Among the regions of the brain, the hippocampus part was the worst affected and showed augmented upregulation of iNOS (inducible nitric oxide synthase) and caspase-3 activity with reduced neuronal count in the CA1 and CA3 subfields of both young and aged rats. Ganesan et al. [17] discussed the most recent trends and applications in Parkinson's disease (PD), including (1) the role of phytoconstituents in decreasing oxidative stress and their bioavailability; (2) the role of nanotechnology and nanodelivery systems in decreasing oxidative stress during PD, and (3) various nano-phytoconstituents compounds (including QT, KF, and RT) and their role in PD. Bagad and Khan [110] prepared an oral drug delivery comprised of QT-loaded poly (n-butyl cyanoacrylate) NPs (QT-pBA-NPs). QT-pBA-NPs were further conjugated with polysorbate-80 (QT-pBA-P80 NPs). Results showed that in vitro release of QT-pBA-NPs and QT-pBA-P80 NPs showed an initial burst release followed by a sustained release as compared to free QT. As compared to free QT, the relative bioavailability of QT-pBA-NPs and QT-pBA-P80 NPs improved the bioavailability of QT by 2.38- and 4.93-fold, respectively. Further, the biodistribution studies in rats indicated that a greater concentration of QT was detected in the brain after the NPs were coated with P-80. Sun et al. [33, 34] prepared PLGA-coated QTNPs (PLGA-QTNPs). The cytotoxicity studies of the PLGA-QTNPs showed dose-dependent behavior on the SH-SY5Y human neuroblastoma cells and inhibited the neurotoxicity of Zn^{2+}-A42 system as well as enhanced the viability of neuron cells. The results from behavioral assays indicated that the injection of PLGA-QTNPs into APP/PS1 mice improved cognition and memory impairments. Thus, the PLGA-QTNPs could be used for the AD. Galho et al. [111] compared the effects of QTNEs and QT suspension (QTS) in a collagenase-induced intracerebral hemorrhage (ICH) in rat model. Animals treated with QTNEs showed significant enhancement in the beam walking and open-field tests. Also, QTNEs reduced the size of the hematoma, preserved the activity of glutathione S-transferase (GST), and enhanced GSH content and the total antioxidant capacity. QTS also improved locomotor activity and enhanced the GSH content and the total antioxidant capacity but comparatively less than QTNEs. Kuo et al. [112] developed QTLs grafted with RMP-7 (bradykinin analog) and lactoferrin (LAC), formed RMP-7-LAC-QT-Ls to infuse the blood-brain barrier (BBB), and release degenerated neurons, for the management of AD. It was observed that surface modification with RMP-7 and LAC improved the ability of QT to cross through BBB without persuading strong toxicity and injuring the tight junction. In addition, RMP-7-LAC-QT-Ls expressively abridged Aβ-induced neurotoxicity and enhanced the viability of SK-N-MC cells (human neuroblastoma cells). Ahmad et al. [113] prepared QTNEs to enhance the bioavailability and evaluated its targeting through intranasal (i.n.) route in the brain. Ultra-performance liquid chromatography-mass spectroscopy (UPLC-ESI-Q-TOF-MS/MS)-based bioanalytical studies showed that the administration of QTNEs through i.n. route improved QT brain bioavailability more as compared to IV. It was observed that QTNE was able to pass through the BBB and enters the brain favorably, decreasing the unwanted systemic effects, so it could be used as a potent agent for the management of cerebral ischemia. Ghosh et al. [114] synthesized QTNCs which efficiently counteracted cerebral ischemia-reperfusion-induced cell death and neurodegeneration in young and aged rats. QTNCs showed higher brain uptake and significant mitochondrial localization post cerebral ischemia-reperfusion. It was observed that QTNEs preserved mitochondrial structural and functional integrity through inhibiting ROS, thus modulating mitochondrial ROS-mediated apoptotic cell death in young and aged rats. Oliveira et al. [115] isolated QT and biapigenin (BA) from *Hypericum perforatum* for potential treatment of neurodegenerative disorders. The isolated compounds were encapsulated into PCL NPs, forming QT/BA-PCL NPs. Results from in vitro assay showed that the QT/BA-PCL NPs exhibited an initial burst release followed by sustained release. QT/BA-PCL NPs protected HepG2cells from inherent compound toxicity at higher concentrations. QT/BA-PCL NPs showed more efficiency in protecting HepG2 cells against tert-butyl hydroperoxide-induced toxicity. Rishitha and Muthuraman [116] evaluated the significance of QT-SLNs in pentylenetetrazole (PTZ)-induced cognitive impairment of *Danio rerio* (zebra fish) model. Results showed that the QT-SLNs reduced the effect in PTZ-induced neurocognitive impairments along with improvement of biochemical variations. So, QT-SLNs, due to its potential AOA and anti-lipid peroxidative and acetylcholinesterase inhibitory actions, could be used as for the management of various neurodegenerative disorders like AD and Parkinson's diseases. Najafabadi et al. [102, 103] fabricated QT-SPIONs coated with dextran by nanoprecipitation method and evaluated their targeting in the plasma and brain of Wistar male rats. Results confirmed that SPIONs as a targeted drug delivery carrier improved the bioavailability of QT in the brain (tenfold) higher as compared to free QT and could preferably be used for the management of neurodegenerative disorders. Also, results showed that biodistribution of QT-SPIONs coated with dextran could possibly be managed in the brain by means of an external magnetic field. Amanzadeh et al. [117] examined the effects of free QT in conjugation with SPIONs (QT-SPIONs) on learning and memory in healthy rats. Also, QT target proteins involved in learning and memory using Morris water maze (MWM) and computational methods were examined. Results of MWM showed an enhancement in learning and memory of rats treated with either QT or QT-SPIONs. Improved learning and memory functions using QT-SPIONs revealed that QT-SPIONs increased the bioavailability of QT. Comparative molecular docking studies showed the greater binding affinity of QT with proteins including RSK2, MSK1, CytC, Cdc42, Apaf1, FADD, and CRK. Katebi et al. [118] combined NGF and QT with various concentrations of SPIONs (NGF/QT-SPIONs) and evaluated their therapeutic efficacy against degenerative diseases such as Parkinson's and Alzheimer's. Results of MTT assay revealed that NGF/QT-SPIONs exhibited no cytotoxicity, and morphological analysis displayed a major growth in neurite length when PC12 cells were incubated in NGF/QT-SPION solution. Also, a higher level of β3-tubulin expression was detected in these cells when treated with both QT and SPIONs. Kuo et al. [119] developed cardiolipin (CL)- and phosphatidic acid (PA)-incorporated liposomes which promoted their ability against hyperphosphorylation of tau protein, and a transactivator of transcription (TAT) peptide, to infuse through BBB for transporting NGF, rosmarinic acid (RA), CUR, and QT (TAT-NGF-RA-CUR-QT-CL/PALs). Results revealed that this optimized TAT-NGF-RA-CUR-QT-CL/PALs enhanced the viability of SK-N-MC cells treated with Aβ-fibrils and prohibited Wistar rat brain from generating hyperphosphorylated tau. TAT-NGF-RA-CUR-QT-CL/PALs also improved the expressions of p-ERK5 and p-cyclic adenosine monophosphate response element-binding protein. Liu et al. [120] synthesized a QT-modified P-80-coated gold (Au)/palladium (Pd) core-shell structures. Results showed that QT-modified shells activated autophagy of SH-SY5Y cells, promoted the fusion of autophagosomes and lysosomes, accelerated the clearance of Aβ, and protected SH-SY5Y cells from Aβ-induced cytotoxicity damage. Additionally, QT modified shells also showed better biocompatibility and high BBB permeability. So, it was concluded that these shells could be used as a potential autophagy inducer to treat AD. Mandic et al. [121] prepared flavonoids (QT/MT) loaded in mesoporous silica NPs (QT/MT-MSNs) and investigated the relationship between flavonoid structure and protective activity toward oxidative stress. QT showed the more enhanced and prolonged protective AOA as compared to MT. These results concreted the way concerning the progress of new and enhanced markers for oxidative stress-associated neurological disorders.

2.6 Cardioprotective Effect

Soloviev et al. [122] assessed the effects of phosphatidylcholine-coated QTLs (P-QTLs) on peroxynitrite (ONOO-)-induced cardiac arrhythmias. Experiments were performed using various experimental models, including isolated rat papillary muscle, Langendorff-perfused rat hearts, and anesthetized animals. It was observed that i.v. injection of ONOO- promoted the presence of a high frequency of arrhythmias in anesthetized rats and decreased MLVP. Also, the P-QTLs restored normal cardiac contractility in both isolated tissues and anesthetized animals. Dugas et al. [123] designed PLGA-coated QT/RES-PNPs (P-QT/RES-PNPs) firmly attached with the surface of balloons (catheters), followed by linkage with the negatively charged bilayer of the vascular wall. P-QT/RES-PNPs showed admirable biocompatibility with less or no cytotoxicity. Also, P-QT/RES-PNP-released QT and RES showed sustained release pattern. Hence, these systems could possibly be used for the management of peripheral artery disease. Long et al. [124] fabricated single-cell analysis platform (SCA) for the evaluation of platelets linked to single human umbilical vein endothelial cell (HUVEC) by consuming *E-selectin* (adhesion molecule) on the injured HUVEC as the marker site and integrated electrochemiluminescence (ECL) with the ultrasensitive AuNPs coated zinc-coadsorbed carbon quantum dot Au-ZnCQDs nanoprobes. The Au-ZnCQDs nanoparticles (AgNPs) exhibited a tenfold signal amplification. The SCA efficiently monitored the linkage between single HUVEC and platelets in the deficiency and existence of collagen activation and also showed a good analytical performance with linear range from 1 to 15 platelets. Simultaneously, HUVEC downregulated the expression of adhesion molecules when treated with QT molecules.

2.7 Hepatoprotective Effect

Mandal et al. [125] formulated QT-doped mannosylated liposomes (QT-mLs) and tested in vivo against carbon tetrachloride (CCl_4)-induced liver oxidative damage in rats. QT-mLs reduced the serum enzyme levels (glutamate pyruvate transaminase and alkaline phosphatase) and hepatocellular conjugated diene levels by CCl_4 induction, whereas the same amount of free QT was found ineffective. It was observed that QT-mLs relapsed the injury produced by CCl_4 in the liver. So, the biochemical alterations produced in the serum and liver in CCl_4-treated rats were prohibited by pretreatment with QT-mLs. Mandal and Das [126] investigated the effects of QT (isolated form plant origin) in free, QTLs and galactosylated liposome (QT-gLs) in the rat induced with CCl_4 hepatotoxicity. Results showed that, among the three different forms of QC tested, only QT-gLs exhibited significant protective effects against CCl_4-induced hepatic oxidative damage. Also, QT-gLs enhanced the endogenous AOA, simultaneously reduced the membrane microviscosity formed in membrane in CCl_4-treated rat liver, hence QT-gLs could efficiently be used in the management of human hepatic diseases. Mandal et al. [127] optimized and evaluated the therapeutic efficacy of QT-gLs, in combating *As*-induced hepatic fibrogenesis. Results of histopathological and histochemical analysis showed that QT-gLs-treated rat models exhibited maximum protections from hepatocellular and fatty metamorphosis, necrosis, Kupffer cell hyperplasia, fibrosis, and deposition of collagen contents. Thus, QT-gLs could be used as an effective therapeutic approach for sodium arsenite (NaAsO2)-induced hepatic fibrogenesis. Ghosh et al. [128, 129] evaluated the therapeutic efficiency of QTLs in combating *As*-induced oxidative damage in hepatocytes and brain cells in rat model. In *As*-induced rat models, antioxidant levels in hepatic and neuronal cells were significantly condensed. QTLs reduced antioxidant levels in the liver and brain expressively in *As*-induced rats. Also, QTLs reduced *As*-induced hepatic hydroxyproline (HP) level and prevented the *As*-induced upregulation of cytochrome c (CYc) expression in the liver and brain. Thus, QTLs exhibited potential therapeutic activities against *As*-induced hepatic and neuronal cell damage in rats. Wan et al. [130] examined the effects of QT on hepatitis and hepatic fibrosis prompted by immunological mechanism. It was observed that QT significantly deteriorated the levels of alanine aminotransferase (ALT) and aspartate aminotransferase (AST), though it evidently reduced the pathologic alterations in the liver. Further, histopathological analyses showed that treatment with QT significantly reduced collagen deposition, pseudolobuli expansion, and hepatic stellate cells activation. Also, QT treatment reduced the nuclear factor kappa-light-chain enhancer of activated B cells (NF-kB), and transforming growth factor ß (TGF ß) release signified that the antifibrotic effects of QT were related with its ability to regulate NF-kB and TGF-b release. Sarkar and Sil [131] investigated the cell signaling mechanism of Fe_2O_3-NP-induced oxidative stress and cytotoxicity in vitro using murine hepatocytes as the working model. Additionally, the cytoprotective activity of QT in this pathophysiology was investigated. Results suggested that Fe_2O_3-NP-induced toxicity and cellular damage of hepatocytes via the activation of oxidative stress-responsive cell signaling and QT played a defensive role in Fe_2O_3-NP-induced cytotoxicity and apoptotic death. Schwingel et al. [132] administered RT (20 mg/kg), RES (100 mg/kg), QT (20 mg/kg), and QTNE (20 mg/kg) to BALB/c mice model, and N-acetylcysteine (NAC) as positive control and studied their therapeutic efficiency against neuro- and hepatoxicity induced by oxaliplatin (OXP). RES, RT, and QTNE prevented mechanical allodynia as compared to the OXP-treated animal group. Also, RES, QT, and QTNE decreased the liver weight, which was induced and increased due to treatment with OXP. Also, these compounds reduced the immunopositivity for the apoptosis marker caspase-3. Most importantly, only QT and QTNE reduced the neutrophil migration measured by myeloperoxidase (MPO) activity. Karthivashan et al. [133] isolated QT, KF, and AG from leaves of *Moringa oleifera* and co-loaded into single flavonosome formulation (QT-KF-AGFs) by using sequential loading technique. The QT-KF-AGFs showed no toxicity against human hepatoma cell line. It was observed that optimized QT-KF-AGFs sustained biological activities including antioxidant and hepatoprotective. Verma et al. [14] prepared QT-loaded p-aminophenyl-1-thio-β-Dgalactopryranoside (PAG)-coated N-isopropyl acrylamide (NIPAAM) polymeric nanoparticles (PAG-NIPAAM-QT-PNPs) and compared its therapeutic efficacy with free QT against CCl_4-mediated liver cirrhosis. As compared to free QT, PAG-NIPAAM-QT-PNPs more effectively restored liver membrane integrity, specified by decreased serum markers (alanine transaminase (ALT), AST, alkaline phosphatase (ALP), and LDH). Additionally, PAG-NIPAAM-QT-PNPs decreased the NF-κB and NOS-2 expression more efficiently as compared to QT. Tzankova et al. [134] fabricated C/A-QTNPs and compared its hepatoprotective and antioxidant effects with free QT against in vitro and in vivo models of hepatotoxicity. In vitro results showed that both free QT and C/A-QTNPs exhibited no cytotoxicity in human hepatoma HepG2 cells. In vivo results showed that C/A-QTNPs significantly reduced the augmented levels of serum transaminases ALT and AST, reduced the lipid peroxidation, and reproduced glutathione level. Furthermore, the protective effects of C/A-QTNPs were greater than free QT; hence, C/A-QTNPs could be used against oxidative stress-induced liver injury. Xu et al. [135] prepared PLGA-QTNPs and examined its anti-inflammatory and anti-oxidative stress effects in high-fat diet-induced kidney failure in C57BL/6 mice model. In vitro results showed that PLGA-QTNPs inhibited the upregulation of Toll-like receptor 4 (TLR4)/nuclear factor-kappa B (NF-κB) and oxidative stress, eventually alleviating podocyte injury. PLGA-QTNPs efficiently restored the metabolic disorders caused by high-fat diet and prevented hepatic injury by inhibition of TLR4/NF-κB and oxidative stress. Abdelhalim et al. [136] prepared QT- and arginine (Arg)-loaded AuNPs (QT-ARG- AuNPs) and evaluated its therapeutic efficiency against hepatotoxicity caused by GNPs. Results showed that QT-ARG-AuNPs considerably induced modifications in the biochemical parameters including increase in malondialdehyde, gamma-glutamyl transferase (GGT), ALT, ALP, and total protein levels and decrease in GSH level. QT-ARG-AuNPs also reduced liver enzymes, and the oxidative stress, consequently, improved the liver damage and hepatotoxicity induced by AuNPs. Eftekhari et al. [137] prepared QTNPs by nanoprecipitation method and its hepatoprotective and AOA against toxicity induced by aflatoxin B1. Results showed that as compared to free QT, QTNPs more efficiently reduced the rate of ROS formation and lipid peroxidation and enhanced cell viability, mitochondrial membrane potential, and GSH level and also exhibited significant hepatoprotective activity by reducing AST, ALT, and ALP levels. So, QTNPs showed potential activities and improved the hepatoprotective effect of QT against the cytotoxic effects of aflatoxin B1. Fadda et al. [138] evaluated the protective activities of QT or idebenone (Id) against the hepatotoxicity induced by titanium dioxide nanoparticles (TiO2-NPs). The results showed that the coadministration of QT or Id to rats treated with TiO2-NPs prominently improved the elevation in hepatic malondialdehyde (MDA), ALT, glucose, tumor necrosis factor-α (TNF-α), interleukin-6 (IL-6), immunoglobin G (IgG), and C-reactive protein (CRP) levels as compared to TiO2-NP-treated rats. Palle and Neerati [139] estimated the effect of QT and QTNPs on the pharmacokinetics of bromocriptine (BRO) in rats. The results specified that QT significantly repressed the CYP3A activity in liver and intestinal microsomes. Further, in vivo studies indicated that the amplified levels of Cmax and AUC were relatively higher in QTNE pretreated group than QT group. In addition, pretreatment with QT and QTNP expressively reduced the mean clearance and volume of distribution of BRO. Suke et al. [140] evaluated the protective effect of QTNCs against chlorpyrifos (CPF)-induced hepatic damage and immune alterations in animals. The results showed that QTNCs expressively changed oxidative stress in the liver tissue as well as liver enzyme parameters in blood and immune responses in CPF-treated rats compared to controls. Administration of QTNCs reduced biochemical and immunological parameters. Hence, QTNCs proved to be beneficial in decreasing and refurbishing hepatic oxidative damage and immune modification sustained by pesticide exposure.

2.8 Lungs Targeted

Baowen et al. [141] assessed the preventative activities of QTLs against bleomycin (BM)-induced pulmonary fibrosis. Results indicated that QTLs reduced the upsurge of total cell counts and macrophage counts in bronchoalveolar lavage fluid (BALF). QTLs also reduced the neutrophil and lymphocyte counts. In addition, as compared to BM-treated groups, QTL-treated groups decreased the TNF-α, IL-1β, and IL-6 in BALF level more significantly. Thus, QTLs reduced the BM-induced pulmonary fibrosis by the inhibition of inflammatory cytokines. Liu et al. [12, 13] tested the protective activities of QTLs against radiation-induced lung injury in a C57BL/6J murine model. QTLs efficiently decreased the MDA level and enhanced SOD and glutathione peroxidase (GSH-PX) content in the pulmonary tissues and decreased the total cell counts and inflammatory cell quantities in the BALF, plasma TNF-α and TGF-β1 concentrations, and the HP level in the lung tissues.

2.9 Skin Targeted

Fahlman and Krol [142] reported that QTLs were able to protect the skin from UV-induced lipid peroxidation. Results showed that the defensive activity of QTLs appeared mainly due to the outcome of scavenging of UV-originated radical species. Naumov et al. [143] prepared lecithin-glycine conjugated dihydroQT (DQT) liposomes (L/G-DQTLs), used for the management of glacial acetic acid-induced chemical burn. Results showed that no inflammatory reaction was detected in the wound section. Planimetry and histological studies indicated that enhanced skin restoration was attended by healing of hair follicles and sebaceous glands. Naumov and Potselueva [144] concluded that L/G-DQTLs decreased inflammatory reactions in the thermal burn associated wounds. L/G-DQTLs stabilized the endogenous antioxidant system and restricted the secondary necrotic zone in the wounds. Also, L/G-DQTLs intensified the skin regeneration abilities and repaired the hairy follicles and sebaceous glands. Tan et al. [145] prepared lecithin-chitosan-coated QTNPs (L/C-QTNPs) as topical delivery system. As compared with QT solution, the L/C-QTNPs exhibited greater permeation ability. Results of morphological evaluation revealed that the interface among constituent's L/C-QTNPs and the skin surface distinctly altered the morphology of the stratum corneum and interrupted the corneocyte layers, hence assisting the permeation and accumulation of QT in the skin. Liu et al. [12, 13] investigated the cellular protective ability of QTLs against UV radiation and also the effect of surfactants (Span 20, Tween 80, and sodium cholate) on the properties of QTLs. Further, QTLs showed high %EE and prolonged drug release. Also, cell viability in UVB-irradiated HaCaT cells was enhanced. The ROS and MDA level were also reduced. The rate of penetration of QTLs were 3.8-fold superior than that of the QT suspension. Caddeo et al. [5] developed QTLs and PEG 400-coated QTLs (P400-QTLs) and investigated their efficiency against 12-O-tetradecanoylphorbol-13-acetate (TPA)-induced skin inflammation. QTL size and lamellarity were intensely affected by the amount of PEG 400. QTLs reduced

TPA-induced tissue injuries, with a perceptible reduction of edema and leukocyte infiltration. Moreover, in vitro studies revealed an immense uptake and dispersion of P400-QTLs through dermal fibroblasts; hence, P400-QTLs could be used for the management of inflammatory skin disorders. Castangia et al. [146] fabricated QT/CUR-loaded liposomes (QT/CUR-Ls) to inhibit and reproduce skin tissue damages on chronic cutaneous pathologies. Also, penetration enhancer (PEG 400 and OramixCG110)-coated QT-CUR liposomes (P400/OCG110-QT/CUR-Ls) were attained. In vitro and in vivo results showed that QT/CUR-Ls and P400/OCG110-QT/CUR-Ls inhibited TPA-induced lesions and inflammation. Also, P400/OCG110-QT/CUR-Ls strongly inhibited the myeloperoxidase activity (used to estimate inflammation). Jangde and Singh [147] prepared QTLs by thin-film hydration method, optimized using response surface methodology (RSM). In vitro results revealed that QTLs enhanced the %EE and prolonged drug release in wound areas. Hatahet et al. [148] prepared QTNCs using inversion technique including some alterations to improve QT loading. QTNCs improved the water solubility of pure QT by more than 5000-fold and showed prolonged release kinetics. Finally, QTNCs exhibited protective effects for THP-1 cells from oxidative stress by exogenous hydrogen peroxide. Jangde et al. [149] developed a multiphase C934/gelatin-hydrogel system incorporated with QTLs (QTLs-C934/G-HG) for the management of wound healing. QTLs-C934/G-HG enhanced wound healing in albino rat with major decrease in wound closure time as compared to conventional dosage form. Hatahet et al. [150] compared the ability of three formulations loaded with QT, including QTLs, QTNCs, and QT-smartCrystals® (QTSCs). Drug loading followed the sequence QTSCs > QTNCs > QTLs. Furthermore, no toxicity was detected in HaCaT cells for QT-loaded formulations. Results of cellular interactions with HaCaT (keratinocytes) and TPH-1 (monocytes) cell lines showed that QTSCs permitted the superficial deposition of QT on top of the skin, hence showing good ability as a sunscreen product. Although QTNCs allowed QT delivery to viable epidermis, it could also be used for the management of skin inflammatory disorders. Seong et al. [151] prepared multilayered liposomes conjugated with chitooligosaccharide (COS) and N-succinyl-chitosan (NSC) and loaded with QT (COS/NSC-QTLs). In vitro results showed that the COS/NSC-QTLs were more stable against surfactants and exhibited a relatively high drug release at pH 5.5 than the QTLs. Raghuwanshi et al. [152] explored that HSP70–1 (heat-shock proteins) played crucial role in psoriasis treatment. MT, QT, and EA were isolated from *Woodfordia fruticosa* (*W. fruticosa*) *flowers*, investigating their anti-psoriatic activity using combination of bioinformatics together with ethnopharmacological approach. Further, *W. fruticosa* extract-conjugated GNPs (*Wf*-GNPs) were synthesized. Synthesized *Wf*-GNPs were amalgamated with Carbopol®934 to form *Wf*-GNPs-C934 gel. Results showed that *Wf*-GNPs-C934 gel decreased serum cytokines (TNF-α, IL-22, and IL-23) levels, epidermal thickness, parakeratosis, and hyperproliferation of keratinocytes.

2.10 Anticancer Activity

2.10.1 Lungs

Yuan et al. [153, 154] developed PEGylated (polyethylene glycol 4000) liposomal quercetin (pL-QT) and investigated in vivo distribution of pL-QT and also antitumor efficiency (in vitro-in vivo). The antitumor activity of pL-QT was assessed in the immunocompetent C57BL/6 N mice bearing LL/2 Lewis lung cancer and in BALB/c mice bearing CT26 colon adenocarcinoma and H22 hepatoma. In vitro results revealed that pL-QT induced apoptosis and in vivo results showed that pL-QT expressively reduced the tumor growth (dose-dependent manner). The liposomal formulation could efficiently accumulate in tumor tissues, extend the circulation time of QT in vivo, efficiently constrain numerous types of tumor growth, and lengthen the residual time of tumor-bearing mice. Verma et al. [155] prepared PLGA-coated Fe_3O_4 magnetic nanoparticles (MNPs) loaded with QT (PLGA-QT MNPs) and investigated its targeting activity against lung cancer cells through intranasal delivery. Results showed that the developed PLGA-QT MNPs were biocompatible and showed no cytotoxicity against treated human lung epithelial cells. Moreover, in vivo studies indicated that intranasal treatment of mice with PLGA-QT MNPs showed greater endurance as measured by GLUT and IL-6 release assays. Also, PLGA-QT MNPs efficiently decreased the number of viable A549 cells; hence, PLGA-QT MNPs could be used as a carrier for targeted delivery against lung cancer. Zhou et al. [156] prepared RGD (short peptide containing arginine-glycine-aspartic acid)-modified NLs containing QT-DSPE-PEG2000 forming DSPE-PEG2000-RGD/QT NLs for treating lung cancer. The encapsulation ratio and drug loading of NLs were 89.2% ± 7.4% and 9.2% ± 1.3%, respectively, and mean diameter was 93.4 ± 7.2 nm. In vitro studies showed a relatively slow QT release profile from NLPs. Moreover, in vivo imaging results showed better tumor targeting capacity of RGD-modified LPs. Hence, DSPE-PEG2000-RGD/QT NLs exhibited a substantial antitumor activity in mice with A549 tumors. Arbain et al. [157] prepared palm oil ester (POE)-based QTNE-formed POE-QTNE and optimized by using D-optimal mixture design (DOMD) for pulmonary delivery. In vitro studies showed that aerosol delivery of the optimized NE had aerosol output, aerosol rate, and median mass aerodynamic diameter of 99.31%, 0.19 g/min and 4.25 μm, respectively. Results of the physical characteristics and efficacy for aerosol delivery of POE-QTNE could be used for lung cancer treatment. Riaz et al. [158] prepared T7 surface-functionalized QT liposomes (T7-QTL) for lung cancer therapy by targeting transferrin receptors, which are overexpressed and restricted to tumor cells. In vitro studies showed that 2% T7-QTL demonstrated improved cytotoxicity (~3-fold) and enhanced apoptosis induction and S-phase cell-cycle seizure. T7-QTL expressively improved the anticancer activity of QT and lifespan of mice in orthotopic lung tumor-bearing mice via pulmonary administration. Arbain et al. [159] prepared QTNE (oil-in-water; O/W) in the presence of POE/ricinoleic acid (oil phase) and optimized by using DOMD for the delivery of QT to lungs. The oral bioavailability was enhanced, and the release of QT was prolonged from the NE formulation when delivered from pulmonary route. The cytotoxicity study confirmed that the optimized QTNE could possibly induce cytotoxicity toward A549 lung cancer cells deprived of distressing the normal cells.

2.10.2 Breast Targeting

Wong and Chiu [160] formulated a novel system of QT-VT NLs for the management of breast cancer. Results showed that the co-loaded VT and QT (1:2) in NLs revealed enhanced antitumor efficiency and reduced toxicity in estrogen receptor (ER), progesterone receptor (PR), and trastuzumab-insensitive xenograft models synergistically. It was concluded that this drug delivery system could further develop into a conceivable clinical management option for HER-2 (human epidermal growth factor receptor-2) overexpressing breast cancer. Kumar et al. [161] prepared dextran (DT)-coated MNPs loaded with QT, formed DT-QT-MNPs, and investigated its targeting activity against breast cancer cells. The results of MTT assay showed that no toxicity was inhibited by DT-MNPs against MCF7 cells, while DT-QT-MNPs enhanced the cytotoxicity considerably as compared with pure QT. The treatment of MCF-7 cells with DT-QT-MNPs showed noticeable alterations in cellular morphology which signified the applicability of DT-QT-MNPs as a noble carrier for the treatment of breast cancer. Zafar et al. [162] developed a Eudragit RL100-based co-loaded nanoparticulate system comprising of mitoxantrone dihydrochloride (MDC) along with QT (for Pgp) and hesperetin (HP, for BCRP) (both efflux transporter inhibitors), for targeted delivery of MDC/hyaluronic acid (HA) and concurrently reversing the MDR in breast cancer. It was observed that IC_{50} values were reduced for inhibitor-doped PNPs in various cell lines. The PNPs also exhibited prolonged release with decreased macrophage uptake and enhanced residence time in blood. The enhanced cellular uptake of HA-coated PNPs in MCF7 cells showed their targeting potential. The HA-coated PNPs along with QT/HP exhibited greater tendency for targeted chemotherapy in CD44 overexpressing MDR breast cancer. Sharma et al. [163] synthesized mPEG-PLA encapsulated QT nanoparticles (mPEG-PLA-QTNPs) for their anticancer efficiency. In vitro results exhibited that mPEG-PLA-QTNPs competently induced apoptosis in triple-negative breast cancer cell line (MDA-MB-231) with greater quantity of QT in cell lysate treated with mPEG-PLA-QTNPs as compared to free QT. In vivo studies revealed that mPEG-PLA-QTNPs restricted the tumor growth of xenograft mice model. The terminal deoxynucleotidyl transferase (TdT) dUTP nick-end labeling (TUNEL) assay revealed more existence of apoptotic cells in mPEG-PLA-QTNP-treated tumors rather than free QT at same dose. Pandey et al. [164] synthesized QT-loaded PLA nanoparticles (PLA-QTNPs) for the establishment of its anticancer efficacy against human breast cancer cells. Results showed that PLA-QTNPs retained sustained delivery of QT from the polymer matrix due to late diffusion and stronger interface. In vitro cytotoxicity study indicated that PLA-QTNPs killed almost ~50% breast cancer cells. So, PLA-QTNPs could be efficiently used as carrier for the management of breast cancer. Lv et al. [165] targeted co-delivery of DOX and QT to tumor cells in the presence of biotin (B)-conjugated PEG- PCL NPs for the management of MDR breast cancer. It was observed that tumor-targeting by DOX-QT-B-NPs showed significant benefits associated to management with the free DOX-QT-NPs, DOX-B-NPs, and QT-B-NPs against MDR breast cancer cells (MCF-7/ADR cells). The DOX-QT-B- NPs significantly improved the in vivo anticancer efficacy of DOX-QT in MDR MCF-7/ADR xenograft model. Suksiriworapong et al. [166] investigated the effect of the different hydrophobic chain lengths of PCL-TPGS copolymers on the properties of NPs and efficacy of QT to get delivered to SKBR3 breast cancer cells. It was observed that longer PCL chains showed greater crystallinity and reduced the crystallinity of TPGS chain, hence enhancing the particle size of PCL-TPGS NPs from 20 to 205 nm, and improved the loading efficiency of QT but simultaneously retarded the release of QT. The PCL-TPGS NPs induced the toxicity of QT against SKBR3 cells almost by 2.9 times as compared to the QT solution; thus, PCL-TPGS NPs loaded with QT might be used for the treatment of breast cancer. Minaei et al. [167] explored the activity of nano-QT (phytosome) in DOX-treated apoptosis. MTT assay showed the antiproliferative effect of QT-based phytosomic formulations. Finally, results revealed that co-treatment with nanoQT-DOX phytosomes enhanced the percent apoptosis significantly against MCF-7 breast cancer cells. Tavano et al. [168] co-encapsulated niosomal delivery system of DOX-CUR and DOX-QT for the management of human breast cancer. Results showed that this formulation resulted in extended blood circulation and instinctively accumulated at the tumor site. Moreover, the in vitro studies showed enhanced anticancer activity of niosomes loaded with a DOX-CUR/QT as useful carriers for the treatment of breast tumor. Balakrishnan et al. [3] developed QT-GNPs which significantly reduced protein expression of vimentin (type III intermediate filament (IF) protein that is expressed in mesenchymal cells), N-cadherin (transmembrane protein), transcriptional factors (Snail, Slug, Twist), MMP-2 (matrix metalloproteinase-2), MMP-9 (matrix metalloproteinase-9), p-EGFR (phosphorylated-epidermal growth factor receptor), VEGFR-2 (vascular endothelial growth factor), p-PI3K (phosphoinositide 3-kinase), Akt (protein kinase B (PKB), also known as Akt), and p-GSK3β (glycogen synthase kinase 3 beta) and improved E-cadherin protein expression in response. Results showed that QT-GNPs repressed relocation and invasion of MCF-7 and MDA-MB-231 cells more expressively than free QT. Also, QT-GNPS-treated HUVECs showed decreased cell viability and inhibited new blood vessel development. It was concluded that QT-GNPs suppressed epithelial-mesenchymal transition (EMT), angiogenesis, and metastasis of the breast cancer cells. Balakrishnan et al. [27] prepared QT-GNPs and evaluated the effects of QT-GNPs in MCF-7 and MDA-MB-231 breast cancer cell lines. It was observed that QT-GNPs showed lower IC_{50} value compared to free QT. QT-GNPs increased apoptosis with amplified nuclear condensation. QT-GNPs upregulated caspase-3, downregulated Bcl-2, and inhibited the epidermal growth factor receptor (EGFR) and its downstream signaling molecules PI3K/Akt/mTOR/GSK-3β. Thus, QT-GNPs induced apoptosis and suppressed EGFR signaling; hence, QT-GNPs could be used for the treatment of breast cancer. Zuo et al. [169] investigated the antitumoral effects of 7-O-geranylquercetin (GQT) combined with survivin siRNA (siSuvi) or IL-10 siRNA (siIL-10) to treat breast cancer. It was observed that the liposomal formulation delivered siRNA to tumor efficiently. GQT- and siSuvi-loaded liposomes stimulated the antiproliferation and pro-apoptosis activities of GQT or siSuvi to MCF-7 cells and decreased the amount of survivin and increased the level of caspase-7 in cells. Hence, synergistic antitumor effect of GQT and siRNAs could be potentially used against breast cancer. Hemati et al. [170] optimized the PEGylated niosomes comprising of DOX/QT (PEG-DOX/QT NiS) and small interfering/silencing RNA (siRNA) to improve the therapeutic activity against gastric, prostate, and breast cancer and human foreskin fibroblast cells. Moreover, PEG-DOX/QT NiS showed greater toxicity against cancer cells when related with blank formulations. Moreover, it was observed that PEG-DOX/QT NiS showed enhanced anticancer activities against tumor cell death.

2.10.3 Ovarian and Colon Targeting

Gao et al. [10] encapsulated QT into mPEG-PCL (QT/mPEG-PCL) for the treatment of ovarian cancer. In vitro studies showed that the QT/mPEG-PCL repressed A2780S ovarian cancer cell growth in a dose-dependent manner. The in vivo studies showed that the administration of QT/mPEG-PCL micelles (IV) expressively repressed the development of developed xenograft A2780S ovarian tumors by triggering apoptosis in cancerous cells and hindering angiogenesis. Xu et al. [171] improved the solubility of QT in water and/or enhanced the bioavailability by encapsulating QT into polymeric micellar (QTPM) formulations. Results showed improved effects of QTPM on inducing cell apoptosis, hindering tumor angiogenesis, and restraining cell proliferation. Further, it was concluded that QTPM showed improved antitumor activity, which could be used as a promising anticancer agent against colon cancer. Martirosyan et al. [172] investigated the effects of QT on Ag-NPs against intestinal/colon cancer treatment. It was observed that Ag-NPs reduced cell viability and decreased thiol group sequences, tempted oxidative/nitrosative stress and lipid peroxidation, and also altered the activity of various antioxidant enzymes. QT-AgNPs showed significant anti-inflammatory effect exhibited in downregulated IL-8 levels. Moreover, QT-loaded AG-NPs with reduced cytotoxicity, oxidative stress, and improved metabolic activity of Caco-2 cells signify the shielding activity of QT against the injurious activities of Ag-NPs. Ren et al. [173] explored the activity and probable primary mechanism of QT-GNPs against antitumor activity in liver cancer cells. Results showed that QT-GNPs suppressed the liver cancer cell proliferation, cell migration, and colony establishment, thus inhibiting liver cancer progression. QT-GNPs also upregulated apoptosis significantly. Results proposed that QT-GNPs showed an antitumor effect as it inactivated caspase/Cyto-c pathway, suppressed AP-2β/hTERT and NF-κB/COX-2, and impeded Akt/ERK1/2 signaling pathways; hence, QT-GNPs could act as a potential agent for targeted liver cancer treatment.

2.10.4 Brain Targeting

Ghosh et al. [174] evaluated the therapeutic efficacy of QT nanocapsules (QTNCs) against *As*-induced ROS-mediated oxidative impairment in brain cells and hepatocytes of a rat model. QTNCs prohibited the *As*-induced reduction in antioxidant levels in the brain and hepatic sites. *As* induced a considerable reduction in brain and hepatic cell membrane microviscosities, and QTNC management ensued protection against such reductions. Ghosh et al. [128, 129] evaluated the therapeutic efficacy of QTLs in eliminating *As* toxicity-induced oxidative injury in brain cells and hepatocytes in rats. Results showed that inorganic *As* most distinctly accumulated in neuronal (6.21 ± 0.090 μg/g tissue) and hepatic (9.32 ± 0.100 μg/g tissue) cells of rats, fed with 12 mg/kg body weight of arsenite. Hu et al. [175] prepared and characterized a new type of 1,2-distearoyl-sn-glycero-3-phosphoethanolamine-N-[methoxy (polyethylene glycol)-2000] (DSPE-PEG2000) polymeric nanoliposomes (NLs) for the brain-targeted delivery of QT and temozolomide (TMZ). Results showed that the QT-TMZ nanoliposomes (QT-TMZ NLs) were evidenced to be more efficient at killing C6GC than free TMZ or QT. It was concluded that QT-TMZ NLs possessed characteristics, such as a greater drug encapsulation ratio, reduced in vitro release rate, slow drug clearance, and extended circulation time in vivo. Wang et al. [16] encapsulated QT into freeze-dried NMs (QT-FD-NMs) which improved the bioavailability and solubility of QT which further could be used for glioma chemotherapy. Results revealed that QT-FD-NMs enhanced the intracellular upregulation into Caco-2 cells and effective permeation through blood-brain barrier (BBB) and cytotoxicity on C6GCs. The in vivo studies showed that QT-FD-NMs amassed in tumor-bearing brain tissues and showed significant antitumor effects.

2.10.5 Liver Targeting

Yuan et al. [153, 154] investigated the biodistribution of QT loaded into PEGylated nanoliposomes (NLs) and its therapeutic efficiency on the development of malignant ascites of hepatocellular carcinoma in BALB/c mice. Results revealed that NLs (100 mg/kg) expressively improved the apoptosis in cancerous cells in malignant ascites, repressed the upsurge of body weight, decreased the permeability of peritoneal capillaries, and sustained the residual time of BALB/c mice bearing tumor as compared with control group. Hence, QT-NLs efficiently amassed in the tumor tissues and inhibited the development of carcinomatous ascites. Mandal et al. [176] formulated galactosylated liposomal QT (GLQT) and investigated its efficacy against hepatocarcinoma in rat model. GLQT suspension was given intravenously (IV) to rats (8.98 μM/kg) once a week for 16 weeks. GLQT injections prevented diethylnitrosamine (DEN)-induced hepatocarcinoma and oxidative damage in rat liver. Dabbagh-Bazarbachi et al. [177] established the potential of QT and epigallocatechin-gallate (EG) as combined liposomal formulation rapidly increased labile zinc (Zn) in mouse hepatocarcinoma Hepa 1–6 cells. Results showed that only the co-encapsulated liposomal formulation activated a quick upsurge of FluoZin-3 (Zn-specific detector) fluorescence inside the nanoliposomes, thus signifying the ionophore action of QT, EG, and clioquinol on lipid membrane system. Mandal et al. [178] investigated the regulatory activity and the efficiency of QTNCs on the inhibition of inflammatory mediators (MMP-13) in DEN-mediated hepatocarcinogenesis in Swiss albino rats. It was observed that the QTNCs significantly inhibited the generation of ROS, upregulation of TNF-α and IL-6, initiation of MMP-13, major oxidative damage, hyperplastic nodules with preneoplastic lesions, in DEN-induced hepatocarcinogenesis in Swiss albino rats. Moussa et al. [22] determined the effects of various drug-doped nanocarriers, liposomes, and micelles (Mic) on the distribution and management efficiency for RF ablation (RFA) in the presence of nanocarriers. Results showed that beads with size of 20 nm had greater deposition and deeper tissue penetration in tumor as well as in liver. Mic-DOX and Mic-QT suppressed periablational HIF-1a or HSP70 rim thickness more than liposomal preparations. Finally, it was concluded that with RFA, Mic (small sized) have greater permeation and more effective local molecular modulation. However, with RFA, DOXLs and QTLs resulted in better intratumoral accretion of drug for prolonged time and decreased tumor growth. Varshosaz et al. [179] formulated quercetin-loaded SLNs conjugated with three different sterols (cholesterol stigmasterol and stigmastanol) and was optimized by D-optimal design for enhanced cellular penetration of QT against targeted hepatocellular carcinoma cells. MTT assay was performed to determine cytotoxicity of QT-SLNs on HepG-2 cells. The IC_{50} values of free QT and phytosterol SLNs were almost six and two times more than QT in cholesterol SLNs, respectively, and triggered more accumulation of QT in HepG2 cells. Bishayee et al. [180] prepared gold nanoparticles (GNPs) with QT and loaded QT-GNPs into PLGA NPs and tested the anticancer activity of the prepared NPs on HepG2 hepatocarcinoma cells. It was observed that PLGA-QT-GNPs efficiently triggered conformational alterations in DNA and modified several proteins associated to epigenetic alterations and cell cycle control. In vitro cytotoxicity assay showed that PLGA-QT-GNPs favorably killed cancerous cells as compared to the normal cells and also restrained actions of cell cycle-related proteins, such as p21WAF, cdk1, and pAkt. Also, PLGA-QT-GNPs prompted apoptosis in HepG2 cells by triggering p53-ROS and induced epigenetic changes that lead to cell cycle arrest. Guan et al. [181] prepared QT-loaded PLGA-TPGS nanoparticles (QT-PLGA-TPGS-NPs) for the treatment of liver cancer. QT-PLGA-NP- and QT/coumarin-6 (C6, fluorescent marker)-loaded PLGA-TPGS NPs (QT-C6-PLGA-TPGS-NPs) showed significant activity for cellular uptake by HepG2 and HCa-F and also enhanced apoptosis of HepG2 cells. It was concluded that QT-PLGA-TPGS-NPs could be used as IV administration efficiently for the management of liver cancer with improved liver targeting. Srisa-Nga et al. [28] co-encapsulated QT and superparamagnetic iron oxide nanoparticles (SPIONs) into mPEG-b-oligo (3-caprolactone) micelles (QT/SPIONs-mPEG-b-OCLM) for the suppression of hepatitis B virus-transmitted hepatocellular carcinoma (HepG2.2.15) cellular growth. It was observed that the SPION masses existing inside the center of micellar formulation enhanced T2 relaxivity value signifying decent MRI sensitivity of QT/SPIONs-mPEG-b-OCLM. The HepG2.2.15 cells incubated with QT/SPIONs-mPEG-b-OCLM exhibited greater cytotoxicity than QT. It was concluded that better anticancer activity of QT/SPIONs-mPEG-b-OCLM could be observed at superior magnetic field and mPEG-b-OCLM could be used as a potential co-delivery of QT/SPIONs for targeting hepatocellular carcinoma.

2.10.6 Skin Targeting

Hu et al. [182] developed a deformable QTLs to increase its distribution into the human skin and to enhance its anti-UVB (ultraviolet B) effect. Results showed that the cell sustainability in UVB-induced HaCaT cells increased when treated with QTLs. Also, the ROS and malondialdehyde (MDA) level were condensed. The permeation rate of QTLs was more (3.8-fold) as compared to QT suspensions. Furthermore, the edema and inflammation were improved by deformable QTL, and hence, the potential of deformable QTL might be used as anti-UVB and anti-skin cancer agents. Sahu et al. [183] prepared QT-loaded ethyl cellulose (biocompatible) by nanoprecipitation technique using ethyl cellulose as polymer. Ex vivo results showed that the amount of QT released from PNPs was enhanced and the amount of QT could be retained in the skin for prolonged period of time, therefore reducing the dosing frequency of QT. Hence, QT-loaded ethyl cellulose PNPs could be used topically and also for the management of skin cancer. Zhu et al. [184] showed that QT-loaded PLGA- D-α-tocopheryl polyethylene glycol succinate (TPGS) nanoparticles (PLGA-TPGS-QTNPs) overcame low aqueous solubility of QT and improved its anti-UVB effect. PLGA-TPGS-QTNPs are proficiently absorbed through the epidermis, reached the dermis, and expressively blocked the UVB irradiation-induced COX-2 upregulation and NF-kB activation in Hacat cell line. It was concluded that PLGA-TPGS-QTNPs efficiently protected UVB-induced injuries and might be used for skin cancer therapy. Caddeo et al. [19, 20] investigated anti-ROS activity co-encapsulated QT and resveratrol (RES) liposomes. The results showed that co-encapsulated QT and RES effectively helped the entry of QT and RES in fibroblasts, and their amalgamation enhanced the anti-ROS activity, thus contributing to hinder the inflammatory effect triggered by 12-O-tetradecanoylphorbol 13-acetate (TPA) on the skin, hence resulted in tissue repair, wound healing and may be used in the management of skin cancer. Sandhu et al. [185] developed surface-modified QT-salicylic acid (SA) NMs by the self-assembly method followed by Quality by Design (QbD) approach for effective management of skin carcinoma. The optimized formulation exhibited improved %permeation and enhanced in vitro cellular uptake, and annexin-V-apoptosis assay showed 60.86% cellular death. Also, the 7,12-dimethylbenz(a) anthracene and histopathology studies showed the enhanced antineoplastic action of QT. George et al. [186] extracted QT from onion peel waste, conjugated it with dialdehyde cellulose (DAC) modified from sugarcane bagasse (SCB) cellulose, and efficiently cross-linked it with chitosan hydrogel. Later on, the hydrogel complexes were inserted with zinc oxide (phyto-synthesized using musk melon seeds) nanoparticles (ZnO NPs). Results showed enhanced drug loading (39%) of QT in the hydrogel complexes in the presence of ZnO-NPs as compared to drug loading without ZnO NPs. The hydrogel complex inhibited the growth of *Staphylococcus aureus* (*S. aureus*) and *Trichophyton rubrum* (*T. rubrum*) strains. Also, the biocompatible hydrogel complex retained anticancer activity against L929 murine fibroblast cells and A431 human skin carcinoma cells.

2.11 Miscellaneous

Barzilai and Rahamimoff [187] prepared QTLs, evaluated their inhibitory activity on the calcium (Ca^{2+})-transport ATPase of synaptosomal vesicles, and compared it with MR and RT. Results showed that QT caused a 50% inhibition of Ca^{2+}-transport at low concentration (50 μM), whereas MR and RT had similar effects at higher concentrations (200 μM). Moreover, at higher concentrations, similar order of effectiveness was observed also for ATP hydrolysis. QT also intensely repressed phosphorylation of membrane proteins by ATP in synaptosomal vesicles. MR and RT showed negligible effect on membrane protein phosphorylation. The order of inhibitory potency of the flavonoids on the Ca^{2+}-transport ATPase from synaptosomal vesicles is as follows: QT > MR > RT. Fawzy et al. [188] studied the interaction of retinoids (retinal, retinol, retinoic acid, and retinol acetate) and flavonoids (QT, RT, MR, and sciadopitysin) with phospholipases A2 (PLA2) so as to evaluate the mechanism of inhibition. In vitro studies showed that both the groups inhibited Ca^{2+}-dependent PLA2 activity of human synovial fluid (HSF) in a dose-dependent manner. It was observed that retinal inhibited neutral active Ca^{2+}-dependent PLA2 from human platelets, plasma, and polymorphonuclear leukocytes and also *Naja mossambica mossambica* venom in a dose-dependent fashion while QT inhibited extracellular PLA2 activities of human plasma, HSF, and *N. m. mossambica* venom in a dose-dependent fashion but not PLA2 activity. Theoharidesi et al. [189] developed a mixture of LT, QT, and RT as liposome system comprised of olive kernel oil (OKO) (LQT-OKO-Ls), which enhanced the absorption, as a dietary supplement for the management of autism spectrum disorders (ASD) in children. LQT-OKO-Ls improved the GI and allergy symptoms in about 75% of children, eye contact and attention in about 50%, social interaction in about 25%, and resumption of speech in about 10%. Rich et al. [190] studied the uptake of QT by CaCo-2/TC7 cells effected by mixed micelles existing in the human small intestine. It was observed that the micelles repressed the transport of QT into the cells due to its lower segregation into simple bile salt micelles, and in case of mixed micelles, it was repressed due to increased bile salt concentration. The affinity of QT reduced in the order egg phosphatidylcholine (PC) = lysoPC > mixed micelles > bile salts.

3 Conclusion and Future Prospective

Flavonoids like QT have got much considerations in recent years due to its numerous pharmacological activities. Prominently, because of its antioxidant effect, it has become a vital nutritional compound with potent therapeutic effects. Reports have evidenced that QT have several biological effects in the prevention and treatment of cancer, infections, inflammations, skin-mediated injuries, hepatic injury, and cardiovascular disorders. In this review, the current nanotechnology-based approaches for QT are discussed. Nano-based drug delivery systems of QT have shown promising effects and have improved poor aqueous solubility and low oral bioavailability. Though little is known about its bioavailability, absorption, and metabolism in humans, scientific facts related with the specific treatments of QT-based drug delivery systems are quite uncertain. Additionally, abundant studies concerned with the therapeutic activities of QT-based drug delivery systems have been accomplished either in vitro or in vivo; still some additional data are required to evaluate its potential pharmacological activities in human beings. Therefore, it is expected that in the future well-designed clinical assays and studies in healthy individuals would be performed to endorse the abovementioned beneficial effects of QT.

References

1. Bennet D, Kim S (2013) A transdermal delivery system to enhance quercetin nanoparticle permeability. J Biomater Sci 24(2):185–209
2. Tasdemir D, Kaiser M, Brun R, Yardley V, Schmidt TJ, Tosun F, Ruedi P (2006) Antitrypanosomal and antileishmanial activities of flavonoids and their analogues: in vitro, in vivo, structure-activity relationship, and quantitative structure-activity relationship studies. Antimicrob Agents Chemother 50(4):1352–1364
3. Balakrishnan S, Bhat FA, Raja Singh P, Mukherjee S, Elumalai P, Das S, Patra CR, Arunakaran J (2016) Gold nanoparticle-conjugated quercetin inhibits epithelial-mesenchymal transition, angiogenesis and invasiveness via EGFR/VEGFR-2-mediated pathway in breast cancer. Cell Prolif 49(6):678–697
4. Beninger CW, Hosfield GL (2003) Antioxidant activity of extracts, condensed tannin fractions, and pure flavonoids from Phaseolus vulgaris L. seed coat color genotypes. J Agric Food Chem 51(27):7879–7883
5. Caddeo C, Diez-Sales O, Pons R, Fernandez-Busquets X, Fadda AM, Manconi M (2014) Topical anti-inflammatory potential of quercetin in lipid-based nanosystems: in vivo and in vitro evaluation. Pharm Res 31(4):959–968
6. Jain S, Mehata MS (2017) Medicinal plant leaf extract and pure flavonoid mediated green synthesis of silver nanoparticles and their enhanced antibacterial property. Sci Rep 7(15867):1–13
7. Kumar VD, Prasad PR, Singh SK (2015a) Development and evaluation of biodegradable polymeric nanoparticles for the effective delivery of quercetin using a quality by design approach. LWT Food Sci Technol 61(2):330–338
8. Rathee S, Kamboj A (2018) Optimization and development of antidiabetic phytosomes by the Box-Behnken design. J Liposome Res 28(2):161–172
9. D'Andrea G (2015) Quercetin: A flavonol with multifaceted therapeutic applications? Fitoterapia 106:256–271
10. Gao X, Wang B, Wei X, Men K, Zheng F, Zhou Y, Zheng Y, Gou M, Huang M, Guo G, Huang N, Qian Z, Wei Y (2012) Anticancer effect and mechanism of polymer micelle-encapsulated quercetin on ovarian cancer. Nanoscale 4(22):7021–7030
11. Balestrin LA, Bidone J, Bortolin RC, Moresco K, Moreira JC, Teixeira HF (2016) Protective effect of a hydrogel containing Achyrocline satureioides extract-loaded nanoemulsion against UV-induced skin damage. J Photochem Photobiol B Biol 163:269–276
12. Liu D, Hu H, Lin Z, Chen D, Zhu Y, Hou S, Shi X (2013a) Quercetin deformable liposome: preparation and efficacy against ultraviolet B induced skin damages in vitro and in vivo. J Photochem Photobiol B Biol 127:8–17
13. Liu H, Xue JX, Li X, Ao R, Lu Y (2013b) Quercetin liposomes protect against radiation-induced pulmonary injury in a murine model. Oncol Lett 6(2):453–459
14. Verma SK, Rastogi S, Arora I, Javed K, Akhtar M, Samim M (2016) Nanoparticle based delivery of quercetin for the treatment of carbon tetrachloride mediated liver cirrhosis in rats. J Biomed Nanotechnol 12(2):274–285
15. Riva A, Ronchi M, Petrangolini G, Bosisio S, Allegrini P (2019) Improved Oral absorption of quercetin from quercetin phytosome(R), a new delivery system based on food grade lecithin. Eur J Drug Metab Pharmacokinet 44(2):169–177
16. Wang G, Wang JJ, Chen XL, Du L, Li F (2016) Quercetin-loaded freeze-dried nanomicelles: Improving absorption and anti-glioma efficiency in vitro and in vivo. J Control Release 235:276–290
17. Ganesan P, Ko HM, Kim IS, Choi DK (2015) Recent trends in the development of nanophytobioactive compounds and delivery systems for their possible role in reducing oxidative stress in Parkinson's disease models. Int J Nanomedicine 10:6757–6772

18. Hussain A, Singh S, Das SS, Anjireddy K, Karpagam S, Shakeel F (2019) Nanomedicines as drug delivery carriers of anti-tubercular drugs: from pathogenesis to infection control. Curr Drug Deliv 16(5):400–429
19. Caddeo C, Diez-Sales O, Pons R, Carbone C, Ennas G, Puglisi G, Fadda AM, Manconi M (2016a) Cross-linked chitosan/liposome hybrid system for the intestinal delivery of quercetin. J Colloid Interface Sci 461:69–78
20. Caddeo C, Nacher A, Vassallo A, Armentano MF, Pons R, Fernandez-Busquets X, Carbone C, Valenti D, Fadda AM, Manconi M (2016b) Effect of quercetin and resveratrol co-incorporated in liposomes against inflammatory/oxidative response associated with skin cancer. Int J Pharm 513(1–2):153–163
21. Wang BL, Gao X, Men K, Qiu J, Yang B, Gou ML, Huang MJ, Huang N, Qian ZY, Zhao X, Wei YQ (2012) Treating acute cystitis with biodegradable micelle-encapsulated quercetin. Int J Nanomedicine 7:2239–2247
22. Moussa M, Goldberg SN, Kumar G, Sawant RR, Levchenko T, Torchilin VP, Ahmed M (2014) Nanodrug-enhanced radiofrequency tumor ablation: effect of micellar or liposomal carrier on drug delivery and treatment efficacy. PLoS One 9(8):e102727
23. Kumar VD, Verma PR, Singh SK, Viswanathan S (2015b) LC-ESI-MS/MS analysis of quercetin in rat plasma after oral administration of biodegradable nanoparticles. Biomed Chromatogr 29(11):1731–1736
24. Chen SQ, Wang C, Tao S, Wang YX, Hu FQ, Yuan H (2018) Rational design of redox-responsive and P-gp-inhibitory lipid nanoparticles with high entrapment of paclitaxel for tumor therapy. Adv Healthc Mater 7(17):e1800485
25. Ojeda-Serna IE, Rocha-Guzman NE, Gallegos-Infante JA, Chairez-Ramirez MH, Rosas-Flores W, Perez-Martinez JD, Gonzalez-Laredo RF (2019) Water-in-oil organogel based emulsions as a tool for increasing bioaccessibility and cell permeability of poorly water-soluble nutraceuticals. Food Res Int 120:415–424
26. Omwenga EO, Hensel A, Shitandi A, Goycoolea FM (2018) Chitosan nanoencapsulation of flavonoids enhances their quorum sensing and biofilm formation inhibitory activities against an E. coli Top 10 biosensor. Colloids Surf B Biointerfaces 164:125–133
27. Balakrishnan S, Mukherjee S, Das S, Bhat FA, Raja Singh P, Patra CR, Arunakaran J (2017) Gold nanoparticles-conjugated quercetin induces apoptosis via inhibition of EGFR/PI3K/Akt-mediated pathway in breast cancer cell lines (MCF-7 and MDA-MB-231). Cell Biochem Funct 35(4):217–231
28. Srisa-Nga K, Mankhetkorn S, Okonogi S, Khonkarn R (2018) Delivery of superparamagnetic polymeric micelles loaded with quercetin to hepatocellular carcinoma cells. J Pharm Sci. https://doi.org/10.1016/j.xphs.2018.08.008
29. Yuan Y, Peng Q, Gurunathan S (2017) Effects of silver nanoparticles on multiple drug-resistant strains of *Staphylococcus aureus* and *Pseudomonas aeruginosa* from mastitis-infected goats: an alternative approach for antimicrobial therapy. Int J Mol Sci 18(3):E569
30. Mittal AK, Kumar S, Banerjee UC (2014) Quercetin and gallic acid mediated synthesis of bimetallic (Ag-Se) nanoparticles and their antitumor and antimicrobial potential. J Colloid Interface Sci 431:194–199
31. Yun J, Lee H, Ko HJ, Woo ER, Lee DG (2015) Fungicidal effect of isoquercitrin via inducing membrane disturbance. Biochim Biophys Acta 1848(2):695–701
32. Bidone J, Argenta DB, Kratz J, Pettenuzzo LF, Horn AP, Koester LC, Bassani VL, Simões CO, Teixeira HF (2015) Antiherpes activity and skin/mucosa distribution of flavonoids from achyrocline satureioides extract incorporated into topical nanoemulsions. Biomed Res Int 8:1–7
33. Sun D, Zhang W, Li N, Zhao Z, Mou Z, Yang E, Wang W (2016a) Silver nanoparticles-quercetin conjugation to siRNA against drug-resistant Bacillus subtilis for effective gene silencing: in vitro and in vivo. Mater Sci Eng C 63:522–534
34. Sun D, Li N, Zhang W, Zhao Z, Mou Z, Huang D, Liu J, Wang W (2016b) Design of PLGA-functionalized quercetin nanoparticles for potential use in Alzheimer's disease. Colloids Surf B Biointerfaces 148:116–129

35. Sun D, Zhang W, Mou Z, Chen Y, Guo F, Yang E, Wang W (2017) Transcriptome analysis reveals silver nanoparticle-decorated quercetin antibacterial molecular mechanism. ACS Appl Mater Interfaces 9(11):10047–10060
36. Nguyen HT, Goycoolea FM (2017) Chitosan/cyclodextrin/TPP nanoparticles loaded with quercetin as novel bacterial quorum sensing inhibitors. Molecules 22(1975):3–23
37. Li F, Jin H, Xiao J, Yin X, Liu X, Li D, Huang Q (2018) The simultaneous loading of catechin and quercetin on chitosan based nanoparticles as effective antioxidant and antibacterial agent. Food Res Int 111:351–360
38. Yu L, Shang F, Chen X, Sun D, Xue T (2018) The anti-biofilm effect of silver-nanoparticle-decorated quercetin nanoparticles on a multi-drug resistant *Escherichia coli* strain isolated from a dairy cow with mastitis. PeerJ 6:e5711
39. Lotha R, SriSundaramoorthy N, Shamprasad BR, Nagarajan S, Sivasubramanian A (2018) Plant nutraceuticals (Quercetrin and Afzelin) capped silver nanoparticles exert potent antibiofilm effect against food borne pathogen Salmonella enterica serovar typhimurium and curtail planktonic growth in zebrafish infection model. Microb Pathog 120:109–118
40. Sowmya C, Lavakumar V, Venkateshan N, Ravichandiran V, Saigopal DVR (2018) Exploration of Phyllanthus acidus mediated silver nanoparticles and its activity against infectious bacterial pathogen. Chem Cent J 12(1):1–9
41. Zhou Y, Tang R (2018) Facile and eco-friendly fabrication of colored and bioactive silk materials using silver nanoparticles synthesized by two flavonoids. Polymers 4(10):1–15
42. Mittra B, Saha A, Chowdhury AR, Pal C, Mandal S, Mukhopadhyay S, Bandyopadhyay S, Majumder HK (2000) Luteolin, an abundant dietary component is a potent anti-leishmanial agent that acts by inducing topoisomerase II-mediated kinetoplast DNA cleavage leading to apoptosis. Mol Med 6(6):527–541
43. Sarkar S, Mandal S, Sinha J, Mukhopadhyay S, Das N, Basu MK (2002) Quercetin: critical evaluation as an antileishmanial agent in vivo in hamsters using different vesicular delivery modes. J Drug Target 10(8):573–578
44. Gupta S, Moulik SP, Lala S, Basu MK, Sanyal SK, Datta S (2005) Designing and testing of an effective oil-in-water microemulsion drug delivery system for in vivo application. Drug Deliv 12:267–273
45. Sen G, Mandal S, Roy SR, Mukhopadhyay S, Biswas T (2005) Therapeutic use of quercetin in the control of infection and anemia associated with visceral leishmaniasis. Free Radic Biol Med 38:1257–1264
46. Sen G, Biswas D, Ray M, Biswas T (2007) Albumin–quercetin combination offers a therapeutic advantage in the prevention of reduced survival of erythrocytes in visceral leishmaniasis. Blood Cell Mol Dis 39:245–254
47. Sen G, Mukhopadhyay S, Ray M, Biswas T (2008) Quercetin interferes with iron metabolism in Leishmania donovani and targets ribonucleotide reductase to exert leishmanicidal activity. J Antimicrob Chemother 61:1066–1075
48. Marin C, Boutaleb-Charki S, Diaz JG, Huertas O, Rosales MJ, Perez-Cordon G, Guitierrez-Sanchez R, Sanchez-Moreno M (2009) Antileishmaniasis Activity of Flavonoids from *Consolida oliWeriana*. J Nat Prod 72(6):1069–1074
49. Fonseca-Silva F, Inacio JD, Canto-Cavalheiro MM, Almeida-Amaral EE (2011) Reactive oxygen species production and mitochondrial dysfunction contribute to quercetin induced death in Leishmania amazonensis. PLoS One 6(2):e14666
50. Das S, Roy P, Mondal S, Bera T, Mukherjee A (2013) One pot synthesis of gold nanoparticles and application in chemotherapy of wild and resistant type visceral leishmaniasis. Colloids Surf B Biointerfaces 107:27–34
51. Fonseca-Silva F, Inacio JD, Canto-Cavalheiro MM, Almeida-Amaral EE (2013) Reactive oxygen species production by quercetin causes the death of Leishmania amazonensis intracellular amastigotes. J Nat Prod 76(8):1505–1508
52. Sousa-Batista AJ, Poletto FS, Philipon C, Guterres SS, Pohlmann AR, Rossi-Bergmann B (2017) Lipid-core nanocapsules increase the oral efficacy of quercetin in cutaneous leishmaniasis. Parasitology 144(13):1769–1774

53. Mukhopadhyay P, Maity S, Mandal S, Chakraborti AS, Prajapati AK, Kundu PP (2018) Preparation, characterization and in vivo evaluation of pH sensitive, safe quercetin-succinylated chitosan-alginate core-shell-corona nanoparticle for diabetes treatment. Carbohydr Polym 182:42–51
54. Ebrahimpour S, Esmaeili A, Beheshti S (2018) Effect of quercetin-conjugated superparamagnetic iron oxide nanoparticles on diabetes-induced learning and memory impairment in rats. Int J Nanomedicine 13:6311–6324
55. Maity BK, Vishvakarma V, Surendran D, Rawat A, Das A, Pramanik S, Arfin N, Maiti S (2018) Spontaneous fluctuations can guide drug design strategies for structurally disordered proteins. Biochemistry 57(28):4206–4213
56. Landi-Librandi AP, de Oliveira CA, Azzolini AE, Kabeya LM, Del Ciampo JO, Bentley MV, Lucisano-Valim YM (2011) In vitro evaluation of the antioxidant activity of liposomal flavonols by the HRP-H2O2-luminol system. J Microencapsul 28(4):258–267
57. Kumari A, Yadav SK, Pakade YB, Kumar V, Singh B, Chaudhary A, Yadav SC (2011) Nanoencapsulation and characterization of Albizia chinensis isolated antioxidant quercitrin on PLA nanoparticles. Colloids Surf B Biointerfaces 82(1):224–232
58. Landi-Librandi AP, Caleiro Seixas Azzolini AE, de Oliveira CA, Lucisano-Valim YM (2012a) Inhibitory activity of liposomal flavonoids during oxidative metabolism of human neutrophils upon stimulation with immune complexes and phorbol ester. Drug Deliv 19(4):177–187
59. Landi-Librandi AP, Chrysostomo TN, Caleiro Seixas Azzolini AE, Marzocchi-Machado CM, de Oliveira CA, Lucisano-Valim YM (2012b) Study of quercetin-loaded liposomes as potential drug carriers: in vitro evaluation of human complement activation. J Liposome Res 22(2):89–99
60. Azofeifa G, Quesada S, Boudard F, Morena M, Cristol JP, Perez AM, Vaillant F, Michel A (2013) Antioxidant and anti-inflammatory in vitro activities of phenolic compounds from tropical highland blackberry (Rubus adenotrichos). J Agric Food Chem 61(24):5798–5804
61. Lee GH, Lee SJ, Jeong SW, Kim HC, Park GY, Lee SG, Choi JH (2016) Antioxidative and antiinflammatory activities of quercetin-loaded silica nanoparticles. Colloids Surf B Biointerfaces 143:511–517
62. Strugala P, Tronina T, Huszcza E, Gabrielska J (2017) Bioactivity in vitro of quercetin glycoside obtained in beauveria bassiana culture and its interaction with liposome membranes. Molecules 22(9). https://doi.org/10.3390/molecules22091520
63. Abd El-Fattah AI, Fathy MM, Ali ZY, El-Garawany AEA, Mohamed EK (2017) Enhanced therapeutic benefit of quercetin-loaded phytosome nanoparticles in ovariectomized rats. Chem Biol Interact 271:30–38
64. Aithal GC, Nayak UY, Mehta C, Narayan R, Gopalkrishna P, Pandiyan S, Garg S (2018) Localized in situ nanoemulgel drug delivery system of quercetin for periodontitis: development and computational simulations. Molecules 23(6). https://doi.org/10.3390/molecules23061363
65. Ozdal ZD, Sahmetlioglu E, Narin I, Cumaoglu A (2019) Synthesis of gold and silver nanoparticles using flavonoid quercetin and their effects on lipopolysaccharide induced inflammatory response in microglial cells. 3 Biotech 9(6):212
66. Ratty AK, Sunamoto J, Das NP (1988) Interaction of flavonoids with 1,1-diphenyl-2-picrylhydrazyl free radical, liposomal membranes and soybean lipoxygenase-1. Biochem Pharmacol 37(6):989–995
67. Terao J, Piskula M, Yao Q (1994) Protective effect of epicatechin, epicatechin gallate, and quercetin on lipid peroxidation in phospholipid bilayers. Arch Biochem Biophys 308(1):278–284
68. Saija A, Scalese M, Lanza M, Marzullo D, Bonina F, Castelli F (1995) Flavonoids as antioxidant agents: importance of their interaction with biomembranes. Free Radic Biol Med 19(4):481–486
69. Gordon MH, Roedig-Penman A (1998) Antioxidant activity of quercetin and myricetin in liposomes. Chem Phys Lipids 97:79–85

70. Liao K, Yin M (2000) Individual and combined antioxidant effects of seven phenolic agents in human erythrocyte membrane ghosts and phosphatidylcholine liposome systems: importance of the partition coefficient. J Agric Food Chem 48(6):2266–2270
71. Shirai M, Moon JH, Tsushida T, Terao J (2001) Inhibitory effect of a quercetin metabolite, quercetin 3-O-beta-D-glucuronide, on lipid peroxidation in liposomal membranes. J Agric Food Chem 49(11):5602–5608
72. Filip R, Ferraro GE (2003) Researching on new species of "Mate": ilex brevicuspis: phytochemical and pharmacology study. Eur J Nutr 42(1):50–54
73. Zou Y, Lu Y, Wei D (2004) Antioxidant activity of a flavonoid-rich extract of Hypericum perforatum L. in vitro. J Agric Food Chem 52(16):5032–5039
74. Murata Y, Osaki K, Shimoishi Y, Baba N, Tada M (2004) Degradation of ethyl docosahexaenoate by gamma-ray irradiation and suppression of this degradation by antioxidants. Biosci Biotechnol Biochem 68(3):743–745
75. Ly TN, Hazama C, Shimoyamada M, Ando H, Kato K, Yamauchi R (2005) Antioxidative compounds from the outer scales of onion. J Agric Food Chem 53(21):8183–8189
76. Gabrielska J, Soczynska-Kordala M, Hladyszowski J, Zylka R, Miskiewicz J, Przestalski S (2006) Antioxidative effect of quercetin and its equimolar mixtures with phenyltin compounds on liposome membranes. J Agric Food Chem 54(20):7735–7746
77. Ghosh D, Ghosh S, Sarkar S, Ghosh A, Das N, Das Saha K, Mandal AK (2010) Quercetin in vesicular delivery systems: evaluation in combating arsenic-induced acute liver toxicity associated gene expression in rat model. Chem Biol Interact 186(1):61–71
78. Liang R, Chen CH, Han RM, Zhang JP, Skibsted LH (2010) Thermodynamic versus kinetic control of antioxidant synergism between beta-carotene and (iso)flavonoids and their glycosides in liposomes. J Agric Food Chem 58(16):9221–9227
79. Bernardy N, Romio AP, Barcelos EI, Pizzol CD, Dora CL, Lemos-Senna E, Araujo PHH, Sayer C (2010) Nanoencapsulation of quercetin via miniemulsion polymerization. J Biomed Nanotechnol 6(2):181–186
80. Sahoo NG, Kakran M, Shaal LA, Li L, Muller RH, Pal M, Tan LP (2011) Preparation and characterization of quercetin nanocrystals. J Pharm Sci 100(6):2379–2390
81. Parmar A, Singh K, Bahadur A, Marangoni G, Bahadur P (2011) Interaction and solubilization of some phenolic antioxidants in Pluronic(R) micelles. Colloids Surf B Biointerfaces 86(2):319–326
82. Weiss-Angeli V, Poletto FS, de Marco SL, Salvador M, da Silveira NP, Guterres SS, Pohlmann AR (2012) Sustained antioxidant activity of quercetin-loaded lipid-core nanocapsules. J Nanosci Nanotechnol 12(3):2874–2880
83. Park SN, Lee MH, Kim SJ, Yu ER (2013) Preparation of quercetin and rutin-loaded ceramide liposomes and drug-releasing effect in liposome-in-hydrogel complex system. Biochem Biophys Res Commun 435(3):361–366
84. Bose S, Du Y, Takhistov P, Michniak-Kohn B (2013) Formulation optimization and topical delivery of quercetin from solid lipid based nanosystems. Int J Pharm 441(1–2):56–66
85. Yagolnik EA, Tarahovsky YS, Klenina IB, Kuznetsova SM, Muzafarov EN, Kim YA (2014) Study of membranotropic and antioxidant activity of flavonoids and their complexes with ferric iron. Biophysics 58(5):646–652
86. Tavano L, Muzzalupo R, Picci N, de Cindio B (2014) Co-encapsulation of antioxidants into niosomal carriers: gastrointestinal release studies for nutraceutical applications. Colloids Surf B Biointerfaces 114:82–88
87. Jeon S, Yoo CY, Park SN (2015) Improved stability and skin permeability of sodium hyaluronate-chitosan multilayered liposomes by Layer-by-Layer electrostatic deposition for quercetin delivery. Colloids Surf B Biointerfaces 129:7–14
88. Linkeviciute A, Misiunas A, Naujalis E, Barauskas J (2015) Preparation and characterization of quercetin-loaded lipid liquid crystalline systems. Colloids Surf B Biointerfaces 128:296–303
89. Gupta P, Authimoolam SP, Hilt JZ, Dziubla TD (2015) Quercetin conjugated poly(beta-amino esters) nanogels for the treatment of cellular oxidative stress. Acta Biomater 27:194–204

90. Zorzi GK, Caregnato F, Moreira JC, Teixeira HF, Carvalho EL (2016) Antioxidant Effect of nanoemulsions containing extract of achyrocline satureioides (Lam) D.C.-asteraceae. AAPS PharmSciTech 17(4):844–850
91. Kaur K, Kumar R, Mehta SK (2016) Formulation of saponin stabilized nanoemulsion by ultrasonic method and its role to protect the degradation of quercitin from UV light. Ultrason Sonochem 31:29–38
92. Rezaei-Sadabady R, Eidi A, Zarghami N, Barzegar A (2016) Intracellular ROS protection efficiency and free radical-scavenging activity of quercetin and quercetin-encapsulated liposomes. Artificial Cells Nanomed Biotechnol 44(1):128–134
93. Altunkaya A, Gokmen V, Skibsted LH (2016) pH dependent antioxidant activity of lettuce (L. sativa) and synergism with added phenolic antioxidants. Food Chem 190:25–32
94. Moretti E, Mazzi L, Bonechi C, Salvatici MC, Iacoponi F, Rossi C, Collodel G (2016) Effect of Quercetin-loaded liposomes on induced oxidative stress in human spermatozoa. Reprod Toxicol 60:140–147
95. Anwer MK, Al-Mansoor MA, Jamil S, Al-Shdefat R, Ansari MN, Shakeel F (2016) Development and evaluation of PLGA polymer based nanoparticles of quercetin. Int J Biol Macromol 92:213–219
96. Aluani D, Tzankova V, Kondeva-Burdina M, Yordanov Y, Nikolova E, Odzhakov F, Apostolov A, Markova T, Yoncheva K (2017) capital IE, cyrillicvaluation of biocompatibility and antioxidant efficiency of chitosan-alginate nanoparticles loaded with quercetin. Int J Biol Macromol 103:771–782
97. Huang M, Su E, Zheng F, Tan C (2017) Encapsulation of flavonoids in liposomal delivery systems: the case of quercetin, kaempferol and luteolin. Food Funct 8(9):3198–3208
98. Singh O, Kaur R, Mahajan RK (2017) Flavonoid-surfactant interactions: A detailed physicochemical study. Spectrochim Acta A Mol Biomol Spectrosc 170:77–88
99. Rodriguez EB, Almeda RA, Vidallon MLP, Reyes CT (2018) Enhanced bioactivity and efficient delivery of quercetin through nanoliposomal encapsulation using rice bran phospholipids. J Sci Food Agric. https://doi.org/10.1002/jsfa.9396
100. Singh S, Kushwah V, Agrawal AK, Jain S (2018) Insulin- and quercetin-loaded liquid crystalline nanoparticles: implications on oral bioavailability, antidiabetic and antioxidant efficacy. Nanomedicine 13(5):521–537
101. Farrag Y, Ide W, Montero B, Rico M, Rodriguez-Llamazares S, Barral L, Bouza R (2018) Preparation of starch nanoparticles loaded with quercetin using nanoprecipitation technique. Int J Biol Macromol 114:426–433
102. Najafabadi ER, Kazemipour N, Esmaeili A, Beheshti S, Nazifi S (2018a) Using superparamagnetic iron oxide nanoparticles to enhance bioavailability of quercetin in the intact rat brain. BMC Pharmacol Toxicol 19(1):59
103. Najafabadi RE, Kazemipour N, Esmaeili A, Beheshti S, Nazifi S (2018b) Quercetin prevents body weight loss due to the using of superparamagnetic iron oxide nanoparticles in rat. Adv Biomed Res 7:8
104. Sarkar S, Das N (2006) Mannosylated liposomal flavonoid in combating age-related ischemia-reperfusion induced oxidative damage in rat brain. Mech Ageing Dev 127(4):391–397
105. Matsuzaki K, Noguch T, Wakabayashi M, Ikeda K, Okada T, Ohashi Y, Hoshino M, Naiki H (2007) Inhibitors of amyloid beta-protein aggregation mediated by GM1-containing raft-like membranes. Biochim Biophys Acta 1768(1):122–130
106. Priprem A, Watanatorn J, Sutthiparinyanont S, Phachonpai W, Muchimapura S (2008) Anxiety and cognitive effects of quercetin liposomes in rats. Nanomedicine 4(1):70–78
107. Rivera F, Costa G, Abin A, Urbanavicius J, Arruti C, Casanova G, Dajas F (2008) Reduction of ischemic brain damage and increase of glutathione by a liposomal preparation of quercetin in permanent focal ischemia in rats. Neurotox Res 13(2):105–114
108. Dhawan S, Kapil R, Singh B (2011) Formulation development and systematic optimization of solid lipid nanoparticles of quercetin for improved brain delivery. J Pharm Pharmacol 63(3):342–351

109. Ghosh A, Sarkar S, Mandal AK, Das N (2013) Neuroprotective role of nanoencapsulated quercetin in combating ischemia-reperfusion induced neuronal damage in young and aged rats. PLoS One 8(4):e57735
110. Bagad M, Khan ZA (2015) Poly(n-butylcyanoacrylate) nanoparticles for oral delivery of quercetin: preparation, characterization, and pharmacokinetics and biodistribution studies in Wistar rats. Int J Nanomedicine 10:3921–3935
111. Galho AR, Cordeiro MF, Ribeiro SA, Marques MS, Antunes MF, Luz DC, Hadrich G, Muccillo-Baisch AL, Barros DM, Lima JV, Dora CL, Horn AP (2016) Protective role of free and quercetin-loaded nanoemulsion against damage induced by intracerebral haemorrhage in rats. Nanotechnology 27(17):175101
112. Kuo YC, Tsao CW (2017) Neuroprotection against apoptosis of SK-N-MC cells using RMP-7 and lactoferrin-grafted liposomes carrying quercetin. Int J Nanomedicine 12:2857–2869
113. Ahmad N, Ahmad R, Naqvi AA, Alam MA, Ashafaq M, Abdur Rub R, Ahmad FJ (2018) Intranasal delivery of quercetin-loaded mucoadhesive nanoemulsion for treatment of cerebral ischaemia. Artificial Cells Nanomed Biotechnol 46(4):717–729
114. Ghosh S, Sarkar S, Choudhury ST, Ghosh T, Das N (2017) Triphenyl phosphonium coated nano-quercetin for oral delivery: neuroprotective effects in attenuating age related global moderate cerebral ischemia reperfusion injury in rats. Nanomedicine 13(8):2439–2450
115. Oliveira AI, Pinho C, Fonte P, Sarmento B, Dias ACP (2018) Development, characterization, antioxidant and hepatoprotective properties of poly(Ɛ-caprolactone) nanoparticles loaded with a neuroprotective fraction of Hypericum perforatum. Int J Biol Macromol 110:185–196
116. Rishitha N, Muthuraman A (2018) Therapeutic evaluation of solid lipid nanoparticle of quercetin in pentylenetetrazole induced cognitive impairment of zebrafish. Life Sci 199:80–87
117. Amanzadeh E, Esmaeili A, Abadi REN, Kazemipour N, Pahlevanneshan Z, Beheshti S (2019) Quercetin conjugated with superparamagnetic iron oxide nanoparticles improves learning and memory better than free quercetin via interacting with proteins involved in LTP. Sci Rep 9(1):6876
118. Katebi S, Esmaeili A, Ghaedi K, Zarrabi A (2019) Superparamagnetic iron oxide nanoparticles combined with NGF and quercetin promote neuronal branching morphogenesis of PC12 cells. Int J Nanomedicine 14:2157–2169
119. Kuo YC, Chen CL, Rajesh R (2019) Optimized liposomes with transactivator of transcription peptide and anti-apoptotic drugs to target hippocampal neurons and prevent tau-hyperphosphorylated neurodegeneration. Acta Biomater 87:207–222
120. Liu Y, Zhou H, Yin T, Gong Y, Yuan G, Chen L, Liu J (2019) Quercetin-modified gold-palladium nanoparticles as a potential autophagy inducer for the treatment of Alzheimer's disease. J Colloid Interface Sci 552:388–400
121. Mandic L, Sadzak A, Strasser V, Baranovic G, Domazet Jurasin D, Sikiric MD, Segota S (2019) Enhanced protection of biological membranes during lipid peroxidation: study of the interactions between flavonoid loaded mesoporous silica nanoparticles and model cell membranes. Int J Mol Sci 20(11). https://doi.org/10.3390/ijms20112709
122. Soloviev A, Stefanov A, Parshikov A, Khromov A, Moibenko A, Kvotchina L, Balavoine G, Geletii Y (2002) Arrhythmogenic peroxynitrite-induced alterations in mammalian heart contractility and its prevention with quercetin-filled liposomes. Cardiovasc Toxicol 2(2):129–140
123. Dugas TR, Brewer G, Longwell M, Fradella T, Braun J, Astete CE, Jennings MH, Sabliov CM (2019) Nanoentrapped polyphenol coating for sustained drug release from a balloon catheter. J Biomed Mater Res B Appl Biomater 107(3):646–651
124. Long D, Shang Y, Qiu Y, Zhou B, Yang P (2018) A single-cell analysis platform for electrochemiluminescent detection of platelets adhesion to endothelial cells based on Au@DL-ZnCQDs nanoprobes. Biosens Bioelectron 102:553–559
125. Mandal AK, Sinha J, Mandal S, Mukhopadhyay S, Das N (2002) Targeting of liposomal flavonoid to liver in combating hepatocellular oxidative damage. Drug Deliv 9(3):181–185
126. Mandal AK, Das N (2005) Sugar coated liposomal flavonoid: a unique formulation in combating carbontetrachloride induced hepatic oxidative damage. J Drug Target 13(5):305–315

127. Mandal AK, Das S, Basu MK, Chakrabarti RN, Das N (2007) Hepatoprotective activity of liposomal flavonoid against arsenite-induced liver fibrosis. J Pharmacol Exp Ther 320(3):994–1001
128. Ghosh A, Mandal AK, Sarkar S, Das N (2011a) Hepatoprotective and neuroprotective activity of liposomal quercetin in combating chronic arsenic induced oxidative damage in liver and brain of rats. Drug Deliv 18(6):451–459
129. Ghosh S, Dungdung SR, Chowdhury ST, Mandal AK, Sarkar S, Ghosh D, Das N (2011b) Encapsulation of the flavonoid quercetin with an arsenic chelator into nanocapsules enables the simultaneous delivery of hydrophobic and hydrophilic drugs with a synergistic effect against chronic arsenic accumulation and oxidative stress. Free Radic Biol Med 51(10):1893–1902
130. Wan Y, Tang MH, Chen XC, Chen LJ, Wei YQ, Wang YS (2014) Inhibitory effect of liposomal quercetin on acute hepatitis and hepatic fibrosis induced by concanavalin A. Braz J Med Biol Res 47(8):655–661
131. Sarkar A, Sil PC (2014) Iron oxide nanoparticles mediated cytotoxicity via PI3K/AKT pathway: role of quercetin. Food Chem Toxicol 71:106–115
132. Schwingel TE, Klein CP, Nicoletti NF, Dora CL, Hadrich G, Bica CG, Lopes TG, da Silva VD, Morrone FB (2014) Effects of the compounds resveratrol, rutin, quercetin, and quercetin nanoemulsion on oxaliplatin-induced hepatotoxicity and neurotoxicity in mice. Naunyn Schmiedeberg's Arch Pharmacol 387(9):837–848
133. Karthivashan G, Masarudin MJ, Kura AU, Abas F, Fakurazi S (2016) Optimization, formulation, and characterization of multiflavonoids-loaded flavanosome by bulk or sequential technique. Int J Nanomedicine 11:3417–3434
134. Tzankova V, Aluani D, Kondeva-Burdina M, Yordanov Y, Odzhakov F, Apostolov A, Yoncheva K (2017) Hepatoprotective and antioxidant activity of quercetin loaded chitosan/alginate particles in vitro and in vivo in a model of paracetamol-induced toxicity. Biomed Pharmacother 92:569–579
135. Xu MX, Wang M, Yang WW (2017) Gold-quercetin nanoparticles prevent metabolic endotoxemia-induced kidney injury by regulating TLR4/NF-κB signaling and Nrf2 pathway in high fat diet fed mice. Int J Nanomed 12:327–345
136. Abdelhalim MAK, Moussa SAA, Qaid HAY (2018) The protective role of quercetin and arginine on gold nanoparticles induced hepatotoxicity in rats. Int J Nanomedicine 13:2821–2825
137. Eftekhari A, Ahmadian E, Panahi-Azar V, Hosseini H, Tabibiazar M, Maleki Dizaj S (2018) Hepatoprotective and free radical scavenging actions of quercetin nanoparticles on aflatoxin B1-induced liver damage: in vitro/in vivo studies. Artificial Cells Nanomed Biotechnol 46(2):411–420
138. Fadda LM, Hagar H, Mohamed AM, Ali HM (2018) Quercetin and idebenone ameliorate oxidative stress, inflammation, DNA damage, and apoptosis induced by titanium dioxide nanoparticles in rat liver. Dose Response 16(4):1–9
139. Palle S, Neerati P (2018) Quercetin nanoparticles alter pharmacokinetics of bromocriptine, reflecting its enhanced inhibitory action on liver and intestinal CYP 3A enzymes in rats. Xenobiotica 48(10):1028–1036
140. Suke SG, Sherekar P, Kahale V, Patil S, Mundhada D, Nanoti VM (2018) Ameliorative effect of nanoencapsulated flavonoid against chlorpyrifos-induced hepatic oxidative damage and immunotoxicity in Wistar rats. J Biochem Mol Toxicol 32(5):e22050
141. Baowen Q, Yulin Z, Xin W, Wenjing X, Hao Z, Zhizhi C, Xingmei D, Xia Z, Yuquan W, Lijuan C (2010) A further investigation concerning correlation between anti-fibrotic effect of liposomal quercetin and inflammatory cytokines in pulmonary fibrosis. Eur J Pharmacol 642(1–3):134–139
142. Fahlman BM, Krol ES (2009) Inhibition of UVA and UVB radiation-induced lipid oxidation by quercetin. J Agric Food Chem 57(12):5301–5305
143. Naumov AA, Shatalin YV, Sukhomlin TK, Potselueva MM (2009) Effect of liposomes containing antioxidant, phospholipid, and amino acid on skin regeneration after chemical burn. Bull Exp Biol Med 147(4):531–536

144. Naumov AA, Potselueva MM (2010) Liposomal form of dihydroquercetin contributes to skin regeneration after thermal burns. Cell Tissue Biol 4(3):240–244
145. Tan Q, Liu W, Guo C, Zhai G (2011) Preparation and evaluation of quercetin-loaded lecithin-chitosan nanoparticles for topical delivery. Int J Nanomedicine 6:1621–1630
146. Castangia I, Nacher A, Caddeo C, Valenti D, Fadda AM, Diez-Sales O, Ruiz-Sauri A, Manconi M (2014) Fabrication of quercetin and curcumin bionanovesicles for the prevention and rapid regeneration of full-thickness skin defects on mice. Acta Biomater 10(3):1292–1300
147. Jangde R, Singh D (2016) Preparation and optimization of quercetin-loaded liposomes for wound healing, using response surface methodology. Artificial Cells Nanomed Biotechnol 44(2):635–641
148. Hatahet T, Morille M, Shamseddin A, Aubert-Pouessel A, Devoisselle JM, Begu S (2017) Dermal quercetin lipid nanocapsules: Influence of the formulation on antioxidant activity and cellular protection against hydrogen peroxide. Int J Pharm 518(1–2):167–176
149. Jangde R, Srivastava S, Singh MR, Singh D (2018) In vitro and In vivo characterization of quercetin loaded multiphase hydrogel for wound healing application. Int J Biol Macromol 115:1211–1217
150. Hatahet T, Morille M, Hommoss A, Devoisselle JM, Muller RH, Begu S (2018) Liposomes, lipid nanocapsules and smartCrystals(R): A comparative study for an effective quercetin delivery to the skin. Int J Pharm 542(1–2):176–185
151. Seong JS, Yun ME, Park SN (2018) Surfactant-stable and pH-sensitive liposomes coated with N-succinyl-chitosan and chitooligosaccharide for delivery of quercetin. Carbohydr Polym 181:659–667
152. Raghuwanshi N, Yadav TC, Srivastava AK, Raj U, Varadwaj P, Pruthi V (2019) Structure-based drug designing and identification of Woodfordia fruticosa inhibitors targeted against heat shock protein (HSP70–1) as suppressor for Imiquimod-induced psoriasis like skin inflammation in mice model. Mater Sci Eng C Mater Biol Appl 95:57–71
153. Yuan ZP, Chen LJ, Fan LY, Tang MH, Yang GL, Yang HS, Du XB, Wang GQ, Yao WX, Zhao QM, Ye B, Wang R, Diao P, Zhang W, Wu HB, Zhao X, Wei YQ (2006a) Liposomal quercetin efficiently suppresses growth of solid tumors in murine models. Clin Cancer Res 12(10):3193–3199
154. Yuan ZP, Chen LJ, Wei YQ, Fan LY, Tang MH, Yang GL (2006b) Nanoliposomal quercetin inhibits formation of malignant ascites of hepatocellular carcinoma. Chin J Cancer 25(8):941–945
155. Verma NK, Crosbie-Staunton K, Satti A, Gallagher S, Ryan KB, Doody T, McAtamney C, MacLoughlin R, Galvin P, Burke CS, Volkov Y, Gun'ko YK (2013) Magnetic core shell nanoparticles for drug delivery by nebulization. J Nanobiotechnol 11:1
156. Zhou X, Liu HY, Zhao H, Wang T (2018) RGD-modified nanoliposomes containing quercetin for lung cancer targeted treatment. OncoTargets Ther 11:5397–5405
157. Arbain NH, Salim N, Wui WT, Basri M, Rahman MBA (2018) Optimization of quercetin loaded palm oil ester based nanoemulsion formulation for pulmonary delivery. J Oleo Sci 67(8):933–940
158. Riaz MK, Zhang X, Wong KH, Chen H, Liu Q, Chen X, Zhang G, Lu A, Yang Z (2019) Pulmonary delivery of transferrin receptors targeting peptide surface-functionalized liposomes augments the chemotherapeutic effect of quercetin in lung cancer therapy. Int J Nanomedicine 14:2879–2902
159. Arbain NH, Salim N, Masoumi HRF, Wong TW, Basri M, Abdul Rahman MB (2019) In vitro evaluation of the inhalable quercetin loaded nanoemulsion for pulmonary delivery. Drug Deliv Transl Res 9(2):497–507
160. Wong MY, Chiu GN (2011) Liposome formulation of co-encapsulated vincristine and quercetin enhanced antitumor activity in a trastuzumab-insensitive breast tumor xenograft model. Nanomedicine 7(6):834–840
161. Kumar SR, Priyatharshni S, Babu VN, Mangalaraj D, Viswanathan C, Kannan S, Ponpandian N (2014) Quercetin conjugated superparamagnetic magnetite nanoparticles for in-vitro

analysis of breast cancer cell lines for chemotherapy applications. J Colloid Interface Sci 436:234–242

162. Zafar S, Negi LM, Verma AK, Kumar V, Tyagi A, Singh P, Iqbal Z, Talegaonkar S (2014) Sterically stabilized polymeric nanoparticles with a combinatorial approach for multi drug resistant cancer: in vitro and in vivo investigations. Int J Pharm 477(1–2):454–468
163. Sharma G, Park J, Sharma AR, Jung JS, Kim H, Chakraborty C, Song DK, Lee SS, Nam JS (2015) Methoxy poly(ethylene glycol)-poly(lactide) nanoparticles encapsulating quercetin act as an effective anticancer agent by inducing apoptosis in breast cancer. Pharm Res 32(2):723–735
164. Pandey SK, Patel DK, Thakur R, Mishra DP, Maiti P, Haldar C (2015) Anti-cancer evaluation of quercetin embedded PLA nanoparticles synthesized by emulsified nanoprecipitation. Int J Biol Macromol 75:521–529
165. Lv L, Liu C, Chen C, Yu X, Chen G, Shi Y, Qin F, Ou J, Qiu K, Li G (2016) Quercetin and doxorubicin co-encapsulated biotin receptor-targeting nanoparticles for minimizing drug resistance in breast cancer. Oncotarget 7(22):32184–32199
166. Suksiriworapong J, Phoca K, Ngamsom S, Sripha K, Moongkarndi P, Junyaprasert VB (2016) Comparison of poly(epsilon-caprolactone) chain lengths of poly(epsilon-caprolactone)-co-d-alpha-tocopheryl-poly(ethylene glycol) 1000 succinate nanoparticles for enhancement of quercetin delivery to SKBR3 breast cancer cells. Eur J Pharm Biopharm 101:15–24
167. Minaei A, Sabzichi M, Ramezani F, Hamishehkar H, Samadi N (2016) Co-delivery with nano-quercetin enhances doxorubicin-mediated cytotoxicity against MCF-7 cells. Mol Biol Rep 43(2):99–105
168. Tavano L, Mauro L, Naimo GD, Bruno L, Picci N, Ando S, Muzzalupo R (2016) Further evolution of multifunctional niosomes based on pluronic surfactant: dual active targeting and drug combination properties. Langmuir 32(35):8926–8933
169. Zuo J, Jiang Y, Zhang E, Chen Y, Liang Z, Zhu J, Zhao Y, Xu H, Liu G, Liu J, Wang W, Zhang S, Zhen Y (2019) Synergistic effects of 7-O-geranylquercetin and siRNAs on the treatment of human breast cancer. Life Sci 227:145–152
170. Hemati M, Haghiralsadat F, Yazdian F, Jafari F, Moradi A, Malekpour-Dehkordi Z (2019) Development and characterization of a novel cationic PEGylated niosome-encapsulated forms of doxorubicin, quercetin and siRNA for the treatment of cancer by using combination therapy. Artificial Cells Nanomed Biotechnol 47(1):1295–1311
171. Xu G, Shi H, Ren L, Gou H, Gong D, Gao X, Huang N (2015) Enhancing the anti-colon cancer activity of quercetin by self-assembled micelles. Int J Nanomedicine 10:2051–2063
172. Martirosyan A, Grintzalis K, Polet M, Laloux L, Schneider YJ (2016) Tuning the inflammatory response to silver nanoparticles via quercetin in Caco-2 (co-)cultures as model of the human intestinal mucosa. Toxicol Lett 253:36–45
173. Ren KW, Li YH, Wu G, Ren JZ, Lu HB, Li ZM, Han XW (2017) Quercetin nanoparticles display antitumor activity via proliferation inhibition and apoptosis induction in liver cancer cells. Int J Oncol 50(4):1299–1311
174. Ghosh A, Mandal AK, Sarkar S, Panda S, Das N (2009) Nanoencapsulation of quercetin enhances its dietary efficacy in combating arsenic-induced oxidative damage in liver and brain of rats. Life Sci 84(3–4):75–80
175. Hu J, Wang J, Wang G, Yao Z, Dang X (2016) Pharmacokinetics and antitumor efficacy of DSPE-PEG2000 polymeric liposomes loaded with quercetin and temozolomide: Analysis of their effectiveness in enhancing the chemosensitization of drug-resistant glioma cells. Int J Mol Med 37(3):690–702
176. Mandal AK, Das S, Mitra M, Chakrabarti RN, Chatterjee M, Das N (2008) Vesicular flavonoid in combating diethylnitrosamine induced hepatocarcinoma in rat model. J Exp Ther Oncol 7(2):123–133
177. Dabbagh-Bazarbachi H, Clergeaud G, Quesada IM, Ortiz M, O'Sullivan CK, Fernandez-Larrea JB (2014) Zinc ionophore activity of quercetin and epigallocatechin-gallate: from Hepa 1-6 cells to a liposome model. J Agric Food Chem 62(32):8085–8093

178. Mandal AK, Ghosh D, Sarkar S, Ghosh A, Swarnakar S, Das N (2014) Nanocapsulated quercetin downregulates rat hepatic MMP-13 and controls diethylnitrosamine-induced carcinoma. Nanomedicine 9(15):2323–2337
179. Varshosaz J, Jafarian A, Salehi G, Zolfaghari B (2014) Comparing different sterol containing solid lipid nanoparticles for targeted delivery of quercetin in hepatocellular carcinoma. J Liposome Res 24(3):191–203
180. Bishayee K, Khuda-Bukhsh AR, Huh SO (2015) PLGA-loaded gold-nanoparticles precipitated with quercetin downregulate HDAC-Akt activities controlling proliferation and activate p53-ROS crosstalk to induce apoptosis in hepatocarcinoma cells. Mol Cells 38(6):518–527
181. Guan X, Gao M, Xu H, Zhang C, Liu H, Lv L, Deng S, Gao D, Tian Y (2016) Quercetin-loaded poly (lactic-co-glycolic acid)-d-alpha-tocopheryl polyethylene glycol 1000 succinate nanoparticles for the targeted treatment of liver cancer. Drug Deliv 23(9):3307–3318
182. Hu F, Bu YZ, Liang R, Duan RM, Wang S, Han RM, Wang P, Ai XC, Zhang JP, Skibsted LH (2013) Quercetin and daidzein beta-apo-14′-carotenoic acid esters as membrane antioxidants. Free Radic Res 47(5):413–421
183. Sahu S, Saraf S, Kaur CD, Saraf S (2013) Biocompatible nanoparticles for sustained topical delivery of anticancer phytoconstituent quercetin. Pak J Biol Sci 16(13):601–609
184. Zhu X, Zeng X, Zhang X, Cao W, Wang Y, Chen H, Wang T, Tsai HI, Zhang R, Chang D, He S, Mei L, Shi X (2016) The effects of quercetin-loaded PLGA-TPGS nanoparticles on ultraviolet B-induced skin damages in vivo. Nanomedicine 12(3):623–632
185. Sandhu PS, Kumar R, Katare OP, Singh B (2017) Surface-tailored nanomixed micelles containing quercetin-salicylic acid physical complex for enhanced cellular and in vivo activities: a quality by design perspective. Nanomedicine 12(11):1281–1303
186. George D, Maheswari PU, Begum K (2019) Synergic formulation of onion peel quercetin loaded chitosan-cellulose hydrogel with green zinc oxide nanoparticles towards controlled release, biocompatibility, antimicrobial and anticancer activity. Int J Biol Macromol 132:784–794
187. Barzilai A, Rahamimoff H (1983) Inhibition of Ca2+–transport ATPase from synaptosomal vesicles by flavonoids. Biochim Biophys Acta Biomembr 730(2):245–254
188. Fawzy AA, Vishwanath BS, Franson RC (1988) Inhibition of human non-pancreatic phospholipases A2 by retinoids and flavonoids. Mechanism of action. Agents Actions 25(3–4):394–400
189. Theoharidesi TC, Asadp S, Panagiotidoui S (2012) A case series ofa luteolin formulation (Neuroprotek®) in children with autism spectrum disorders. Int J Immunopathol Pharmacol 25(2):317–323
190. Rich GT, Buchweitz M, Winterbone MS, Kroon PA, Wilde PJ (2017) Towards an understanding of the low bioavailability of quercetin: a study of its interaction with intestinal lipids. Nutrients 9(2). https://doi.org/10.3390/nu9020111

Polyphenol-Loaded Nanomedicines Against Skin Aging

Deepika Singh, Mahfooz ur Rahman, and Vikas Kumar

1 Introduction

The skin serves as an outer barrier and is subjected to multiple oxidative stresses, including ultraviolet radiation and pollutants. In the skin, reactive oxygen species (ROS) and additional free radicals are produced as a reaction to these oxidative as assaults [1].

The skin is accountable in maintaining the equilibrium between pro-oxidants and antioxidants. It is composed of a multitude of lipophilic substances, i.e. carotenoids, and hydrophilic substances, i.e. uric acid, and is sheltered from the harmful impacts of free radicals [2].

Various antioxidant enzymes (catalase) and endogenous (glutathione) and exogenous (vitamin C) molecules which act as a first defence are reduced in reactive oxygen species. When biomolecules are oxidized, biological protective systems repair or replace them. However, biomolecules are gradually oxidized irreversibly, and the accumulation of these biomolecules changes the biological functions over time, ultimately leading to aging and age-related diseases [3].

Due to enhanced exposure to exogenous sources of ROS, oxidative stress impairs the equilibrium between oxidants and antioxidants. Lessening ROS-induced skin harm is possible only by the administration of topical antioxidant, as it can promote the antioxidant status of the skin.

Aging is an extremely complicated method characterized by successive occurrences that encourage changes in an individual organism's ordinary functioning over time. These changes occur in the organismic function at all levels starting

D. Singh (✉) · M. u. Rahman
Natural Product Drug Discovery Laboratory, Department of Pharmaceutical Sciences, Sam Higginbottom University of Agriculture, Technology and Sciences, Allahabad, India

V. Kumar
Department of Pharmaceutical Sciences, Shalom Institute of Health & Allied Sciences, SHUATS, Allahabad, Uttar Pradesh, India

© Springer Nature Singapore Pte Ltd. 2020

M. Rahman et al. (eds.), *Nanomedicine for Bioactives*,
https://doi.org/10.1007/978-981-15-1664-1_10

with minor changes in the manufacturing of cellular protein and alterations of cellular macro-features such as the existence and distribution of signal receptors. Aging modifications also happen on a bigger scale and include a widespread reduction in function effectiveness in whole organs or systems, which in turn may lead to disease development [4].

In the age-related factors, the real aetiology is commonly multifactorial and naturally complex, and prospective modifications generally occur. Different factors related to aging, i.e. Alzheimer diseases, cancer, and heart diseases, occur due to basic biological processes [5].

From vascular plants or tracheophytes and numerous marine creatures, polyphenols compounds are available as secondary molecule and belong to the family of ubiquitous compound. Natural polyphenols are precious compounds found in radical oxygen species in crops, fruits, legumes, chocolate, tea, wine, and aquatic creatures. These phytocompounds are available from easy molecules to complicated constructions which have one or more hydroxy functions in common with the presence of benzenic cycles. These molecules plays an important role in the development, reproduction, resistant against pathogen and disease. These capabilities make polyphenols useful in cosmetic formulations for either the therapy of different illnesses such as inflammation and cancer or for anti-aging reasons. These phytochemicals, in specific, contribute significantly to plant colour and organoleptic characteristics, but in case of aquatic creatures, such as microalgae and cyanobacteria, polyphenols work against UV exposure in the antioxidant reaction [6]. In reality, Ignat et al. have shown that microalgae and cyanobacteria contain various classes of flavonoids, such as flavonols, anthocyanidins, flavones, catechins, and phenolic compounds in spite of the reality that these animals have different evolution than land-dwelling crops [7].

Polyphenols show beneficial impacts in the treatment of inflammation, control the cell-signal transduction, and stimulate the pathways of detoxification. Polyphenols are found to be effective in health promotion, moderating the risks of chronic illness (governed by the absorption in systemic circulation and metabolism), and vigorously interacting with the microbiota. Generally they have low bioavailability when consider to external environment and possesses different molecular weight and chemical structure, little stability in gastrointestinal, comprehensive stage II metabolism, low hydrosolubility and fast elimination. As a result, there are restricted clinical applications of PPH [8].

2 Classification of Polyphenols

There are various number of polyphenolic compounds attached together in different assemblies and depend on the fundamental differences in the skeleton of the chemical compounds, i.e. methylation, oxidation, glycosylation, hydroxylation, and potential relation with other compounds. Polyphenols are categorized into four classes:

- Flavonoids
- Coumarins

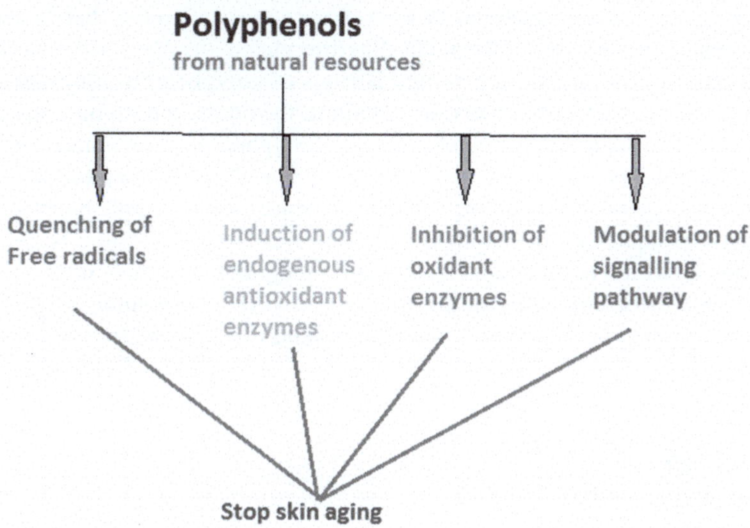

Fig. 1 The use of polyphenol against skin aging

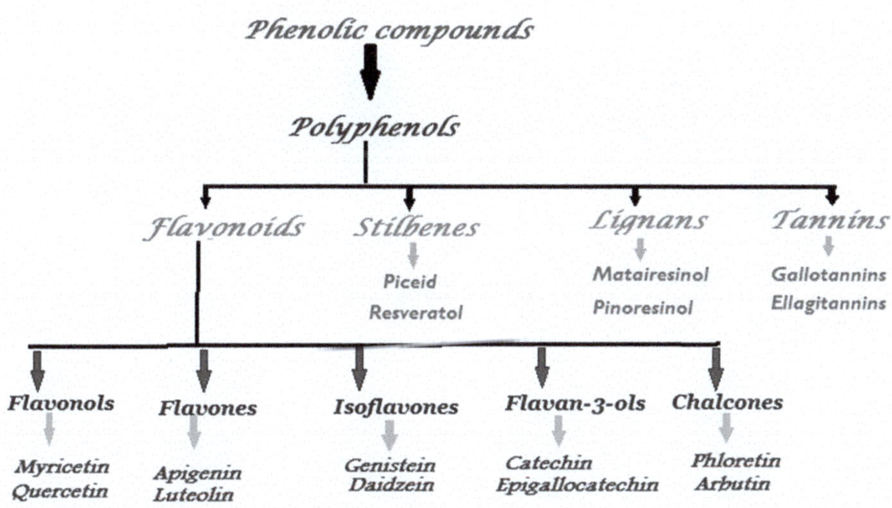

Fig. 2 Classification of polyphenols

- Stilbenes
- Tannins

Flavonoids are composed of C15 compounds with the framework of C6-C3-C6 and a group of three carbons linking two benzene rings together. The group C3 structure determines the classification of the compounds. Typical flavonoids show A-, B-, and C-ring on the left hand with the A-ring. The A-ring is derived from molecule of three malonyl-CoA which is condensed, and the other B-ring is derived from molecule p-coumaroyl-CoA. Literature accounted that in most of the flavonoids, A-ring is either meta-dihydroxylated or meta-trihydroxylated. Different groups attached on the A- and B-rings of polyphenolic compounds distinguish the anthocyanidins, catechins, flavones, flavonols, flavanones, and isoflavones together with the structure of the ring-C. Polyphenolic compounds stilbenes, having a structure of C6-C2-C6, are chemically diarylethene in nature, displayed on each carbon atom of the double bond with a core ethane double bond replaced with phenyl groups. Tannins comprise a group of compounds with a broad structural diversity which have a capacity to bind, precipitate proteins, and form complex structure. Molecules of tannins typically involve at minimum 12 groups of hydroxyl and phenyls (5 groups) which act to bind the proteins (protein binders). They are categorized into three subclasses, depending on the form of easy unit connection [9]:

- Condensed tannins
- Hydrolysable tannins
- Complex tannins

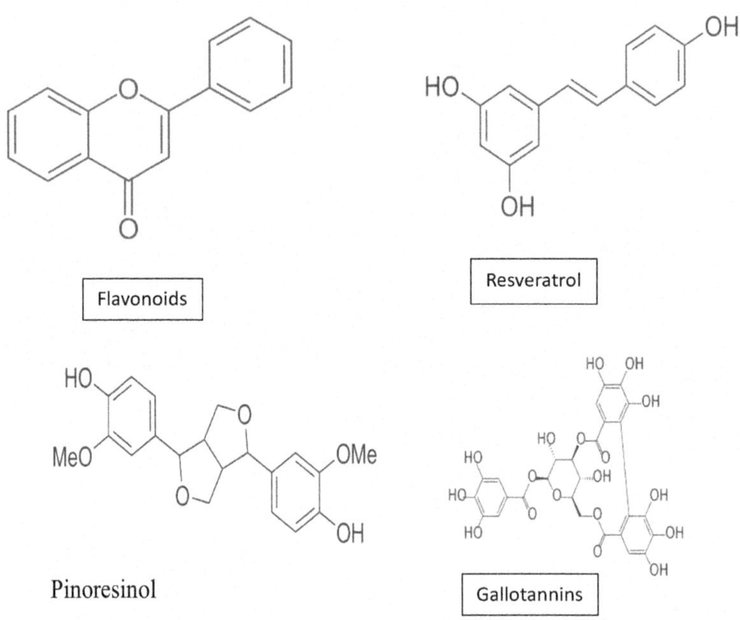

Fig. 3 Polyphenol compounds

3 Antioxidants

Homologue of vitamin E, α-tocopherol is a lipid soluble, active, and membrane-bound antioxidant of the various enzymatic and non-enzymatic antioxidants in the skin. Tocopherol acts mainly through two mechanism:

- It gives one of its electrons to a free radical as an antioxidant and maintain its stability.
- They are helpful in slowing down the collagen breakdown process.

If the endogenous antioxidants are inadequate to decrease oxidative stress, then in various cosmetic formulations, antioxidants from various sources can be introduced to the skin [10].

In order to study the physiology and function of the skin, retinol was intensively investigated. Retinol is stored or converted into retinoic acid in the skin as a retinyl ester. Retinoic acid is the biologically active form of retinoid, but retinols produce less irritation and develop effects similar to retinoic acid. Because of this quality, the use of retinol makes it more favourable in cosmetic products. Researchers also not provided a wide data on the use of trans-retinol regarding its quantitative penetration and skin distribution although it is widely used in the cosmetics sector [11].

The most potent antioxidants found in plants are flavonoids, a class of polyphenols, and catechin belonging to this class was found plentifully in the extract of green tea. The other polyphenolic compound is resveratrol, regarded as a major antioxidant in the stilbene family, discovered abundantly in the skin of white wine, cranberries, grapes, dark chocolate, nuts, and blue berries [12]. These compounds demonstrate isomerism (geometric isomerism), but several biological impacts were exhibits by trans-resveratrol, including anticancer, anti-aging, and antioxidant activity. In dermal products, trans-resveratrol is a remarkable medicine to be incorporated due to its better property such as short half-life, easy metabolism, good bioavailability, and good solubility [13].

Another best polyphenol compound is curcumin which provides colour to the rhizomes of *Curcuma longa*. It is a most frequently used drug in dermatological products and traditional medicines [14].

Among others, the most widely used antioxidants for topical use are vitamin C, coenzyme Q10, and quercetin. Topical antioxidants when administered through the skin can boost antioxidant capacity and decrease skin damage caused by reactive oxygen species. The physical and chemical characteristics of polyphenols and other antioxidants make them excellent candidates in the treatment of skin diseases or cosmetics, but due to their poor water solubility, it is difficult to make their application in some cases. Various tremendous antioxidants, i.e., vitamin C, due to their positive impacts on the skin, are used in dermal application and cosmeceuticals [15]. However, in heterogeneous structures, vitamin C has bad chemical stability and may be subject to dismutation reactions. One approach to prevent this sort of inconvenience is to use distinct nanocarriers to formulate these antioxidants as nano.

The administration of phenolic compounds therefore involves the formulation of a completed protective product capable of maintaining the polyphenol's structural integrity until consumption or administration, masking its flavour, increasing its water solubility and bioavailability, and transmitting it exactly to a physiological goal.

4 Polyphenol Bioavailability

While describing the term bioavailability in case of polyphenol, it plays a major role and is an important element to play a role in managing diseases. However, the function of polyphenols to determine the aetiology of human illness and therapy is barely decisive, and sometimes questions arise on how the polyphenolic compounds work or what the mechanisms of action within the body are. And while some researchers have disputes about the mechanism of action, epidemiological studies already indicate that polyphenols have positive impacts on health. However, in order to preserve a correct view on the emerging potential health advantages, the real implications of diseases need to be recognized continuously [16].

Low bioavailability of polyphenols is primarily because of short absorption following consumption in the gastrointestinal (GI) tract of human, biotransformation of polyphenol rings within the intestine, and fast body clearance from the body. Various polyphenols (PPHs) are particularly accessible as in the form of glycosylated compounds, and glycosylated PPHs downregulate the diffusion process across obstacles in the gastrointestinal tract of humans. Some PPHs are not absorbed directly from enterocytes because of low bioavailability in the acidic condition of the stomach and alkaline situation of the small intestine [17]. In addition, PPH is extensively transformed in the enterocytes of the small intestine through phase II pathways, predominantly methylation, glucuronidation, and sulphation, and then metabolized in the liver to facilitate their rapid excretion. Several approaches have been utilized to improve chemical stability or permeability of polyphenols for dermal application. Various chemical additives are added while formulating PPH which makes it a suitable candidate in the cosmetic preparation:

- Dissolving agent: enhance solubility
- Decreasing agent: preserve the structure
- Complexing agent: interact with the stage I and/or II enzyme inhibitors to prevent biotransformation of polyphenols [18]

5 Nanotechnology

Nanotechnology-based delivery systems were created to overcome these constraints and improve therapeutic applications of polyphenols, and nanoencapsulation was a successful approach among all [19]. Nanocarriers are an exciting means among the present stabilization methods. The use of nanoformulated polyphenol instead of free compounds makes it a successful candidate for the delivery of drugs. Nanoencapsulation is now an efficient strategy to improving solubility, minimizing the process of degradation, reducing toxicity, and controlling polyphenol's active absorption and biological reaction. To prevent these disadvantages, nanodistribution schemes have been created that can retain the structural integrity of the bioactive molecules. There are presently various nanocarrier techniques available, and nanoencapsulated materials are widely used in the food, pharmaceutical, and cosmetic industries but also in many other areas such as personal care, agricultural products, veterinary medicine, industrial chemicals, biotechnology, biomedicine, and sensors [20].

In the field of encapsulation, various nanovector agents, e.g. strong dispersion, nanoparticles, ethosomes, niosomes, liposomes and cyclodextrins, have appeared to enhance the bioavailability, solubility, delivery to site of action, and distribution of polyphenol. In the inner framework (core-shell or matrix), the structure and physical condition of the active substance encapsulated may vary [21]. Additionally, the encapsulation efficiency of polyphenol-encapsulated nanoparticles of at least 60% (the proportion of drug trapping effectiveness) is calculated as follows:

$$\text{Encapsulation efficience (\%)} = \text{mass of drug in final product/total mass of drugs} \times 100$$

6 Nanoencapsulation System

Nanotechnology is a branch of science which involves the development of units, size varying in diameter from 1 to 1000 nm, and end products which are obtained to display characteristics or different phenomena which attribute to its size. Encapsulation of nanomaterials relates to several techniques based on principles of chemistry, physical and physiochemistry. There are three types of nanoencapsulation system:

- Physical procedures (e.g. method of air suspension, pan coating techniques).
- By chemical method nanoencapsulation: it involves in situ polymerization method and interfacial techniques that need polymerization of monomer at the interface of two immiscible substances by adding an internal cross-linker.
- Physiochemical procedures involve nanoprecipitation method, phase separation and coacervation, and melt dispersion method. These approaches produce drug nanosuspensions or nanoparticles of nanometer size which are stable via approaches of small particle size. The suitability of this procedure is to improve drug loading capability, persistence at the target locations and the impact of permeation and retention.

Several studies have recently analysed the delivery of polyphenol-loaded nanoparticles which depends on the characteristic of polymers, i.e. biodegradable and biocompatible in nature [22].

7 Different Types of Nanocarriers

In recent years, nanocarriers have considered and appeared as a promising agent for drug delivery scheme that offers numerous benefits when compared to conventional drugs. In the literature, the utilization of nanocarriers in cosmetics and its benefits have been reviewed.

It has several advantages over passive delivery agent:

- Enhanced surface area
- Larger stability
- Enhanced solubility
- Protection from degradation
- Enhanced loading of drug
- Decreased drug irritation
- Enhanced skin permeation

Various types of nanocarriers are as follows:

- Liposomes: Liposomes are self-closed vesicles, spherical in shape in which lipid bilayer is composed of synthetic or natural phospholipids, colloidal in dimensions, and in the sequester part of the solvent, phospholipid bilayers freely float into the interior.
- Nanocapsules: Nanocapsules are vesicular systems at the nanoscale level (10–1000 nm), and inner liquid core is encapsulated by polymeric membrane.
- Solid lipid nanoparticles: SLC are submicron colloidal carriers which are made up of physiological lipid distributed in aqueous solution of surfactant or in water.
- Nanocrystals: They are the masses which are made up of thousands of atoms and conglomerate into a bunch and range from 10 to 400 nm in size. Nanocrystals are found to be effective in the carriage of the drugs which are poorly soluble.
- Niosomes: They are vesicles which are made up of nonionic surfactants and are used to encapsulate hydrophilic, lipophilic, and amphiphilic drug moieties. They have advantages over liposomes due to its high entrapment efficiency, better chemical stability, and improved penetration.

Many products are available in the market as lipid nanocarrier for cosmetic applications.

Nowadays, lipid-based nanodrugs have been scrutinized for dermal application which are comprised of nanoemulsions, nanostructured lipid carriers, and strong lipid nanoparticles.

SLN contains strong lipids and remains lipids at ambient temperature, while NLC consists of solid and liquid lipid mixtures. Second generation of lipid nanoparticles known as nanostructured lipid carriers are stable to leak, and drug loading capacity is high.

Benefits of SLN and NLC over other nanocarriers:

- Usage of physiological lipids in the structure
- To avoid organic solvents while in preparation
- Chances of producing focused lipid suspensions
- Accessibility of proven proper scale-up processes [23]

Topical application of SLN and NLC structures with polyphenol compounds such as retinoic acid, tocopherol, and vitamin A has been researched.

The study outlined a latest summary of the most representative molecules such as ellagic acid, resveratrol, coenzyme Q10, epigallocatechin-3-gallate, ellagitannins, curcumin, and quercetin in physicochemical nanoencapsulated polyphenol compounds.

Table 1 Nanocarrier encapsulated polyphenolic compounds

Polyphenolic compounds	Nano system	System components	References
Epigallocatechin-3-gallate (EGCG)	Polyester nanoparticles	PLGA biodegradable nanoparticles	Italia et al. [24]
	Serum albumin nanoparticles	Bovine serum albumin (BSA)	Zu et al. [25]
	Carbohydrate matrix	Chitosan nanoparticles	Dube et al. [26]
Quercetin (QC)	Solid lipid nanoparticles	Soya lecithin, Tween-80, and PEG	Li et al. [27]
	Cyclodextrins	2-(Hydroxy-propyl)-β-cyclodextrins (HP-β-CD)	Mercader-Ros et al. [28]
Resveratrol (RE)	Micelles	PCL-PEG	Lu et al. [29]
	Liposomes	Distearoylphosphatidylcholine (DSPC)	Basavaraj and Betageri [30]
	Solid lipid nanoparticles	Cetyl palmitate and polysorbate 60	Neves et al. [31]
Ellagitannins	Carbohydrate matrix	Gelatin nanoparticles	Li et al. [32]
Ellagic acid	Carbohydrate matrix	Chitosan nanoparticles	Arulmozhi et al. [33]
	Polyester nanoparticles	PLGA nanoparticles	Bala et al. [34]
Oleuropein	Serum albumin nanoparticles	Bovine serum albumin (BSA)	Mowlazadeh et al. [35]
	Polyester nanoparticles	PLGA-mPEG	Guan et al. [36]
Curcumin (CU)	Solid lipid nanoparticles	Dioctyl sodium sulfosuccinate and Poloxamer 188	Li et al. [37]
	Serum albumin nanoparticles	Human serum albumin	Kim et al. [38]
Q10	Ultra-small nanostructured lipid carrier (NLC)		Lohan et al. [39]

8 Different Aspects of Nanocarriers Encapsulated Polyphenolic Compounds in Skin Aging

8.1 Nanocarriers for Resveratrol Administration

Resveratrol-loaded solid lipid nanoparticles (SLN) were developed for application on the skin and uptake of cellular components, and transportation and keratinocyte internalization were considered. Within short time or 15 min, nanoparticles crossed the cell membrane and the SLN-encapsulated drug showed greater intracellular delivery, solubility, and solution compared to resveratrol in [40].

Nanocarriers such as liposomes and niosomes increase the residence time of drug in the different layers of the skin (stratum corneum and epidermis), decreasing the systemic absorption of drug. Niosomes are economic in cost and have high chemical stability and benefits over liposomes. Resveratrol encapsulated niosomes showed greater delivery of resveratrol in deeper layers of the skin relative to liposomes. Niosomes consist of various proportions of different polymers such as glycerol monooleate (PEC), cholesterol, or polypolyglyceryl-3 dioleate (PLU). They consist of PEC (30 mg/mL) and cholesterol (30 mg/mL) which made it possible to accumulate the largest quantity of resveratrol in the stratum corneum [41].

Another option is the concurrent encapsulation in lipid-core nanocapsules of resveratrol with curcumin, as previously suggested. Studies of skin penetration showed enhanced resveratrol delivery in deeper layers of the skin at co-delivery compared to one individual polyphenol encapsulation [42].

8.2 Nanocarriers for Quercetin Delivery

Quercetin has been shown to be the largest antioxidant activity flavonoids, characterized by various processes. In addition, quercetin's topical application reduces skin harm caused by UV radiation exposure. However, quercetin's skin protective activity is inadequate because of its bad penetration in percutaneous layer, and this limits the compound's quantity to reach its target site.

Quercetin nanostructured lipid carriers were created using an emulsification method based on solvent (chloroform/acetone) and assessed for dermal application. The significant drawback of manufacturing technique while formulating a solvent-free, strong lipid-based quercetin nanosystem is that the use of organic solvents produces harmful solvents which are caused by the residues of solvent [43].

8.3 Delivery of Nanocarriers for Coenzyme Q10 (CoQ10)

In humans, compound CoQ10 is a lipophilic cell antioxidant which is synthesized endogenously in the body. CoQ10 acts as an antioxidant and works in the skin epidermis layer with ten times concentration compared to in the dermis. Because of its capacity to decrease in vivo aging caused by sunlight and ascribed to enhance the components manufacturing of basal membrane, protection from oxidative stress, proliferation of fibroplast, this quality makes its excellent candidate to use these agent in cosmeceuticals [44] .

8.4 Nanocarriers for Vitamin C

Because of its biological activity, vitamin C was used in cosmetic products as a non-enzymatic soluble antioxidant. Vitamin C can safeguard biomolecules against oxidative degradation, by its capacity to decrease ROS, quenching, and free radicals. Vitamin C, however, is volatile and can readily be oxidized under aerobic circumstances, with increases in the rate [45].

8.5 Nanocarriers for Curcumin Delivery

Curcumin is one of the most significant phytopolyphenolic compounds which are composed of three curcuminoids and extracted from the rhizomes of the plant *Curcuma longa* (common name turmeric). It is a type of spice and used as curries in South Asian cuisine. Ayurveda and Chinese medicine suggested its application or the therapy in wound healing, dermal preparation, and inflammation. It is a powerful free radical scavenger that inhibits the peroxidation of lipids and quenches singlet oxygen, hydroxyl radicals, and superoxide anions. Free curcumin, though, degrades quickly when exposed to sunlight, decreasing its antioxidant ability [46].

To prevent this impact, one approach is to encapsulate it into biodegradable, secure, cheap, and commercially available polymeric nanospheres that protect against photodegradation and retain its antioxidant capacity.

8.6 Nanocarriers for Tocopherol Delivery

It is well reflected in the literature that tocopherol is a lipid-soluble, effective, and membrane-bound compound and well-existing antioxidant in the skin. Literature reported the decreasing constituents of tocopherol in the stratum corneum of the skin, and continuous effects of UV lights on the skin also diminish the antioxidant levels in the skin along with the time [47].

To stop aging, tocopherol is applied topically on the skin's upper layer to regenerate the antioxidant levels and safeguard against oxidative stress.

An extremely viscous skin irritant and light-sensitive fluid is synthetically accessible tocopherol. This property makes the incorporation of tocopherol in topical/cosmetic formulations hard. Another alternative is to formulate tocopherol in nanocarriers to create an aqueous formulation that is non-irritating, stable, and cosmetically attractive. The findings established that in comparison with nanoemulsions, nanostructured lipid carriers (NLC) provide higher release of tocopherol; hence, it is concluded that NLC is an excellent nanocarrier for dermal preparation and cosmetic products [48].

9 Conclusion

Phenolic compounds are strong compounds which are derived from plants and are used in a variety of diseases such as diabetes, skin aging, cancer, inflammation, etc. Due to its limited solubility, stability, and poor bioavailability, it is not used in pharmaceutical industries, food, or cosmetics. In the recent times, approaches regarding the drug delivery provides the uses of active pharmaceutical ingredient make it a fascinating means to prospective their activity. Nanocarrier are also preferred when formulating the polyphenol compounds for cosmetic purpose or as anti-aging agent. Nowadays, lipid core nanocapsules appear to have better potential in skin aging and in making polyphenol into a usable compound. As food additives and nutritional supplements, use of plant-derived polyphenolic compounds should be increased in industry in drugs or for cosmetic uses.

Conflict of Interest None.

References

1. Byrd AL, Belkaid Y, Segre JA (2018) The human skin microbiome. Nat Rev Microbiol 16:143–155
2. Shukla T, Upmanyu N, Agrawal M, Saraf S, Saraf S, Alexander A (2018) Biomedical applications of microemulsion through dermal and transdermal route. Biomed Pharmacother 108:1477–1494
3. Jimenez-Escrig A (2006) Polyphenol and carotenoid protection in biological systems through the modulation of antioxidant enzymes. Curr Enzym Inhib. https://doi.org/10.2174/157340806777934793
4. Kammeyer A, Luiten RM (2015) Oxidation events and skin aging. Ageing Res Rev 21:16–29
5. Masaki H (2010) Role of antioxidants in the skin: anti-aging effects. J Dermatol Sci 58(2):85–90
6. Petti S, Scully C (2009) Polyphenols, oral health and disease: a review. J Dent 37(6):413–423
7. Ignat I, Volf I, Popa VI (2011) A critical review of methods for characterisation of polyphenolic compounds in fruits and vegetables. Food Chem 126(4):1821–1835
8. Williamson G, Manach C (2005) Bioavailability and bioefficacy of polyphenols in humans. II. Review of 93 intervention studies. Am J Clin Nutr 81(1 Suppl):243S–255S
9. Handique JG, Baruah JB (2002) Polyphenolic compounds: an overview. React Funct Polym. https://doi.org/10.1016/S1381-5148(02)00091-3
10. Inoue T, Tatemori S, Muranaka N, Hirahara Y, Homma S, Nakane T et al (2012) The identification of vitamin e homologues in medicinal plant samples using ESI(+)-LC-MS3. J Agric Food Chem 60(38):9581–9588
11. Romieu I, Parra S, Hernández JF, Madrigal H, Willett W, Hernández M (1999) Questionnaire assessment of antioxidants and retinol intakes in Mexican women. Arch Med Res 30(3):224–239
12. Yilmaz Y, Toledo RT (2004) Major flavonoids in grape seeds and skins: antioxidant capacity of catechin, epicatechin, and gallic acid. J Agric Food Chem 52(2):255–260

13. Buryanovskyy L, Fu Y, Boyd M, Ma Y, Hsieh TC, Wu JM et al (2004) Crystal structure of quinone reductase 2 in complex with resveratrol. In: Biochemistry
14. Bisht K, Wagner KH, Bulmer AC (2010) Curcumin, resveratrol and flavonoids as anti-inflammatory, cyto- and DNA-protective dietary compounds. Toxicology 278(1):88–100
15. Bouayed J, Hoffmann L, Bohn T (2011) Total phenolics, flavonoids, anthocyanins and antioxidant activity following simulated gastro-intestinal digestion and dialysis of apple varieties: bioaccessibility and potential uptake. Food Chem 128(1):14–21
16. Scalbert A, Williamson G (2000) Dietary intake and bioavailability of polyphenols. J Nutr 130(8S Suppl):2073S–2085S. https://doi.org/10.1093/jn/130.8.2073S
17. Bohn T (2014) Dietary factors affecting polyphenol bioavailability. Nutr Rev 72(7):429–452
18. Landete JM (2012) Updated knowledge about polyphenols: functions, bioavailability, metabolism, and health. Crit Rev Food Sci Nutr 52(10):936–948
19. Ahmed W, Elhissi A, Subramani K (2012) Introduction to Nanotechnology. In: Nanobiomaterials in clinical dentistry. Elsevier, Amsterdam
20. Bilia AR, Isacchi B, Righeschi C, Guccione C, Bergonzi MC (2014) Flavonoids loaded in nanocarriers: an opportunity to increase oral bioavailability and bioefficacy. Food Nutr Sci. https://doi.org/10.4236/fns.2014.513132
21. Hu B, Liu X, Zhang C, Zeng X (2017) Food macromolecule based nanodelivery systems for enhancing the bioavailability of polyphenols. J Food Drug Anal 25(1):3–15
22. Pinto Reis C, Neufeld RJ, Ribeiro AJ, Veiga F (2006) Nanoencapsulation I. Methods for preparation of drug-loaded polymeric nanoparticles. Nanomedicine 2(1):8–21
23. Vinardell MP, Mitjans M (2015) Nanocarriers for delivery of antioxidants on the skin. Cosmetics. https://doi.org/10.3390/cosmetics2040342
24. Italia JL, Datta P, Ankola DD, Kumar MNVR (2008) Nanoparticles enhance per oral bioavailability of poorly available molecules: epigallocatechin gallate nanoparticles ameliorates cyclosporine induced nephrotoxicity in rats at three times lower dose than oral solution. J Biomed Nanotechnol. https://doi.org/10.1166/jbn.2008.341
25. Zu YG, Yuan S, Zhao XH, Zhang Y, Zhang XN, Jiang R (2009) Preparation, activity and targeting ability evaluation in vitro on folate mediated epigallocatechin-3-gallate albumin nanoparticles. Yao Xue Xue Bao 44(5):525–531
26. Dube A, Ng K, Nicolazzo JA, Larson I (2010) Effective use of reducing agents and nanoparticle encapsulation in stabilizing catechins in alkaline solution. Food Chem. https://doi.org/10.1016/j.foodchem.2010.03.027
27. Li HL, Bin ZX, Ma YK, Zhai GX, Li LB, Lou HX (2009) Enhancement of gastrointestinal absorption of quercetin by solid lipid nanoparticles. J Control Release 133(3):238–244
28. Mercader-Ros MT, Lucas-Abellán C, Fortea MI, Gabaldón JA, Núñez-Delicado E (2010) Effect of HP-β-cyclodextrins complexation on the antioxidant activity of flavonols. Food Chem. https://doi.org/10.1016/j.foodchem.2009.05.061
29. Lu X, Ji C, Xu H, Li X, Ding H, Ye M et al (2009) Resveratrol-loaded polymeric micelles protect cells from Aβ-induced oxidative stress. Int J Pharm 375(1–2):89–96
30. Basavaraj S, Betageri GV (2014) Improved oral delivery of resveratrol using proliposomal formulation: investigation of various factors contributing to prolonged absorption of unmetabolized resveratrol. Expert Opin Drug Deliv 11(4):493–503
31. Neves AR, Lúcio M, Martins S, JLC L, Reis S (2013) Novel resveratrol nanodelivery systems based on lipid nanoparticles to enhance its oral bioavailability. Int J Nanomedicine 8:177–187
32. Li Z, Percival SS, Bonard S, Gu L (2011) Fabrication of nanoparticles using partially purified pomegranate ellagitannins and gelatin and their apoptotic effects. Mol Nutr Food Res 55(7):1096–1103
33. Arulmozhi V, Pandian K, Mirunalini S (2013) Ellagic acid encapsulated chitosan nanoparticles for drug delivery system in human oral cancer cell line (KB). Colloids Surf B Biointerfaces 110:313–320
34. Bala I, Bhardwaj V, Hariharan S, Kharade SV, Roy N, Kumar MNVR (2006) Sustained release nanoparticulate formulation containing antioxidant-ellagic acid as potential prophylaxis system for oral administration. J Drug Target 14(1):27–34

35. Mowlazadeh S, Ebrahimpourmoghaddam S, Purkhosrow A, Mohammadi S (2012) Magnetic bovine serum albumin-based nanoparticles as potential controlled release drug delivery systems. Proc 4th, Int Conf Nanostructures 12–14 March, 2012, Kish Island, IR Iran
36. Guan Q, Sun S, Li X, Lv S, Xu T, Sun J et al (2016) Preparation, in vitro and in vivo evaluation of mPEG-PLGA nanoparticles co-loaded with syringopicroside and hydroxytyrosol. J Mater Sci Mater Med 27(2):24
37. Li C, Zhang Y, Su T, Feng L, Long Y, Chen Z (2012) Silica-coated flexible liposomes as a nanohybrid delivery system for enhanced oral bioavailability of curcumin. Int J Nanomedicine 7:5995–6002
38. Kim TH, Jiang HH, Youn YS, Park CW, Tak KK, Lee S et al (2011) Preparation and characterization of water-soluble albumin-bound curcumin nanoparticles with improved antitumor activity. Int J Pharm 403(1–2):285–291
39. Lohan SB, Bauersachs S, Ahlberg S, Baisaeng N, Keck CM, Müller RH et al (2015) Ultra-small lipid nanoparticles promote the penetration of coenzyme Q10 in skin cells and counteract oxidative stress. Eur J Pharm Biopharm 89:201–207
40. Carlotti ME, Sapino S, Ugazio E, Gallarate M, Morel S (2012) Resveratrol in solid lipid nanoparticles. J Dispersion Sci Technol 33:465–471. https://doi.org/10.1080/01932691.2010.548274
41. Caddeo C, Manconi M, Fadda AM, Lai F, Lampis S, Diez-Sales O et al (2013) Nanocarriers for antioxidant resveratrol: formulation approach, vesicle self-assembly and stability evaluation. Colloids Surf B Biointerfaces 111:327–332
42. Molpeceres J, Aberturas MR, Chacon M, Berges L, Guzman M, De Souza TD et al (2015) Skin penetration behavior of lipid-core nanocapsules for simultaneous delivery of resveratrol and curcumin. Eur J Pharm Sci 78:204–213
43. Guo CY, Yang CF, Li QL, Tan Q, Xi YW, Liu WN et al (2012) Development of a Quercetin-loaded nanostructured lipid carrier formulation for topical delivery. Int J Pharm 430(1–2):292–298
44. Schwarz JC, Baisaeng N, Hoppel M, Löw M, Keck CM, Valenta C (2013) Ultra-small NLC for improved dermal delivery of coenyzme Q10. Int J Pharm 447(1–2):213–217
45. Janesirisakule S, Sinthusake T, Wanichwecharungruang S (2013) Nanocarrier with self-antioxidative property for stabilizing and delivering ascorbyl palmitate into skin. J Pharm Sci 102(8):2770–2779
46. El-Refaie WM, Elnaggar YSR, El-Massik MA, Abdallah OY (2015) Novel curcumin-loaded gel-core hyaluosomes with promising burn-wound healing potential: development, in-vitro appraisal and in-vivo studies. Int J Pharm 486(1–2):88–98
47. Fernandes KS, Silva AHM, Mendanha SA, Rezende KR, Alonso A (2013) Antioxidant effect of 4-nerolidylcatechol and α-tocopherol in erythrocyte ghost membranes and phospholipid bilayers. Braz J Med Biol Res 46(9):780–788
48. Pinto F, de Barros DPC, Fonseca LP (2018) Design of multifunctional nanostructured lipid carriers enriched with α-tocopherol using vegetable oils. Ind Crops Prod. https://doi.org/10.1016/j.indcrop.2018.03.042

New Paradigms in the Treatment of Skin Infections: Lipid Nanocarriers to the Rescue

Karan Razdan, V. R. Sinha, and Kamalinder K. Singh

1 Introduction

The hike in the incidence of skin infections due to various microorganisms has led to enormous increase in morbidity and health-care troubles. Management of skin infections has always been a difficult task due to low efficacy of conventional topical formulations, short residence time at the site leading to high frequency of application, lack of patient compliance coupled with antimicrobial resistance (AMR), a major challenge in the prevailing scenario [1, 2]. Also, physico-chemical properties of drugs, viz. pKa, molecular size and weight, solubility and partition coefficient, affect the delivery across the skin barrier [3]. The skin is the largest body organ, composed of three layers: epidermis, dermis and hypodermis. The stratum corneum (SC), the outermost layer of the epidermis, is made up of keratinocytes embedded in lipid matrix comprising of cholesterol, ceramides and fatty acids. This makes SC a very effective barrier to delivery of anti-infectives across the skin, thus reducing their therapeutic efficacy [4].

Treatment of skin diseases by topical application of drug molecules is a viable and appealing strategy as it allows for reduction in drug dosage and subsequent

K. Razdan
School of Pharmacy and Biomedical Sciences, Faculty of Clinical and Biomedical Sciences, University of Central Lancashire, Preston, UK

Pharmaceutics Division, University Institute of Pharmaceutical Sciences, UGC-Centre of Advanced Study, Panjab University, Chandigarh, India

V. R. Sinha
Pharmaceutics Division, University Institute of Pharmaceutical Sciences, UGC-Centre of Advanced Study, Panjab University, Chandigarh, India

K. K. Singh (✉)
School of Pharmacy and Biomedical Sciences, Faculty of Clinical and Biomedical Sciences, University of Central Lancashire, Preston, UK
e-mail: ksingh1@uclan.ac.uk

© Springer Nature Singapore Pte Ltd. 2020

M. Rahman et al. (eds.), *Nanomedicine for Bioactives*,
https://doi.org/10.1007/978-981-15-1664-1_11

systemic side effects associated with oral or parenteral administration [5]. Skin infections in particular require constant drug exposure so as to ensure complete elimination of causative microbes. Moreover, it also reduces the incidence of developing resistance by microorganisms as fluctuating systemic levels and elimination of drug lead to lesser availability of drug to infected skin site.

The issue pertaining to low permeability of most drug molecules across the skin has led to research in finding the strategies to enhance the transport through the skin. One such strategy is to use chemical permeation enhancers such as azones, ethanol, terpenes, etc., which promote the transport of drugs by fluidizing or partially extracting the skin lipids or by enhancing the drug partitioning [6, 7]. Another approach is the use of physical permeation enhancers such as iontophoresis, sonophoresis, electroporation, microneedles, thermal ablation and microwaves, either alone or in combination. These approaches circumvent the SC and deliver the drug directly into target site [8–12]. Both physical and chemical enhancers have demonstrated the potential to successfully permeate different types of drug molecules across the skin, but most physical enhancers can cause pain and/or skin damage and are also costly with poor patient compliance, whereas chemical enhancers can cause skin irritation and permanent damage to skin.

So a third approach to enhance skin permeability is the use of nanocarrier systems. Nanocarrier-enabled delivery systems can enhance the transport of drugs across the skin by increasing drug solubility, enhancing skin penetration and improving drug distribution in the skin. Lipid-based nanocarriers (LNs) have demonstrated substantial potential for dermal drug delivery. LNs contain components which are biocompatible and biodegradable in nature [13–15]. These nanocarriers can act by various mechanisms, viz. (1) adsorption onto the skin and fusion with skin cells leading to transfer of drug entrapped in the vesicles to the skin and (2) enhancing penetration of drug through SC by altering the fluidity of SC lipids and changing intercellular lipid lamellae [14, 16, 17]. The peculiarity of these systems resides in the fact that the biomimetic nature of lipid bilayer structure of the vesicles allows for the fusion with cells of infectious microorganism, thus delivering anti-infective drug directly to the target site [18].

This chapter aims to provide comprehensive overview of various lipid nanocarriers for topical drug delivery to overcome the current challenges in the treatment of skin infections.

2 Skin Structure and Permeation Mechanisms

The skin is the largest body organ, accounting for 10% of body mass and 2 m^2 of body surface area. One of the major functions of the skin is to act as a barrier to entry of microorganisms [19].

The skin is made up of three chief layers: the outermost layer known as the epidermis (50–150 μm thick), dermis about 250 μm thick and underlying hypodermis or subcutaneous fat. The epidermis consists of five layers, out of which SC is the outermost part (15–20 μm thick) and acts as a rate-controlling barrier to diffusion of most molecules [20]. It is composed of corneocytes (terminally differentiated keratinocytes) which are embedded in lamellae of the lipid matrix. This peculiar arrangement presents the barrier and reduces the transport of molecules, specifically those larger than 500 Da [21, 22]. The viable epidermis (excluding the stratum corneum) is composed of keratinocytes, melanocytes, Langerhans cell and Merkle cells [23]. The dermis is present beneath the viable epidermis and is composed of collagen and elastin fibres. It contains nerve supply, vascular and lymphatic network for exchange of fluid and nutrients [24]. The last layer of skin is the hypodermis, composed of adipocytes (fat cells) linked to collagen fibres. Hypodermis forms a thermal barrier and acts as a mechanical cushion for the body. Skin appendages such as pilosebaceous glands, sweat glands and hair follicles emerge either from the dermis or the hypodermis.

A molecule can pass through the skin via two pathways—(1) transepidermal route (diffusion across skin layers) and (2) appendageal route (via sweat ducts or hair follicles)—as depicted in Fig. 1. In transepidermal route, the molecule may transverse transcellularly or intercellularly. In the former, the molecule has to cross alternate cell layers and extracellular matrix (ECM), while in the latter the molecule steers through the tortuous path within the ECM. The appendageal or shunt pathway includes transport through sweat ducts and hair follicles. Recent reports have shown transappendageal route to be an effective pathway for permeation of topical formulations [4].

3 Types of Skin Infections and Their Aetiology

3.1 Bacterial

3.1.1 Impetigo

Impetigo is a communicable, bacterial infection affecting the superficial skin caused by *Staphylococcus aureus* and haemolytic *Streptococcus pyogenes*. The typical presentation is yellowish cold crust on the face, arms or legs. It mostly affects children, but other risk factors such as diabetes mellitus and dermatitis can increase the risk of contracting the infection in the elderly [25].

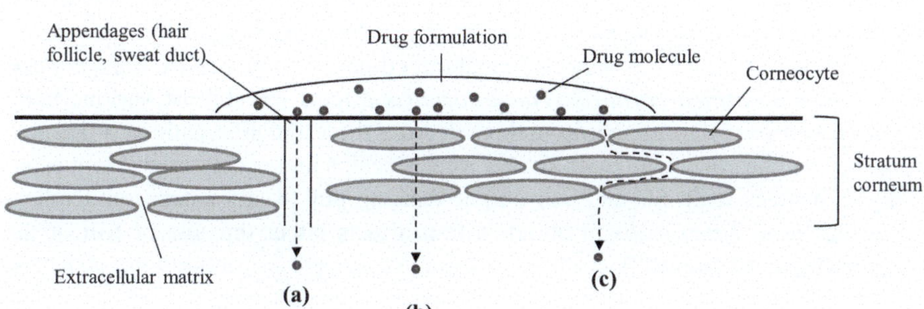

Fig. 1 Drug permeation routes through the skin (stratum corneum depicted): (**a**) the appendageal or shunt route, (**b**) the transcellular, and (**c**) the intercellular route

3.1.2 Ecthyma

Ecthyma is an exacerbated form of impetigo involving the deep skin. It is characterized by crusted sores masking the ulcers beneath the crusts. It also leads to erosion of the skin into the dermis and most commonly affects the leg and thigh areas. Ecthyma usually starts as a vesicle progressing to ulcerations. Bacteria involved in this infection are *S. aureus* and *S. pyogenes* [25].

3.1.3 Cellulitis and Erysipelas

Cellulitis is a bacterial skin infection specifically affecting the dermis and subcutaneous fatty tissue. It particularly affects the lower limbs and often occurs on legs afflicted by venous ulcerations or lymphoedema. The microbes make their way through a break in the skin. There is pain and slight swelling with diffused erythema. Erysipelas is a subcategory of cellulitis which is characterised by abrupt onset of painful, well-marginated and swollen erythematous area, usually in the lower limbs. There may be severe inflammation and area may blister. The major causative organism of cellulitis is Group A beta-hemolytic *Staphylococcus aureus* along with other microbes like *P. aeruginosa*, *Escherichia coli* and *Klebsiella* species. On the contrary, erysipelas is caused more commonly by beta-hemolytic *Streptococcus* [26].

3.1.4 Folliculitis, Furunculosis and Carbunculosis

Folliculitis is a cutaneous infection and inflammation of hair follicles present anywhere on the skin except on palms and soles. Furunculosis or a boil is deep folliculitis, leading to swelling due to accumulation of pus and dead tissue, and individual boils cluster together to cause carbunculosis. The infection is caused mostly by coagulase-positive *S. aureus* strains. Other factors which can contribute are diabetes mellitus, obesity and intake of glucocorticoids [26].

3.2 Fungal

3.2.1 Dermatophytosis (Ringworm)

Dermatophytosis is an infection of the hair, nails or skin caused by *Trichophyton*, *Microsporum* and *Epidermophyton* species. These fungi have high affinity for keratinized tissue. This includes various fungal infections, viz. *tinea pedis*, *tinea capitis*, *tinea cruris* and *tinea corporis*. *Tinea corporis* affects the skin of various body parts. *Tine cruris* is prominent in males affecting the groin area with well-defined, itchy, red patches, whereas *tinea capitis* affects scalps and occurs in children especially. It invades scalp SC and hair cuticle causing pink, scaly patches and subsequent hair loss. *Tinea pedis* or athlete's foot affects either the sole of feet or the space between the toes [27].

3.2.2 Candidiasis

Candidiasis is caused by the yeast *Candida albicans* and is the common cause of vulvovaginitis in pregnant women and in women taking oral contraceptives. It also occurs frequently in immunocompromised patients. Acute inflammation of perineum takes place leading to itchy, reddish and scaly skin [27].

3.2.3 Pityriasis Versicolor

Also called as tinea versicolour, this fungal infection is caused by *Pityrosporum ovale Malassezia furfur*. This fungus becomes pathogenic when there is compromised immunity or enhanced sebum secretion. The infection is characterized by pale, scaly macules over the skin of back and chest in young adults. The fungus interferes with normal skin pigmentation, by filtering sunrays and inhibiting tyrosinase leading to discoloured patches on the skin [28].

3.3 Viral

3.3.1 Herpes Simplex

It is caused by *herpes simplex* virus (HSV), a small DNA virus which is of two types usually, HSV 1 and HSV 2. Type 1 affects the genitalia whereas Type 2 infects the face and oropharynx. The onset of infection is very displeasing with severe inflammation of the mouth and lips along with pyrexia in infants. The lesions around the mouth or on the lips start as grouped, tender papules and then coalesce to form a crusted erosion. The virus infects by coming in direct contact with mucosa or abraded skin [29].

3.3.2 Herpes Zoster

Herpes zoster also known as shingles is an acute dermatitis affecting 10–20% of adults but also occurs in immunocompromised patients. The disease is not acquired but surfaces due to reactivation of a virus residing dormant in a posterior root ganglion of a spinal nerve. It is characterized by grouped vesicles within the dermatome accompanied by fever and malaise [29].

3.3.3 Warts

Warts are caused by different types of human papillomavirus (HPV). They are most likely caught by direct contact of the skin with wart virus-containing horny debris. There are usually little black dots near the surface of the wart, representing thrombosed capillaries in elongated dermal papillae. There is epidermal thickening, with particular increase in the granular cell layer, which also shows a characteristic basophilic stippled appearance [30].

3.4 Cutaneous Leishmaniasis

Cutaneous leishmaniasis (CL) is caused by various species of parasitic protozoa belonging to the genus *Leishmania* and is spread by infected female sandfly (genus *Phlebotomus*). If ulcerative lesions occur at site of the bite, then it is localized CL, whereas multiple non-ulcerative nodules encompass diffused CL. The first symptom of infection is a small erythema which develops into a papule and then nodule which ulcerates over a time of 2 weeks to convert into a lesion [31]. Figure 2 depicts the various types of skin infections.

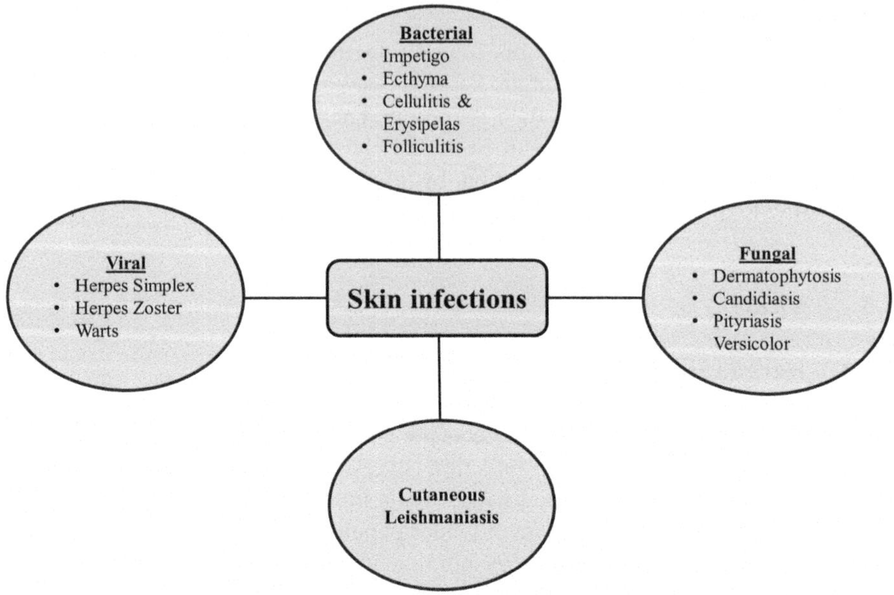

Fig. 2 Schematic illustration of various skin infections

4 Current Challenges in Skin Infection Therapy and Need of Nanotechnology

Conventional topical preparations such as ointments and creams act on the outer skin surface providing a concentrated layer of active moiety on the skin surface. The incorporated drug partitions from these formulations into the skin. These vehicles require high drug concentrations for efficacious treatment due to low efficiency of these systems, which can lead to toxicity and skin irritation. Issues of greasiness and stickiness associated with conventional preparations affords patient non-compliance [32, 33]. These systems also exhibit low skin penetration and non-targeted delivery. Most bacterial and fungal infections are deep seated and require penetration of drug to the infection site. Nanotechnology can aid in alleviating the pitfalls associated with conventional topical formulations by delivering anti-infectives to target site in a precise and controlled manner [34]. Nanocarriers are able to encapsulate anti-infectives that may not be bioavailable otherwise due to poor solubility or inability to maintain intimate contact with pathogen. Nanocarriers can act as a depot of encased drug by sustaining its release over a longer period of time and allowing constant contact with the target pathogen [35, 36]. As described earlier in Sect. 2, intercellular pathway is one of the most significant routes to penetrate SC. In this regard, lipid nanocarriers can serve as perfect carriers due to their similarity in lipid composition with that of the skin lipids and can efficiently interact with them. LNs can also impart chemical stability to bioactives and synthetic anti-infectives that are sensitive to light and oxidation [37]. Hence, LNs can overcome the shortcomings associated with conventional preparations and provide an effective topical therapy for skin infections.

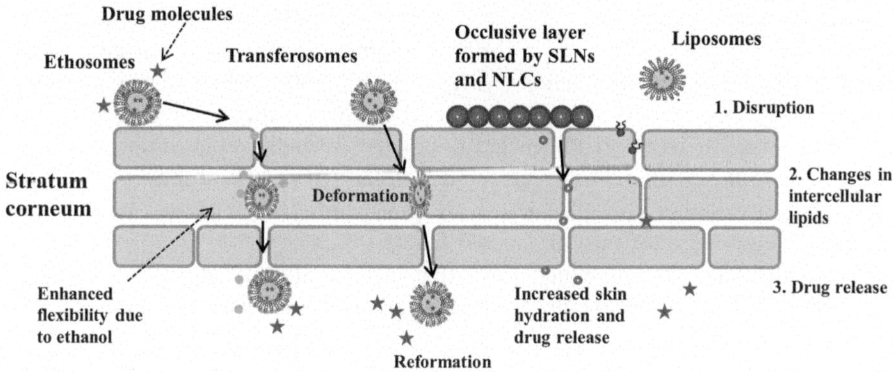

Fig. 3 Schematic depiction of skin permeation mechanisms of various lipid nanocarriers

4.1 Lipid Nanocarriers and Their Application to Skin Infections

Lipid-based nanocarriers to be used for topical treatment of skin infections are composed of biodegradable and biocompatible lipids and surfactants that provide optimum and controlled drug release and efficient contact with target microbes. Various lipid-based systems have been put to use for the treatment of viral, fungal and bacterial infections. These systems are regarded as efficient, safe and attractive delivery strategies for anti-infective pharmaceuticals.

Lipid nanocarriers can be classified into vesicular-, particulate- and emulsion-based carriers. The following section details about the chief representatives of the lipid-based systems used for topical treatment and their application in therapy of skin infections. Mechanism of skin permeation of different LNs is depicted in Fig. 3.

4.2 Liposomes

Liposomes were discovered by Alec Bangham more than 50 years ago [38], and the term is derived from two Greek words: *lipos* (fat) and *soma* (body). These are composed of one or more bilayers of phospholipids enclosing an aqueous compartment(s). Phospholipids can be neutral or surface charged, and cholesterol is added to provide rigidity to the membrane. Liposomes can incorporate both hydrophilic and hydrophobic drugs [39].

Liposomes seem to deliver drugs to the skin by virtue of their biomimetic nature, and in 1980 Mezei and Gulasekharam reported enhanced deposition of liposomal triamcinolone acetonide in the epidermis and dermis [40]. It was in 1988 that first topical liposomal formulation of econazole was approved for topical dermatomycosis [41]. Since then, liposomes for topical drug delivery of anti-infectives have been extensively investigated. Liposomes make an ideal delivery system for anti-infectives due to mimetic nature of lipid bilayer to cell membrane which allows for fusion with bacteria. Thus, encapsulated anti-infective drug can make intimate contact with bacterial cell warranting high concentration at the infection site [42].

The ability to deliver drug depends on physical and chemical characteristics of liposomes such as lipid composition, membrane rigidity, size, surface charge and drug-loading capacity. Generally, liposomes ≥600 nm in size are unable to release drug deep into skin layers and tend to form a film after drying on the SC. Vesicles with sizes ≤100 nm get deposited in viable epidermis as well as deep down in the dermis. However, the debate is still open and matter of further investigation whether these vesicles can pass through the skin, intact or the released drug penetrates into the deeper layers [43, 44].

There are many studies which show that incorporation of anti-infective drugs into liposomes has been advantageous. Madan et al. [45] prepared liposomes of curcumin, lauric acid and azithromycin and incorporated into the gel to evaluate their effect against *P. acnes*-induced infection in rat ear model. Co-application of liposomal gel of curcumin and lauric acid (1:1) displayed significant increase in antibacterial effect against both macrolide-sensitive and macrolide-resistant strains of *P. acnes* in comparison with the individual treatment. Moosavian et al. [46] developed stearyl amine tethered liposomes loaded with meglumine antimonate (MA) and evaluated their efficacy in treatment of *L. major*-induced cutaneous leishmaniasis (CL) in BALB/c mice in comparison with MA-cream and control liposomes. Liposomal MA containing stearyl amine showed higher skin permeation and higher toxicity against promastigotes and amastigotes of *L. major* as compared to liposomal MA. Lesion size was significantly reduced. Kavian et al. [47] also formulated liposomes of miltefosine (ML) in varying concentrations of ML (0.5, 1, 2 and 4%) for topical treatment of CL and evaluated the efficacy in *L. major*-induced CL in BALB/c mice. Liposomal ML 4% showed highest skin retention (82%) and significantly lowered lesion parasite burden as compared to ML cream and other liposomes at 12 weeks postinfection.

Study by Rukavina et al. [48] showed enhanced antibacterial activity of azithromycin (AZT)-loaded liposomes against methicillin-resistant *Staphylococcus aureus* (MRSA) to treat local skin infections. They prepared conventional liposomes (CLs), deformable liposomes (DLs), propylene glycol-containing liposomes (PGLs) and cationic liposomes (CATLs). All liposomes demonstrated higher skin retention of AZT in comparison with free AZT solution with CATLs showing highest retention (54%), while DLs were retained mostly inside the skin (77%) owing to their elastic nature. All liposomal formulations showed excellent biocompatibility with keratinocytes and fibroblasts. CATLs, DLs and PGLs demonstrated 32-fold lower MIC and MBIC than AZT solution.

Chhibber et al. [49] evaluated the liposomal phage cocktail in comparison with free phage cocktail (*S. aureus*-specific lytic phages MR-5 and MR-10) for their therapeutic potential in resolving diabetic wound infection in mice. Entrapment of phage cocktail in liposomes leads to better persistence of phage at wound site. A 2-log increase in phage titer was observed with liposomal formulation at the wound site, and rate of resolution of infection also increased. Hsu et al. [50] evaluated the follicular delivery and anti-MRSA activity of chloramphenicol (Cm) liposomes. The liposomes prepared were made malleable by incorporation of dimyristoylphosphatidylcholine (DMPC) or deoxycholic acid (DA). Malleable liposomes showed increased follicular uptake of Cm up to twofold in comparison with free Cm solution. The optimized formulation exhibited both intra- and extracellular anti-MRSA activity and can be useful in the treatment of diseases such as furuncles and folliculitis. Tanriverdi et al. [51] also evaluated the antifungal potential of terbinafine HCl (TBF-HCl)-loaded liposomes incorporated in pullulan and Eudragit films for treatment of onychomycosis. It was found that TBF-HCl-loaded liposome film formulation had better antifungal activity and at the end of 10 days, drug content in nail plate was found to be more than MIC of TBF-HCl for onychomycosis.

Nunes et al. [52] investigated the effect of usnic acid-loaded liposomes (UAL) incorporated into gelatin-based membrane on burn wound healing in pigs in comparison with marketed products. The membranes showed controlled release with increased deposition of collagen in comparison with free UA. UAL treatment was found to be as efficacious as marketed duoDerm® dressing in the development and maturation of granulation tissue and scar repair. Tanriverdi and Ozer [53] attempted first time to overcome nail plate barrier for the treatment of onychomycosis by employing terbinafine HCl entrapped in vesicular carriers. The accumulated drug in nail plate was in the range of treatment concentrations, and highest accumulation was seen with liposomal gel to provide sustained release of terbinafine. Gupta et al. [54] prepared fluconazole-loaded liposomal and niosomal gel for topical treatment of cutaneous candidiasis induced in immunosuppressed albino rats. Liposomal and niosomal gel showed 2.18-fold and 1.85-fold skin retention, respectively, in comparison with plain drug-loaded gel. Liposomal gel was found to be superior in comparison with niosomal gel as liposomal gel successfully treated five out of six animals as evident from negative culture, whereas in latter case, only three out of six got cured.

4.3 Ethosomes

Ethosomes are flexible liposomes composed of phosphatidylcholine and ethanol (up to 50%). These vesicular delivery systems were pioneered by Touitou et al. [55]. These are malleable and soft vesicles that can penetrate intact through the human skin. This nature can be attributed to the presence of ethanol which confers fluidity and elasticity by lowering the transition temperature of phospholipids. Ethanol can permeate through the skin carrying drug along with into the deeper layers, and it also lends negative surface charge that inhibits the aggregation of vesicles [14].

Shetty et al. [56] prepared ethosomal gel of clove oil and evaluated its efficacy in treatment of cutaneous candidiasis. The average particle size of all formulations ranged between 2 and 10 μm. The optimized formulation was found to be non-irritant to the skin and exhibited satisfactory antifungal activity in comparison with pure clove oil. To avoid the side effects associated with oral voriconazole (VRC) therapy, Faisal et al. [57] prepared ethosomal formulations of VRC and assessed the permeability of VRC through rat abdominal skin. Ethosomes were found to have a size range of 420–600 nm and up to 46% encapsulation efficiency (EE). In vitro antifungal study against *Aspergillus flavus* demonstrated that ethosomal VRC achieved same zones of inhibition as achieved by DMSO solution of VRC, at two-fold lower concentration. The ethosomal VRC showed sixfold higher skin deposition in comparison with hydroalcoholic solution of VRC.

Griseofulvin (GRF) is the most commonly prescribed drug for treatment of dermatophytoses and childhood mycoses, but its poor solubility has limited its topical application. Marto et al. [58] developed ethosomes of GRF by using soybean phosphatidylcholine, ethanol and water and characterized for particle size, charge and permeability was assessed using newborn pig skin. Ethosomes exhibited excellent retention in SC with negligible permeation into deeper layer. The formulation was non-cytotoxic in HaCaT cell lines and showed enhanced antifungal activity. Iizhar et al. [59] prepared ethosomal gel of TBF-HCl comprising of soya phosphatidylcholine, ethanol and propylene glycol. Optimized formulation showed EE of 69.1% and particle size of 248 nm. Transdermal flux of ethosomal gel across rat abdominal skin in ex vivo study was found to be 144.61 μg/cm^2/h, whereas 1% marketed cream of TBF-HCl showed flux of 121.6 μg/cm^2/h. This study demonstrated that ethosomal gel can improve patient compliance by virtue of increased residence time at target site.

Deep dermal and subdermal infections caused by *S. aureus* require systemic antibiotic treatment. Godin et al. [60] investigated new approach by formulating ethosomes of erythromycin (EMC) for ascertaining its efficiency in the treatment of deep dermal *S. aureus* infection in mice, in comparison with i.p. EMC and hydroethanolic EMC solution. Ethosomal EMC-treated group showed normal skin structure and no bacterial growth, while hydroethanolic EMC-treated group developed deep dermal abscesses, and *S. aureus* colonies were also present. Topical ethosomal therapy was found to be as effective as systemic EMC treatment.

4.4 Transfersomes

Conventional liposomes have narrow utility in dermal drug delivery. Cevc and Blume [61] introduced ultradeformable liposomes called transferosomes. In addition to phospholipids and cholesterol, transfersomes contain an edge activator (usually a surfactant) to confer elasticity and flexibility. Commonly used edge activators are sodium cholate, polysorbates and sorbitan esters and dipotassium glycyrrhizinate [62, 63]. These activators are usually a single chain surfactant which cause destabilization of lipid bilayer of vesicle and increase the deformability of bilayer by reduction in interfacial tension [64, 65]. These are highly adaptable complexes which change shape in response to stress and avoid vesicle rupture. Transferosomes can enter intact into the SC under the driving force of transcutaneous hydration gradient occurring naturally. After gaining entry into SC, vesicles alter the intercellular lipid lamellae and promote the permeation of free drug molecules [61]. Various studies have demonstrated the potential of transferosomes in topical treatment of skin infections.

Miconazole nitrate (MCN) is a broad-spectrum antifungal drug used for the treatment of cutaneous fungal infections. Due to poor skin permeability, conventional MCN formulations afford patient non-compliance. Qushawy et al. [66]

prepared and evaluated transferosomal gel of MCN for the treatment of cutaneous candidiasis in in vivo rat model in comparison with marketed cream (Daktarin® 2%). Thin film hydration method was used for the preparation, while Span 80 and Tween 80 were used as edge activators. Transferosomal gel showed higher steady-state flux (85.968 μg/cm²/h) than Daktarin® 2% cream (72.488 μg/cm²/h) after 24 h. Zone of inhibition was 55 mm with MNC transferosomal gel, whereas with marketed cream, it was 50 mm. In in vivo pharmacodynamic studies, group treated with MCN transferosomal gel showed normal skin with slight inflammation, whereas in Daktarin® 2% cream group, scars were still present. Pandit et al. [67] also demonstrated higher efficacy of MCN-loaded transferosomes in comparison with conventional liposomes and drug suspension in water. Better skin penetration of transfersomes was confirmed by fluorescence microscopy, and in vivo studies showed early recovery from infection in comparison with conventional liposomes.

Chhibber et al. [68] utilized transfersomes as carrier for delivery of methicillin-resistant *S. aureus* (MRSA) phage cocktail to treat soft tissue infection in rats in comparison with free phage cocktail. Transferosomes showed EE of 93% and vesicle size of 450 nm. Thigh infection in rats resolved within 7 days as compared to 20-day period for untreated animals. Unlike free phages, the phage cocktail entrapped in transferosomes protected all animals even when there was 12-h delay in administration after the infection. The enhanced recovery can be attributed to higher phage infectivity due to deeper transfersome penetration.

Abdellatif et al. [69] developed various vesicular carriers of sertaconazole nitrate (STN) and evaluated its ability to treat deep dermal fungal infection. Nanovesicles were found to be unilamellar and spherical in shape. Amongst various vesicular systems prepared, transferosomal system containing 3% soya PC and 0.15% sodium deoxycholate showed highest flux (645 μg/cm²/h), 3.5-fold higher than commercial cream (Dermofix®). Confocal microscopy confirmed penetration of transfersomes to viable epidermis. Transferosomal gel of STN showed lower colony count after 10 days of treatment in immunosuppressed rat model of fungal skin infection than Dermofix®, indicating better efficacy.

Li et al. [70] formulated transferosomes/nanoliposomes loaded with daptomycin (DAP) for topical application using response surface methodology. In vitro and in vivo studies demonstrated potent bacteriostatic activity against *S. aureus*. Transferosomes showed high skin permeation of DAP in 12 h (661.15 ± 88.65 μg/cm²). DAP was detected in various skin layers and even in the underlying structures. This study showed the potential to have clinical application.

4.5 Niosomes

Niosomes were first developed and patented by L'Oreal in the 1970s. These are vesicular systems made up of non-ionic surfactants, cholesterol and fatty alcohols. Most commonly used surfactants include alkyl ethers, alkyl esters, polysorbates and sorbitan esters. Charge inducers such as stearyl amine and dicetyl phosphate provide positive and negative charge, respectively, to vesicles and enhance entrapment efficiency [71, 72]. Niosomes are non-toxic and non-immunogenic due to their composition. Hydrophilic-lipophilic balance (HLB) of a surfactant affects entrapment of drug in niosomes: HLB of 8.6 yields niosomes with highest EE, whereas EE decreases as HLB decreases below 8.6. HLB in range of 14–17 is not suitable for niosome formation [73]. In comparison with liposomes, niosomes are physically and chemically stable and cost-effective and do not need special storage conditions.

Chitosan gel-embedded moxifloxacin-loaded niosomal system was prepared by Sohrabi et al. [74] for topical antimicrobial treatment. The optimized formulation showed particle size of 290 nm and zeta potential of −23 mV, 73% drug entrapment and 47% drug release in 8 h. The hybrid system showed sustained-release behaviour and superior antibacterial activity against *S. aureus*.

Alomrani et al. [75] prepared itraconazole (ITZ)-loaded niosomes using various non-ionic surfactants to enhance its skin penetration. Various surfactants explored included sorbitan esters (Span 60, Span 80 and Span 85), polysorbates (Tween 60, Tween 80 and Tween 85) and Brij 35. Niosomes prepared with Tween 80 showed the highest skin permeation [75].

4.6 Solid Lipid Nanoparticles (SLNs) and Nanostructured Lipid Carriers (NLCs)

SLNs were developed in the 1990s as an alternative drug delivery system to polymeric nanoparticles and liposomes [76]. These are colloidal systems made up of physiological lipids, stabilized by suitable surfactant. The lipid remains solid at body and skin temperature. Both hydrophilic and lipophilic molecules can be incorporated in the matrix [77, 78]. Lipids used include triglycerides, partial glycerides, fatty acids or waxes or mixtures thereof. SLN formulations typically incorporate 0.1–30% w/w of solid lipid and 0.5–5% w/w of surfactants [14]. SLNs are spherical in shape with mean particle size in range of 50–1000 nm. Advantages of SLNs include their good biocompatibility, low toxicity, scale up feasibility and organic solvent-free preparation process. Due to high surface area and small particle size, SLNs make an intimate contact with SC to form an occlusive film [14, 79]. But major pitfall of SLNs is their low drug loading and expulsion of drug during storage. This led to birth of second-generation lipid carriers, the NLCs, put forth by Muller et al. [80]. NLCs are prepared by precise mixing of solid and liquid lipids, leading to nanostructures with enhanced drug loading and release profiles. Blending of lipid molecules of different confirmations and size leads to formation of imperfect crystalline matrix, which can accommodate greater amount of drug. This increases drug loading capacity and minimizes expulsion of drug during storage [81, 82]. NLC preparation involves mixing of solid and liquid lipid in a ratio of 70:30 to a ratio of 99.9:0.1, and surfactant concentration varies from 0.5 to 5% w/v [83].

Fluconazole (FLZ)-loaded SLNs were prepared by El-Housiny et al. [84] and incorporated into Carbopol gel for enhanced treatment of pityriasis versicolor (PV). SLNs were formulated by employing high shear homogenization technique using

Compritol 888 ATO and Precirol ATO 5 as solid lipids, whereas Cremophor RH40 and Poloxamer 407 were used as surfactants. The particles were spherical in shape with EE in range of 55.49–83.04%. A randomized controlled clinical trial of optimized formulations was also undertaken in 30 patients suffering from PV in comparison with marketed cream Candistan® 1%. Clinical trial demonstrated significant improvement in curing the PV in comparison with marketed cream (1.4-fold, healing %; 4-fold, complete eradication). Also the formulations were found to be free from any adverse effects. Kakadia and Conway [85] developed triclosan (TSN)-loaded SLNs comprising glyceryl behenate and glyceryl palmitostearate as solid lipids and Tween 80 and Transcutol® P as S_{mix}. Formulations exhibited mean particle size in range of 192.23–231.72 nm and negative zeta potential. % EE was in range of 77.17–82.19%. SLN formulations showed 4.5-fold higher permeation rate of TSN in comparison with control solution.

Vaghasiya et al. [86] prepared and evaluated TBF-HCl-loaded SLN gel for skin targeting. SLNs were prepared using solvent-injection technique and optimized by Design of Experiments (DoE). The optimized formulation showed % EE of 73.74% and particle size of 300 nm. Drug release followed Higuchi model. SLN-based gel showed 40.57% skin retention of drug, whereas it was only 29.43% with marketed product Daskil®, after 8 h. This could be attributed to enhanced contact with corneocytes and skin occlusion. In vivo pharmacodynamic study in albino rats demonstrated the superiority of optimized formulation over Daskil®, as it took 18 days for full recovery for SLN gel-treated group as compared to 24.8 days taken by Daskil®-treated group. Bhalekar et al. [87] prepared MCN-loaded SLNs using Compritol 888 ATO as lipid phase, propylene glycol, Tween 80 and glyceryl monostearate as surfactants. Particles showed mean size in range of 244–766 nm and % EE between 80% and 100%. The SLNs were incorporated into Carbopol gel, and ex vivo skin permeation through the human cadaver skin showed higher skin deposition (57%) for MCN-loaded SLNs in comparison with marketed gel (30%). Jain et al. [88] also developed MCN-loaded SLN gel employing solvent injection method and compared its release with MCN suspension and hydrogel. MCN-loaded SLN gel demonstrated tenfold higher skin retention in comparison with MCN suspension and hydrogel as evidenced by tape stripping experiments.

Lv et al. [89] developed SLNs of penciclovir for topical application by using double emulsion (w/o/w) technique. The particles were spherical in shape and exhibited EE of 92.40% and zeta potential of −25 mV. SLNs of penciclovir showed twofold higher skin penetration than marketed cream through excised rat skin after 12 h of administration. Apparent morphology of SC was found to be changed due to interaction between SLNs and skin surface as evidenced by microscopic study. HSV is a common viral disease in humans causing mucocutaneous lesions. Lai et al. [90] formulated *Artemisia arborescens* essential oil (proven to possess activity against HSV-1)-loaded SLNs and evaluated the permeation through newborn pig skin. The formulation was found to be stable even after 2 years, and skin permeation study demonstrated accumulation of oil in the skin, whereas control solution of oil showed higher permeation and low retention.

Increased incidence of resistance of *Candida* species to conventional drugs presents a great challenge for treatment of fungal skin infections. Essential oils (EO) have been reported to enhance the effectiveness of conventional drugs. Carbone et al. [91] developed Mediterranean EO (*R. officinalis*, *Lavandula intermedia* and *Origanum vulgare*)-loaded NLCs for combined delivery of clotrimazole optimized using DoE approach. NLCs were characterized for particle size, polydispersity and long-term stability. Particle size was found to be <100 nm and broad-size distribution was observed (PDI < 0.15). On basis of in vitro biosafety studies in HaCaT and A431 (tumoral cell line), *Lavandula* and *Rosmarinus* oils were selected. Calorimetric studies on biomembrane models showed the potential of oils to permeabilize the lipidic membranes, thus affording antifungal activity. In vitro antifungal activity against *C. albicans*, *C. krusei* and *C. parapsilosis* demonstrated superiority and synergy of clotrimazole and EOs when incorporated into NLC formulations. Khezri et al. [92] prepared rosemary EO (REO)-loaded NLCs and evaluated their in vitro antibacterial activity and wound-healing potential in full-thickness wound infection animal model. The wounds were infected with 10^7 CFU of *S. aureus* and *P. aeruginosa*. Animals were treated with REO and REO-NLCs in comparison with mupirocin. REO-NLCs were found to be superior to mupirocin and pure oil in reducing bacterial load in wound and wound size. They also promoted vascularization, collagen production and re-epithelialization. Ghodrati et al. [93] also conducted similar study by preparing peppermint EO-loaded NLCs and evaluated them for in vitro antibacterial activity and in vivo wound healing in full-thickness wound infection model in mice. PEO-NLC formulation increased collagen deposition, wound contraction rate and re-epithelialization in comparison with control.

Alalaiwe et al. [94] prepared and evaluated oxacillin-loaded cationic NLCs for anti-MRSA activity against clinical isolates for treatment of cutaneous infections. Soya ethyl morpholinium ethosulphate was used to impart positive surface charge to NLCs. The zeta potential and diameter of lipid particles were found to be 19 mV and 177 nm, respectively. The combination of NLC and oxacillin significantly reduced biofilm thickness from 31.2 to 13 μm, which was more effective than NLC (18.2 μm) or oxacillin (25.2 μm) alone. Significant reduction in intracellular MRSA burden in THP-1 cells was observed with oxacillin-loaded NLCs due to bacterial membrane disruption and enhanced oxacillin transport into cytoplasm.

Gaba et al. [95] developed TBF-HCl loaded NLC gel using high-pressure homogenization employing glyceryl monostearate as solid lipid, Labrasol as liquid lipid and Pluronic F127 as surfactant for treatment of topical fungal infections. Particles were spherical in shape with mean diameter of 128 nm. 92.60% of drug was released from NLCs in 24 h as compared to 82.82% from marketed formulation. NLCs showed 86.35% skin permeation, while marketed formulation could afford 69.41% skin permeation. In vivo fungal infection model in rats demonstrated the superiority of NLC-based gel as it showed significant reduction in fungal burden after 5 days (771 ± 41.79 CFU) as compared to marketed formulation (1558 ± 140.52 CFU). Song et al. [96] developed VRC-loaded NLC-based gel for treatment of mycotic skin infections. The NLC formulation was prepared by high-pressure homogenization using Precirol ATO 5 as solid lipid and Labrafil 1944 CS and Tween 80 as surfactant. The particle size of nanoparticles was ~210 nm and zeta potential was −30 mV. The NLC gel showed 2.8-fold higher flux and greater deposition than conventional cream in skin permeation study across mice skin. This study demonstrates the potential of novel NLC formulation in the treatment of fungal skin infections in comparison with oral therapy with systemic side effects.

4.7 Microemulsions (ME) and Nanoemulsions (NE)

Hoar and Schulman [97] first time introduced the structure of ME when transparent solution was obtained after addition of co-surfactant into conventional emulsion. MEs are colloidal dispersions with globule size ranging from 10 to 100 nm. They are formed spontaneously upon combining oil, water and mixture of surfactant and co-surfactant. MEs are an effective drug delivery system for topical applications due to their thermodynamic stability, ease of preparation and high solubilization capacity. MEs adhere to skin surface due to small droplet size and enhance drug permeation [98]. Microemulsion can be of three types: oil-in-water (o/w), water-in-oil (w/o) and bicontinuous structures [99]. MEs themselves exert antimicrobial action due to breach of structural integrity of cell membrane and eventual cell death [100].

Nanoemulsions are kinetically stable submicron-sized emulsions with droplet size in range of 20–500 nm. NEs require high energy input for preparation as amount of surfactant used is less as compared to MEs. In terms of cutaneous delivery, NEs offer unique advantages: high physical stability, skin friendly due to low surfactant amount, high drug loading and ease of preparation [101]. NEs can obtain higher penetration rates and accumulation in the skin than lipid nanoparticles [102]. These are low viscosity systems, but higher viscosity requirement can be met by incorporation of gelling agents [103].

There are various reports which prove the potential of MEs and NEs for topical delivery of anti-infectives. Talaat et al. [104] investigated for the first time the potential of terconazole (Tr) in the treatment of cutaneous mycosis by incorporation into lecithin ME-based lipogels (LMBGs). LMBGs showed enhanced permeability of Tr through animal skin as compared to conventional gels and were found to be non-irritant to the skin. Kaur et al. [105] prepared w/o ME of acyclovir using isopropyl myristate, Tween 20, Span 20, water and DMSO and evaluated the antiviral activity in murine model of zosteriform cutaneous HSV-1 infection. Optimized formulation was found to have particle size of 198.2 nm and pH between 6.8 and 7.0. ME formulation showed higher skin retention (43.06 μg/cm^2) as compared to marketed cream (41.81 μg/cm^2) and conventional PEG ointment (22.49 μg/cm^2). Optimized formulation was able to completely inhibit the development of skin lesions when applied 24 h after infection vis-à-vis marketed cream. Alam et al. [106] prepared ketoconazole 2% (KTZ) and clotrimazole 1% (CTZ) ME using clove oil (6%) as oil phase and Tween 20 as surfactant (25–30%). Developed ME formulations were found to be non-irritant to the skin, and average zone of inhibitions against *Candida albicans* were about 2.5–3 times greater than commercial cream. In vivo antifungal activity demonstrated that in comparison with CTZ cream, CTZ ME-treated group showed scabbing of skin lesion after 3 days, and scabs started to erode after 5 days.

Pseudolaric acid B (PAB) possesses potent fungicidal activity against *Candida albicans* and other *Candida* species. The poor water solubility however limits the topical application, thus necessitating the need for enhanced topical delivery. Wan et al. [107] prepared ME-based gel of PAB using isopropyl myristate (IPM) as oil phase, Cremophor EL as surfactant and Transcutol P as co-surfactant. ME was incorporated into Carbopol gel, and particle size, pH and rheological assessments were carried out. Permeation study through rat skin demonstrated that ME significantly enhanced PAB retention and permeation through the skin, regardless if it was incorporated into gel or not. In vitro antifungal test revealed that 8 mg/g of PAB ME gel showed higher activity than 20 mg/g MCN cream.

Burn wounds are an ideal spot for bacterial colonization and multiplication. Chhibber et al. [108] developed phospholipid-based ME-gel of fusidic acid (2%) for treatment of methicillin-sensitive *S. aureus* (MSSA)-infected burn wound. IPM was used as oil phase and S_{mix} comprised of phospholipid/Tween80/ethanol in ratio of 1:2:1. In vivo burn wound infection was induced in BALB/c mice with MSSA ATCC 25923. The optimized formulation showed complete wound healing in 7 days while conventional cream took 10 days.

Ramteke et al. [109] formulated NE using cinnamon oil as an active ingredient as well as oil phase, and usnic acid was added as drug to evaluate potential synergistic antifungal activity using candidiasis model in rats and dermatophytosis model in guinea pigs. The cinnamon oil and usnic acid (CUN)-blended NE showed maximum zone of inhibition (1.19 cm) which was comparable to fluconazole (1 mg/ml), i.e. 1.82 cm. CUN showed significant antifungal activity in cutaneous candidiasis model with log CFU per infected site to be 1.36 as compared to untreated control group (3.91). However, in dermatophytosis model in guinea pigs, CUN gel showed highest clinical efficacy (41.3%) in comparison with ciclopirox (38%) and cinnamon oil-usnic acid (CUS) gel (35.6%).

Thakur et al. [110] prepared positively charged bilayered NE of fusidic acid (FA) for topical treatment of MRSA burn wound infection. The developed system was characterized for globule size, rheology and cell viability assay, in vitro antibacterial activity and ex vivo studies. The formulations showed globule size in range of 20.24–110.62 nm. FA-NE demonstrated exceptional antibacterial activity against MRSA, showing >100-fold reduction in MIC in comparison with ethanolic solution of FA. The formulations were found to be safe and non-toxic when studied in HaCaT cell lines even up to 100 μg/ml after 72 h of treatment. FA-NE showed 2.76-fold higher skin retention and complete eradication of bacteria within 7 days as compared to marketed cream, which took 15 days for eradication of infection clearly establishing superiority of the FA-NE.

Sosa et al. [111] developed NE of amphotericin B (AmB) for topical treatment of cutaneous fungal infections and aspergillosis. Pseudoternary phase diagram was constructed to select best formulation ingredients which were chosen on the basis of AmB solubility and compatibility with the skin. One hundred per cent of AmB was found to be released after 75 h demonstrating sustained release. The developed AmB NE exhibited good skin tolerability and antifungal activity.

Thomas et al. [112] prepared curcumin-loaded NE-based chitosan gel using triacetin and propylene glycol as oil phase, Tween 80 as surfactant and PEG-400 as co-surfactant for wound-healing purpose. The optimized NE showed droplet size of 84.03 nm and highest skin permeation (46.07%). NE incorporated into chitosan (2%) gel exhibited lower skin permeation (31.25%) than NE, thus enhancing skin retention. NE-based gel led to complete wound closure in 12 days, whereas untreated group was healed in 18 days. Cao et al. [113] developed NB-201, a novel membrane-disruptive NE compound for the treatment of MRSA-infected skin wounds. NB-201 was evaluated in in vivo murine and porcine infection models. NB-201 topical application reduced bacterial load in MRSA-infected wounds and was found to be non-toxic to the skin. It also inhibited epidermal and deep dermal inflammation. The study demonstrated effective MRSA killing irrespective of genetic basis of antibiotic resistance.

Many reports have demonstrated the anti-herpetic activity of pentyl gallate (PGE), but the major drawback is its poor aqueous solubility, thus limiting topical application. Kelmann et al. [114] incorporated PGE into o/w NE and evaluated it for in vitro skin penetration and anti-herpetic activity against HSV-1. The NE formulations showed particle size in range of 124.8–143.7 nm and loading efficiency >99%. PGE was able to reach deep into the dermis more effectively from NE-F4. Thus, the PGE NE formulation exhibited potential for treatment of herpes labialis infection.

Karri et al. [115] developed NE-based gel of TBF-HCl to improve its permeability by high-pressure homogenization (Gel-P) and high-speed homogenization (Gel-S). Droplet size obtained with former method was less than 2 nm in radius, while it was <10 nm in latter. Gel-P showed higher permeation (51.19%) as compared to Gel-S (31.72%) and marketed cream (19.78%) in in vitro permeation studies conducted using pork skin. *Trichophyton mentagrophytes* infection induced in Wistar rats was cured in 3 days with NE-based gels of TBF-HCl as compared to 14 days taken by marketed cream. This study highlights the potential of NE-based formulations in efficient treatment of skin infections.

5 Conclusion

Dermatological diseases are both burdensome and a stigma for people around the world. Skin infections in particular represent major category of dermatological illnesses accounting for considerable health and financial burden. The rise of antimicrobial resistance has made treatment of infectious diseases further challenging. Conventional topical formulations require longer administration periods and have failed to achieve desired efficacy due to poor skin permeation and retention. The advent of lipid-based drug delivery systems has revolutionized the dermatology industry, and these systems have been used for effective delivery of anti-infectives. The success of lipid nanocarriers can be attributed to their various advantages such as targeted and controlled drug delivery, eradication of infectious pathogen by making intimate contact and reduced toxicity. Fungisome®, a liposomal preparation of AmB, represents commercial success of lipid nanocarrier-based technology in the category of anti-infectives.

However, still consistent and advanced efforts are required to provide safe and effective topical delivery of anti-infective-loaded lipid nanocarriers. Regulatory and safety guidelines need to be developed. Despite extensive research in this area, a lot has yet to be achieved for moving from bench to bed side.

Acknowledgement The author, Karan Razdan, acknowledges the Department of Science and Technology, New Delhi, India, for their financial assistance as DST INSPIRE JRF (IF170172) and is grateful to Commonwealth Scholarship Commission, UK for their support for carrying out the research work at University of Central Lancashire, UK.

References

1. Lam P, Lee K, Wong R, Cheng G, Bian Z, Chui C et al (2018) Recent advances on topical antimicrobials for skin and soft tissue infections and their safety concerns. Crit Rev Microbiol 44(1):40–78
2. Thakur K, Sharma G, Singh B, Chhibber S, Katare OP (2018a) Current state of nanomedicines in the treatment of topical infectious disorders. Recent Pat Antiinfect Drug Discov 13(2):127–150
3. Roberts M, Mohammed Y, Pastore M, Namjoshi S, Yousef S, Alinaghi A et al (2017) Topical and cutaneous delivery using nanosystems. J Control Release 247:86–105
4. Desai P, Patlolla RR, Singh M (2010) Interaction of nanoparticles and cell-penetrating peptides with skin for transdermal drug delivery. Mol Membr Biol 27(7):247–259
5. Zhang L, Jiang Y, Ding Y, Povey M, York D (2007) Investigation into the antibacterial behaviour of suspensions of ZnO nanoparticles (ZnO nanofluids). J Nanopart Res 9(3):479–489
6. Lane ME (2013) Skin penetration enhancers. Int J Pharm 447(1–2):12–21
7. Williams A, Barry B (2012) Penetration enhancers. Adv Drug Deliv Rev 64:128–137
8. Azagury A, Khoury L, Enden G, Kost J (2014) Ultrasound mediated transdermal drug delivery. Adv Drug Deliv Rev 72:127–143
9. Han T, Das DB (2015) Potential of combined ultrasound and microneedles for enhanced transdermal drug permeation: a review. Eur J Pharm Biopharm 89:312–328
10. Nawaz A, Wong TW (2017) Microwave as skin permeation enhancer for transdermal drug delivery of chitosan-5-fluorouracil nanoparticles. Carbohydr Polym 157:906–919
11. Shahzad Y, Louw R, Gerber M, Du Plessis J (2015) Breaching the skin barrier through temperature modulations. J Control Release 202:1–13
12. Wang Y, Thakur R, Fan Q, Michniak B (2005) Transdermal iontophoresis: combination strategies to improve transdermal iontophoretic drug delivery. Eur J Pharm Biopharm 60(2):179–191
13. Garcês A, Amaral M, Lobo JS, Silva A (2018) Formulations based on solid lipid nanoparticles (SLN) and nanostructured lipid carriers (NLC) for cutaneous use: A review. Eur J Pharm Sci 112:159–167
14. Jain S, Patel N, Shah MK, Khatri P, Vora N (2017) Recent advances in lipid-based vesicles and particulate carriers for topical and transdermal application. J Pharm Sci 106(2):423–445
15. Sala M, Diab R, Elaissari A, Fessi H (2018) Lipid nanocarriers as skin drug delivery systems: Properties, mechanisms of skin interactions and medical applications. Int J Pharm 535(1–2):1–17

16. Al-mahallawi AM, Abdelbary AA, Aburahma MH (2015) Investigating the potential of employing bilosomes as a novel vesicular carrier for transdermal delivery of tenoxicam. Int J Pharm 485(1–2):329–340
17. Barua S, Mitragotri S (2014) Challenges associated with penetration of nanoparticles across cell and tissue barriers: a review of current status and future prospects. Nano Today 9(2):223–243
18. Baker C, Pradhan A, Pakstis L, Pochan DJ, Shah SI (2005) Synthesis and antibacterial properties of silver nanoparticles. J Nanosci Nanotechnol 5(2):244–249
19. McGrath J, Eady R, Pope F (2004) Anatomy and organization of human skin. In: Rook's textbook of dermatology, vol 10. Wiley, Hoboken, p 9781444317633
20. Agache P (2004) Metrology of the stratum corneum. In: Humbert et al (eds) Measuring the skin. Springer International Publishing, Switzerland, pp 101–111
21. Gray G, White R, Williams R, Yardleymrc H (1982) Lipid composition of the superficial stratum corneum cells of pig epidermis. Br J Dermatol 106(1):59–63
22. Wertz PW, Swartzendruber DC, Kitko DJ, Madison KC, Downing DT (1989) The role of the corneocyte lipid envelopes in cohesion of the stratum corneum. J Investig Dermatol 93(1):169–172
23. Gentilhomme E, Neveux Y (2004) Epidermal physiology. In: Humbert et al (eds) Measuring the skin. Springer International Publishing, Switzerland, p 165
24. Burgeson RE, Christiano AM (1997) The dermal—epidermal junction. Curr Opin Cell Biol 9(5):651–658
25. Bisno AL, Stevens DL (1996) Streptococcal infections of skin and soft tissues. N Engl J Med 334(4):240–246
26. Callahan EF, Adal KA, Tomecki KJ (2000) Cutaneous (non-HIV) infections. Dermatol Clin 18(3):497–508
27. Hay R (1990) Overview of studies of fluconazole in oropharyngeal candidiasis. Rev Infect Dis 12(Supplement_3):S334–S3S7
28. Drake LA, Dinehart SM, Farmer ER, Goltz RW, Graham GF, Hordinsky MK et al (1996) Guidelines of care for superficial mycotic infectionsof the skin: Pityriasis (tinea) versicolor. J Am Acad Dermatol 34(2):287–289
29. Memar O, Tyring SK (1995) Cutaneous viral infections. J Am Acad Dermatol 33(2):279–287
30. Sterling J, Handfield-Jones S, Hudson P (2001) Guidelines for the management of cutaneous warts. Br J Dermatol 144(1):4–11
31. Reithinger R, Dujardin J-C, Louzir H, Pirmez C, Alexander B, Brooker S (2007) Cutaneous leishmaniasis. Lancet Infect Dis 7(9):581–596
32. Pople PV, Singh KK (2010) Targeting tacrolimus to deeper layers of skin with improved safety for treatment of atopic dermatitis. Int J Pharm 398(1–2):165–178
33. Teichmann A, Heuschkel S, Jacobi U, Presse G, Neubert RH, Sterry W et al (2007) Comparison of stratum corneum penetration and localization of a lipophilic model drug applied in an o/w microemulsion and an amphiphilic cream. Eur J Pharm Biopharm 67(3):699–706
34. Jadhav M, Kalhapure RS, Rambharose S, Mocktar C, Singh S, Kodama T et al (2018) Novel lipids with three C18-fatty acid chains and an amino acid head group for pH-responsive and sustained antibiotic delivery. Chem Phys Lipids 212:12–25
35. Hasanpouri A, Lotfipour F, Ghanbarzadeh S, Hamishehkar H (2018) Improvement of dermal delivery of tetracycline using vesicular nanostructures. Res Pharm Sci 13(5):385
36. Zhang L, Pornpattananangkul D, Hu C-M, Huang C-M (2010) Development of nanoparticles for antimicrobial drug delivery. Curr Med Chem 17(6):585–594
37. Luan L, Chi Z, Liu C (2019) Chinese white wax solid lipid nanoparticles as a novel nanocarrier of curcumin for inhibiting the formation of Staphylococcus aureus biofilms. Nano 9(5):763
38. Bangham AD, Horne R (1964) Negative staining of phospholipids and their structural modification by surface-active agents as observed in the electron microscope. J Mol Biol 8(5):660–IN10

39. Rahman M, Alam K, Beg S, Anwar F, Kumar V (2019) Liposomes as topical drug delivery systems: state of the arts. In: Biomedical applications of nanoparticles. Elsevier, Amsterdam, pp 149–161
40. Mezei M, Gulasekharam V (1980) Liposomes-a selective drug delivery system for the topical route of administration I. Lotion dosage form. Life Sci 26(18):1473–1477
41. Egbaria K, Weiner N (1990) Liposomes as a topical drug delivery system. Adv Drug Deliv Rev 5(3):287–300
42. Rukavina Z, Vanić Ž (2016) Current trends in development of liposomes for targeting bacterial biofilms. Pharmaceutics 8(2):18
43. Boakye CH, Patel K, Singh M (2015) Doxorubicin liposomes as an investigative model to study the skin permeation of nanocarriers. Int J Pharm 489(1–2):106–116
44. Verma DD, Verma S, Blume G, Fahr A (2003) Particle size of liposomes influences dermal delivery of substances into skin. Int J Pharm 258(1–2):141–151
45. Madan S, Nehate C, Barman TK, Rathore AS, Koul V (2019) Design, preparation, and evaluation of liposomal gel formulations for treatment of acne: in vitro and in vivo studies. Drug Dev Ind Pharm 45(3):395–404
46. Moosavian SA, Fallah M, Jaafari MR (2019) The activity of encapsulated meglumine antimoniate in stearylamine-bearing liposomes against cutaneous leishmaniasis in BALB/c mice. Exp Parasitol 200:30–35
47. Kavian Z, Alavizade SH, Golmohamadzadeh S, Badiee A, Khamesipour A, Jaafari MR (2019) Development of topical liposomes containing miltefosine for the treatment of Leishmania major infection in susceptible BALB/c mice. Acta Trop. https://doi.org/10.1016/j.actatropica.2019.05.018
48. Rukavina Z, Klarić MŠ, Filipović-Grčić J, Lovrić J, Vanić Ž (2018) Azithromycin-loaded liposomes for enhanced topical treatment of methicillin-resistant Staphyloccocus aureus (MRSA) infections. Int J Pharm 553(1–2):109–119
49. Chhibber S, Kaur J, Kaur S (2018) Liposome Entrapment of Bacteriophages Improves Wound Healing in a Diabetic Mouse MRSA Infection. Front Microbiol 9:561
50. Hsu CY, Yang SC, Sung CT, Weng YH, Fang JY (2017) Anti-MRSA malleable liposomes carrying chloramphenicol for ameliorating hair follicle targeting. Int J Nanomedicine 12:8227–8238
51. Tanriverdi ST, Hilmioglu Polat S, Yesim Metin D, Kandiloglu G, Ozer O (2016) Terbinafine hydrochloride loaded liposome film formulation for treatment of onychomycosis: in vitro and in vivo evaluation. J Liposome Res 26(2):163–173
52. Nunes PS, Rabelo AS, Souza JC, Santana BV, da Silva TM, Serafini MR et al (2016) Gelatin-based membrane containing usnic acid-loaded liposome improves dermal burn healing in a porcine model. Int J Pharm 513(1–2):473–482
53. Tanriverdi ST, Ozer O (2013) Novel topical formulations of Terbinafine-HCl for treatment of onychomycosis. Eur J Pharm Sci 48(4–5):628–636
54. Gupta M, Goyal AK, Paliwal SR, Paliwal R, Mishra N, Vaidya B et al (2010) Development and characterization of effective topical liposomal system for localized treatment of cutaneous candidiasis. J Liposome Res 20(4):341–350
55. Touitou E, Dayan N, Bergelson L, Godin B, Eliaz M (2000) Ethosomes - novel vesicular carriers for enhanced delivery: characterization and skin penetration properties. J Control Release 65(3):403–418
56. Shetty S, Jose J, Kumar L, Charyulu RN (2019) Novel ethosomal gel of clove oil for the treatment of cutaneous candidiasis. J Cosmet Dermatol 18(3):862–869
57. Faisal W, Soliman GM, Hamdan AM (2018) Enhanced skin deposition and delivery of voriconazole using ethosomal preparations. J Liposome Res 28(1):14–21
58. Marto J, Vitor C, Guerreiro A, Severino C, Eleuterio C, Ascenso A et al (2016) Ethosomes for enhanced skin delivery of griseofulvin. Colloids Surf B Biointerfaces 146:616–623
59. Iizhar SA, Syed IA, Satar R, Ansari SA (2016) In vitro assessment of pharmaceutical potential of ethosomes entrapped with terbinafine hydrochloride. J Adv Res 7(3):453–461

60. Godin B, Touitou E, Rubinstein E, Athamna A, Athamna M (2005) A new approach for treatment of deep skin infections by an ethosomal antibiotic preparation: an in vivo study. J Antimicrob Chemother 55(6):989–994
61. Cevc G, Blume G (1992) Lipid vesicles penetrate into intact skin owing to the transdermal osmotic gradients and hydration force. Biochim Biophys Acta 1104(1):226–232
62. Cevc G, Blume G, Schätzlein A, Gebauer D, Paul A (1996) The skin: a pathway for systemic treatment with patches and lipid-based agent carriers. Adv Drug Deliv Rev 18(3):349–378
63. Trotta M, Peira E, Carlotti ME, Gallarate M (2004) Deformable liposomes for dermal administration of methotrexate. Int J Pharm 270(1–2):119–125
64. Elsayed MM, Abdallah OY, Naggar VF, Khalafallah NM (2007) Lipid vesicles for skin delivery of drugs: reviewing three decades of research. Int J Pharm 332(1–2):1–16
65. Jain S, Jain P, Umamaheshwari R, Jain N (2003) Transfersomes—a novel vesicular carrier for enhanced transdermal delivery: development, characterization, and performance evaluation. Drug Dev Ind Pharm 29(9):1013–1026
66. Qushawy M, Nasr A, Abd-Alhaseeb M, Swidan S (2018) Design, optimization and characterization of a transfersomal gel using miconazole nitrate for the treatment of candida skin infections. Pharmaceutics 10(1). https://doi.org/10.3390/pharmaceutics10010026
67. Pandit J, Garg M, Jain NK (2014) Miconazole nitrate bearing ultraflexible liposomes for the treatment of fungal infection. J Liposome Res 24(2):163–169
68. Chhibber S, Shukla A, Kaur S (2017) Transfersomal phage cocktail is an effective treatment against methicillin-resistant Staphylococcus aureus-mediated skin and soft tissue infections. Antimicrob Agents Chemother 61(10):e02146–e02116
69. Abdellatif MM, Khalil IA, Khalil MAF (2017) Sertaconazole nitrate loaded nanovesicular systems for targeting skin fungal infection: In-vitro, ex-vivo and in-vivo evaluation. Int J Pharm 527(1–2):1–11
70. Li C, Zhang X, Huang X, Wang X, Liao G, Chen Z (2013) Preparation and characterization of flexible nanoliposomes loaded with daptomycin, a novel antibiotic, for topical skin therapy. Int J Nanomedicine 8:1285–1292
71. Ashtikar M, Nagarsekar K, Fahr A (2016) Transdermal delivery from liposomal formulations–Evolution of the technology over the last three decades. J Control Release 242:126–140
72. Gangwar M, Singh R, Goel R, Nath G (2012) Recent advances in various emerging vescicular systems: An overview. Asian Pac J Trop Biomed 2(2):S1176–S1S88
73. Kumar GP, Rajeshwarrao P (2011) Nonionic surfactant vesicular systems for effective drug delivery—an overview. Acta Pharm Sin B 1(4):208–219
74. Sohrabi S, Haeri A, Mahboubi A, Mortazavi A, Dadashzadeh S (2016) Chitosan gel-embedded moxifloxacin niosomes: an efficient antimicrobial hybrid system for burn infection. Int J Biol Macromol 85:625–633
75. Alomrani AH, Al-Agamy MH, Badran MM (2015) In vitro skin penetration and antimycotic activity of itraconazole loaded niosomes: various non-ionic surfactants. J Drug Deliv Sci Technol 28:37–45
76. Müller RH, Mäder K, Gohla S (2000) Solid lipid nanoparticles (SLN) for controlled drug delivery–a review of the state of the art. Eur J Pharm Biopharm 50(1):161–177
77. Doktorovová S, Kovačević AB, Garcia ML, Souto EB (2016) Preclinical safety of solid lipid nanoparticles and nanostructured lipid carriers: current evidence from in vitro and in vivo evaluation. Eur J Pharm Biopharm 108:235–252
78. Guimarães KL, Ré MI (2011) Lipid nanoparticles as carriers for cosmetic ingredients: the first (SLN) and the second generation (NLC). In: Nanocosmetics and nanomedicines. Springer, New York, pp 101–122
79. López-García R, Ganem-Rondero A (2015) Solid lipid nanoparticles (SLN) and nanostructured lipid carriers (NLC): occlusive effect and penetration enhancement ability. J Cosmet Dermatol Sci Appl 5(02):62
80. Muller RH, Radtke M, Wissing SA (2002) Nanostructured lipid matrices for improved microencapsulation of drugs. Int J Pharm 242(1–2):121–128

81. Montenegro L, Lai F, Offerta A, Sarpietro MG, Micicche L, Maccioni AM et al (2016) From nanoemulsions to nanostructured lipid carriers: A relevant development in dermal delivery of drugs and cosmetics. J Drug Deliv Sci Technol 32:100–112
82. Müller RH, Alexiev U, Sinambela P, Keck CM (2016) Nanostructured lipid carriers (NLC): the second generation of solid lipid nanoparticles. In: Percutaneous penetration enhancers chemical methods in penetration enhancement. Springer, New York, pp 161–185
83. Beloqui A, Solinís MÁ, Rodríguez-Gascón A, Almeida AJ, Préat V (2016) Nanostructured lipid carriers: Promising drug delivery systems for future clinics. Nanomedicine 12(1):143–161
84. El-Housiny S, Shams Eldeen MA, El-Attar YA, Salem HA, Attia D, Bendas ER et al (2018) Fluconazole-loaded solid lipid nanoparticles topical gel for treatment of pityriasis versicolor: formulation and clinical study. Drug Deliv 25(1):78–90
85. Kakadia PG, Conway BR (2018) Solid lipid nanoparticles for targeted delivery of triclosan into skin for infection prevention. J Microencapsul 35(7–8):695–704
86. Vaghasiya H, Kumar A, Sawant K (2013) Development of solid lipid nanoparticles based controlled release system for topical delivery of terbinafine hydrochloride. Eur J Pharm Sci 49(2):311–322
87. Bhalekar MR, Pokharkar V, Madgulkar A, Patil N, Patil N (2009) Preparation and evaluation of miconazole nitrate-loaded solid lipid nanoparticles for topical delivery. AAPS PharmSciTech 10(1):289–296
88. Jain S, Jain S, Khare P, Gulbake A, Bansal D, Jain SK (2010) Design and development of solid lipid nanoparticles for topical delivery of an anti-fungal agent. Drug Deliv 17(6):443–451
89. Lv Q, Yu A, Xi Y, Li H, Song Z, Cui J et al (2009) Development and evaluation of penciclovir-loaded solid lipid nanoparticles for topical delivery. Int J Pharm 372(1–2):191–198
90. Lai F, Sinico C, De Logu A, Zaru M, Müller RH, Fadda AM (2007) SLN as a topical delivery system for Artemisia arborescens essential oil: in vitro antiviral activity and skin permeation study. Int J Nanomedicine 2(3):419
91. Carbone C, Teixeira M, Sousa M, Martins-Gomes C, Silva AM, Souto EMB et al (2019) Clotrimazole-loaded mediterranean essential oils NLC: a synergic treatment of candida skin infections. Pharmaceutics 11(5):231
92. Khezri K, Farahpour MR, Mounesi RS (2019) Accelerated infected wound healing by topical application of encapsulated Rosemary essential oil into nanostructured lipid carriers. Artif Cells Nanomed Biotechnol 47(1):980–988
93. Ghodrati M, Farahpour MR, Hamishehkar H (2019) Encapsulation of Peppermint essential oil in nanostructured lipid carriers: In-vitro antibacterial activity and accelerative effect on infected wound healing. Colloids Surf A Physicochem Eng Asp 564:161–169
94. Alalaiwe A, Wang P-W, Lu P-L, Chen Y-P, Fang J-Y, Yang S-C (2018) Synergistic anti-MRSA activity of cationic nanostructured lipid carriers in combination with oxacillin for cutaneous application. Front Microbiol 9:1493
95. Gaba B, Fazil M, Khan S, Ali A, Baboota S, Ali J (2015) Nanostructured lipid carrier system for topical delivery of terbinafine hydrochloride. Bull Fac Pharm Cairo Univ 53(2):147–159
96. Song SH, Lee KM, Kang JB, Lee SG, Kang MJ, Choi YW (2014) Improved skin delivery of voriconazole with a nanostructured lipid carrier-based hydrogel formulation. Chem Pharm Bull 62(8):793–798
97. Hoar T, Schulman J (1943) Transparent water-in-oil dispersions: the oleopathic hydro-micelle. Nature 152(3847):102
98. Kreilgaard M, Kemme MJ, Burggraaf J, Schoemaker RC, Cohen AF (2001) Influence of a microemulsion vehicle on cutaneous bioequivalence of a lipophilic model drug assessed by microdialysis and pharmacodynamics. Pharm Res 18(5):593–599
99. Shukla A, Janich M, Jahn K, Neubert RH (2003) Microemulsions for dermal drug delivery studied by dynamic light scattering: effect of interparticle interactions in oil-in-water micro-emulsions. J Pharm Sci 92(4):730–738
100. Al-Adham I, Khalil E, Al-Hmoud N, Kierans M, Collier P (2000) Microemulsions are membrane-active, antimicrobial, self-preserving systems. J Appl Microbiol 89(1):32–39

101. Solans C, Izquierdo P, Nolla J, Azemar N, Garcia-Celma M (2005) Nano-emulsions. Curr Opin Colloid Interface Sci 10(3–4):102–110
102. Calderilla-Fajardo S, Cazares-Delgadillo J, Villalobos-Garcia R, Quintanar-Guerrero D, Ganem-Quintanar A, Robles R (2006) Influence of sucrose esters on the in vivo percutaneous penetration of octyl methoxycinnamate formulated in nanocapsules, nanoemulsion, and emulsion. Drug Dev Ind Pharm 32(1):107–113
103. Mou D, Chen H, Du D, Mao C, Wan J, Xu H et al (2008) Hydrogel-thickened nanoemulsion system for topical delivery of lipophilic drugs. Int J Pharm 353(1–2):270–276
104. Talaat SM, Elnaggar YS, Abdalla OY (2019) Lecithin microemulsion lipogels versus conventional gels for skin targeting of terconazole: in vitro, ex vivo, and in vivo investigation. AAPS PharmSciTech 20(4):161
105. Kaur A, Sharma G, Gupta V, Ratho RK, Katare OP (2018) Enhanced acyclovir delivery using w/o type microemulsion: preclinical assessment of antiviral activity using murine model of zosteriform cutaneous HSV-1 infection. Artif Cells Nanomed Biotechnol 46(2):346–354
106. Alam MA, Al-Janoobi FI, Alzahrani KA, Al-Agamy MH, Abdelgalil AA, Al-Mohizea AM (2017) In-vitro efficacies of topical microemulsions of clotrimazole and ketoconazole; and in-vivo performance of clotrimazole microemulsion. J Drug Deliv Sci Technol 39:408–416
107. Wan T, Xu T, Pan J, Qin M, Pan W, Zhang G et al (2015) Microemulsion based gel for topical dermal delivery of pseudolaric acid B: in vitro and in vivo evaluation. Int J Pharm 493(1–2):111–120
108. Chhibber T, Wadhwa S, Chadha P, Sharma G, Katare OP (2015) Phospholipid structured microemulsion as effective carrier system with potential in methicillin sensitive Staphylococcus aureus (MSSA) involved burn wound infection. J Drug Target 23(10):943–952
109. Ramteke P, Pandey AC, Pandey H (2019) Evaluation of antifungal activity of blended cinnamon oil and usnic acid nanoemulsion using candidiasis and dermatophytosis models. Biocatal Agric Biotechnol 18:101062
110. Thakur K, Sharma G, Singh B, Jain A, Tyagi R, Chhibber S et al (2018b) Cationic-bilayered nanoemulsion of fusidic acid: an investigation on eradication of methicillin-resistant Staphylococcus aureus 33591 infection in burn wound. Nanomedicine 13(8):825–847
111. Sosa L, Clares B, Alvarado HL, Bozal N, Domenech O, Calpena AC (2017) Amphotericin B releasing topical nanoemulsion for the treatment of candidiasis and aspergillosis. Nanomedicine 13(7):2303–2312
112. Thomas L, Zakir F, Mirza MA, Anwer MK, Ahmad FJ, Iqbal Z (2017) Development of curcumin loaded chitosan polymer based nanoemulsion gel: In vitro, ex vivo evaluation and in vivo wound healing studies. Int J Biol Macromol 101:569–579
113. Cao Z, Spilker T, Fan Y, Kalikin LM, Ciotti S, LiPuma JJ et al (2017) Nanoemulsion is an effective antimicrobial for methicillin-resistant Staphylococcus aureus in infected wounds. Nanomedicine 12(10):1177–1185
114. Kelmann RG, Colombo M, Lopes SCDA, Nunes RJ, Pistore M, Agnol DD et al (2016) Pentyl gallate nanoemulsions as potential topical treatment of herpes labialis. J Pharm Sci 105(7):2194–2203
115. Karri VNR, Raman SK, Kuppusamy G, Mulukutla S, Ramaswamy S, Malayandi R (2015) Terbinafine hydrochloride loaded nanoemulsion based gel for topical application. J Pharm Investig 45(1):79–89

Lipid Nanocarriers for Dermal Delivery of Lutein

Shammy Jindal, Avichal Kumar, Kamya Goyal, Rajendra Awasthi, and Giriraj T. Kulkarni

1 Introduction

Lutein (β, ε-carotene-3, 3′-diol) is a naturally occurring carotenoid. It is a lipid-soluble pigment present in vegetables such as spinach and kale. It is an oxygenated derivative of carotenoids known as xanthophylls. It is a yellow pigment present in the macular region of the eye retina. This carotenoid is not produced in the human body; it is obtained from food sources. Out of 20 different antioxidants, xanthophyll is the major antioxidant found in soma [1]. Oral bioavailability of lutein is about 10–15%. Acceptable daily intake of carotenoid is 2 mg/kg weight (equivalent to 220 mg/day for a 60-kg person).

Zeaxanthin is the most common carotenoid alcohol found in nature associated with xanthophyll cycle. It is present in high concentrations within the macular area of the human eye. It protects the macula from blue light. It improves visual acuity known for its importance for up vision [2]. Oxidation of the lens is a major cause of cataracts. Zeaxanthin and lutein are the most common antioxidants present in the eye, which nutrients neutralize free radicals. Thus, it reduces the risk of chronic eye disorders, including cataracts and age-related macular degeneration.

Human skin also contains lutein and carotenoids. The skin needs antioxidants to protect it by reducing and counteracting free radical production [3]. Xanthophyll plays an important role in developing and maintaining a healthy skin cell layer by inhibiting the UV radiations from sunlight. It reduces the cell damage and hence

S. Jindal (✉)
Laureate Institute of Pharmacy, Kathog, Jawalamukhi, Himachal Pradesh, India

Amity Institute of Pharmacy, Amity University Uttar Pradesh, Noida, Uttar Pradesh, India

A. Kumar · K. Goyal
Laureate Institute of Pharmacy, Kathog, Jawalamukhi, Himachal Pradesh, India

R. Awasthi (✉) · G. T. Kulkarni
Amity Institute of Pharmacy, Amity University Uttar Pradesh, Noida, Uttar Pradesh, India

© Springer Nature Singapore Pte Ltd. 2020

M. Rahman et al. (eds.), *Nanomedicine for Bioactives*,
https://doi.org/10.1007/978-981-15-1664-1_12

less inflammation is produced in skin layers. It has been observed that the highest exposure to UV radiations leads to generation of carcinoma. Carotenoids like lutein have therapeutic value that reduces the risk of carcinoma [4].

Lutein is lipophilic in nature and has a poor aqueous solubility. Nanocarrier formulations are expected to overcome the limitations associated with the dermal delivery of lutein. A brief account of physicochemical properties of lutein is presented in Table 1. Chemical structure of lutein and its derivative zeaxanthin is presented in Fig. 1.

In a double-blind, 12-week placebo-controlled study, oral and topical administration of lutein and zeaxanthin alone or in combination showed significant antioxidant activity in the skin. However, the highest degree of antioxidant protection was obtained in subjects that received a combination of lutein and zeaxanthin. An oral dose of lutein provided better protection than topical application. Lutein and zeaxanthin also found to decrease the lipid peroxidation and increase skin hydration, surface lipids, and skin elasticity [6].

Table 1 A brief account of physicochemical properties of lutein [5]

Name of drug	Lutein
Chemical class	Xanthophyll (carotenoid)
Appearance	Yellowish-red/or yellowish-orange crystalline powder
Molecular weight	568.87 g/mol
Melting point	176–179°C
Molecular formula	$C_{40}H_{56}O_2$
Half-life	10–15 days
Solubility	Highly lipophilic, practically insoluble in water, soluble in methanol, ethanol, ethyl acetate, methylene chloride, n-hexane, and tetrahydrofuran

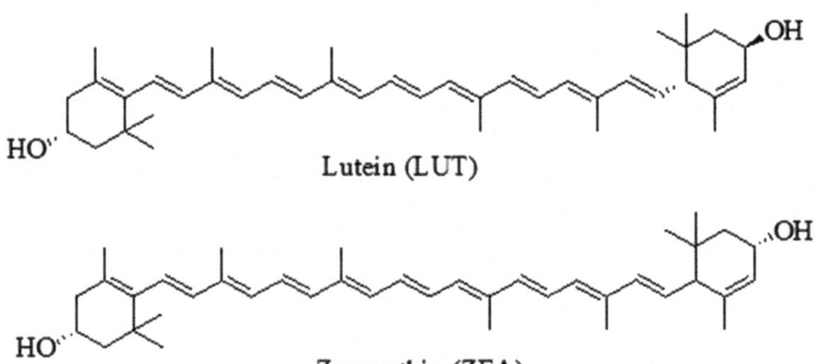

Fig. 1 Chemical structure of lutein and its derivative zeaxanthin containing nine conjugated double bonds in the polyene chain characterized by the presence of two hydroxyl groups

Lutein-loaded nanocarriers synthesized by high-pressure homogenization have shown excellent in vitro dissolution and in vitro penetration profile in fresh pig ear skin. The results showed that the bioactive molecule remained in the skin and is not absorbed systemically. Also the formulation protected the bioactive against UV degradation [7].

The application of natural carotenoids includes both nutraceutical and biological process supplements. It protects the skin as a protective shield from any kind of injury as well as helps in the inhibition of chemical reaction that prevents cholesterol synthesis. It defends inflammatory diseases and reduces aging of respiratory organ [8, 9]. Additionally, it is also used as an orange-yellowish coloring agent in food and liquid products [10].

Due to the increased number of skin disorders, attention has been focused on the topical delivery of natural bioactives using nanotechnology [11–14]. The nanoparticles as well as liposomal delivery systems have limitations in industrial applications (e.g., physical stabilities of formulations) and many polymers have a shortage of specific accepted regulative standing levels and lack of massive scale production effective values (e.g., compound nanoparticles) [15, 16]. Among numerous skin delivery systems investigated, lipid nanocarriers attracted a wide attention of formulation scientists due to their unique functional significance and mode of drug delivery [2]. Solid lipid nanoparticles (SLNs) and nanostructured lipid carriers (NLCs) are widely used as delivery carriers for carotenoids. In cosmetic dermal product, the NLCs are already in the market. SLNs and NLCs have the ability to improve the chemical stability and delivery of pharmacologically active molecules and cosmeceuticals [17, 18]. Several types of nanocarriers have been reported for drug delivery. This chapter recapitulates therapeutic applications of lutein and its delivery via SLNs and NLCs.

2 Structure of Skin

It is the largest organ present in the human body. Total area of adult human skin is approximately 2 square meters. Protecting body from the external environment, maintaining water loss by sweating, controlling cooling effects by water evaporation, and creating a barrier for unwanted foreign body absorption are the primary functions of skin. Anatomically, skin consists of three layers—epidermis, dermis, and subcutaneous fat tissues [19, 20] (Fig. 2).

2.1 Epidermis Layer

The cuticle is separated into two regions: the dead cuticle or stratum corneum covered with viable cuticle. The epidermal layer is composed of keratinized stratified squamous epithelium and water [19]. Normally, this layer consists of about 10–25 layers of dead, elongated, fully keratinized corneocytes. The outmost layer of skin is known as horny layer, which represents the physiological state of skin and has

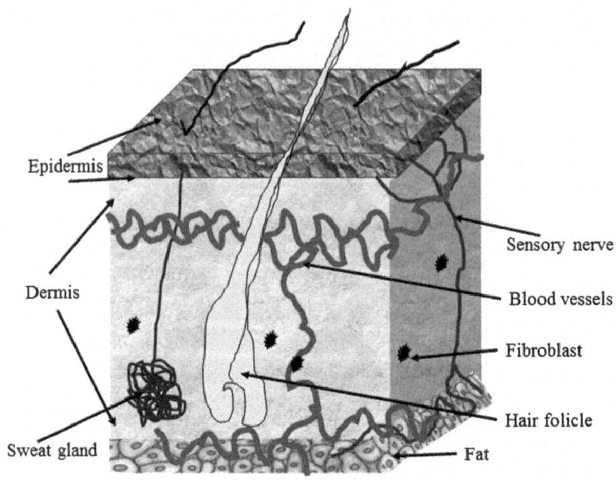

Fig. 2 Schematic representation of anatomical structure of human skin

protecting functions [21, 22]. This layer of skin does not contain blood vessels within it (i.e., it is avascular). Based on the number of skin layers it is classified into two classes, i.e., thin skin and thick skin. A thin skin has four layers which are stratum basale, stratum spinosum, stratum granulosum, and stratum corneum (from deep to superficial). Thick skin has a fifth layer, called the stratum lucidum. This layer is located between stratum granulosum and stratum corneum. Thick skin is found only on the soles of the feet and the palms of the hands.

2.2 Dermis Layer

This layer is the major structural element of skin which has three components, namely fibrous matrix, cellular, and diffuse and filamentous matrix. Basically, it is composed of fibrous, an inter-fiber gel of glycosaminoglycans, different salts, and water. This layer contains different kinds of blood vessels, nerve endings, follicle, oleaginous glands, and sweat glands embedded at specific intervals. The vascular network present in depth within the dermis plays an important role in supplying nutrition to the skin, repairing processes, immunological signals, heat sensitivity regulation, etc. [23].

2.3 Subcutaneous "Fat" Tissue Layer

Subcutaneous layer (also called hypodermis) serves as a depot for the storage of fat. The albuminoid present in between the fat cells provides the linkage. The subcutaneous fat tissue acts as a dielectric of heat and prevents heat loss [24]. A hypodermic injection is given through the subcutaneous tissue.

3 Methods of Skin Penetration

There are two possible routes of drug penetration across the intact skin, namely the transepidermal (intracellular and intercellular) and transappendageal (sweat ducts, hair follicles, and sebaceous glands) pathways.

3.1 Transepidermal Pathway

The passage of bioactives takes through the stratum corneum, which acts as a multicellular and multilayered barrier. Depending on the mode of permeation, transepidermal pathway is further classified as intercellular or intracellular penetration. Intercellular transport allows the diffusion of nonpolar or lipophilic molecules through the intercellular spaces of continuous lipid matrix via a tortuous pathway. However, in case of intracellular transport the transport of polar or hydrophilic solutes takes place through corneocytes (terminally differentiated keratinocytes) [19].

3.2 Transappendageal Pathway

Transappendageal drug diffusion could affect the overall skin permeation and local distribution of drugs to the skin appendages and may influence the percutaneous fate of topically administered bioactive. In transappendageal pathway, the penetrating substances diffuse through the corneum *via* a shunt pathway through the hair follicles or sweat glands. The hair follicles play an important role for this kind of pathway [25] (Fig. 3). This route is also known as the shunt route.

4 Principle of Skin Permeation

The most simplistic skin permeation approach is passive permeation, which is explained *via* Fick's first law of diffusion. According to the Fick's first law, the rate ($\mathrm{d}Q/\mathrm{d}t$) of a solute permeation through the skin membrane having unit area (A) in one direction is directly proportional to the gradient of concentration (dc/dx) across the skin membrane. The flux of permeation (J) can be defined by a mathematical equation as follows:

$$J = \frac{\mathrm{d}Q}{\mathrm{d}t \times A}$$

According to the equation, the flux of permeation and the concentration gradient across the skin membrane are directly proportional to each other. The coefficient of diffusion (D) can be represented by the following equation:

$$D = \frac{BT}{6\pi hR}$$

where B is Boltzmann constant, T is temperature, h is viscosity, and R is radius of the solute. Hence, the equation can be represented as

$$J = D\frac{A(\mathrm{C1} - \mathrm{C2})}{h}$$

where C1 and C2 represent the concentration level across the membrane and h is the membrane thickness. The partition coefficient (K) can be represented as the following equation:

$$K = \frac{\mathrm{C1}}{\mathrm{Cd}} = \frac{\mathrm{C2}}{\mathrm{Cr}}$$

where Cd and Cr represent different concentrations present in the donor and receptor compartments (Fig. 4). When we introduce the factor of partition coefficient, the equation can be written as follows:

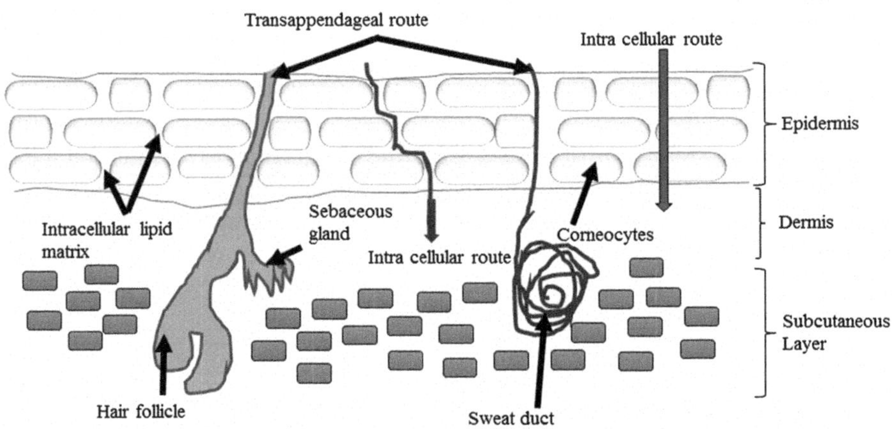

Fig. 3 Schematic presentation of pathways for percutaneous absorption

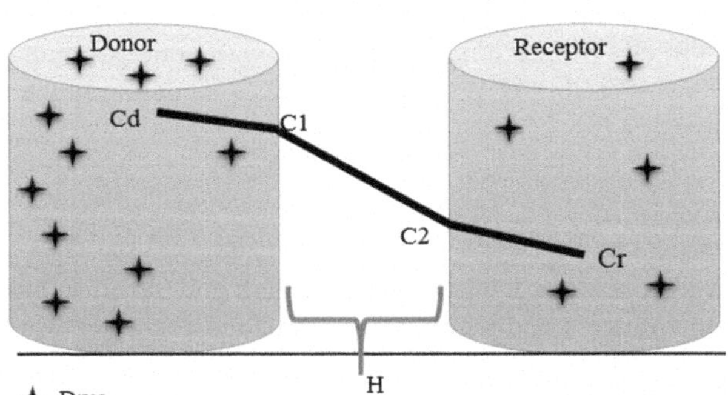

Fig. 4 Permeation of drug molecule from donor compartment to receptor compartment across the concentration gradient

$$J = \frac{\mathrm{DAK}\left(\mathrm{Cd} - \mathrm{Cr}\right)}{h}$$

According to the concepts, the passive diffusion of lutein or other drug formulations is directly dependent on the concentration gradient, factor of temperature, viscosity of the solute, medium or drug delivery system, and molecular size or particle size of the drug as well as type of delivery system used [23, 26].

5 Carrier for Lutein Delivery

During the past 2–3 decades, various delivery carriers such as polymeric nanoparticles, pH-responsive polymeric micelles, liposomes, and nanoemulsion have been reported for the effective delivery of lutein. The nanoparticles are known as the mixture of the solid particles having a range from 10 to 1000 nm, within which the bioactive molecules are entrapped. However, commercialization of these carriers is not easy due to their physical stability problems (particularly for liposomes), lack of regulatory approval of many polymers, and costly production of polymeric nanoparticles. NLCs and SLNs are expected to overcome the limitations associated with topical delivery of lutein. Lipid-based nanoparticles are classified into two categories known as SLN and NLC. The solid lipid nanoparticles have been synthesized to overcome a number of restrictions of mixture carriers such as liposomes, emulsions, and different kinds of compound nanoparticles. NLCs are commercially available in cosmetic dermal delivery of lutein. However, SLNs containing pure solid lipid may form a relatively perfect crystal lattice and leave a limited space for drug entrapment. This may lead to the leaching of drug during storage due to the polymorphic changes in lipid phase. This limitation can overcome by incorporating solid fatty material to form SLNs. The inner core of SLNs is made up of a mixture of liquid and solid lipid, which is thermodynamically stable during storage at room temperature. These lipid NLs have been recommended for oral, parenteral, and topical administration. The SLNs are modified into the NLCs that enhance the loading capability. Three basic structural models have been developed for NLCs. These lipid NLCs have a significant role in drug delivery and cosmetic research. These nanoformulations are nontoxic and site specific [27]. A schematic presentation of the historical development of various colloidal carrier systems is given in Fig. 5. SLNs combine the benefits of different mixture carriers and also over a number of limitations associated with different individual carriers. The SLNs also enhance the bioavailability of bioactive molecules [28] (Table 2). These nanocarriers are free from organic solvents, well acceptable by human skin, and are easy to scale up by existing technologies.

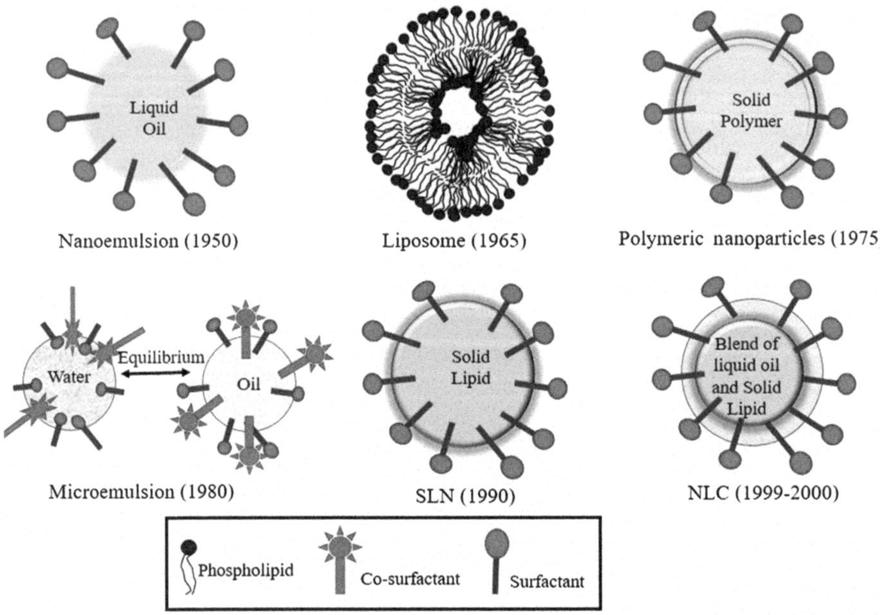

Fig. 5 Schematic presentation of historical development of different colloidal carrier systems

Table 2 Comparison of solid lipid nanoparticles and nanostructured lipid carriers

Parameters	Solid lipid nanoparticles	Nanostructured lipid carriers
Composition	SLNs are composed of 1% w/w to 30% w/w solid lipid dispersed in an aqueous medium and if necessary stabilized with preferably 0.5% w/w to 5% w/w surfactant	Blends of solid lipids and liquid lipids (oils), preferably in a weight ratio of 70:30 up to a ratio of 90:10
Microstructure	Highly ordered lipid matrix	Less ordered lipid matrix
Preferred method of preparation	High-pressure homogenization	High-pressure homogenization
Drug loading	Due to its high degree of order, the number of imperfections in the crystal lattice is reduced leading to drug expulsion	By blending solid and liquid lipids, a less ordered lipid matrix is created with higher drug load potential
Occlusivity	Higher	Lower
Mechanism of enhancing skin permeation	Due to small particle size and consequently higher surface area, these nanoparticles achieve close contact with superficial junction of corneocyte clusters and channels of stratum corneum	Similar to SLNs. Due to small particle size and consequently higher surface area, these nanoparticles achieve close contact with superficial junction of corneocyte clusters and channels of stratum corneum

6 Ingredients of Nanostructure Lipid Carriers

NLCs are derived from the o/w nanoemulsion system. These nanoformulations are made up of biocompatible and biodegradable lipids and/or fatty acids, which are generally recognized as safe. NLCs are composed of various components such as surfactants, lipids, and water. In addition, NLCs do not require organic solvents. A specific part of the oil content in o/w emulsion can be replaced with a lipid (solid) resulting as a matrix of solid particle of this carrier system having specific temperature conditions to obtain the blends for the matrix of particles [29].

6.1 Lipids

The selection of acceptable lipid mixture is a critical step in the production of NLCs with specific physicochemical characteristics. The selection criteria of an appropriate lipid mix are mentioned as below:

- The solubility index of the pharmaceutically active molecule in a given matrix of lipid is important. The unfavorable solubility profile of bioactive molecules into the lipid nanocarrier may invariably influence the percentage drug loading, percentage encapsulation efficiency, and consequent properties of the synthesized NLCs.
- Lipids should have the ability to provide particles within the nanostructure matrix scale.
- The lipids should not result in production of any toxic residues. Normally used lipids in the preparation of NLCs are mono-, di-, and triacylglycerols, waxes, fatty acids, etc. [30].

6.1.1 Liquid Lipids

The medium-chain triglycerides (MCTs) and monounsaturated fatty acid are most commonly used liquid lipids in NLC preparations. The medium-chain triglycerides are a novel category of saturated lipids, which also act as a wetting agent. MCTs are food-grade lipids, which are prepared by modified hydrolysis and fractionation of natural oils like coconut or nut oils. The MCTs are famed to have quicker predigestion as compared to long-chain triglycerides, e.g., oils, and have high stability against reactions.

6.1.2 Solid Lipids

The most commonly used solid lipids in NLC preparations are radical behenate, radical palmitostearate, radical monostearate/monostearin, cetyl palmitate, saturated fatty acid, etc. These fats are commercially accessible and, except cetyl palmitate, approved as GRAS for direct human use in food products [28] (Table 3).

6.2 Emulsifying Agents

Selection of an emulsifying agent is critical in the formulation of lipid nanoemulsion carriers. These agents influence the formation feasibility and the stability of the nanodispersions. Emulsifying agents used in the formulation of lipid nanocarriers are broadly classified into two types: surfactants or small amphiphilic molecules, and polymers, such as proteins/peptides and poly(vinyl alcohol). Polysaccharides have also been investigated for their emulsifying properties. Proteins and polysaccharides have the advantage of being a natural ingredient. However, polysaccharides do not get more popularity as an emulsifying agent as these do not actively adsorb onto the oil/water interface in o/w-type emulsions. Most of the surfactants used are anionic (e.g., oleates, lactic acid esters, and phospholipids) or nonionic compounds (e.g., aliphatic alcohols, monoacylglycerols, and their esters). The kinetic stability of nanoemulsion system depends on the nature and concentration of the emulsifier which coats the droplet surface. There are mainly two types of emulsifiers: (i) polymers, such as proteins/peptides and poly(vinyl alcohol), and (ii) small, amphiphilic molecules or surfactants. A list of emulsifying agents that have been reported for the synthesis of nanoemulsions is presented in Table 4 [31].

Drug entrapment or incorporation can be explained by three basic models, namely solid solution model, drug-enriched shell model, and drug-enriched core model (Table 5 and Fig. 6).

An overview of three different structures for NLCs is presented in Fig. 7. The structure in the first category is distorted sort within which the liquid fats (oil) and solids are mixed. The characteristic controlled conditions within the process of crystallization result in a very disordered structure. The second category structure is

Table 3 Lipids used for preparation of nanocarrier [28]

Type of lipids	Examples
Triacylglycerols	Tricaprin, trilaurin, trimyristin, tripalmitin, and tristearin
Acylglycerols	Glycerol monostearate, glycerol behenate, and glycerol palmitostearate
Fatty acids	Stearic acid, palmitic acid, decanoic acid, and behenic acid
Waxes	Cetyl palmitate

Table 4 List of surfactants used for preparation of lipid nanocarrier [28]

Type of surfactant	Examples
Phospholipids	Soy lecithin, egg lecithin, and phosphatidylcholine
Ethylene oxide/propylene oxide copolymers	Poloxamer 188, poloxamer 182, poloxamer 407, and poloxamine 908
Sorbitan ethylene oxide/ propylene	Polysorbate 20, polysorbate 60, and polysorbate 80
Alkylaryl polyether alcohol polymers	Tyloxapol
Bile salts	Sodium cholate, sodium glycol cholate, sodium tauro cholate, and sodium taurodeoxycholate
Alcohols	Ethanol and butanol

formless sort (a noncrystalline matrix). This may not have a crystalline structure and hence lead to the expulsion of loaded amorphous drug. The formation of unit-shaped crystals during cooling and use of a bound lipid mixture may prevent expulsion of loaded drug. In the third category, the drug solubility in liquid lipid is higher than in solid lipid. Hence, it controls the degradation by solid lipid. This type of NLCs forms water/oil/water (w/o/w) emulsion [33] (Table 6).

7 Preparation of Nanoparticles

There are different techniques for the preparation of LNP. Usually, the SLNs are prepared using homogenization at elevated or low temperatures (including hot homogenization and cold homogenization), solvent emulsification, evaporation or diffusion, critical fluid (supercritical fluid extraction of emulsions), ultrasonication or high-speed homogenization, spray drying, etc. [34].

Table 5 Three models of drug incorporation into solid lipid nanoparticles [32]

Solid solution model	Core-shell model (drug-enriched shell)	Core-shell model (drug-enriched core)
• Formation of this model in cold homogenization technique • Using no drug-solubilizing surfactant • Drug dispersed in lipid matrix • There is a strong interaction between lipid and drug	• Formation of this model in hot homogenization technique • Formation of lipid core at recrystallization temperature of lipid • Cooling of the obtained dispersion leads to re-partitioning of the drug to the lipid phase • Concentration of drug in surrounding membrane	• Dispersion cooling leads to a supersaturation of the drug which is dissolved in the lipid • Precipitation of drug in melted lipid • Finally, further cooling leads to recrystallization of the lipid • Formation of drug-enriched core

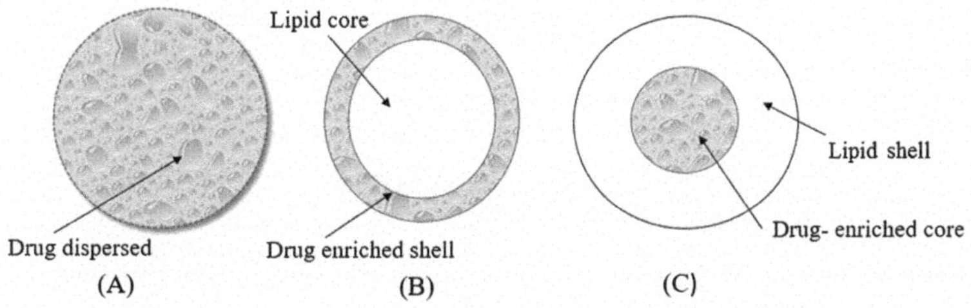

Fig. 6 Schematic representation of drug incorporation models of solid lipid nanoparticles: (**a**) solid solution model, (**b**) core-shell model (drug-enriched shell), (**c**) core-shell model (drug-enriched core)

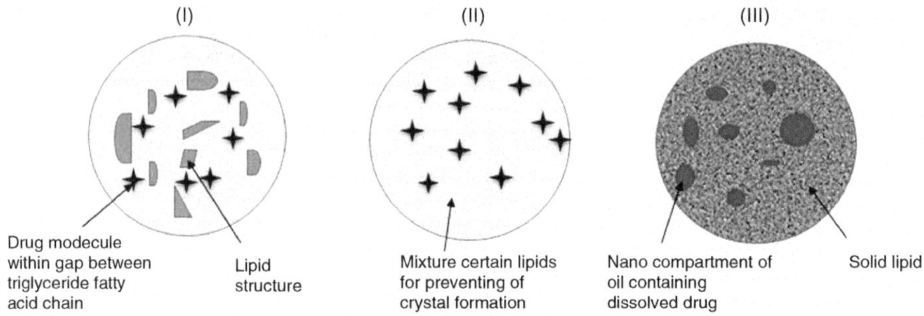

Fig. 7 Structures of nanostructure lipid carriers: Class I: imperfect type, class II: formless type, class III: multiple type

7.1 High-Pressure Homogenization

This method is based on the cavitation approach, where the nanoscale drug suspension is synthesized by applying high-pressure air jet milling (~100–2000 bar) in the presence of an aqueous solution of surfactant. The approach is free from contaminants. The crystal structure may get changed due to the high pressure applied. The high-pressure homogenization method is also known as emulsion homogenization. High-pressure homogenization approach is broadly classified into two categories, namely cold homogenization and hot homogenization.

7.2 High-Pressure Hot Homogenization

During hot homogenization, the system temperature needs to be maintained above the melting point of lipid. High-shear mixing of drug-loaded lipid melts and an aqueous emulsifier phase leads to the formation of a pre-emulsion. This pre-emulsification is carried out at temperatures above the melting point of lipid. A decrease in viscosity of the inner phase at elevated temperature leads to decrease in particle size. This elevated temperature might be unfavorable for certain heat-sensitive drugs. Hot homogenization may lead to the drug distribution into the aqueous phase during homogenization, temperature-induced drug degradation, and complexity of the crystallization. Increasing number of cycles or homogenization pressure may also increase particle size due to the high kinetic energy of particles. Nanoparticle synthesis by hot homogenization technique is illustrated in Fig. 8.

7.3 High-Pressure Cold Homogenization

To overcome the drawbacks of hot homogenization cold homogenization has been developed where cooling, grinding, and dispersing in a cold surfactant solution are carried out for drug-lipid pre-suspension preparation. In this technique a pre-suspension is prepared and homogenization of this pre-suspension is carried out below room temperature. The nanosize range of lipid particles is achieved due to the gravitation force to form solid lipid nanoparticles.

Table 6 Types of different nanostructure lipid carriers [31]

NLC type	Nature of matrix	Comments	Diagram
Imperfect	Imperfectly structured solid matrix	Spatially different lipids are mixed creating imperfections in the crystal order of lipid nanoparticles; show high-drug payload	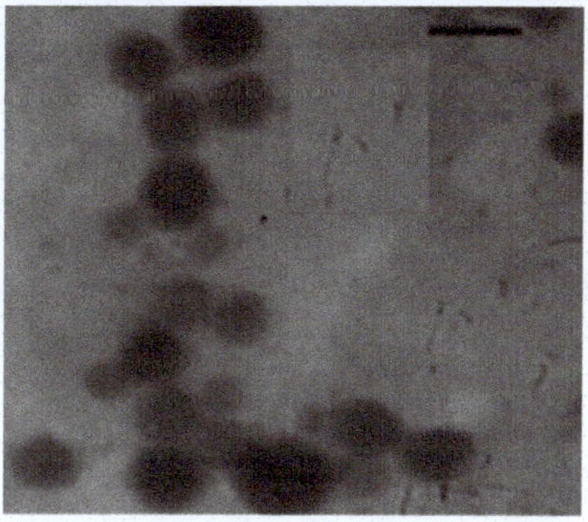
Amorphous	Structureless solid amorphous matrix	Formed by mixing solid lipids with special lipids like hydroxyoctacosanyl hydroxystearate, isopropyl myristate, or medium-chain triglycerides such as miglyol 812, thus preventing drug repulsion, moderate drug payload	
Multiple	Multiple oil in fat in water	Solubility of drug in lipophilic phase decreases during the cooling process after homogenization and crystallization process during storage	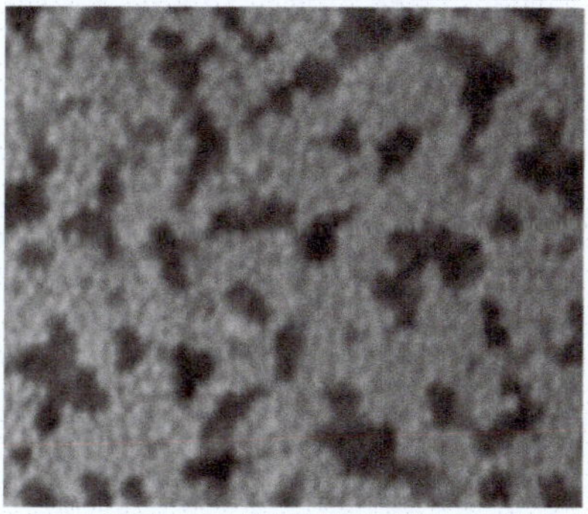

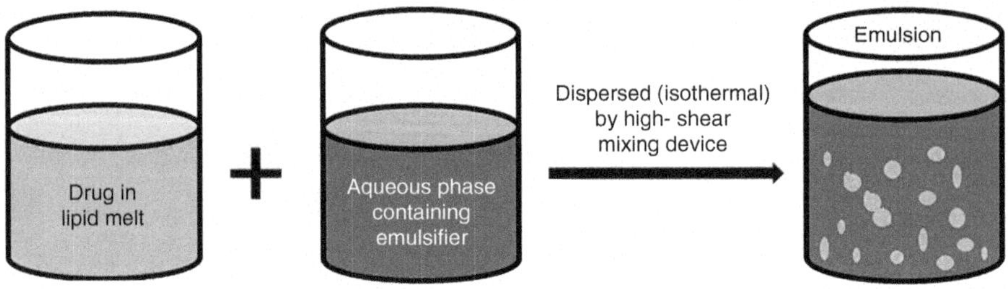

Fig. 8 Hot homogenization technique for production of solid lipid nanoparticles

7.4 Solvent Emulsification or Evaporation Process

This is the most widely investigated technique for the production of nanoparticles where the emulsification may be carried as a single step or double step. The organic solvent is evaporated after formation of stable nanoformulation. In single-emulsion solvent evaporation technique, a water-soluble solvent is used to dissolve the drug and polymer. This is added to lipid-lipid-PEG aqueous solution and homogenized to form oil/water emulsion system. Here, the drug used is hydrophobic in nature. Double-emulsion solvent evaporation technique involves the emulsification of aqueous drug solution in an organic phase containing lipid and polymer. The system is continuously homogenized to obtain a water/oil emulsion. The formed primary emulsion is re-emulsified to form a water/oil/water emulsion. The re-emulsification is carried out in aqueous lipid-PEG solution at high-speed homogenization. In the double-emulsion solvent evaporation technique, the drug used is hydrophilic in nature.

7.5 Supercritical Fluid Extraction of Emulsions

A supercritical fluid substance does not condense or evaporate to form a liquid or gas at a point where the temperature and pressure reach above its critical points. Such substances do not have surface tension and show properties between those of liquids and gases. Carbon dioxide is the example of the most investigated supercritical fluid for the synthesis of nanoparticles under certain conditions of heat and pressure applied. At the supercritical point, carbon dioxide gas converts from liquid phase to supercritical phase. The bioactive is dissolved in supercritical fluid to formulate nanoparticle and sprayed through a nozzle for rapid expansion of bioactive containing supercritical fluid. The supercritical fluid gets evaporated and finally nanoparticles are produced.

7.6 Ultrasonication or High-Speed Homogenization

Ultrasonication or high-speed homogenization techniques are recommended for the production of solid lipid nanoparticles. A combination of both of these techniques leads to the formation of smaller particles. Briefly, a drug and molten solid lipid are mixed. The resultant mixture is added to an aqueous solution of surfactant. High-speed

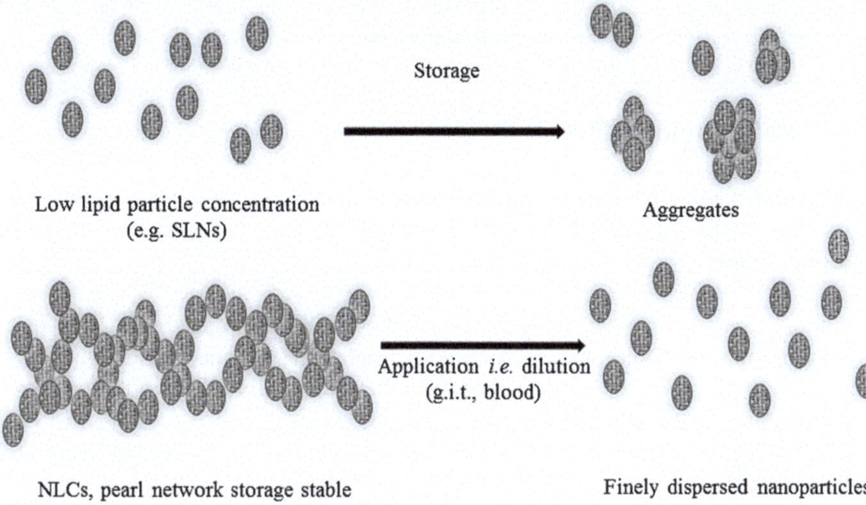

Fig. 9 Stabilization effect in highly concentrated lipid particle dispersions

homogenization (8000 rpm) is carried out for 15 min. The formed pre-emulsion is ultra-sonicated at 0 °C using an ultrasonicator. The nanoparticles are collected, filtered, and stored at 4 °C. The method is easy and simple, and does not require shear stress. However, metal contamination due to ultrasonication and physical instability during storage such as particle growth upon storage are the major limitations associated with this technique.

7.7 Spray Drying

This is a single-step technique used for the production of stable nanoparticles from a liquid phase. In this technique drug and a preformed polymer are dissolved in water and spray-dried. Spray drying of the mixture occurs due to the atomization followed by solvent evaporation at 30–55 °C. This leads to the production of nanoparticles. However, high temperatures and high shear forces may result in aggregation of particle [35].

8 Stability of Nanostructured Lipid Dispersions

NLCs may contain mixture structures (mixed micelles, liposomal particles, nano-emulsions, etc.) that contribute to their stability. These nanocarriers mainly have stability issues during their storage, which may lead to their particle size improvement, drug expulsion from the lipid matrix, etc. The physical stability can be observed using different techniques such as zeta potential determination, particle size analysis, and thermal analysis. Various studies have indicated that SLNs are physically stable and decomposed over a period of 1 year [36] (Fig. 9).

Different approaches have been reported to improve the stability of NLCs such as used of polyethylene glycol, lyophilization, etc.

8.1 Polyethylene Glycol

Surface modification by coating with a hydrophilic substance such as polyethylene glycol has below-mentioned advantages in respect to the stability of nanoformulations:

- Helping to give best physical stability and dispersibility properties of colloids
- Developing the colloid presence in blood circulation for general uses
- Improving the stabilization properties of colloids in the body
- Developing the transport of mixture across the tissue
- Controlling the interaction of colloids with membrane for specific delivery and drug targeting
- Enhancing the biocompatibility and reducing thrombogenicity of drug carriers

8.2 Lyophilization

Lyophilization is another efficient approach to improve the stability of NLCs by dehydration. However, the freeze-dried SLNs not containing cryoprotectant can lead to the aggregation of particles [37].

9 Characterization of Lipid Nanocarriers and Nanoemulsion

9.1 Particle Size Analysis

Light microscopy is the commonly used technique to examine particle aggregation. The mean particle size can be examined using an electron microscope, photon correlation microscopy, Malvern mastersizer, light scattering method, or freeze fracture electron microscopy. Surface morphology can be examined by transmission electron microscopy (specific for liquid sample), optical microscopy, or scanning electron microscopy. Dimensional analysis of the particle is determined by using energy-dispersive X-ray spectroscopy.

9.2 Chemical Stability

Chemical stability of lutein and xanthophyll loaded in nanocarrier is determined to identify the validity of production method. The proportion of remaining xanthophyll in every preparation is determined using an ultraviolet radiation photometer.

9.3 Determination of Zeta Potential

The zeta potential (ζ) value is used for checking quantification of the charge present on the surface of the particles. The stability of a colloidal dispersion depends on the zeta potential value of system. The nano zeta-sizer measures the cataphoretic quality of the particles in specific electrical field. The rate of hydration can be estimated in a Neubauer's chamber [27].

9.4 Drug Entrapment Efficiency

The amount of drug entrapped within the particle is determined by extracting the entrapped drug in a suitable solvent and analyzing the resultant drug-containing solution by the appropriate assay method such as UV spectrophotometry of high-performance liquid chromatography. The amount of unentrapped drug can be estimated by gel filtration, exhaustive dialysis, and ultracentrifugation. The content of the drug can also be estimated by energy-dispersive X-ray spectroscopy.

9.5 Photostability Study

For the determination of photostability, irradiation is done by employing a solar simulator and arc lamp equipped with a xenon lamp graduated with a radiometer. The lamp is graduated before every detection process. After completion of irradiation, every tested sample is recovered from the used cuvette by a combination of THF:H_2O (9:1) and analyzed using a suitable spectrophotometric technique [27, 28].

9.6 In Vitro Drug Release Study

Drug release profile from the lipid nanocarriers is commonly examined using dialysis bag method. In this method a dialysis bag is washed and soaked in dissolution media. The nanocarrier suspension is filled into the soaked bag. The drug-loaded nanocarrier-containing bag is then placed in the measured amount of suitable buffer solution. The system is constantly shaken at controlled temperature. The aliquots are collected at the predetermined time interval and equal volume of fresh dissolution media maintained at same temperature is added after each sampling to maintain complete sink condition. After suitable dilution, the sample solution is analyzed for the drug content using an appropriate analysis technique.

10 Therapeutic Applications of Lutein

10.1 Hyperglycemia-Induced Retinal Disorders

Lutein, the most abundant antioxidant carotenoid in green leafy vegetables and eggs, is used as a dietary supplement to enhance eye health. Carotenoids exhibit anti-senescent property by suppressing generation of reactive oxygen species (ROS) through SIRT1 upregulation in retinal pigment epithelium cells exposed to hyperglycemia mimicking diabetic conditions. Thus, lutein could be a therapeutic intervention for treating hyperglycemia-induced retinal disorders [38].

Lutein inhibits hyperglycemia-induced cellular senescence in ARPE-19 cells by inducement of SIRT1 expression, thereby modulating cellular ROS generation. This upregulation of SIRT1, that influences epigenetic pathways, could underlie several functions of carotenoid as well as its antioxidant activity and ability to forestall cellular senescence. Thus, carotenoids are recommended as a potential therapeutic agent for hyperglycemia-induced retinal disorders, together with a far better understanding of the clinical effectiveness of carotenoid as a dietary supplement [39, 40].

10.2 Antioxidant Activity

The primary immediate defense against environmental skin harm depends on the antioxidant capability of the skin. However, this defense system is compromised by moderate exposure to ultraviolet light. Therefore, bolstering the antioxidant defense system of the skin could be a doubtless necessary strategy for reducing environmentally evoked skin harm. These xanthophyll carotenoids can be administered orally and/or topically.

The combined oral and topical administration of lutein and zeaxanthin provides the best degree of inhibitor protection. Oral and topical administration of such antioxidants severally conjointly provides important activity within the skin [40].

10.3 Anti-ulcerogenic Activity

Gastric ulcer may lead to chronic ulceration in the stomach if it persists for about 10–20 years due to the perennial episodes of healing and re-exacerbation. It is affecting up to 10% of the world's population. Medicines used in ulcer treatment act by blocking the histamine receptor, inhibiting proton pump, and affecting the mucosal barrier (antacid, cimetidine, omeprazole). To seek out a solution for the side effects as a result of long-run treatments, the eye-protecting nutrient lutein has great importance. The structure of lutein (3, 3′-dihydroxy-β, ε-carotene) is characterized by the presence of a hydroxyl group attached to each of the two terminal β-ionone rings within the molecule. This xanthophyll is more hydrophilic than other carotenoids. The hydrophilic nature of lutein allows it to react with singlet oxygen generated within the water section than the nonpolar carotenoids. The antiulcer activity of lutein is also due to the inhibition of oxidative stress made by alcohol by lutein administration [41].

10.4 Antitumor Activity

Cancer is the second leading cause of death. In 2018, 9.6 million deaths are reported globally due to cancer. To elucidate the antitumor activity of natural carotenoids, in vitro and in vivo models have been reported during the past three decades. Most of the in vitro anticancer studies on natural carotenoids have been reported using cancer derived from prostate (PC-3), breast (MCF-7), and blood-forming (HL-60) cells. Lutein has been reported to act as a ROS inducer for HeLa cells, whereas it acts as a normalizer for normal cells. The lutein-induced apoptosis of HeLa cells is associated with the increased ROS generation, loss of mitochondrial membrane potential, and upregulation of Bax and caspase-3-mediated pathway, resulting in fragmentation of nuclei DNA. Overall, lutein can be a prospective candidate for chemoprevention and chemotherapy of human cervical carcinoma and different chronic diseases. Cooperative effects of lutein with known anticancer medication can be examined as an effective chemotherapy for human cervical carcinoma [42].

10.5 Prevention of Selenite-Induced Cataract

Cataract is the clouding of the lens in the eye which results in decrease in vision. About half of the world population blindness is due to cataract, for which surgery is the only viable treatment. In a study, selenite-induced cataract in Crl:WI rats was significantly reduced after 7 days of topical application or oral administration of lutein-loaded nanoparticles. Cataract severity was considerably reduced in rats treated with lutein-containing nanoparticles applied topically when compared to the animals treated with free lutein or other lutein formulations [37].

11 Safety Versus Toxicity Issues

Till date, sufficient safety data for nutrient supplements is not available. Further clinical trials are warranted to ensure the safety aspects of nutrient supplementation in disease prevention. The data concerning safety and toxicity of xanthophyll and its formulations should be available. More attention needs to identify the stage at which oxidative damage starts and the antioxidant supplement may be important [40, 43]. Toxicity study for 28 days in Wistar rats fed with 773 mg/kg/day of purified crystalline lutein did not show significant hematological, clinical, histopathological, or biochemical changes [44]. The xanthophyll toxicity information is limited and additional toxicity-centered studies should be performed to access their therapeutic values.

12 Clinical Trials on Lutein-Based Products

Thirty different types of carotenoid are present in the nature. The human inhibitor defense system also includes a combination of different accelerator and nonenzymatic antioxidants [18]. A list of clinical investigations based on lutein supplements by different organizations is presented in Table 7. Table 8 presents a list of commercially available cosmeceuticals containing lutein. Commercially available nutraceuticals containing lutein are listed in Table 9.

With the numbers of marketed requirements, the food as well as beverage companies are also focusing on products with lutein in order to meet the growing consumer need.

13 Conclusion

This chapter recapitulates the basic information of lutein, its chemistry parameters, and nanotechnology approaches utilized in the synthesis of topical formulations for lutein. Research is focused on finding ways of improving therapeutic efficacy of natural bioactives such as lutein by developing the novel nanoformulations. The drawbacks associated with the conventional drug delivery approaches can be overcome by utilizing advanced nanocarriers. Special attention has been given to the use of nanoparticles in pharmaceuticals and cosmeceuticals for the delivery of natural bioactives due to their unique physicochemical properties, affinity to complex with hydrophilic and hydrophobic substances, chemical stability, local delivery, and biodegradable character. Based on the commercially available products, literature evidences, and status of clinical trials, lutein is considered as a safe and perfect naturally occurring nutrient for the prevention of macular degeneration in the aging population.

Table 7 Lutein under different phases of the clinical trials

Institute/company	Name of the study	Clinical trials identifier/phase	Number of enrolled participants
Oregon Health and Science University	Dietary lutein and the retinopathy of prematurity	NCT00672230/ Not Applicable	212
Peking University	Effects of lutein supplementation on age-related macular degeneration	NCT01048476/ Phase II	120
Maastricht University Medical Center in collaboration with Newtricious	Effect of lutein-enriched-egg beverage on age-related macular degeneration (AMD)	NCT00902408/ Not Applicable	100
Peking University	Effects of lutein supplementation on oxidative stress in healthy subjects	NCT01056094/ Phase II	30
Maastricht University Medical Center in collaboration with Cognis Deutschland GmbH & Co. KG	The influence of lutein supplements on age-related macular degeneration	NCT01042860/ Not Applicable	80
National Institutes of Health Clinical Center (CC) sponsored by the National Eye Institute (NEI)	Lutein/zeaxanthin and omega-3 supplementation in persons over age 60	NCT00121589/ Phase I	40
National Eye Institute (NEI) in collaboration with the National Heart, Lung, and Blood Institute (NHLBI)	Age-related eye disease study 2 (AREDS2)	NCT00345176/ Phase III	4203
National Eye Institute (NEI)	Effects of antioxidants on human macular pigments	NCT00718653/ Not Applicable	40
National Institutes of Health Clinical Center (CC) (sponsored by the National Eye Institute)	Lutein for age related macular degeneration	NCT00006202/ Phase II	45
Wyeth	Effect of lutein in S-26 gold on growth and safety (S-26 Gold L)	NCT00913406/ Not Applicable	232
Medical University of Vienna	Effects of lutein supplementation on macular pigment optical density and visual acuity in patients with age-related macular degeneration	NCT00879671/ Not Applicable	126

(continued)

Table 7 (continued)

Institute/company	Name of the study	Clinical trials identifier/phase	Number of enrolled participants
Oregon Health and Science University in collaboration with Oregon Partnership for Alzheimer's Research	Lutein and Alzheimer's disease study (LAD)	NCT00596024/ Phase II	3
University of Jena	Long-term effects of lutein/ zeaxanthin and omega-3 supplementation on optical density of AMD patients (LUTEGA)	NCT00763659/ Not Applicable	172
University of Jena	Long-term effects of lutein/ zeaxanthin and omega-3 supplementation for two more years (LUTEGA 2)	NCT01648660/ Not Applicable	80
National Center for Complementary and Alternative Medicine (NCCAM)	Effects of lutein in retinitis pigmentosa	NCT00029289/ Phase II	--
The Ludwig Boltzmann Institute of Retinology and Biomicroscopic Laser Surgery	Rheohemapheresis and lutein supplementation in the non-exsudative Form of age-related macular degeneration (ReLux-AMD)	NCT00443911	20
University Hospital Inselspital, Berne	Enhancement of macular pigment density by oral lutein supplementation (EMPOLS)	NCT00563979/ Not Applicable	80
University Hospital, Bordeaux in collaboration with Laboratoires Thea	Lutein influence on macula of persons issued from AMD parents (LIMPIA)	NCT01269697/ Phase III	120
The Lowy Medical Research Institute Limited	MPOD in macular telangiectasia following supplementation of lutein and zeaxanthin (MacTEL-Supp)	NCT00568828	12

Table 8 Commercially available cosmeceuticals containing lutein

Company	Brand name/products
Ecco Bella Botanicals	Day cream Age antidote day cream Eye nutrient cream Night rebuilder cream
CosMed	Physician's complex: antiaging super C-plus serum
HDS Cosmetics Lab, Inc.	DDF silky C serum and DDF daily matte, SPF 15 DDF-doctor's dermatologic formula retinol energizing moisturizer
Petite Marie Ltd	Petite Marie organics new age gold cream

Table 9 List of commercially available nutraceuticals containing lutein

Company	Brand name/product
Bioelements Inc.	Lutein indoor protective day cream LUTEIN ofta® LUTEIN tablet as a dietary supplement for nursing mothers: 10 mg LUTEIN mamma® capsules: 10 mg LUTEIN ofta® gocce drops for oral use Lutein for the development and defense of the eyesight (28 mg) MACULAR oil Dressing based on extra-virgin olive oil with LUTEIN added
NEOOX group-Sooft Italia	LUTEIN omega-3: contains lutein, polyunsaturated omega-3 fatty acids (DHA + EPA), vitamins, zinc, and copper: capsule 10 mg MACULAR® active Supplement specific for macular protection: capsule 10 mg MACULAR® B forte To protect against the risk of vascular disorders and oxidation damage in ARMD: tablet 10 mg MYOOPS® components improve retinal metabolism and the functionality of rod and cone cells (tablet and liquid): 6 mg
Nutrina Co Inc. USA	Nutrina lutein capsules: 1 mg
Alcon Inc. Switzerland	ICaps® lutein and zeaxanthin tablets: 0.5 mg Ocuvite Lutein Eye Vitamins and Mineral Supplement tablets and capsules: 0.6 mg
Bausch & Lomb	Ocuvite eye vitamin and mineral supp for adults 50 and over soft gels Lutein 6 MG 50ct (6 mg) PreserVision® soft gel lutein eye vitamins (5 mg)
Quest	Quest eye nutrient tablets: 0.6 mg Quest lutein capsule: 10 mg
GNC Natural Brands R	Natural Brand™ Lutein capsule Natural Brand™ Lutein soft gel capsule GNC Herbal Plus® Standardized Bilberry Plus Lutein TWINLAB® OcuGuard® Plus
Nature Plus	Lutein RX-eye® Vcaps™ Ultrasource of life® with lutein tablet SunScape® Vcaps™ Ultralutein® softgel Miracle Essential™ tablets
Pure Encapsulation, USA	Macular support formula Lutein 20 mg
Vitamin Health, Inc.	Viteyes® AREDS formula eye vitamins Viteyes Plus Lutein Beta-Carotene Free Viteyes AREDS Advanced
Wyeth Nutrition	Procal Gold 3 with lutein powder Promil Gold 2 with lutein powder Promise Gold 4 with lutein powder S-26 Gold 1 with lutein powder S-26 RTF Gold with lutein solution

References

1. Heyang JI, Fei XI, Jiang C, Yaping ZH, Lin HE (2009) Nanoencapsulation of lutein with hydroxypropylmethyl cellulose phthalate by supercritical antisolvent. Chin J Chem Eng 17(4):672–677
2. Landrum JOT, Bone RA, Joa H, D Kilburn MA, Moore LL, Sprague KE (1997) A one year study of the macular pigment: the effect of 140 days of a lutein supplement. Exp Eye Res 65(1):57–62
3. Roberts RL, Green J, Lewis B (2009) Lutein and zeaxanthin in eye and skin health. Clin Dermatol 27(2):195–201
4. Sommerburg O, Keunen JE, Bird AC, van Kuijk FJ (1998) Fruits and vegetables that are sources for lutein and zeaxanthin: the macular pigment in human eyes. Br J Ophthalmol 82(8):907–910
5. Araus KA, Casado V, del Valle JM, Robert PS, Juan C (2019) Cosolvent effect of ethanol on the solubility of lutein in supercritical carbon dioxide. J Supercrit Fluids 143:205–210
6. Palombo P, Fabrizi G, Ruocco V, Ruocco E, Fluhr J, Roberts R, Morganti P (2007) Beneficial long-term effects of combined oral/topical antioxidant treatment with the carotenoids lutein and zeaxanthin on human skin: a double-blind, placebo-controlled study. Skin Pharmacol Physiol 20(4):199–210
7. Mitri K, Shegokar R, Gohla S, Anselmi C, Müller RH (2011) Lipid nanocarriers for dermal delivery of lutein: preparation, characterization, stability and performance. Int J Pharm 414(1–2):267–275
8. Alves-Rodrigues A, Shao A (2004) The science behind lutein. Toxicol Lett 150(1):57–83
9. Chen H, Yao Y (2017) Phytoglycogen to increase lutein solubility and its permeation through Caco-2 monolayer. Food Res Int 97:258–264
10. Lacatusu I, Mitrea E, Badea N, Stan R, Oprea O, Meghea A (2013) Lipid nanoparticles based on omega-3 fatty acids as effective carriers for lutein delivery. Preparation and *in vitro* characterization studies. J Funct Foods 5(3):1260–1269
11. Apanasenko IE, Selyutina OY, Polyakov NE, Suntsova LP, Meteleva ES, Dushkin AV, Vachali P, Bernstein PS (2015) Solubilization and stabilization of macular carotenoids by water soluble oligosaccharides and polysaccharides. Arch Biochem Biophys 572:58–65
12. Henry LK, Catignani GL, Schwartz SJ (1998) Oxidative degradation kinetics of lycopene, lutein, and 9-cis and all-trans β-carotene. J Am Oil Chem Soc 75(7):823–829
13. Rai VK, Mishra N, Yadav KS, Yadav NP (2018) Nanoemulsion as pharmaceutical carrier for dermal and transdermal drug delivery: formulation development, stability issues, basic considerations and applications. J Control Release 270:203–225
14. Zhao C, Shen X, Guo M (2018) Stability of lutein encapsulated whey protein nano-emulsion during storage. PLoS One 13(2):e0192511
15. Shakeel F, Shafiq S, Haq N, Alanazi FK, Alsarra IA (2012) Nanoemulsions as potential vehicles for transdermal and dermal delivery of hydrophobic compounds: an overview. Expert Opin Drug Deliv 9(8):953–974
16. Tamjidi F, Shahedi M, Varshosaz J, Nasirpour A (2013) Nanostructured lipid carriers (NLC): a potential delivery system for bioactive food molecules. Innovative Food Sci Emerg Technol 19:29–43
17. Brum AA, dos Santos PP, da Silva MM, Paese K, Guterres SS, Costa TM, Pohlmann AR, Jablonski A, Flôres SH, de Oliveira Rios A (2017) Lutein-loaded lipid-core nanocapsules: Physicochemical characterization and stability evaluation. Colloids Surf A Physicochem Eng Asp 522:477–484
18. Madaan T, Choudhary AN, Gyenwalee S, Thomas S, Mishra H, Tariq M, Vohora D, Talegaonkar S (2017) Lutein, a versatile phyto-nutraceutical: an insight on pharmacology, therapeutic indications, challenges and recent advances in drug delivery. Pharm Nutr 5(2):64–75

19. Erdő F, Hashimoto N, Karvaly G, Nakamichi N, Kato Y (2016) Critical evaluation and methodological positioning of the transdermal microdialysis technique. A review. J Control Release 233:147–161
20. Feingold KR, Grunfeld C (2012) Psoriasis: it's more than just the skin. J Lipid Res 53(8):1427–1429
21. Berenson GS, Burch GE (1951) Studies of diffusion of water through dead human skin: the effect of different environmental states and of chemical alterations of the Epidermis1. Am J Trop Med Hyg 1(6):842–853
22. Nemanic MK, Elias PM (1980) In situ precipitation: a novel cytochemical technique for visualization of permeability pathways in mammalian stratum corneum. J Histochem Cytochem 28(6):573–578
23. Jain S, Patel N, Shah MK, Khatri P, Vora N (2017) Recent advances in lipid-based vesicles and particulate carriers for topical and transdermal application. J Pharm Sci 106(2):423–445
24. Scheuplein RJ, Blank IH (1971) Permeability of the skin. Physiol Rev 51(4):702–747
25. Sinko PJ (2006) Martin's physical pharmacy and pharmaceutical sciences: physical chemical and biopharmaceutical principles in the pharmaceutical sciences, 6th edn Lippincott Williams & Wilkins, Philadelphia, PA.
26. Elias PM, Friend DS (1975) The permeability barrier in mammalian epidermis. J Cell Biol 65(1):180–191
27. Jaiswal P, Gidwani B, Vyas A (2016) Nanostructured lipid carriers and their current application in targeted drug delivery. Artificial Cells Nanomed Biotechnol 44(1):27–40
28. Mehnert W, Mäder K (2012) Solid lipid nanoparticles: production, characterization and applications. Adv Drug Deliv Rev 64:83–101
29. Araújo J, Nikolic S, Egea MA, Souto EB, Garcia ML (2011) Nanostructured lipid carriers for triamcinolone acetonide delivery to the posterior segment of the eye. Colloids Surf B: Biointerfaces 88(1):150–157
30. Schäfer-Korting M, Mehnert W, Korting HC (2007) Lipid nanoparticles for improved topical application of drugs for skin diseases. Adv Drug Deliv Rev 59(6):427–443
31. Westesen K, Bunjes H, Koch MH (1997) Physicochemical characterization of lipid nanoparticles and evaluation of their drug loading capacity and sustained release potential. J Control Release 48(2–3):223–236
32. Naseri N, Valizadeh H, Zakeri-Milani P (2015) Solid lipid nanoparticles and nanostructured lipid carriers: structure, preparation and application. Adv Pharm Bull 5(3):305
33. Kaur S, Nautyal U, Singh R, Singh S, Devi A (2015) Nanostructure lipid carrier (NLC): the new generation of lipid nanoparticles. Asian Pac J Health Sci 2(2):76–93
34. Ranganathan A, Manabe Y, Sugawara T, Hirata T, Shivanna N, Baskaran V (2019) Poly (d, l-lactide-co-glycolide)-phospholipid nanocarrier for efficient delivery of macular pigment lutein: absorption pharmacokinetics in mice and antiproliferative effect in Hep G2 cells. Drug Deliv Transl Res 9(1):178–191
35. Puglia C, Bonina F (2012) Lipid nanoparticles as novel delivery systems for cosmetics and dermal pharmaceuticals. Expert Opin Drug Deliv 9(4):429–441
36. Garud A, Singh D, Garud N (2012) Solid lipid nanoparticles (SLN): method, characterization and applications. Int Curr Pharm J 1(11):384–393
37. Bodoki E, Vostinaru O, Samoila O, Dinte E, Bodoki AE, Swetledge S, Astete CE, Sabliov CM (2019) Topical nanodelivery system of lutein for the prevention of selenite-induced cataract. Nanomedicine 15(1):188–197
38. Lakshminarayana R, Aruna G, Sangeetha RK, Bhaskar N, Divakar S, Baskaran V (2008) Possible degradation/biotransformation of lutein *in vitro* and *in vivo*: isolation and structural elucidation of lutein metabolites by HPLC and LC-MS (atmospheric pressure chemical ionization). Free Radic Biol Med 45(7):982–993
39. Hwang JS, Han SG, Lee CH, Seo HG (2018) Lutein suppresses hyperglycemia-induced premature senescence of retinal pigment epithelial cells by upregulating SIRT1. J Food Biochem 42(3):e12495
40. Johnson EJ (2002) The role of carotenoids in human health. Nutr Clin Care 5(2):56–65

41. Sindhu ER, Kuttan R (2012) Carotenoid lutein protects rats from gastric ulcer induced by ethanol. J Basic Clin Physiol Pharmacol 23(1):33–37
42. Gansukh E, Mya KK, Jung M, Keum YS, Kim DH, Saini RK (2019) Lutein derived from marigold (Tagetes erecta) petals triggers ROS generation and activates Bax and caspase-3 mediated apoptosis of human cervical carcinoma (HeLa) cells. Food Chem Toxicol 127:11–18
43. Jacques PF (1999) The potential preventive effects of vitamins for cataract and age-related macular degeneration. Int J Vitam Nutr Res 69(3):198–205
44. Kruger CL, Murphy M, DeFreitas Z, Pfannkuch F, Heimbach J (2002) An innovative approach to the determination of safety for a dietary ingredient derived from a new source: case study using a crystalline lutein product. Food Chem Toxicol 40(11):1535–1549

Bioactive Flavonoid Apigenin and Its Nanoformulations: A Promising Hope for Diabetes and Cancer

Biswajit Mukherjee, Soumyabrata Banerjee, Laboni Mondal, Shreyasi Chakraborty, Deepayan Chanda, and Jayawardane Athukorallage Chamindika Perera

Abbreviations

ADA	American Diabetes Association
AMP	Adenosine monophosphate
AMPK	Adenosine monophosphate kinase
ATP	Adenosine triphosphate
BDNF	Brain-derived neurotrophic factor
BMI	Basal metabolic index
CDK2	Cyclin-dependent kinase 2
CDK4	Cyclin-dependent kinase 4
CDK6	Cyclin-dependent kinase 6
CNP	Carbon nanopowder
COX 2	Cyclooxygenase 2
CREB	Cyclic-AMP response element binding protein
CSC	Cancer stem cells
DM	Diabetes mellitus
DPP	Dipeptidyl peptidase
ERK	Extracellular-signal-regulated kinase
GAD	Glutamic acid decarboxylase
GBM	Glioblastoma multiforme
GDM	Gestational diabetes mellitus
GLP	Glucagon-like peptide
GLUT	Glucose transporter
GLUT4	Glucose transporter type 4
GM-CSF	Granulocyte-macrophage colony-stimulating factor
GnRH	Gonadotropin releasing hormone

B. Mukherjee (✉) · S. Banerjee · L. Mondal · S. Chakraborty · D. Chanda · J. A. C. Perera
Department of Pharmaceutical Technology, Jadavpur University, Kolkata, India
e-mail: biswajit.mukherjee@jadavpuruniversity.in

© Springer Nature Singapore Pte Ltd. 2020
M. Rahman et al. (eds.), *Nanomedicine for Bioactives*,
https://doi.org/10.1007/978-981-15-1664-1_13

GTP	Glutathione peroxidase
HbA1c	Glycated haemoglobin
IAA	Insulin autoantibodies
IDF	International Diabetes Federation
IL	Interleukin
IRS	Insulin receptor substrate
K-ATP	ATP-sensitive potassium channel
LNCaP	Androgen-sensitive human prostate adenocarcinoma
NF-κb	Nuclear factor-Kb
NO	Nitric oxide
$p21^{WAF1/CIP1}$	Cyclin-dependent kinase inhibitor
$p27^{KIP1}$	Universal cyclin-dependent kinase inhibitor
PI3K/Akt	Phosphatidylinositol 3-kinase (PI3K) and protein kinase B (Akt)
PPAR	Proliferator-activated receptor
PPARγ	Peroxisome proliferator-activated receptor gamma
PSC	Pancreatic stem cell
SCC-25	Squamous cell carcinoma cell line
SD	Solid dispersion
SFCs	Sphere-forming cells
SGLT	Sodium/glucose transporter
SOD	Superoxide dismutase
TNF	Tumour necrosis factor
TNFα	Tumour necrosis factor alpha
UDP	Uridine diphosphate
ZnT	Zinc transporter protein
β-cells	Beta cells

1 Introduction

Apigenin is a potential bioactive flavone with polyphenolic structure, widely present in celery, parsley, red peppers, onions, oranges, tea, white sprouts, chamomile, mint, ginkgo biloba, etc. In foods, apigenin is present as glycoside conjugates and various acetylated derivatives which are water soluble than its pure form having poor solubility in either water or many of the organic solvents. Despite being a potential bioactive compound, apigenin requires a suitable drug delivery system to improve its poor solubility and low bioavailability. Apigenin formulations have shown its increased solubility, bioavailability and absorption into the living system. Diabetes, a metabolic disorder and the fourth major cause of death, has been associated consistently with the risk factors of several forms of cancer, the second major cause of death. Although diabetes and cancer have many common risk factors, the correlation between these two diseases is not clear till date. Recent findings depict various functions of apigenin such as cell cycle arrest, triggering apoptosis of the cancerous cells, suppressing cell migration in cancer showing the anticancer activity and again improving islet of beta cell damage with the attenuation of elevated blood glucose level resulting in beneficial effects on diabetes, which have made it to explore further as an essential therapeutic tool against these diseases. Further, antioxidant and anti-inflammatory activities of apigenin make it more promising to explore its application in human health care. Modern nanoformulation strategies offer certain advantages over the traditional formulations to combat these diseases more efficiently. Recently, the ligand-conjugated apigenin-loaded nanocarriers have shown its potency to reach the target organs, opening a new window of targeted cancer therapy with apigenin. This novel class of therapeutics has also increased the solubility, sustained drug release and reduced the toxicity of the pure compound in both in vivo and in vitro. Thus, it suggests the potential of apigenin nanoformulations for development as a promising diabetic or cancer chemopreventive therapeutic options in the near future. This chapter emphasises on the different nanoformulations of apigenin, its molecular mechanisms of action and their formulary variations along with their challenges as an ideal replacement for the ongoing chemotherapeutics in the treatment of these two major diseases.

2 Different Types of Flavonoids

Flavonoids are an important class of natural products; particularly, they belong to a class of plant secondary metabolites having a polyphenolic structure, widely found in fruits, vegetables and certain beverages [1]. Many of these natural products were found to be potential for human health care. Flavonoids are useful nutraceuticals which show antioxidant, anti-inflammatory and anticancer properties, and they have the ability to change key cellular enzyme function [2, 3].

Flavonoids are divided into different subgroups depending on the position of the carbon of C ring on which B ring is attached and the degree of unsaturation of flavonoids and oxidation depends on B ring (Fig. 1). In brief, flavonoids in which B ring is attached to the third carbon of C ring are called as isoflavones. Similarly, when B ring is attached to the fourth carbon atom of the C ring (Fig. 2), the flavonoids are called as neoflavonoids. Thus, on the basis of the structural features of the C ring, flavonoids have several classes such as flavones, flavonols, flavanones, flavanonols, flavanols or catechins, anthocyanins and chalcones [4].

2.1 Flavonols

Flavonols are the unique flavonoids with a ketone group which is the building block of all the proanthocyanins. Compared with flavones, flavonols have a hydroxyl group in the third carbon position of the C ring. Flavonols are found in high concentration in variety of fruits and vegetables. Flavonols are very diverse in methylation, hydroxylation and the different glycosylation patterns [4].

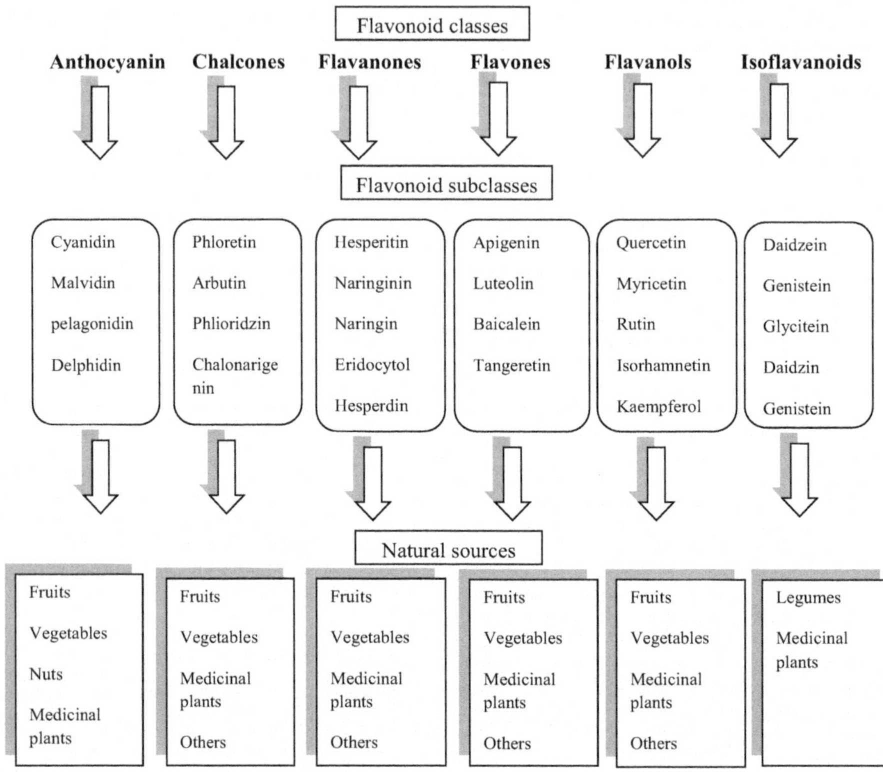

Fig. 1 Different classes and subclasses of flavonoids and their natural sources

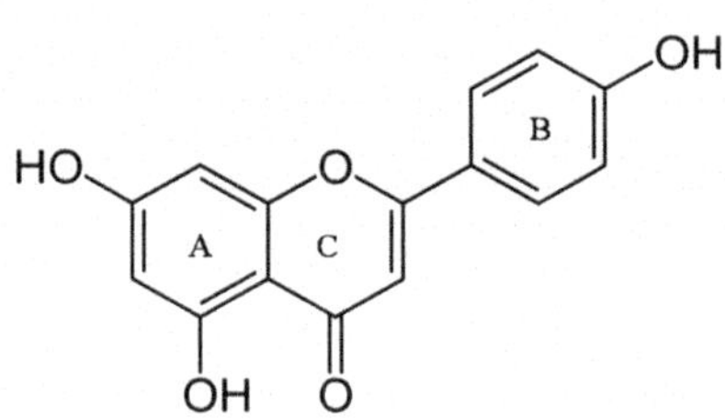

Fig. 2 Structure of apigenin; molecular formula, $C_{15}H_{10}O_5$; IUPAC name, 5,7-dihydroxy-2-(4-hydroxyphenyl)chromen-4-one

2.2 Flavanones

Flavanones are another important class which is generally present in rinds of citrus fruits such as oranges, lemons and grapes. Hesperetin, naringenin and eriodictyol belong to this class of flavonoids. Flavanones are also called as dihydroxyflavones. Over the last few decades, the number of flavanones has been isolated and found to have significant potential in health care as evidenced from the experimental data [5].

2.3 Isoflavonoids

Isoflavonoids are a large and distinctive subgroup of flavonoids. Isoflavonoids are abundantly found in soya beans and other leguminous plants. Some isoflavonoids are present in microbes. They are also found to play an important role as precursors for the development of phytoalexins during plant microbe interactions. Isoflavonoids have the ability to fight a number of diseases. The genistein and daidzein which are found in soya bean are commonly regarded to be phytoestrogens because of their oestrogenic activity in vivo. Genistein have the ability to modulate several metabolic and hormonal pathways [6].

2.4 Neoflavonoids

Neoflavonoids, the polyphenolic compounds, are the first isolated flavonoids from *Calophyllum inophyllum* seeds. It is also found in the bark and timber of Sri Lankan endemic plant *Mesua thwaitesii* [7].

2.5 Anthocyanins

Anthocyanins are pigments responsible for colours in different plants, flowers and fruits. Cyanidin, delphinidin, malvidin, pelargonidin and peonidinare are some of the most commonly studied anthocyanins. They are abundantly present in the outer cell layer of various fruits such as cranberries, black currants, red grapes, merlot grapes, raspberries, strawberries, blueberries, bilberries and blackberries. Anthocyanins have health benefits, so they are widely used in food industry as nutraceuticals; the colour of the anthocyanin mostly depends on pH [4].

2.6 Chalcones

Chalcones are characterised by the absence of ring 'C' of the basic flavonoid skeleton structure (Fig. 2) and referred as open-chain flavonoids. The most common chalcones are phloridzin, arbutin, phloretin and chalconaringenin. They are present in significant amounts in tomatoes, pears, strawberries, bearberries and certain wheat products. Chalcones and their derivatives have gathered considerable attention because of numerous nutritional and biological activities [8, 9].

2.7 Flavone

Flavones belong to one of the important subgroups of flavonoids. Flavones are widely present in leaves, flowers and fruits as glycosides. Flavones are mostly found in celery, parsley and ginkgo biloba. Apigenin belongs to this subclass of flavonoids [3, 4].

3 Apigenin: An Important Naturally Obtained Flavone

Apigenin is flavone (Fig. 2) which is obtained from the plant *Matricaria recutita* belonging to the family Asteraceae. This is a bioflavonoid compound that is mostly found in herbs and also in plants. It is a dimer found in nature. Some of the popular and abundant sources include chamomile tea [10], grapefruits, onions, oranges and some species of parsley [11]. It is also found in higher amount in celery, yarrow, cilantro, foxglove, liquorice, spearmint, basil and oregano [12, 13].

3.1 Pharmacokinetic Behaviours of Apigenin

Apigenin upon iv injection get rapidly metabolised via UDP-glucuronosyltransferase UGT1A1 and released into serum as glucoside and sulphate conjugates [13, 14]. It has half-life of 91.8 h. Apigenin appears in blood after 24 h taken by oral route [14]. It is mostly excreted via urine in the form of sulphate and glucoside conjugates.

3.2 Bioactivity of Apigenin

3.2.1 Activity on Gamma-Aminobutyric Acid (GABA) Receptor

Apigenin shows its antianxiety property by acting on benzodiazepine receptor. Normal dosage (3–10 mg/kg body weight) does not show any muscle relaxant or sedative property. But it shows sedation by increasing 3–10-fold of the above-mentioned dose (i.e. 30–100 mg/kg body weight) [14].

3.2.2 Neuroprotection

Apigenin, in the form of biapigenin (3,8″-biapigenin), can exert a neuroprotective effects by reducing the oxidative stress-induced reduction in mitochondrial membrane lipid peroxidation and transmembrane electric potential in the neural mitochondria [15]. A study by Zhao et al. [16, 17] reported the neuroprotective effects of apigenin in Alzheimer's disease (AD) mice model through the mechanism of superoxide anion scavenging and improved SOD and GTP enzyme activity. Additionally, they showed the restoration of neurotrophic ERK/CREB/BDNF pathway in the cerebral cortex which emphasises the representation of apigenin as an alternative medication for the therapy and prevention of neurodegenerative diseases.

3.2.3 Inflammation and Immunology

Apigenin gives anti-inflammatory property via suppressing the induction of NO synthase and COX 2 enzymes in macrophages via lipopolysaccharide influence. Apigenin also has inhibitory effects on interleukin-4 production [18, 19]. It may also inhibit TNFα elevations via interference with NF-κb transcription [20].

3.2.4 Antidiabetic Activity

Apigenin upon oral administration (0.78 mg/kg body weight) for 10 days was shown to reverse the decrease in hepatic antioxidants in alloxan-induced insulin-dependent diabetic mice, confirming the free radical scavenging activity [21]. In streptozotocin-induced diabetic rats, the intraperitoneal (ip) apigenin administration showed a prevalent anti-hyperglycaemic effect [22]. In clonal β-cells, apigenin treatment attenuated 2-deoxy-D-ribose-induced apoptosis through its antioxidant effect by controlling the mitochondrial membrane potential [23]. In human THP-1 monotypic cells, apigenin inhibits TNFα- and IL-1β-induced activation of NF-κB [24], and in HepG2 hepatocytes, this has improved AMPK phosphorylation [25]. Apigenin was found to be 200-fold more potent than metformin, a well-known activator of AMPK. In HepG2 cells exposed to high glucose, apigenin decreases the phosphorylation and delayed the lipid accumulation [25], supporting the beneficial effects of apigenin on hyperlipidaemia and diabetes by regulating AMPK-dependent energy metabolism. Another antiobesity and antidiabetic flavone, luteolin, found in high amount in vegetables and fruits such as onion leaves, cabbage, broccoli, celery, parsley, carrots, peppers, apple skins and chrysanthemum flowers also showed its antidiabetic potential [25–27]. In primary mouse adipose cells and 3T3-L1 adipocytes, luteolin was reported to potentiate insulin action and enhance the expression and transcriptional activation of PPARγ target genes [27].

Apigenin and two glucoranoside glycosides of apigenin, from the plant *Cephalotaxus sinensis* of the plum yew family, have been shown to exert beneficial effects in diabetes by potentiating GLUT4 response to insulin [28]. Apigenin as dietary flavonoids decrease gluconeogenesis and glucose output of the liver. Apigenin also increase glycogenesis and regulate carbohydrate metabolism. It improves glucose homeostasis and decreases insulin resistance. Apigenin also reduces interleukin-1 beta. Thus, it improves inflammation, beta cell function and insulin action. Apigenin also enhances phosphorylation of insulin receptor substrate (IRS) and thereby enhances the activity of phosphatidylinositol-3-kinase pathway. Thereby, it increases insulin sensitivity and glucose utilisation, which decreases glycosylated haemoglobin. On the other hand, apigenin decreases 3-hydroxy-3-methylglutaryl-coenzyme reductase enzyme activity and downregulates sterol regulatory binding protein synthesis, resulting in decrease in fatty acid and very-low-density lipoprotein synthesis from the liver. Thus, it causes a reduction in cholesterol synthesis and triglyceride levels, which improves glucose homeostasis and decreases insulin resistance and glycosylated haemoglobin levels [28].

3.2.5 Apigenin in Cancer Therapy

Carcinogenesis is a multistage complex process with a series of genetic and epigenetic changes. This leads to initiation, promotion and progression of the disease. The obvious strategies to treat cancer are the elimination of tumour cells by triggering cell apoptosis or by inhibiting cancer cell proliferation by inducing cell cycle arrest. The current strategies of treating the disease include the inhibition of tumour cell migration and its invasion and stimulation of the immune response of patients. Recent researches are showing its promise towards the use of apigenin in cancer therapy demonstrating all these antitumor activities with different tumour types in vitro and in vivo [29, 30].

Induction of Apoptosis

Apoptosis is the process of programmed cell death involving energy-dependent cascade events and different morphological characteristics [29]. It is induced by two core pathways: (1) the extrinsic, known as death receptor pathway, and (2) the intrinsic, known as mitochondrial pathway. It is a complex process which removes nonfunctional, undesirable cells under normal physiological conditions. Absence of this programmed process is one of the most important characteristics of cancer cells making them different from normal cells. Thus, triggering apoptosis in those cancer cells by targeting apoptotic pathways with chemotherapy is a widely used promising modern-day strategy to treat cancer. In this smart strategy, apigenin is proving itself as a precious one by boosting the hope few folds to the cancer patients with its effectiveness for triggering apoptosis via either the intrinsic or extrinsic pathway in human cancer cells [29].

Modulation of the Cell Cycle

Uncontrolled and rapid cell division is another characteristic of cancer. A number of natural compounds found to arrest cell cycle have showed their effectiveness by suppressing cancer in vitro, in vivo and in clinical studies [31, 32]. Apigenin inhibits cancer cell proliferation by modulating the cell cycle and inhibits the cell cycle phase at G_2/M or G_0/G_1 checkpoint. It can also induce cell cycle arrest at G_0/G_1 or S checkpoint. In human prostate cancer, it has been found that in LNCaP cells this apigenin resulted in G_1 arrest of cell cycle progression. Apigenin treatment also markedly reduces the protein expression of cyclin D1, D2 and E and their activating partners CDK2, CDK4 and CDK6 and increases expression of p21$^{WAF1/CIP1}$ and p27^{KIP1} concomitantly. The induction of p1^{W}2$^{AF1/CIP1}$ appears to be transcriptionally upregulated and p53-dependent [31]. In an oral squamous cell carcinoma cell line (SCC-25), apigenin treatment caused cell cycle arrest at both G_0/G_1 and G_2/M checkpoints, associated with decreased expression of cyclin D1 and E, and inactivation of CDK1 [33]. Interestingly, apigenin exposure induces G_2/M arrest in imatinib-sensitive human K562 myelogenous leukaemia cells while arresting imatinib-resistant K562/IMA3 (imatinib-resistant version of K562) cells in S phase especially at a dosage of 100 μM apigenin [32]. Taken together, these observations suggest that apigenin modulates cell cycle progression in a dose-dependent and/or cell line-specific manner which is very promising to build up new therapeutic strategy of cancer with a modulation of cell cycle at the different phases.

Induction of Autophagy

Autophagy is another programmed cell death process characterised by the appropriation of cytoplasmic material into vacuoles for bulk degradation by the lysosomal enzymes. It is a dynamic process in which the cell digests its own cytoplasmic materials within the lysosomes which then results in the sequestration and degradation of macromolecules [34]. There are growing evidences about the relationship between autophagy and cancer, which is complex and contradictory [35, 36]. In some cases, autophagy can serve as a cell survival pathway by providing recycled metabolic substrates and maintaining energy homeostasis during starvation; in other reports, it can cause cell death, either in collaboration with apoptosis or as a backup mechanism. Autophagy triggered by apigenin was first observed in erythroleukaemia TF1 cells without apoptosis [37, 38]. Since then, more evidences have been presented in favour of apigenin showing its capability to induce autophagy which serves a tumour suppressive or tumour protective role under different circumstances [39, 40]. Apigenin exerts its chemopreventive action by inducing autophagy in human keratinocytes via activation of AMPK [41]. In human breast cancer T47D and MDA-MB-231 cells, apigenin exposure triggers the cell apoptosis and autophagy with the accumulation of acidic vesicular organelles (AVOs) and LC3-II, a marker of Atg5/Atg7-dependent autophagy [42]. Similarly, in human colon cancer cells (HCT116), apigenin concomitantly caused apoptosis and autophagy [43]. Hence, it may be speculated with the existing observations that the autophagy plays an important role in apigenin-induced apoptosis in cancer cells [42, 43].

Apigenin on Cancer Stem Cells

Apigenin shows its significant and selective cell cytotoxicity against various types of cancer cells with low or no toxicity to normal cells. The cancer stem cells (CSCs) are closely associated with drug resistance, metastasis and the recurrence of cancer with time. In the head and neck squamous carcinoma cells (HN-8, HN-30 and HSC-3), apigenin significantly downregulates the stem cell markers (CD44, NANOG and CD105) and abolished the hypoxia-induced increase in CD44(+) cells, CD105(+) cells and STRO-1(+) cells [44]. In human glioblastoma cells, apigenin has been found to inhibit both the self-renewal capacity (such as cell growth and clonogenicity) and the invasiveness of GBM stemlike cells by inhibiting the activation of c-Met signalling pathway [44]. In addition, in CD44 (+) prostate CSCs of PC3 cells, apigenin inhibits cell survival and proliferation in a dose-dependent manner by inducing extrinsic cell apoptosis and increasing cell cycle arrest. Apigenin is also found to suppress the stem cell migration and adhesion by downregulating matrix metallopeptidase-2 and matrix metallopeptidase-9, Snail and Slug. Meanwhile, the apigenin treatment can significantly downregulate stemness marker Oct3/Oct4 protein expression by downregulation of PI3K/Akt and NF-κB signalling pathways [45]. Sphere-forming cells (SFCs) have self-renewal capacity and found to possess stem-like cell properties. Apigenin downregulates and inhibits the self-renewal capacity of SFCs in SKOV3 and HeLa cells [46, 47]. Interestingly, it has been experimentally evidenced that apigenin synergistically enhanced PI3K/Akt inhibitor-induced apoptosis in CD34(+) CD38(−) leukaemia cells by targeting CK2 without harming healthy haematopoietic stem cells [48]. Apigenin shows its anticancer effects by inhibiting the self-renewal capacity of CSCs. This evidence further demonstrated the effective anticancer activities of apigenin.

These experimental observations gleefully encourage the scientific community about the effectiveness, efficiency and potential of apigenin to counteract the different kinds of disease including two major diseases, diabetes and cancer, which share their risk factors with each other [49]. The pharmaceutical scientists are continuously trying to overcome different hurdles by their immense and reputed research work to use apigenin as a medicine in clinical purpose in the near future. Different formulations have been evolved to deliver apigenin in a disease-specific manner, especially acquiring the unique pharmaceutical technologies to reach the target tissue of the individuals. In this circumstance, we now will go in details about the different approaches about those formulations of apigenin in the very next section.

4 Different Formulations of Apigenin

Aqueous solubility, bioavailability and antioxidant activity of apigenin were improved by developing liposome. These phospholipid-based formulations were shown to retain all the functional activities of loaded apigenin while improving the solubility around 36 times than the pure drug. The liposome loaded with apigenin was efficient to restore liver function enzymes in a rat model [50]. Apigenin-loaded self-microemulsifying drug delivery system was another approach to overcome the problems associated with solubility of apigenin by increasing the solubility about 7500 folds [17]. In addition, nanosuspensions of apigenin were developed by utilising combined technology of bead milling and high-pressure homogenisation. Almost doubled antioxidant property of the apigenin nanocrystals was reported. The researchers also have reported that these smart crystals can be used in dermal applications after incorporation in gel to serve as a potent antioxidant agent [51]. Enhanced bioavailability and dissolution were reported for the developed nanocrystals of apigenin than the coarse apigenin powder by Zhang et al. [52]. Supercritical antisolvent technique was applied to formulate the nanocrystals that resulted in an improved pharmacokinetic profile of apigenin than in the coarse powder after a single oral dose. In another study by Ding et al. [53], novel apigenin-loaded carbon nanotubes were developed by solid dispersion (SD), and solubility criteria was compared with that of free apigenin. These carbon nanopowder (CNP)-based solid dispersions (SDs) showed promises for improvement of bioavailability of apigenin with improved drug safety [53]. Antibacterial property of apigenin was estimated for the liposomal apigenin, and the activity of the formulation was compared to that of free-drug. Nanorange distearoylphosphatidylcholine liposomes were developed by Banerjee et al. [54]. Bacterial membrane perturbation occurred at high extent after effective adherence and fusion of this apigenin liposomes to the bacteria. Liposomes delivered apigenin at a high concentration to the bacterial cells which was claimed for generation of reactive oxygen species (ROS) that ultimately augment the antibacterial activity of apigenin. Marked reduction in inflammation and fibrosis-related mRNAs through application of apigenin-loaded PEGylated PLGA nanoparticles in chronic pancreatitis was reported by Feng et al. [55]. They found that mRNA expression associated with PSC pro-inflammation- and pro-fibrosis was reduced by apigenin loaded nanoparticles in vitro. Mice treated with these nanocarriers showed longer presence of it in circulation than the free-apigenin. Apigenin-loaded ethosomes were compared to liposomes, and deformable liposomes showed a therapeutic potential for effective skin deposition and transdermal flux. These novel formulations showed higher in vitro and in vivo skin targeting which may be an active tool for the treatment of ultraviolet B (UVB)-induced skin inflammation [56].

Pulmonary delivery of bovine serum albumin nanoparticles containing apigenin (BSA-Api-NPs) was investigated as inhalable formulation of apigenin to control the respiratory diseases to utilise the antioxidant and anti-inflammatory properties of apigenin in the lung. Dispersity enhancers such as lactose monohydrate and L-leucine were used along with the powdered nanoparticles to dispense through a next-generation impactor and compared to an excipient-free formulation. The formulations delivered to pulmonary region through this technique were reported to preserve the potent antioxidant property of apigenin which concluded the successful use of apigenin-loaded BSA nanoparticles against lung injury [57]. To improve the solubility and bioavailability of apigenin, the apigenin-containing nanoparticles were developed using a technique called liquid antisolvent precipitation method, and the developed nanoparticles were subjected to in vitro characterisation. Oral bioavailability of these nanoparticles was found to be higher than that of free drug. Pluronic P123- and Solutol HS 15-based polymeric micelles were formulated by Zhai et al. [58] to incorporate apigenin as active constituent. The primary objective of this study was to improve the solubility of native apigenin. They also showed greater cytotoxic potential of these micellar formulations against HepG2 and MCF 7 cells over pure apigenin [58]. Apigenin was successfully loaded to the PLGA polymeric matrix, and antitumor activity was seen in vitro in chemically induced hepatocellular carcinoma-bearing rat model. This study showed different pharmacokinetic parameters of apigenin nanoparticles. In vivo biodistribution of the nanocarriers was also studied after labelling the polymeric nano-devices with radio-isotope [59]. PLGA nanoparticle loaded with apigenin as active constituent was developed for treating skin melanoma. The formulation actively showed to target A375 cells in vitro. Overall, the researchers were successful to show the potential of apigenin nanoparticle in skin melanoma [60]. In another study, Dutta et al. [61] have shown the efficacy of colorectal cancer-targeted aptamer conjugated PLGA nanoparticles loaded with apigenin. These active nanocarriers were successful to deliver apigenin to the site of action [61].

In another study, $CD133^{+}$/ALDH1-high cancer stem cells (CSCs) were isolated from various human prostate cancer cells such as 22Rv1, DU145 and PC3 cells and treated with the optimised apigenin-loaded polyethylene glycol nanoparticles which showed greater extent of time-dependent cell viability and sphere formation when compared to that of non-CSCs counterparts. Exposure of nanoparticles in cancer stem cells in prostate cancer was reported to downregulate β-catenin, c-Myc, SOX2 and Oct4 proteins. The overexpressed CSCs were successfully targeted and killed by the optimised apigenin-loaded nanoparticles. The study has shown promises of apigenin-loaded nanoparticles to abolish tumour recurrence and/or metastasis in prostate cancer [62]. Higher extent of inhibition in cell viability in chronic myeloid leukaemia cells (K562 cells) was observed through the treatment of nanogel formulation of apigenin. Apigenin was incorporated in chitosan–stearate nanogel which was further optimised in terms of different in vitro studies. The toxic effects of the nanoparticles were evaluated and compared to those of control group which showed greater toxicity in case of nanogel formulation of apigenin [63]. Nanoencapsulation of apigenin in PLGA nanoparticles was found to be effective to enhance anticarcinogenic effect than the free-drug against benzo(a)pyrene and ultraviolet B-mediated skin cancer and mitochondrial dysfunction in mouse model. Reduced tumour tissue damage and decreased frequency of chromosomal aberrations were found to be associated with the apigenin nanoparticles. Mitochondrial apoptosis mediated by increased ROS accumulation was also reported to be effective in combating skin cancer by the treatment with apigenin nanoparticles [60]. Polymer-based nanoparticles containing apigenin released entrapped apigenin for a prolonged period of time. These flavone-loaded polymeric nanoparticles were shown to inhibit tumour cell adhesion by suppressing the expression of endothelial cell adhesion molecule. In comparison to free-apigenin, this drug delivery system provided prolonged localised release of potent drug and emphasised the targeted anti-metastatic therapies [64]. Graphene oxide-coated nanotised apigenin was used to produce synergistic effect along with paclitaxel in ovarian cancer and also to reduce the dose and dose-dependent toxicity of paclitaxel. Modified Hummers method was used to produce the nanoparticles. Improved ROS accumulation, mitochondrial depolarisation, suppression of SOD and cell cycle arrest were collectively assisted to induce apoptosis by the apigenin nanoformulation with paclitaxel when treated in SKOV3 ovarian cancer cells. Synergistic antiproliferative effect was observed that was higher than both the group of nanotised apigenin and paclitaxel in SKOV3 cells reported to be occurring through pro- and anti-apoptotic gene modulation which may be responsible for the reduction of side effects of paclitaxel [65]. Improved skin permeability and better stability of apigenin was studied by formulating the apigenin as transfersomes which has been effective in skin cancer therapy.

5 Conclusion and Future Perspective

Apigenin has shown its potential in diabetes and cancer. The molecular basis of efficacy of the flavonoid to modulate those dreadful diseases is being explored well. Several proposed mechanistic approaches including the modulation of apoptotic pathway, autophagy and self-renewal of cells and modulation of glucose metabolism to control blood glucose level of hyperglycaemia by altering several significant proteins including GLUT proteins, various enzymes and even IRS provide support for such claim. This chapter also covers therapeutic effects of apigenin nanoformulations in cancer and diabetes. It is evident from different reports that bioavailability of apigenin has been improved in form of its nanocarriers which in turn also provide sustained drug release for a prolonged duration. Further, apigenin nanoformulations also showed promise to target cancer cells. However, different preclinical and clinical studies are warranted to establish apigenin nanoformulation as potential agent to treat diabetes and cancer.

References

1. Burak M, Imen Y (1999) Flavonoids and their antioxidant properties. Turkiye Klin Tip Bil Derg 19:296–304
2. Castañeda-Ovando A, Pacheco-Hernández ML, Páez-Hernández ME, Rodríguez JA, Galán-Vidal CA (2009) Chemical studies of anthocyanins: a review. Food Chem 113:859–871
3. Panche AN, Diwan AD, Chandra SR (2016) Flavonoids: an overview. J Nutr Sci 5:e47. (1–15)
4. Iwashina T (2013) Flavonoid properties of five families newly incorporated into the order caryophyllales (review). Bull Natl Mus Nat Sci 39:25–51
5. Matthies A, Clavel T, Gütschow M, Engst W, Haller D, Blaut M, Braune A (2008) Conversion of daidzein and genistein by an anaerobic bacterium newly isolated from the mouse intestine. Appl Envrion Microbiol 74:4847–4852
6. Linuma M, Tanaka T, Hamada K, Mizuno M, Asai F, Peher G, Kraus L (1987) Revised structure of neoflavone in Coutarea hexandra. Phytochemistry 26:3096–3097
7. Garazd M, Garazd Y, Khilya V (2003) Neoflavones. 1. Natural distribution and spectral and biological properties. Chem Nat Comp 39:54–121
8. López-Lázaro M (2009) Distribution and biological activities of the flavonoid luteolin. Mini Rev Med Chem 9:31–59
9. Manach C, Scalbert A, Morand C, Rémésy C, Jiménez L (2004) Polyphenols: food sources and bioavailability. Am J Clin Nutr 79:727–747
10. McKay DL, Blumberg JB (2006) A review of the bioactivity and potential health benefits of chamomile tea (Matricaria recutita L.). Phytother Res 20:519–530
11. Shukla S, Gupta S (2010) Apigenin: a promising molecule for cancer prevention. Pharm Res 27:962–978
12. Birt DF, Hendrich S, Wang W (2001) Dietary agents in cancer prevention: flavonoids and isoflavonoids. Pharmacol Ther 90:157–177
13. Patel D, Shukla S, Gupta S (2007) Apigenin and cancer chemoprevention: progress, potential and promise (review). Int J Oncol 30:233–245
14. Gradolatto A, Basly JP, Berges R, Teyssier C, Chagnon MC, Siess MH, Canivenc-Lavier MC (2005) Pharmacokinetics and metabolism of apigenin in female and male rats after a single oral administration. Drug Metab Dispos 33:49–54
15. Silva B, Oliveira PJ, Dias A, Malva JO (2008) Quercetin, Kaempferol and Biapigenin from Hypericum perforatum are neuroprotective against excitotoxic insults. Neurotox Res 13:265–279
16. Zhao L, Wang JL, Liu R, Li XX, Li JF, Zhang L (2013a) Neuroprotective, anti-amyloidogenic and neurotrophic effects of apigenin in an Alzheimer's disease mouse model. Molecules 18:9949–9965
17. Zhao L, Zhang L, Meng L, Wang J, Zhai G (2013b) Design and evaluation of a self-microemulsifying drug delivery system for apigenin. Drug Dev Ind Pharm 39:662–669
18. Kawai M, Hirano T, Higa S, Arimitsu J, Maruta M, Kuwahara Y, Ohkawara T, Hagihara K, Yamadori T, Shima Y, Ogata A, Kawase I, Tanaka T (2007) Flavonoids and related compounds as anti-allergic substances. Allergol Int 56:113–123
19. Yano S, Umeda D, Yamashita T, Ninomiya Y, Sumida M, Fujimura Y, Yamada K, Tachibana H (2007) Dietary flavones suppresses IgE and Th2 cytokines in OVA-immunized BALB/c mice. Eur J Nutr 46:257–263
20. Panés J, Gerritsen ME, Anderson DC, Miyasaka M, Granger DN (1996) Apigenin inhibits tumor necrosis factor-induced intercellular adhesion molecule-1 upregulation in vivo. Microcirculation 3:279–286
21. Panda S, Kar A (2007) Apigenin (41,5,7-trihydroxyflavone) regulates hyperglycaemia, thyroid dysfunction and lipid peroxidation in alloxan-induced diabetic mice. J Pharm Pharmacol 59:1543–1548

22. Rauter AP, Martins A, Borges C, Mota-Filipe H, Pinto R, Sepodes B, Justino J (2010) Antihyperglycaemic and protective effects of flavonoids on streptozotocin-induced diabetic rats. Phytother Res 24:S133–S138
23. Suh KS, Oh S, Woo JT, Kim SW, Kim JW, Kim YS, Chon S (2012) Apigenin attenuates 2-deoxy-d-ribose-induced oxidative cell damage in HIT-T15 pancreatic. β-cells. Biol Pharm Bull 35:121–126
24. Zang M, Xu S, Maitland-Toolan KA, Zuccollo A, Hou X, Jiang B, Wierzbicki M, Verbeuren TJ, Cohen RA (2006) Polyphenols stimulate amp-activated protein kinase, lower lipids, and inhibit accelerated atherosclerosis in diabetic ldl receptor-deficient mice. Diabetes 55:2180–2191
25. Neuhouser ML (2004) Dietary flavonoids and cancer risk: evidence from human population studies. Nutr Cancer 50:1–7
26. Gates MA, Tworoger SS, Hecht JL, de Vivo I, Rosner B, Hankinson SE (2007) A prospective study of dietary flavonoid intake and incidence of epithelial ovarian cancer. Int J Cancer 121:2225–2232
27. Miean KH, Mohamed S (2001) Flavonoid (myricetin, quercetin, kaempferol, luteolin, and apigenin) content of edible tropical plants. J Agric Food Chem 49:3106–3112
28. Li W, Dai RJ, Yu YH, Li L, Wu CM, Luan WW, Meng WW, Zhang XS, Deng YL (2007) Antihyperglycemic effect of Cephalotaxus sinensis leaves and GLUT-4 translocation facilitating activity of its flavonoid constituents. Biol Pharm Bull 30:1123–1129
29. Elmore S (2007) Apoptosis: a review of programmed cell death. Toxicol Pathol 35:495–516. https://doi.org/10.1080/01926230701320337
30. Kim B, Jung N, Lee S, Sohng JK, Jung HJ (2016) Apigenin inhibits cancer stem cell-like phenotypes in human glioblastoma cells via suppression of c-Met signaling. Phytother Res 30:1833–1840. https://doi.org/10.1002/ptr.5689
31. Gupta S, Afaq F, Mukhtar H (2002) Involvement of nuclear factor-kappa B, Bax and Bcl-2 in induction of cell cycle arrest and apoptosis by apigenin in human prostate carcinoma cells. Oncogene 21:3727–3738. https://doi.org/10.1038/sj.onc.1205474
32. Solmaz S, Adan Gokbulut A, Cincin B, Ozdogu H, Boga C, Cakmakoglu B, Kozanoglu I, Baran Y (2014) Therapeutic potential of apigenin, a plant flavonoid, for imatinib-sensitive and resistant chronic myeloid leukemia cells. Nutr Cancer 66:599–612. https://doi.org/10.1080/01635581.2014.894099
33. Maggioni D, Garavello W, Rigolio R, Pignataro L, Gaini R, Nicolini G (2013) Apigenin impairs oral squamous cell carcinoma growth in vitro inducing cell cycle arrest and apoptosis. Int J Oncol 43:1675–1682. https://doi.org/10.3892/ijo.2013.2072
34. Korolchuk VI, Rubinsztein DC (2011) Regulation of autophagy by lysosomal positioning. Autophagy 7:927–928. https://doi.org/10.4161/auto.7.8.15862
35. Wilde L, Tanson K, Curry J, Martinez-Outschoorn U (2018) Autophagy in cancer: a complex relationship. Biochem J 475:1939–1954
36. Yang ZJ, Chee CE, Huang S, Sinicrope FA (2011) The role of autophagy in cancer: therapeutic implications. Mol Cancer Ther 10:1533–1541
37. Chaabane W, User SD, El-Gazzah M, Jaksik R, Sajjadi E, Rzeszowska-Wolny J, Los MJ (2013) Autophagy, apoptosis, mitoptosis and necrosis: interdependence between those pathways and effects on cancer. Arch Immunol Ther Exp 61:43–58. https://doi.org/10.1007/s00005-012-0205-y
38. Ruela-de-Sousa RR, Fuhler GM, Blom N, Ferreira CV, Aoyama H, Peppelenbosch MP (2010) Cytotoxicity of apigenin on leukemia cell lines: implications for prevention and therapy. CellDeath Dis 1:e19. https://doi.org/10.1038/cddis.2009.18
39. Salmani JMM, Zhang XP, Jacob JA, Chen BA (2017) Apigenin's anticancer properties and molecular mechanisms of action: recent advances and future prospectives. Chin J Nat Med 15:321–329
40. Sung B, Chung HY, Kim ND (2016) Role of apigenin in cancer prevention via the induction of apoptosis and autophagy. J Cancer Prev 21:216–226. https://doi.org/10.15430/JCP.2016.21.4.216

41. Tong X, Smith KA, Pelling JC (2012) Apigenin, a chemopreventive bioflavonoid, induces AMP-activated protein kinase activation in human keratinocytes. Mol Carcinog 51:268–279. https://doi.org/10.1002/mc.20793
42. Cao X, Liu B, Cao W, Zhang W, Zhang F, Zhao H, Meng R, Zhang L, Niu R, Hao X, Zhang B (2013) Autophagy inhibition enhances apigenin-induced apoptosis in human breast cancer cells. Chin J Cancer Res 25:212–222
43. Lee Y, Sung B, Kang YJ, Kim DH, Jang JY, Hwang SY, Kim M, Lim HS, Yoon JH, Chung HY, Kim ND (2014) Apigenin-induced apoptosis is enhanced by inhibition of autophagy formation in HCT116 human colon cancer cells. Int J Oncol 44:1599–1606. https://doi.org/10.3892/ijo.2014.2339
44. Ketkaew Y, Osathanon T, Pavasant P, Sooampon S (2017) Apigenin inhibited hypoxia induced stem cell marker expression in a head and neck squamous cell carcinoma cell line. Arch Oral Biol 74:69–74. https://doi.org/10.1016/j.archoralbio.2016.11.010
45. Erdogan S, Doganlar O, Doganlar ZB, Serttas R, Turkekul K, Dibirdik I, Bilir A (2016) The flavonoid apigenin reduces prostate cancer CD44(+) stem cell survival and migration through PI3K/Akt/NF-kappaB signaling. Life Sci 162:77–86. https://doi.org/10.1016/j.lfs.2016.08.019
46. Liu J, Cao XC, Xiao Q, Quan MF (2015) Apigenin inhibits HeLa sphere-forming cells through inactivation of casein kinase 2alpha. Mol Med Rep 11:665–669. https://doi.org/10.3892/mmr.2014.2720
47. Tang AQ, Cao XC, Tian L, He L, Liu F (2015) Apigenin inhibits the self-renewal capacity of human ovarian cancer SKOV3derived sphere-forming cells. Mol Med Rep 11:2221–2226. https://doi.org/10.3892/mmr.2014.2974
48. Cheong JW, Min YH, Eom JI, Kim SJ, Jeung HK, Kim JS (2010) Inhibition of CK2{alpha} and PI3K/Akt synergistically induces apoptosis of CD34+CD38− leukaemia cells while sparing haematopoietic stem cells. Anticancer Res 30:4625–4634
49. Edward G, David MH, Michael CA, Richard MB, Susan MG, Laurel AH, Michael P, Judith GR, Douglas Y (2010) Diabetes and cancer: a consensus report. Diabetes Care 33:1674–1685
50. Telange DR, Patil AT, Pethe AM, Fegade H, Anand S, Davec VS (2017) Formulation and characterization of an apigenin-phospholipid phytosome (APLC) for improved solubility, in vivo bioavailability, and antioxidant potential. Eur J Pharm Sci 108:36–49
51. Shaal L, Shegokar R, Müller RH (2011) Pharmaceutical nanotechnology production and characterization of antioxidant apigenin nanocrystals as a novel UV skin protective formulation. Int J Pharm 420:133–140
52. Zhang J, Huang Y, Liu D, Gao Y, Qian S (2013) Preparation of apigenin nanocrystals using supercritical antisolvent process for dissolution and bioavailability enhancement. Eur J Pharm Sci 48:740–747
53. Ding S, Zhang Z, Song J, Cheng X, Jiang J, Jia X (2014) Enhanced bioavailability of apigenin via preparation of a carbon nanopowder solid dispersion. Int J Nanomedicine 9:2327–2333
54. Banerjee K, Banerjee S, Das S, Mandal M (2015) Probing the potential of apigenin liposomes in enhancing bacterial membrane perturbation and integrity loss. J Colloid Interface Sci 453:48–59
55. Feng W, Guo H, Xue T, Wang X, Tang C, Ying B, Gongb H, Cui G (2015) Anti-inflammation and anti-fibrosis with PEGylated, apigenin loaded PLGA nanoparticles in chronic pancreatitis disease. RSC Adv 5:83628–83635
56. Li-Na S, Zhanga YT, Wangb Q, Xub L, Fenga N-P (2014) Pharmaceutical nanotechnology enhanced in vitro and in vivo skin deposition of apigenin delivered using ethosomes. Int J Pharm 460:280–288
57. Pápay ZE, Kósa A, Böddi B, Merchant Z, Saleem IY, Zariwala MG, Klebovich I, Somavarapu S, Antal I (2017) Study on the pulmonary delivery system of apigenin-loaded albumin nanocarriers with antioxidant activity. J Aerosol Med Pulm Drug Deliv 30:274–288
58. Zhai Y, Guoa S, Liu C, Yang C, Dou J, Li L, Zhaia G (2013) Preparation and in vitro evaluation of apigenin-loaded polymeric micelles. Colloids Surf A Physicochem Eng Asp 429:24–30

59. Bhattacharya S, Mondal L, Mukherjee B, Dutta L, Ehsan I, Debnath MC, Gaonkar RH, Pal MM, Majumdar S (2018) Apigenin loaded nanoparticle delayed development of hepatocellular carcinoma in rats. Nanomedicine 14:1905–1917
60. Das S, Das J, Samadder A, Paul A, Khuda-Bukhsh AR (2013) Strategic formulation of apigenin-loaded PLGA nanoparticles for intracellular trafficking, DNA targeting and improved therapeutic effects in skin melanoma in vitro. Toxicol Lett 223:124–138
61. Dutta D, Chakraborty A, Mukherjee B, Gupta S (2018) Aptamer-conjugated apigenin nanoparticles to target colorectal carcinoma: a promising safe alternative of colorectal cancer chemotherapy. ACS Appl Bio Mater 1:1538–1556
62. Kanwal R, Shukla S, Exner AA, Tzakos AG, Sanjay G (2018) Apigenin nanoparticle suppresses sphere formation in CD133+/ALDH1high prostate cancer stem cells through downregulation of stem cell markers. AACR Cancer Res 78
63. Samadian N, Hashemi M (2018) Effects of apigenin and apigenin- loaded nanogel on inductionof apoptosis in human chronic myeloid leukemia cells. Galen Med J 7:e1008. https://doi.org/10.22086/gmj.v0i0.1008
64. Cochran DB, Gray LN, Anderson KW, Dziubla TD (2016) Degradable poly(apigenin) polymerinhibits tumor cell adhesion to vascular endothelial cells. J Biomed Mater Res Part B Appl Biomater 104B:1438–1447
65. Pal MK, Jaiswar SP, Dwivedi A, Goyal S, Dwivedi VN, Pathak AK, Kumar V, Sankhwar PL, Ray RS (2017) Synergistic effect of graphene oxide coated nanotised apigenin with paclitaxel (GO-NA/PTX): a ROS dependent mitochondrial mediated apoptosis in ovarian cancer. Anticancer Agents Med Chem 17:1721–1732

Therapeutic Potential of "Aegeline," an Important Phytochemical of *Aegle marmelos*: Current Health Perspectives for the Treatment of Disease

Dinesh Kumar Patel, Kanika Patel, Mahfoozur Rahman, and Sudhir Chaudhary

Abbreviations

TNBS	2,4,6-Trinitrobenzene sulfonic acid
hMSCs	Human mesenchymal stem cells
HFD	High-fat diet
GLUT4	Glucose transporter type 4
LDL	Low-density lipoprotein
STZ	Streptozotocin
CYP	Cytochrome P450 enzymes
ROS	Reactive oxygen species
NMR	Nuclear magnetic resonance
HR-ESI/MS	High-resolution electrospray ionization mass spectrometry
UHPLC	Ultrahigh-performance liquid chromatography
PDA	Photodiode array detector

1 Introduction

Natural products have been used as food material for mankind since ancient times and provided some of the most important drugs and therapeutic agents called drugs for centuries for preventive and curative purpose. Ayurveda, homoeopathy, and Siddha also describe the importance of these plant-based phytochemical and drugs

D. K. Patel (✉) · K. Patel · M. Rahman
Department of Pharmaceutical Sciences, Shalom Institute of Health & Allied Sciences, SHUATS, Allahabad, Uttar Pradesh, India

S. Chaudhary
BBS Institute of Pharmaceutical & Allied Sciences, Greater Noida, Uttar Pradesh, India

© Springer Nature Singapore Pte Ltd. 2020
M. Rahman et al. (eds.), *Nanomedicine for Bioactives*,
https://doi.org/10.1007/978-981-15-1664-1_14

in the indigenous system of medicine for the treatment of disorders. For maintaining proper health and combating diseases, most of the population in the world uses plant-derived drugs as their first line of defense. Plants contain numerous phytochemicals such as terpenes, terpenoids, tannins, alkaloids, coumarins, flavonoids, polysaccharides, glycosides, gum, and phenols. Evidence also suggests that plant-derived medicines are more valuable and acceptable to the human body compared to the allopathic drugs [1, 2]. Herbal medicines also play an important role in the human civilization as they have been used for different purposes from the ancient times on beliefs of its less side effect and economic values until the synthetic drugs are developed in the world [3–5]. Phytochemicals present in various herbs, fruits, vegetables, oil, seeds, etc. play an important role in the human life as some of the best-ever drugs used by human being are derived from natural sources. Various plant extracts have been claimed for the treatment of numerous disorders, and these activities can be further improved by preparation of nanoparticles. Nanoparticles have various beneficial applications in sensing devices, medicine, microelectronics, pharmacology, and new drug delivery systems. AgNPs are one of the best examples among various nanoparticles which have diverse biomedical applications. Chemical, physical, and photochemical methods are available for the synthesis of nanoparticles. Nanoparticles are categorized as, for example, gold and silver nanoparticles, carbon nanoparticles, and semiconductor nanoparticles. Gold and silver nanoparticles provide superior material properties with functional versatility, which have been used as potential tools for medical imaging too. Due to its safety, simplicity, low-cost, and eco-friendly nature, nowadays, green synthesis of nanoparticles using different plant parts and components is the better approach in this modern age. Successful synthesis of AgNPs using various plant parts such as *Dillenia indica* fruit, *Acacia concinna* fruit, *Syzygium cumini* fruit, *Mimusops elengi* fruit, *Lagerstroemia speciosa* fruit, *Emblica officinalis* fruit, *Eucalyptus globulus* leaf, *Azadirachta indica* leaf, *Ocimum sanctum* leaf, *Arbutus unedo* leaf, *Stevia rebaudiana* leaf, *Nicotiana tobaccum* leaf, and *Ficus benghalensis* leaf has been mentioned in the literature [6].

2 Importance of *Aegle marmelos*

Aegle marmelos is also known as Indian quince, Bengal quince, holy fruit, Bael, or Bilva in Hindi and golden apple in English. The fruit is a rich source of riboflavin and vitamin C. In Ayurvedic medicine, the bark and fruit are used for dysentery and other intestinal disorders. The crude extracts of *Aegle marmelos* have antidiabetic, anti-inflammatory, analgesic, anti-ulcer, antimicrobial, antihyperglycemic, anti-dyslipidemic, antidiarrheal, antifungal, gastric mucosal protective, anti-malarial, antibacterial, antiparasitic, antioxidant, anticancer, antiviral, radioprotective, cardioprotective, antiasthmatic, and antispermatogenic potentials. Since very ancient times, herbal tea with boiling water of dried Bael flower was used as popular beverage in Sri Lanka, and distillation of flowers has been also used as tonic for the stomach and intestine. *Aegle marmelos* is also present in one of the best-selling products Dashamoola which is widely used in various Ayurvedic formulations such as Dashmularishta, Dashamoola Kashayam, and Dasamulakatutrayadi Kashayam [7–9]. *Aegle marmelos* is a tropical fruit native to Southeast countries mainly grown in India, Pakistan, Sri Lanka, and Bangladesh. Seeds of unripe Bael fruits contain gum and are highly branched in nature with terminal units of galactose, galacturonic acid, rhamnose, and arabinose. The branching heteropolysaccharide feature allows for easy preparation of sulfated compounds that would be difficult to obtain by a synthetic approach. *Aegle marmelos* fruit pulp contains various health-beneficial components including carotenoids, alkaloids, coumarins, flavonoids, pectins, phenolics, tannins, terpenoids, and many pure phytochemicals such as aegelin, aurapten, cinenol, citral, citronellal, cuminaldehyde, eugenol, fagarine, lupeol, vangetin, marmelide, marmelosin, marmesinin, marmin, psoralen, and skimmianine (Fig. 1) [10–14]. *Aegle marmelos* is used in traditional medicine for the treatment of dysentery, fever, diabetes, asthma, heart problems, ophthalmic, hemorrhoids, and urinary problems in humans. The fruit of *Aegle marmelos* is prescribed in the treatment of tuberculosis and hepatitis and used against stomach complications. The chemical constituents of the essential oil of *Aegle marmelos* recorded promising antifungal and antimicrobial activities. Methanolic extract and mucilage of *Aegle marmelos* fruits were recently reported to have hypoglycemic effect. *Aegle marmelos* is widely used in Ayurvedic and folk medicines to cure peptic ulcers, tuberculosis, headache, hypertension, dysentery, diabetes, etc. This plant possesses antimicrobial, antidiarrheal, antidiabetic, anti-ulcerative colitis, hepatoprotective, cardioprotective, and radioprotective properties [15]. *Aegle marmelos* stem and root bark are used to treat ulcers, dyspepsia, and stomachalgia. The unripe fruit is used for diarrhea, dysentery, ulcers, and hypoglycemia. The fresh flowers allay thirst and vomiting and is useful in dysentery, whereas seeds were reported to possess antimicrobial, anthelmintic, and hypoglycemic activities [12].

3 Pharmacological Usage of *Aegle marmelos*

Earlier reports have shown antioxidant and nitric oxide scavenging properties of *Aegle marmelos* fruit pulp aqueous extract. Preclinical studies of *Aegle marmelos* whole plant extract and fruit pulp extract show hepatoprotective effect. The observed hepatoprotective effects were attributed to enhancement of antioxidant levels, preventing hyperproliferation in the liver [13]. Recent studies demonstrate the curative effects of the ethanolic extract of Bilva plants against 2,4,6-trinitrobenzene sulfonic acid (TNBS)-induced colitis in rats [9]. The aqueous decoction of the leaves has been reported to have a significant hypoglycemic effect and helps in the regeneration of a damaged pancreas. Extract of the leaves has been reported to have chemopreventive potential especially against chemical carcinogenesis. Aqueous and alcoholic extracts of *Aegle marmelos* leaves reported to have cardiotonic effect [15].

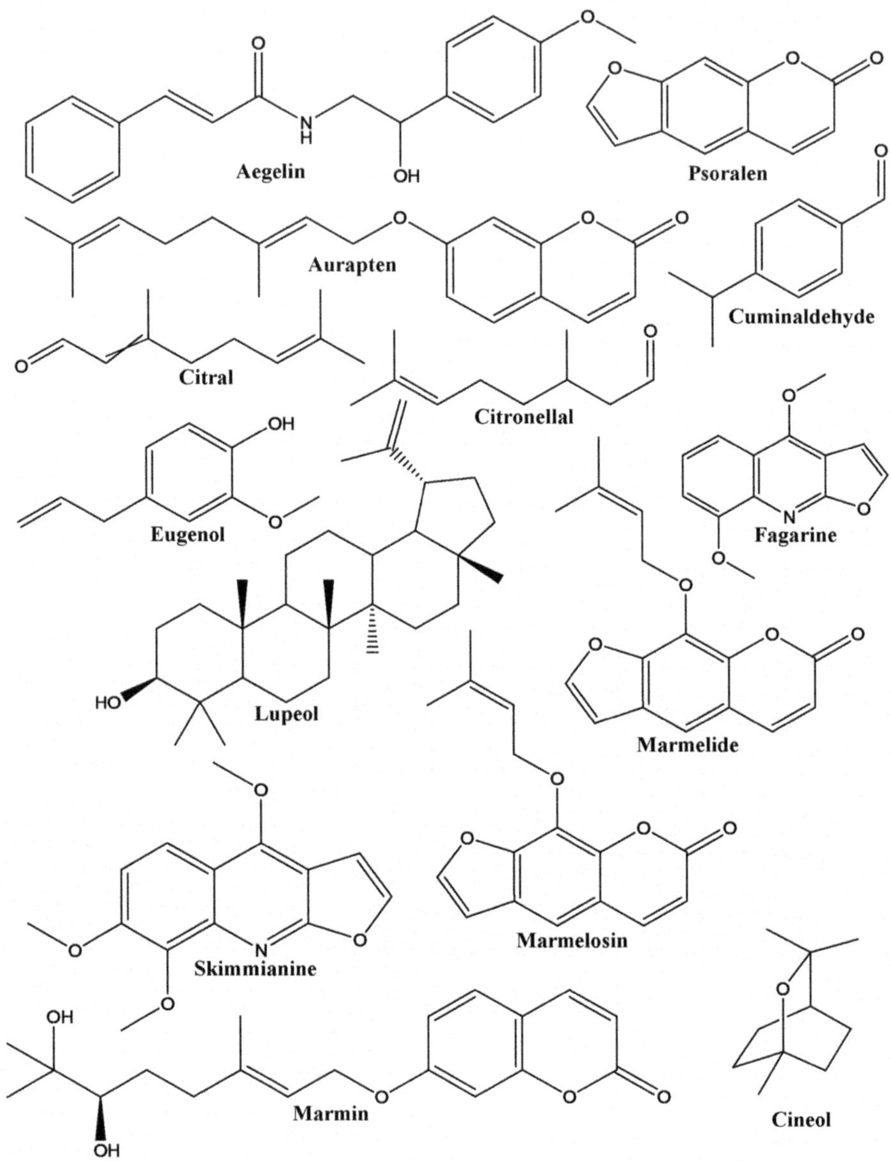

Fig. 1 Chemical structure of various phytocompounds present in *Aegle marmelos*

3.1 Therapeutic Interventions for Insulin Resistance and Type 2 Diabetes

Leaves of *Aegle marmelos* are being used since ages in Indian Ayurvedic medicine for the treatment of diabetes mellitus. Aegeline is an alkaloidal-amide, isolated from the leaves of *Aegle marmelos*, and has already been shown to be responsible for antihyperglycemic as well as antidyslipidemic activity in the validated animal models of type 2 diabetes mellitus [16]. β3-AR activity of novel N-acyl-1-amino-3-arylopropanol synthetic mimics of aegeline and its beneficial effect in insulin resistance have been investigated. Based on 3D pharmacophore modeling of known β3-AR agonist, 20 synthetic derivatives of aegeline from the literature were screened. The most active compound was selected, and β3-AR activity was further validated in white and brown adipocytes differentiated from human mesenchymal stem cells (hMSCs). Insulin resistance model developed in hMSC-derived adipocytes was used to study the insulin-sensitizing property. Eight-week HFD-fed C57BL6 mice were given 50 mg/Kg of the selected compound, and metabolic phenotyping was done to evaluate its antidiabetic effect. The compound 10C was found to be the most active and specific β3-AR agonist. The compound 10C also improved insulin sensitivity and glucose tolerance in 8-week HFD-fed C57BL6 mice. This study signified the use of β3-AR agonist as therapeutic interventions for insulin resistance and type 2 diabetes [17].

3.2 Impact of Aegeline on Obesity and Its Associated Metabolic Disorders

Natural amino alcohol "aegeline" isolated from the plant *Aegle marmelos* had antidiabetic and antidyslipidemic potential. Here, a series of novel amino alcohol and thiazolidinedione hybrid molecules have been synthesized and their antiadipogenic activity studied. As a result, we have identified a potent hybrid compound 12c as an inhibitor of adipocyte differentiation. The compound 12c inhibits lipid accumulation and adipogenesis in 3T3-L1 preadipocyte cell line. Study proved that compound 12c showed potential ability to inhibit adipocyte differentiation which can be used therapeutically for the treatment of obesity and its associated metabolic disorders [18]. Aegeline is an alkaloidal-amide, isolated from the leaves of *Aegle marmelos*, and has shown antihyperglycemic as well as antidyslipidemic activities in the validated animal models of type 2 diabetes mellitus. Aegeline enhanced GLUT4 translocation-mediated 2-deoxy-glucose uptake in both time- and concentration-dependent manner. These findings suggest that aegeline stimulates the glucose transport through Akt and Ras-related C3 botulinum toxin substrate 1-dependent distinct parallel pathways and has cytoskeletal roles via stimulation of the PI3-kinase-Rac1-PAK1-cofilin pathway in the skeletal muscle cells [15]. Aegeline isolated from the leaves of *Aegle marmelos* act as an antihyperlipidemic and antihyperglycemic agent. New N-acyl-1-amino-2-alcohols (N-acrylated-1-amino-2-phenylethanol and N-acylated-1-amino-3-aryloxypropanols) via Ritter reaction have been synthesized and screened for their in vivo antihyperlipidemic activity in Triton-induced hyperlipidemia model, LDL oxidation, and antioxidant activity. Few synthesized compounds showed good antihyperlipidemic activity, LDL oxidation, and antioxidant activity which was compared with marketed antidyslipidemic drug [19]. n-Hexane, dichloromethane, ethyl acetate, and methanol extracts of the leaves of *Aegle marmelos* at the concentrations of 25, 50, 75, and 100 μg/ml, respectively, for adipogenesis inhibition in the adipocytes have been screened. The most active dichloromethane extract was selected for further isolation, and 14 compounds were isolated from dichloromethane extract of *Aegle marmelos* leaves. The compounds were screened for the adipogenesis inhibition at two concentrations, and it was found out that of the 14 compounds aegeline was found to be active which signified adipogenesis inhibition by *Aegle marmelos* as one of the pathways for its anti-obesity effect [20]. A series of alkaloidal-amides [N-(2-hydroxy-2-p-tolylethyl)-amides and N-(2-oxo-2-p-tolylethyl)-amide derivatives] related to aegeline were synthesized and screened for their in vivo antihyperlipidemic activity in Triton-induced hyperlipidemia model. The synthetic compounds 4, 17, and 20 showed equipotent activity to the aegeline. These compounds also showed strong antioxidant activity, which support their antihyperlipidemic activity [21]. From the leaves of *Aegle marmelos*, an alkaloidal-amide, aegeline 2, was isolated and found to have antihyperglycemic activity as evidenced by lowering the blood glucose levels in sucrose-challenged streptozotocin-induced diabetic rats (STZ-S) model at the dose of 100 mg/kg body weight. Aegeline 2 has also significantly decreased the plasma triglyceride levels, total cholesterol, and free fatty acids accompanied with increase in HDL-C and HDL-C/TC ratio in dyslipidemic hamster model at the dose of 50 mg/kg body weight [22].

3.3 Interaction of Aegeline with Histamine-1 Receptor

Affinities of six active compounds including aegeline of *Aegle marmelos* as antihistamines in histamine H1 receptor have been investigated in comparison to cetirizine, diphenhydramine, and chlorpheniramine as ligands. Molecular docking determines the interaction of ligand binding to its receptor, and aegeline was found to be potential as antihistamine agents, especially as histamine H1 receptor antagonists by interacting with amino acid residues, Asp107, Lys179, Lys191, Asn198, and Trp428, of histamine H1 receptor [23]. Effects of aegeline on the histamine release from mast cell using rat basophilic leukemia (RBL-2H3) cell line and rat peritoneal mast cells (RPMCs) are investigated. Aegeline inhibited the histamine release from RBL-2H3 cells induced by DNP(24)-BSA. The inhibitory effects of aegeline on the histamine release from mast cells depended on the type of mast cell and intracellular Ca(2+) signaling events or downstream process of intracellular Ca(2+) signaling in mast cells [24].

3.4 Inhibition of CYP3A4 and CYP1A2 Through Aegeline

Inhibitory potential of methanolic extract of *A. marmelos* fruit and its constituent aegeline towards major cytochrome P450 enzymes (CYP3A4, 2D6, 1A2, 2C9, and 2C19) using human liver microsomes and recombinant CYPs was evaluated. Aegeline was a very weak inhibitor of CYP3A4 and showed no inhibition for CYP1A2 isoform, and further, no significant inhibition of recombinant CYP2D6, 2C9, and 2C19 was seen with aegeline [25].

3.5 Role of Aegeline on Onset of Apoptosis

Library of natural products consisting of 394 compounds was screened using yeast cells that express either human α-synuclein or human Bax to find out small molecule that would mimic ySec22p, and it was found that aegeline overcame apoptosis induced by both α-synuclein and Bax in yeast. Besides, aegeline also prevented growth block in cells expressing the more toxic A53T α-synuclein mutant. Restoration of cell growth occurred through inhibition of increased ROS levels, mitochondrial membrane potential loss, and nuclear DNA fragmentation, characteristics of apoptosis manifested in α-synuclein or Bax-expressing cells. These results highlight the importance of yeast systems to identify rapidly molecules that may prevent the onset of apoptosis that occurs in Parkinson's disease [26].

4 Determination of Aegeline from Different Parts of the *Aegle marmelos*

Various analytical techniques are available for the determination of pure chemical present in the plant samples with different principles. A fast UHPLC-PDA method was developed for the simultaneous determination of aegeline and other coumarins from the root, leaf, fruit, and bark of *Aegle marmelos*. The developed UHPLC method is simple, fast, and suitable for quality control of aegeline from *Aegle marmelos* and other commercial dietary supplements [27]. Two new cytotoxic furoquinoline alkaloids were isolated from the $CHCl_3$ extract of leaves of *Aegle marmelos* and aegeline from the pet ether extract. The structures were elucidated through [1]D and [2]D NMR spectroscopy and HR-ESI/MS [28]. A colorimetric method based on the reduction of the tetrazolium salt XTT to a water-soluble formazan proved to be a reliable and convenient method for the screening of methanolic extracts of 1059 plants and bioassay-guided fractionation of the alkaloid aegeline from *Sarcorhachis naranjoana* [29].

5 Conclusion

During ancient times, medicinal plants are the only source to treat disorders, and their important contribution in primary healthcare system has been well known. Because of their low costs and effectiveness, people of developing countries rely on traditional system of medicine for their primary healthcare needs. Studies on traditional medicinal plants have become increasingly valuable in the development of healthcare and conservation programs in different parts of the world [30]. *Aegle marmelos* is a popular tree in India and other Southeast Asian countries. The fruit is usually consumed as dried, fresh, or juice and is reported to have a high nutritional value [25]. *Aegle marmelos* is a tropical fruit native to Southeast countries mainly grown in India, Pakistan, Sri Lanka, and Bangladesh. Since very ancient times, herbal tea with boiling water of dried Bael flower was used as popular beverage in Sri Lanka, and distillation of flowers has been also used as tonic for the stomach and intestine. *Aegle marmelos* fruit pulp contains various health-beneficial phytocomponents including phenolics, carotenoids, alkaloids, and flavonoids. The crude extracts of *Aegle marmelos* have antidiabetic, anti-inflammatory and analgesic, anti-ulcer, antimicrobial, antihyperglycemic, antidyslipidemic, antidiarrheal, oral hypoglycemic, antifungal, gastric mucosal protective, anti-malarial, antibacterial, antiparasitic, antioxidant, anticancer, antiviral, radioprotective, cardioprotective, antiasthmatic, and antispermatogenic agents. More than 100 bioactive compounds including skimmianine, aegelin, lupeol, cinenol, citral, citronellal, cuminaldehyde, eugenol, marmesinin, marmelosin, luvangetin, aurapten, psoralen, marmelide, fagarine, marmin, and tannin have been present and proved to have different biological potential. Earlier reports have shown antioxidant and nitric oxide scavenging properties of *Aegle marmelos* fruit pulp aqueous extract. Recent studies demonstrate the curative effects of the ethanolic extract of Bilva plants against its antibacterial and antioxidant properties. The aqueous decoction of the leaves has been reported to have a significant hypoglycemic effect and helps in regeneration of a damaged pancreas. Leaves of *Aegle marmelos* are being used since ages in Indian Ayurvedic medicine for the treatment of diabetes mellitus. From these above facts, it was concluded that *Aegle marmelos* is one of the better fruits or drugs that is there in the nature through which we can treat numerous disorders; however, its nutritional aspect is also one of the choices. The presented data of the *Aegle marmelos* and aegelin in this book chapter will be beneficial to all researchers, scientists, and scholars to explore more biological and pharmaceutical importance of this one of the holy trees of the Hindu community.

References

1. Patel K, Kumar V, Verma A, Rahman M, Patel DK (2017) β-sitosterol: bioactive compounds in foods, their role in health promotion and disease prevention "a concise report of its phytopharmaceutical importance". Curr Tradit Med 3(3):168–177
2. Patel K, Patel DK (2016a) Medicinal importance, pharmacological activities, and analytical aspects of hispidulin: a concise report. J Tradit Complement Med 7(3):360–366
3. Patel K, Patel DK (2016b) Medicinal significance, pharmacological activities, and analytical aspects of ricinine: a concise report. J Coast Life Med 4(8):663–667
4. Patel K, Mishra R, Patel DK (2016) A review on phytopharmaceutical importance of asiaticoside. J Coast Life Med 4(12):1000–1007
5. Patel K, Kumar V, Rahman M, Verma A, Patel DK (2018) New insights into the medicinal importance, physiological functions and bioanalytical aspects of an important bioactive compound of foods 'Hyperin': health benefits of the past, the present, the future. Beni-Suef Univ J Basic Appl Sci 7(1):31–42
6. Devi M, Devi S, Sharma V, Rana N, Bhatia RK, Kumar A (2019) Green synthesis of silver nanoparticles using methanolic fruit extract of Aegle marmelos and their antimicrobial potential against human bacterial pathogens. J Tradit Complement Med. https://doi.org/10.1016/j.jtcme.2019.04.007
7. Chandrasekara A, Daugelaite J, Shahidi F (2018) DNA scission and LDL cholesterol oxidation inhibition and antioxidant activities of Bael (Aegle marmelos) flower extracts. J Tradit Complement Med 8(3):428–435
8. Patel P, Asdaq SMB (2010) Immunomodulatory activity of methanolic fruit extract of Aegle marmelos in experimental animals. Saudi Pharm J 18(3):161–165
9. Rajaram A, Vanaja GR, Vyakaranam P, Rachamallu A, Reddy GV, Anilkumar K et al (2018) Anti-inflammatory profile of Aegle marmelos (L) Correa (Bilva) with special reference to young roots grown in different parts of India. J Ayurveda Integr Med 9(2):90–98
10. Azmi L, Shukla I, Goutam A, Allauddin RCV, Jawaid T et al (2019) In vitro wound healing activity of 1-hydroxy-5,7-dimethoxy-2-naphthalene-carboxaldehyde (HDNC) and other isolates of Aegle marmelos L.: enhances keratinocytes motility via Wnt/β-catenin and RAS-ERK pathways. Saudi Pharm J 27(4):532–539
11. Jindal M, Rana V, Kumar V, Sapra B, Tiwary AK (2017) Synthesis, physico-chemical and biomedical applications of sulfated *Aegle marmelos* gum: green chemistry approach. Arab J Chem 10:S2151–S2159
12. Lanjhiyanaa S, Patra KC, Ahirwara D, Rana AC, Lanjhiyana SK (2012) A validated HPTLC Method for Simultaneous estimation of two marker compounds in *Aegle marmelos* (L.) Corr., (Rutaceae) root bark. Der Pharm Lett 4(1):92–97
13. Rathee D, Kamboj A, Sachdev RK, Sidhu S (2018) Hepatoprotective effect of *Aegle marmelos* augmented with piperine co-administration in paracetamol model. Rev Bras Farmacogn 28(1):65–72
14. Wijewardana RMNA, Nawarathne SB, Wickramasinghe I, Gunawardane CR, Wasala WMCB, Thilakarathne BMKS (2016) Retention of physicochemical and antioxidant properties of Dehydrated Bael (*Aegle Marmelos*) and palmyra (*Borassus Flabellifer*) fruit powders. Procedia Food Sci 6:170–175
15. Ibrahim NA, Mohammed MMD, Aly HF, Ali SA, Al-Hady D-A (2018) Efficiency of the leaves and fruits of *Aegle marmelos* methanol extract (L.) Correa and their relative hepatotoxicity induced by CCL4 and identification of their active constituents by using LC/MS/MS. Toxicol Rep 5:1161–1168
16. Gautam S, Ishrat N, Singh R, Narender T, Srivastava AK (2015) Aegeline from *Aegle marmelos* stimulates glucose transport via Akt and Rac1 signaling, and contributes to a cytoskeletal rearrangement through PI3K/Rac1. Eur J Pharmacol 762:419–429

17. Rajan S, Satish S, Shankar K, Pandeti S, Varshney S, Srivastava A et al (2018) Aegeline inspired synthesis of novel β3-AR agonist improves insulin sensitivity in vitro and in vivo models of insulin resistance. Metabolism 85:1–13
18. Satish S, Srivastava A, Yadav P, Varshney S, Choudhary R, Balaramnavar VM et al (2018) Aegeline inspired synthesis of novel amino alcohol and thiazolidinedione hybrids with antiadipogenic activity in 3T3-L1 cells. Eur J Med Chem 143:780–791
19. Sarkar S, Sonkar R, Bhatia G, Tadigoppula N (2014) Synthesis of new N-acryl-1-amino-2-phenylethanol and N-acyl-1-amino-3-aryloxypropanols and evaluation of their antihyperlipidemic, LDL-oxidation and antioxidant activity. Eur J Med Chem 80:135–144
20. Karmase A, Jagtap S, Bhutani KK (2013) Anti adipogenic activity of *Aegle marmelos* Correa. Phytomedicine 20(14):1267–1271
21. Narender T, Rajendar K, Sarkar S, Singh V, Chaturvedi U, Khanna A et al (2011) Synthesis of novel N-(2-hydroxy-2-p-tolylethyl)-amide and N-(2-oxo-2-p-tolylethyl)-amide derivatives and their antidyslipidemic and antioxidant activity. Bioorg Med Chem Lett 21(21):6393–6397
22. Narender T, Shweta S, Tiwari P, Papi Reddy K, Khaliq T, Prathipati P et al (2007) Antihyperglycemic and antidyslipidemic agent from *Aegle marmelos*. Bioorg Med Chem Lett 17(6):1808–1811
23. Nugroho AE, Agistia DD, Tegar M, Purnomo H (2013) Interaction of active compounds from *Aegle marmelos* CORREA with histamine-1 receptor. Bioinformation 9(8):383–387
24. Nugroho AE, Riyanto S, Sukari MA, Maeyama K (2011) Effects of aegeline, a main alkaloid of *Aegle marmelos* Correa leaves, on the histamine release from mast cells. Pak J Pharm Sci 24(3):359–367
25. Manda VK, Avula B, Chittiboyina AG, Khan IA, Walker LA, Khan SI (2016) Inhibition of CYP3A4 and CYP1A2 by *Aegle marmelos* and its constituents. Xenobiotica [Internet] 46(2):117–125
26. Derf A, Sharma A, Bharate SB, Chaudhuri B (2019) Aegeline, a natural product from the plant *Aegle marmelos*, mimics the yeast SNARE protein Sec22p in suppressing α-synuclein and Bax toxicity in yeast. Bioorg Med Chem Lett 29(3):454–460
27. Avula B, Chittiboyina A, Wang Y-H, Sagi S, Raman V, Wang M et al (2016) Simultaneous determination of Aegeline and six coumarins from different parts of the plant *Aegle marmelos* using UHPLC-PDA-MS and chiral separation of aegeline enantiomers using HPLC-ToF-MS. Planta Med 82(06):580–588
28. Mohammed MMD, Ibrahim NA, El-Sakhawy FS, Mohamed KM, Deabes DAH (2016) Two new cytotoxic furoquinoline alkaloids isolated from Aegle marmelos (Linn.) Correa. Nat Prod Res 30(22):2559–2566
29. Williams C (2003) Hydrosoluble formazan XTT: its application to natural products drug discovery for Leishmania. J Microbiol Methods 55(3):813–816
30. Krupa J, Sureshkumar J, Silambarasan R, Priyadarshini K, Ayyanar M (2019) Integration of traditional herbal medicines among the indigenous communities in Thiruvarur District of Tamil Nadu. India J Ayurveda Integr Med 10(1):32–37

Epigallocatechin-3-Gallate-Loaded Nanocarriers for Health Benefits

Nupur Garg and Farhan Jalees Ahmad

1 Introduction

In recent years, there has been a major shift towards consumption of natural products for prevention or treatment of an ailment or, simply, for benefit of health. Natural components that are rich in bioactive materials like flavonoids, polyphenols, and catechins have been used as ingredients of several supplements and therapies for health sustainability, prevention, and cure of diseases. Green tea, containing all these bioactive components, is the most popular among them and is widely accepted for its health-improving activities. Of the tea catechins, epigallocatechin-3-gallate (EGCG) is widely known and the most abundant constituent especially in green tea constituting more than 50% of the total catechins present [1].

The chemical structure of EGCG contains a flavanol core (flavan-3-ols) along with a gallocatechol group and a gallate ester. The chemical structure is shown in Fig. 1 [2]. These two gallocatechol rings display the potent antioxidant and chelating properties to EGCG [3]. Each of the gallocatechol rings is efficient in capturing free radicals from the environment directly.

EGCG is a biologically active and potential component that has a direct effect on human health and diseases associated. Studies have demonstrated that EGCG has a stronger antioxidant capacity compared with the other green tea catechins, and it is also shown that the efficiency of EGCG is more than that of vitamins E and C in radical scavenging activity [4]. Several preclinical studies have shown its effects as antioxidant, antiaging, antitumorigenic, antidiabetic, antileukemic, anti-Alzheimer, and anti-obesity and in cardiovascular diseases, metabolic health, several types of cancers, atherosclerosis, multiple sclerosis, ulcerative colitis, skin and ocular diseases, etc. [5–11]. They have illustrated in detail the pathways and the mechanisms

N. Garg (✉) · F. J. Ahmad
Nanomedicine Research Lab, School of Pharmaceutical Education & Research, Jamia Hamdard, New Delhi, India
e-mail: nupursgarg_sch@jamiahamdard.ac.in

© Springer Nature Singapore Pte Ltd. 2020

M. Rahman et al. (eds.), *Nanomedicine for Bioactives*,
https://doi.org/10.1007/978-981-15-1664-1_15

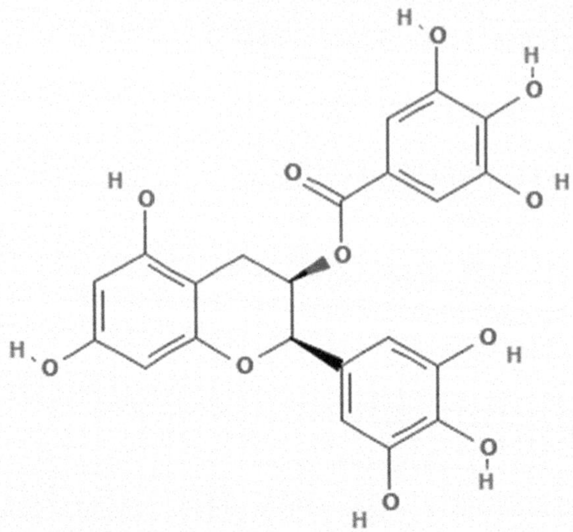

Fig. 1 Chemical structure of epigallocatechin-3-gallate [2]

associated. Numerous clinical studies to evaluate the preventive or therapeutic or both effects are currently undergoing and have established the therapeutic potential of EGCG [6, 10]. Due to this, the study of EGCG is of great importance because this compound seems to be useful in the prevention and treatment of various diseases.

In this chapter, therapeutic potential of the EGCG in different diseases and disorders are reviewed. Also, their practical applications in EGCG-loaded nanocarriers for different therapeutic applications have been considered and explained by reviewing the existing literature.

1.1 Therapeutic Applications of EGCG

Numerous potential mechanisms have been proposed to explain the health benefits of EGCG, including radical scavenging activity, induction of topoisomerase I and II DNA complexes, suppression of adipocyte differentiation, regulation of the tumor-suppressor mRNAs, inhibition of hepatocyte growth factor receptor activity, and inhibition of Iκb kinase activity [7, 12–14]. This implies that this bioactive compound is a part of several receptor pathways and mechanisms that define its role in the disease prevention and treatment. The major therapeutic properties of EGCG are represented diagrammatically in Fig. 2.

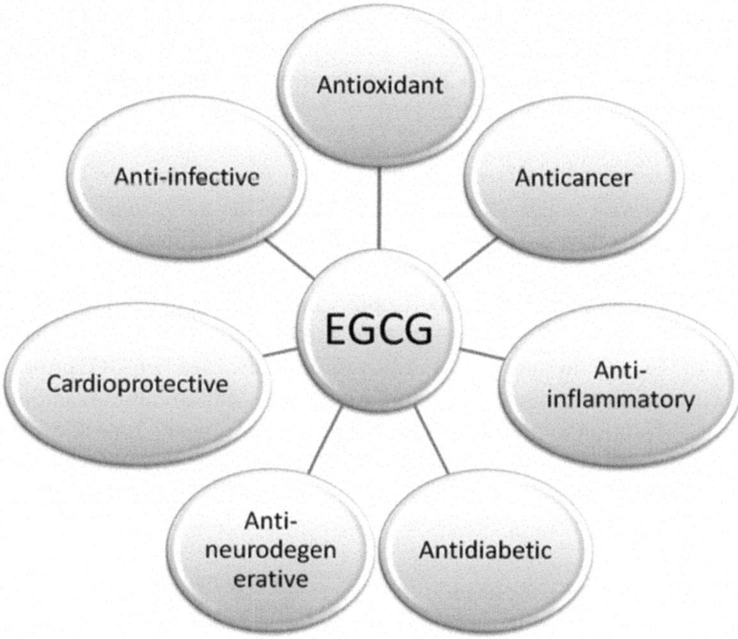

Fig. 2 Therapeutic applications of EGCG

1.1.1 Cancer

The scope of exploring the anticancer property of EGCG has been the focal point for majority of the research. The findings about its antioxidant and radical scavenging properties have shown EGCG to be a promising molecule against cancer. Studies suggest that it can limit the tumor proliferation by inhibiting angiogenesis [13]. EGCG effectively interacts with specific targets that are involved in the regulation of crucial steps of cell proliferation, survival, and metastatic spread. The targets involve several proteins like the transmembrane receptor 67LR, Pin1, TGFR-II, and metalloproteinases (MMPs) which regulate EGCG-dependent inhibition of nuclear factor-kappaB (NF-κB) and cellular invasion [1]. Shankar et al. found that EGCG inhibits tumor growth, angiogenesis, and metastasis in pancreatic cancer that are results of inhibition of PI3K/AKT and ERK pathways and activation of FKHRL1/FOXO3a [14]. Previous studies report that EGCG promotes a reduction in the migration and metastatic state of tumor accomplishing a more reliable and efficient chemotherapy [15]. As EGCG possess anti-inflammatory and antioxidant effects, it can also be useful in combination therapies to reduce the side effects.

1.1.2 Antioxidant Effect

The chemical structure of EGCG itself suggests its antioxidant potential as the presence of phenol rings acts as electron acceptors, hence, free radical scavengers, which prevent the production of reactive oxygen species (ROS), thereby reducing the effect of oxidative stress [3]. They chelate the metal ions present in the body, especially the iron ion, which produce free radicals, hence reducing their production. It is reported that EGCG can effectively inhibit protein tyrosine nitration induced by oxidative stress in blood platelet and can improve the function of mitochondria reducing the oxidative DNA damage [16].

1.1.3 Cardiovascular Diseases

EGCG exhibits a wide range of therapeutic properties including anti-atherosclerosis, anticardiac hypertrophy, and anti-myocardial infarction. These therapeutic effects are mainly associated with the inhibition of LDL cholesterol (anti-atherosclerosis), inhibition of NF-κB (anticardiac hypertrophy), and inhibition of MPO activity (anti-myocardial infarction) [17]. Several in vitro and in vivo studies have demonstrated the effect of EGCG in reducing cardiovascular disorder. EGCG improves the systemic circulation and dilates the capillaries, lowers inflammation, and interferes with the fat absorption and digestion. EGCG can also reduce cholesterol and initiates its excretion through the bile [18, 19]. This catechin can also modulate the process of platelet formation, from macrophage recruiting to macrophage uptake of cholesterol [20].

1.1.4 Neurological Diseases

The common clinical features of neurological diseases like Parkinson's disease (PD) and Alzheimer's disease (AD) may attribute towards the oxidative damage of neurons and accumulation of iron in cerebral areas [21]. The neuroprotective effect of the EGCG relates to its antioxidant, anti-inflammatory, and iron chelating properties [19]. Moreover, EGCG, unlike other compounds, is permeable to the blood-brain barrier (BBB) [18]. In neuronal cells, this compound enhances cell survival stimulation and inhibits the cell death signals, leading to an improvement in neuronal health. Studies also confirm that EGCG promotes the degree of cognition, lowers anxiety, and induces an overall increase in the cerebral activity and calmness [22].

1.1.5 Diabetes

EGCG is said to prevent or treat diabetes mellitus by several therapeutic effects like increasing insulin secretion, inhibiting insulin resistance, regulating glucose uptake, and maintaining glucose level through multiple pathways that signal each role [23]. EGCG is shown to resist starch hydrolysis by binding to the active site of α-amylase and α-glucosidase. EGCG was given in high-fat diet and streptozotocin (STZ)-induced type-2 diabetes that improved the glucose homeostasis, and through PXR/CAR receptor-mediated activation, it slowed the process of gluconeogenesis and lipogenesis in the liver [24]. Treatment of HepG2 cells with EGCG has shown to improve insulin resistance by increased glucose uptake and decreased glucose content. It also stimulates the glucose transporter 2 (GLUT2) protein and its downstream proteins, peroxisome proliferator-activated receptor coactivator (PGC)-1β, in EGCG-treated human hepatoma cells [25].

1.1.6 Inflammatory Disorders

Inflammatory conditions like rheumatoid arthritis, ulcerative colitis, and osteoporosis have degenerative effect on body due to bone and cartilage erosion and fragility of tissues and joints. Studies have shown that current therapies using anti-inflammatory drugs combined with analgesics are inefficient in some cases and tend to lose efficacy in body with time [26]. Therefore, new therapies are explored, and EGCG may act as a potential candidate due to its potent antioxidant and anti-inflammatory activities. In cartilage cell cultures, EGCG is reported to show specific inhibition COX-2 expression and activity induced by IL-1β that stimulates the nitric oxide synthase [27].

1.1.7 Obesity

Green tea catechins are widely popular for their weight loss effects, although improving the lifestyle, including diet modification, and stress-free living are considered as the most effective treatments of obesity [28]. However, drugs and supplements may be used in some cases to help in weight reduction as obesity may pose negative effects on health leading to other conditions like diabetes and arteriosclerosis. As stated earlier, EGCG directly acts on the lipid digestion process by inhibiting the phospholipase A2 which is highly relevant in the management of weight [29]. Moreover, EGCG enhances the lipid metabolism, consequently leading to fat burn. EGCG can also interfere with the digestion of starch by inhibition of α-amylase [30]. Besides that, the ingestion of EGCG during a weight loss program is very useful because its administration is strongly linked with improving circulation, reactive oxygen species or free radical scavenging, and mood enhancement [22].

1.2 Limitations of EGCG as Potential Drug Candidate

In vitro studies have shown that the effective concentrations of EGCG range from 1 to 100 μmol/L. However, the peak plasma levels of tea catechins were usually in the low micromolar range in human subjects or animals after the oral administration of EGCG [31, 32], which happened to have lower effective concentration in in vitro tests. This causes limited low gastrointestinal tract stability and limited intestinal membrane permeability and, hence, limited therapeutic efficacy [33, 34]. Also, its therapeutic potential is limited due to poor systemic absorption after oral administration, including low absorption and bioavailability, less biodistribution, high first-pass metabolism, and low tissue accumulation, hence, low targeting efficacy. The inconsistency between in vitro biological activity and low in vivo absorption studies can also be caused due to its low metabolic stability. EGCG is chemically unstable under physiological conditions and metabolized rapidly through conjugation biotransformation [35]. This means it is highly sensitive towards heat and light and has short half-life. Also, EGCG shows its effect at relatively high concentration at the target-related molecular site. It is shown that taking EGCG content equivalent to 8–16 cups of green tea daily may mitigate the poor bioavailability of EGCG [20, 21]. Daily intakes as high as 800 mg of EGCG extract for up to 4 weeks have been used safely and effectively. This implies that systemic availability of EGCG increases at higher doses; however, it is never recommended to use excessive dose of catechins for improving the efficacy [36]. Despite being antioxidants, they can also generate reactive oxygen species (ROS) [37]. EGCG supplements have been associated with serious side effects, such as liver and kidney failure, dizziness, low blood sugar, and anemia.

1.3 Role of Efficient Delivery Systems

In short, EGCG exhibits some major limitations including low intestinal permeability, degradation during processing or within the gastrointestinal tract, and alkaline conditions, leading to decreased bioavailability. For these reasons, many studies have tried to improve functionality of EGCG and overcome its limitations through delivery systems, to increase the stability and bioavailability. Nanocarriers are believed to be a novel drug delivery system that can be engineered to harness site-specific targeting of drugs, to have a high drug-loading capacity, demonstrating enhanced pharmacokinetics and safe and effective drug delivery, hence, bioavailability of therapeutics.

2 Nanocarriers as Delivery Systems

Nanocarriers are colloidal drug carrier systems having particle size in the nanoscale range typically <500 nm [38]. Nanocarriers have been extensively investigated in the past few decades as they showed great promise in the area of drug delivery. Nanocarriers have a very high surface-area-to-volume ratio, due to which they possess distinct physicochemical properties and bioactivity of drugs. This results in improved pharmacokinetics and biodistribution parameters, improved solubility and stability, decreased toxicities, controlled release, and site-specific delivery of therapeutic. In addition, the composition, shapes, sizes, surface properties, etc. can be modified to obtain desired physiochemical properties [39]. The main aim of drug-loaded nanocarriers is target-specific delivery to ensure maximum therapeutic effect and least side effects to treat a disease effectively. In the case of EGCG, use of nanotechnology plays a crucial role in overcoming the limitations as a bioactive molecule. Several studies reported the sustained release and enhanced bioavailability of EGCG-loaded nanocarriers at much lower doses than conventional formulations. Carriers like lipids [40], proteins [41], and carbohydrate [42] can be used for encapsulation of EGCG and show positive effects on the bioavailability of EGCG, by enhancing solubility, preventing its degradation in the intestine, and increasing the permeation, thereby increasing concentration in the blood stream. Several nanosystems have been used in EGCG delivery including polymeric nanoparticles (NPs); solid-lipid nanoparticles (SLNs); liposomes; metallic nanoparticles including gold, platinum, palladium, and iron nanoparticles; protein-/peptide-based nanocarriers; and inorganic nanocarriers that have proved to be more potent and efficient [38].

2.1 Lipid Nanocarriers

The bioavailability of an orally administered drug majorly depends on its solubility in the GIT and its permeability across cell membranes. Many drugs, like EGCG, having fair aqueous solubility still have low bioavailability reason being poor permeability. The only way to increase the permeability of drugs so as to make them reach the tissues is to entrap them in lipid-based nanocarriers. Lipid-based nanocarriers are composed of physiological lipids that are biocompatible and biodegradable; hence, they are well tolerated and nontoxic [43]. For the past decade, researchers have focused their studies on new approaches of drug delivery and targeting using novel lipid-based nanocarriers including liposomes, niosomes, ethosomes, transfersomes, solid-lipid nanoparticles (SLNs), nanostructured lipid carriers (NLCs), and lipid nanoemulsions for safe and effective delivery of drugs. For natural drugs like EGCG, many studies have been reported that use lipids as nanocarriers for enhanced, targeted, and safe effects. Few of such studies are compiled in Table 1.

There are two types of lipid nanocarriers: solid-lipid nanoparticles (SLNs) and nanostructured lipid carriers (NLCs). Solid-lipid nanoparticles commonly use the solid lipid as a matrix material for drug encapsulation. A variety of lipids can be selected depending on the compatibility and target of the drug, from monoglycerides to triglycerides and lipid acids. However, the crystallized firm structure of SLN matrix generally leads to limited drug-loading capacity and even drug expulsion. To improve this, nanostructured lipid carriers (NLCs) are prepared by introducing a liquid lipid into the solid matrix. Liquid lipids such as oleic acid, miglyol, and several oils provide more efficient loading [51]. For EGCG nanocarrier formulations, both lipid nanocarriers have been successfully tried to extract the desired therapeutic effect of EGCG.

Dzulhi et al. [44] formulated EGCG-loaded SLNs using the solvent emulsification evaporation method and reported an in vitro increased penetration through stratum corneum for cosmetic purposes. Fangueiro et al. [45] developed cationic lipid nanoparticles using cetyltrimethylammonium bromide (CTAB) as the cation and evaluated the biopharmacokinetic parameters of the formulation for ocular pathology related to antioxidant and anti-inflammatory pathways. In vitro, in vivo, and ex vivo studies showed prolonged release, nonirritancy, safety, and better permeation to anterior and posterior segments of the eye. Shtay et al. evaluated the gut stability and storage conditions of EGCG SLNs by using food-grade lipids and found that the drug is protected under adverse conditions and have prolonged stability. Zhang et al. published two studies [46, 47] that used NLCs loaded with EGCG to increase the gut stability and to induce site-specific anti-inflammatory effect by increasing accumulation in macrophages. Smith and coworkers reported the significant enhancement of EGCG's oral bioavailability by lipid nanoparticles in vivo

Table 1 Lipid nanocarriers for EGCG

Nanocarrier	Particle size (nm)	In vitro/in vivo results	Reference
SLN (glyceryl monostearate, lecithin, poloxamer 188)	155 ± 10.04	Increased penetration in skin layer	Dzulhi et al. [44]
Cationic lipid nanocarrier (soy phosphatidylcholine, poloxamer 188, transcutol P)	<150	Enhanced rabbit eye permeability, nonirritancy, and safety, in vitro, in vivo	Fangueiro et al. [45]
SLN (cocoa butter, Na stearoyl-2-lactylate, and glycerides)	108–122	Food grade with prolonged storage period under adverse conditions	Shtay et al. [38]
NLCs (glyceryl tridecanoate, glyceryl tripalmitate, soy lecithin and Kolliphor HS15, and chitosan)	50	High stability in both acidic and neutral environments. In vitro studies performed in THP-1-derived macrophages showed a decrease in inflammation and accumulation of cholesterol	Zhang et al. [46]
NLCs (phosphatidylcholine, Kolliphor HS15, and alpha-tocopherol acetate)	108	Increased EGCG stability. Enhanced accumulation inside macrophages and decrease in the production of MCP-1	Zhang et al. [47]
Phosphatidylcholine, phosphatidylethanolamine, phosphatidic acid, and phosphatidylinositol	30–80	Enhanced EGCG oral bioavailability in vivo and induction of α-secretase activity in neuronal cells in vitro	Smith et al. [48]
NLCs (cetyl palmitate/ phospholipon 80, sesame oil, and Tween-80)	126–167	Photodegradation of EGCG under UVA radiation	Chen et al. [49]
SLN (glycerol monostearate, stearic acid, soya lecithin, and Pluronic F68)	157	Enhanced stability in physiological fluids. Increased induction of cell death in breast cancer cells MDA-MB-231 and prostate cancer cells DU-145	Radhakrishnan et al. [50]

compared to free EGCG and increased systemic absorption. EGCG-loaded nanocarriers were also able to promote α-secretase activity in neuronal cells, thereby proving to have an important role in neurodegenerative diseases [48]. Chen et al. [49] developed lipid nanoparticles for EGCG encapsulation, for skin care applications for UVA radiation protection. Radhakrishnan and coworkers produced SLN for delivery of EGCG to enhance its stability at physiological conditions, sustained release, and anticancer effects. In vitro studies using breast cancer cells MDA-MB-231 and prostate cancer (PCa) cells DU-145 resulted in a sharp decrease in cell viability and higher levels of apoptosis as compared to free EGCG.

Table 2 Liposomes as EGCG carrier

Components	Particle size (nm)	Study results	Reference
Mg-poloxamer modified	201.9 ± 9.4	Superior antioxidant activity	Minnelli et al. [53]
Lecithin-cholesterol-Tween 80	111.10 ± 0.52	Increased antioxidant activity	Chen et al. [54]
Soy PC-cholesterol	105 ± 13	Increased dermal intracellular uptake	Marwah et al. [55]
Dextran-sulfate-DCMC	189.8	In vitro digestion stability	Zou et al. [56]
Egg lecithin and cholesterol	89–93	Improved in vitro and in vivo anti-MRSA activity	Karger Publishers [57]
PhC and cholesterol	180	Desired release profile in the gastrointestinal fluids	Luo et al. [58]
Sorbitan monostearate and cholesterol	100	Higher stability in neutral pH and higher Caco-2 cell permeability	Song et al. [59]
Cholesterol, phosphatidylcholine, and Tween-80	126.7 ± 4.3	Synergistic effects between EGCG and paclitaxel in inducing apoptosis in MDA-MB-231 cells	Ramdass et al. [60]

2.2 Nanoliposomes

Liposomes are amphiphilic spherical vesicles formed of hydrophobic phospholipid bilayer surrounding an aqueous core. Liposomes are, therefore, versatile natural systems and can entrap and distribute different lipophilic and hydrophilic drugs. The surface modifications, flexibility, and size-adjustable nature of liposomes to reach the specific target sites provide a suitable method of delivery [39]. Burst release of drug from liposomes occurs with the help of environmental factors, such as temperature, pH, light, and enzymes through modifications on the cell membrane that cause the disruption of membrane integrity [52]. Despite the potential of liposomes as nanocarriers, few studies have utilized them for EGCG delivery for therapeutic applications which are compiled in Table 2.

Minnelli et al. prepared modified EGCG anionic nanoliposomes using poloxamer 407 in order to maximize EGCG internalization and its metabolic stability so as to achieve the concentration effective enough to act as antioxidant in stress and oxidative damage [53].

Recently, Chen and coworkers coencapsulated EGCG and quercetin to obtain synergistic and optimized antioxidant effect through in vitro assay [54]. Studies by Marwah on a deformable liposomal preparation of EGCG deliver a formulation that passes through skin to increase the uptake by dermal cells in treatment of skin cancer by using EGCG as potent compound. They demonstrated the localization of liposomes intracellularly within human dermal fibroblast and keratinocyte cells [55]. Gharib et al. formulated EGCG-loaded nanoliposomes using egg lecithin and cholesterol and evaluated their activity against methicillin-resistant *Staphylococcus aureus* (MRSA). The results showed enhanced effect against MRSA burn wound infections compared to free EGCG both in vitro and in vivo. It established that EGCG is an efficient antibacterial agent [57]. Zou et al. developed a dextran sulfate-coated chitosan nanoliposomal delivery system that gave a good in vitro stability in gastrointestinal conditions which has been a major drawback with EGCG formulations [56]. Luo et al. [58] synthesized liposomes made of phosphatidylcholine (PhC) and cholesterol with a view to prevent carcinogenesis. The liposome demonstrated high stability in simulated gastric conditions and enhanced inhibitory effect of EGCG on tumor cell viability at a higher concentration. Song et al. developed liposomes using nonionic surfactants and cholesterol to improve the gastric absorption of EGCG. They also investigated the permeability of Caco-2 cell monolayers to the EGCG-loaded liposomes compared with free EGCG and found an increased apparent permeability of EGCG. The formulation exhibited better stability and lower toxicity [59]. These promising results encourage the investigation of a liposome-based nanoformulation to improve the bioavailability of EGCG by increasing its intestinal permeability. Another study by Ramadass et al. [60] co-loaded cholesterol and phosphatidylcholine liposome with EGCG and paclitaxel for breast cancer therapy. The nanoformulation had a synergistic effect between the two compounds and promoted breast cancer cell apoptosis in MDA-MB-231 cells.

2.3 Polymeric Nanocarriers

Among several types of nanocarriers, polymeric nanocomplexes have widely been employed for target delivery of the drug. The major reason is the ease of production, versatile nature, compatibility, flexibility in modifications, nontoxic, controlled and target-specific drug release, and availability of wide range of polymers. They can be formed using both natural and synthetic polymers [61]. In polymeric nanocarriers, the drug is entrapped in the matrix so that it can be protected from the outside environment. The outer layer of the polymer can be modified and functionalized with markers to target some specific site or to improve the pharmacokinetic profile of the drug, therefore representing an optimal alternative for the administration of drugs. These nanosystems show a high loading capacity of chemically different drugs [62]. Recent studies report the use of polymeric nanocarriers to deliver EGCG in vitro and in vivo with encouraging results as summarized in the Table 3.

Several biocompatible polymers have been employed to formulate epigallocatechin-3-gallate-loaded nanocarriers by researchers. Chitosan is one such polymer which is widely used to develop nanocarriers for a number of indications like neurodegenerative diseases, inflammatory disorders, chemoprevention, and what not. Hong et al. [63] synthesized self-assembled EGCG nanoparticles using chitosan and polyaspartic acid for studying the effectiveness of EGCG against rabbit atherosclerosis. Most recently, Cano et al. [70] have explored another therapeutic application of EGCG, to develop EGCG nanoparticles for anti-epilepsy therapy. They used PEGylated poly(lactic-co-glycolic acid) (PLGA) for protection and effective and sustained brain delivery of EGCG. Studies showed reduction in neuroinflammation and number of epileptic seizures. Sanna et al. used biocompatible

Table 3 Polymeric nanocarriers for EGCG

Components	Particle size (nm)	Study results	Reference
Chitosan and PAA	100	Higher stability in body fluids and improved in vivo activity against atherosclerosis	Hong et al. [63]
PLGA-PEG-ligand copolymer	209.13 ± 1.74	Higher affinity towards target antigen on prostate cancer cells in vivo	Sanna et al. [64]
Folic acid and PEG	185.0 ± 13.5 142.7 ± 7.2	High stability, inhibited MCF-7 cell proliferation	Zeng et al. [65]
PEGylated PLGA EGCG/AA NPS	124.8 ± 5.2	Enhanced therapeutic efficacy in Alzheimer's disease mice model	Cano et al. [66]
PLGA-PVA	91.3 ± 0.8	Lower proteinuria excretion of nephritic syndrome in mice	Zhang and Zhang [67]
PLGA-PVA	317.8	Reduction of neurobehavioral deficits in Alzheimer's pathology in rats	Singh et al. [68]
Fucose-chitosan/gelatin	244 ± 8	Inhibition of *H. pylori* growth and gastric inflammation	Lin et al. [69]
PEGylated PLGA	168.5 ± 9.9	Effective anticonvulsant activity in vivo	Cano et al. [70]
Gelatin-hyaluronic acid	<350	Anti-inflammatory effect on rabbit dry-eye syndrome	Huang et al. [71]
PLGA	<1 µm	Enhanced efficacy in inflammation treatment	Wu et al. [72]

polymers PLGA-PEG, for EGCG loading functionalized with a specific ligand that targets prostate-specific membrane antigen (PSMA), a well-known transmembrane protein that is overexpressed on prostate cancer (PCa) epithelial cells. The ligand-based EGCG nanocarrier specifically binds PSMA leading to enhanced antiproliferative activity, thus increasing the effectiveness of EGCG in vitro and in vivo [64]. In another study, Zeng et al. [65] synthesized two types of particles using folic acid and PEG polymers with EGCG with folic acid as outer shell to improve the targeting ability for recognizing folate receptors that are overexpressed in cancer cells. In vitro studies using MCF-7 cells showed the increased effectiveness of EGCG nanoparticles compared to free EGCG by reduction in cell number and enhanced cellular uptake. Singh et al. [68] used PLGA nanoparticles for EGCG loading to study the effect on Alzheimer's disease pathology in aluminum chloride-induced neurodeficit behavior in rats. The results showed the biochemical, immunohistochemical, and protein levels were greatly reduced in nanoEGCG-treated rat. Similar study done by Cano et al. [66] used similar EGCG nanocarrier for oral treatment of APPswe/PS1dE9(APP/PS1) mice model of Alzheimer's disease. The morphological changes and enhanced spatial learning and memory in mice proposed the stabilization of EGCG in NPs, and a destabilized blood-brain barrier accounted to enhanced therapeutic EGCG concentrations in the brain. A study by Zhang and Zhang [67] was done to improve the oral bioavailability of EGCG by using pH-sensitive polymer complex containing poly(lactic-co-glycolic acid) (PLGA) and poly(vinyl alcohol)

(PVA) for demonstrating the effect on chronic kidney disease. The results showed that the nephrotic syndrome in rats treated with EGCG NPs was significantly lower than those treated with EGCG powder, and kidney pathology scores in EGCG NP-treated rats were also significantly lower than EGCG powder-treated rats. Another therapeutic potential of EGCG as anti-inflammatory agent is explored by Huang et al. [71] and Lin et al. [69] who used gelatin-based polymers for enhanced effect in eye disorders in rabbit model and in gastric disorders against *Helicobacter pylori* as antibacterial, respectively.

These studies effectively suggest the use of polymers as EGCG nanocarriers as the most efficient, safe, target-specific, flexible, and easy-to-prepare method to deliver EGCG to its full potential to a variety of disorders and conditions.

2.4 Metallic Nanocarriers

Metallic nanoparticles find immense applications in the biomedical and pharmaceutical area. Metallic nanoparticles like gold, silver, selenium, and iron oxide are considered a better option than other nanoparticle formulations because of their biocompatibility, eco-friendly, easy preparation, chemical stability, unique optical properties, and ease of surface modification as they can bind to variety of biomolecules. Furthermore, it is shown that conjugation of ligands to metallic NPs by interactions increases their stability, by protecting them from enzymatic degradation [73]. Gold nanoparticles are made from a gold core encapsulated in a film of functional ligands, which are capable of protecting and transporting drugs and targeting molecules. Several studies, as presented in Table 4, have described gold nanoparticles as effective delivery vehicles of EGCG. EGCG has a capacity to combine with metals, forming complexes by the gallate ring. Therefore, it is feasible to synthesize

Table 4 Metallic nanocarrier-loaded EGCG

Nanocarrier	Particle size (nm)	Study results	Reference
EGCG-AuNP	25	Increased gut stability and intracellular uptake, inhibition of NF-κB growth of prostate tumor	Chavva et al. [73]
EGCG-Au-MTX/dox nanoclusters	50	Target delivery to FA receptors overexpressed cancer cells in vitro and in vivo	Wu et al. [74]
EGCG-gelatin/dox gold conjugates	35–85	Enhanced cellular uptake and improved apoptosis in PC-3 cells	Tsai et al. [75]
Gold	68	High cytotoxicity ex vivo	Chen et al. [76]
Selenium	29	Increased inhibition of Aβ fibrillation and protection of PC-12 against Aβ-mediated toxicity	Zhang [77]
B6-ME-NP SPIONS	80	Enhanced accumulation in PD lesions and increased dopaminergic neurons	Li et al. [78]

its metal nanocarrier complex [79]. Chavva et al. [73] formulated EGCG-gold nanoparticles (AuNPs) of around 25 nm size that cause an increased intracellular uptake in case of prostate tumor. AuNPs at this size give reddish color, while an increased size turns the colloid blue due to the red shift in the wavelength. This is an added advantage for AuNPs to detect the relatable size along with its own tumor selective chemotoxicity. In vivo studies by Chavva and team had shown the inhibition in growth of prostate tumor through several signaling pathways including inhibition of the nuclear translocation and transcriptional activity of nuclear factor-kappaB (NF-κB). A similar study by Tsai et al. [75] reflects the use of EGCG-gelatin/doxorubicin (dox) conjugate-coated gold nanoparticles that are formed in layered assembly, for prostate cancer treatment. This complex helped in fluorescence imaging and inhibition of prostate cancer cell growth in vivo. Recently, Wu et al. [74] biosynthesized a nanocluster of dual-drug-loaded EGCG and methotrexate/doxorubicin-coated AuNPs. These flower-shaped nanoclusters have targeted synergistic effects as shown in the in vitro and in vivo studies. They could effectively target and kill cancer cells through identifying folic acid receptor overexpression. Zhang et al. [77] attached EGCG on the surface of selenium nanoparticles coated with Tet-1 peptide to target the pathway included in Alzheimer's disease. This nanocomplex can inhibit Aβ fibrillation into nontoxic aggregates. Also, Tet-1 peptides can significantly improve the cellular uptake of the nanocarrier in PC-12 cells responsible for the signaling pathways. More recently, superparamagnetic iron oxide nanocubes (SPIONs) were loaded in the nanoparticles for diagnostic purpose to trace the accumulation of EGCG in the brain in case of Parkinson's disease. This study was done by Li and coworkers [78]. It was shown that the amount of EGCG accumulation in the lesions was markedly enhanced and traced via magnetic resonance imaging, and most importantly, the dopaminergic neurons were increased.

2.5 Miscellaneous Nanocarriers

Apart from the nanocarriers described above, researchers have explored other different types of delivery vehicles for EGCG. They have tried to develop conjugates, dendrimers, peptides, transferosomes, etc. to obtain desired therapeutic efficacy with maximum stability and potency and minimum toxicity and dose. For example, Avadhani et al. [80] developed transferosomes of EGCG with hyaluronic acid to obtain a synergistic protective effect from UV radiations and to minimize skin damage. The preparation proved to be effective to be used in sunscreen creams to prevent skin aging and damage. Another study by Shetty and team [81] used peptide dendrimers for transdermal delivery of EGCG and evaluated their ability to permeate through skin for its antioxidant property. Shafiei et al. [82] formulated nanohybrids of EGCG with Ca/Al NO_3 double hydroxides as inorganic matrix for the effective delivery of EGCG to PC-3 cells present in tumor with controlled release in vitro.

3 Conclusion

The researchers have time and again showed that green tea catechins, especially EGCG, possess solutions to several health problems including various types of cancers, cardiovascular disorders, neurodegenerative diseases like Alzheimer's and Parkinson's diseases, inflammatory conditions, diabetes, and most importantly oxidative damage that is thought to be a common step in the pathways of several other disorders. Being a natural compound, it is nontoxic and has no side effects per se. Also, it is having good solubility. Attributed to its several physicochemical limitations like low stability, high metabolism, low permeation, and low site specificity, hence, low bioavailability and need of large doses to reach therapeutic concentration [83], there is a need of delivery system that can exploit and control the potential of EGCG by protecting it from degradation and metabolism and making it reach target site through site-specific functionalized delivery. Nanocarriers, providing the same, make excellent delivery vehicles that have an added advantage of their nanoscale size by which they can easily travel and permeate to the delivery site. Different types of nanocarriers like lipid nanoparticles, liposomes, polymeric nanoparticles, and metallic nanoparticles have been tried successfully for loading EGCG to obtain desired therapeutic efficacy against several diseases both in vitro and in vivo. Most of the nanocarriers like lipid and polymeric nanocarriers are biocompatible, hence, nontoxic and have several beneficial properties like flexibility for modifications, controlled and sustained release of drug, protection from degradation, and what not. Nanoparticles have proven to be promising delivery vehicles for natural products specifically EGCG with better bioavailability and stability results. To summarize, nanocarriers could efficiently enhance the therapeutic potential of EGCG in the treatment of numerous disorders including cancer and neurodegenerative, metabolic, skin, cardiovascular, and ocular diseases.

4 Future Perspectives

Nanotechnology comes with a major limitation, its cost of fabrication and application. To make the research reach commercial aspect, it is important to lower the cost of nanocarriers as delivery vehicles. Lipid nanocarriers, being cheap and efficient, show promising future as delivery vehicles. Such type of nanocarriers can have high loading capacity which is important in reduction of cost and toxicity. This is due to the nature of this hydrophilic molecule to combine with the lipids. Also, high bioavailability is inversely proportional to dose required. The drug-loaded nanocarriers have marked the importance of nanotechnology in clinical applications, and EGCG combined with the nanosystems holds a promising future for advancements in the therapeutic applications of natural products.

References

1. Negri A, Naponelli V, Rizzi F, Bettuzzi S (2018) Molecular targets of epigallocatechin—gallate (EGCG): a special focus on signal transduction and cancer. Nutrients 10(12). https://doi.org/10.3390/nu10121936
2. PubChem (-)-Epigallocatechin gallate [Online]. Available https://pubchem.ncbi.nlm.nih.gov/compound/65064. Accessed 26 June 2019
3. Braicu C, Ladomery MR, Chedea VS, Irimie A, Berindan-Neagoe I (2013) The relationship between the structure and biological actions of green tea catechins. Food Chem 141(3):3282–3289
4. Rice-Evans CA, Miller NJ, Bolwell PG, Bramley PM, Pridham JB (1995) The relative antioxidant activities of plant-derived polyphenolic flavonoids. Free Radic Res 22(4):375–383
5. Chu C et al (2017) "Green tea extracts epigallocatechin-3-gallate for different treatments: Figure 1." [Online]. Available https://www.hindawi.com/journals/bmri/2017/5615647/fig1/. Accessed 16 June 2019]
6. Du G-J et al (2012) Epigallocatechin gallate (EGCG) Is the most effective cancer chemopreventive polyphenol in green tea. Nutrients 4(11):1679–1691
7. Higdon JV, Frei B (2003) Tea catechins and polyphenols: health effects, metabolism, and antioxidant functions. Crit Rev Food Sci Nutr 43(1):89–143
8. Khan N, Mukhtar H (2018) Tea polyphenols in promotion of human health. Nutrients 11(1). https://doi.org/10.3390/nu11010039
9. Machado ECFA, Ambrosano L, Lage R, Abdalla BMZ, Costa A (2017) Nutraceuticals for healthy skin aging. In: Nutrition and functional foods for healthy aging. Elsevier, Amsterdam, pp 273–281
10. Salimi A, Pourahmad J (2018) Role of natural compounds in prevention and treatment of chronic lymphocytic leukemia. In: Polyphenols: prevention and treatment of human disease. Elsevier, Amsterdam, pp 195–203
11. Wolfram S (2007) Effects of green tea and EGCG on cardiovascular and metabolic health. J Am Coll Nutr 26(4):373S–388S
12. Granja A, Frias I, Neves AR, Pinheiro M, Reis S (2017a) Therapeutic potential of epigallocatechin gallate nanodelivery systems. BioMed Res Int 2017:5813793
13. "Retracted: Epigallocatechin-3-gallate inhibits photocarcinogenesis through inhibition of angiogenic factors and activation of CD8+ T cells in tumors - Mantena - 2005 - Photochemistry and Photobiology - Wiley Online Library." [Online]. Available https://onlinelibrary.wiley.com/doi/abs/10.1562/2005-04-11-RA-487. Accessed 20 June 2019
14. Shankar S, Marsh L, Srivastava RK (2013) EGCG inhibits growth of human pancreatic tumors orthotopically implanted in Balb C nude mice through modulation of FKHRL1/FOXO3a and neuropilin | SpringerLink." [Online]. Available: https://link.springer.com/article/10.1007/s11010-012-1448-y. Accessed 20 June 2019
15. Granja A, Frias I, Neves AR, Pinheiro M, Reis S (2017b) Therapeutic potential of epigallocatechin gallate nanodelivery systems. Biomed Res Int 2017:1–15
16. Meng Q (2008) Regulating the age-related oxidative damage, mitochondrial integrity, and antioxidative enzyme activity in Fischer 344 rats by supplementation of the antioxidant epigallocatechin-3-gallate | Rejuvenation Research. [Online]. Available: https://www.liebertpub.com/doi/10.1089/rej.2007.0645. Accessed 20 June 2019
17. Eng QY, Thanikachalam PV, Ramamurthy S (2018) Molecular understanding of epigallocatechin gallate (EGCG) in cardiovascular and metabolic diseases. J Ethnopharmacol 210:296–310
18. Giunta B et al (2010) Fish Oil enhances anti-amyloidogenic properties of Green Tea EGCG in Tg2576 mice. Neurosci Lett 471(3):134–138
19. Lee JW et al (2009) Green tea (−)-epigallocatechin-3-gallate inhibits beta-amyloid-induced cognitive dysfunction through modification of secretase activity via inhibition of ERK and NF-kappaB pathways in mice. J Nutr 139(10):1987–1993

20. Koo SI, Noh SK (2007) Green tea as inhibitor of the intestinal absorption of lipids: potential mechanism for its lipid-lowering effect. J Nutr Biochem 18(3):179–183
21. Skovronsky DM, Lee VM-Y, Trojanowski JQ (2006) Neurodegenerative diseases: new concepts of pathogenesis and their therapeutic implications. Annu Rev Pathol 1:151–170
22. Scholey A et al (2012) Acute neurocognitive effects of epigallocatechin gallate (EGCG). Appetite 58(2):767–770
23. Park J-H, Bae J-H, Im S-S, Song D-K (2014) Green tea and type 2 diabetes. Integr Med Res 3(1):4–10
24. Li X, Li S, Chen M, Wang J, Xie B, Sun Z (2018a) (-)-Epigallocatechin-3-gallate (EGCG) inhibits starch digestion and improves glucose homeostasis through direct or indirect activation of PXR/CAR-mediated phase II metabolism in diabetic mice. Food Funct 9(9):4651–4663
25. Zhang Q et al (2018) "Epigallocatechin gallate improves insulin resistance in HepG2 cells through alleviating inflammation and lipotoxicity," *Diabetes Res.* Clin Pract 142:363–373
26. Firestein GS (2003) Evolving concepts of rheumatoid arthritis | Nature. [Online]. Available https://www.nature.com/articles/nature01661. Accessed 22 June 2019
27. Singh R, Ahmed S, Islam N, Goldberg VM, Haqqi TM (2002) Epigallocatechin-3-gallate inhibits interleukin-1β–induced expression of nitric oxide synthase and production of nitric oxide in human chondrocytes: suppression of nuclear factor κB activation by degradation of the inhibitor of nuclear factor κB. Arthritis Rheum 46(8):2079–2086
28. Thielecke F, Boschmann M (2009) The potential role of green tea catechins in the prevention of the metabolic syndrome – a review. Phytochemistry 70(1):11–24
29. Moon H-S, Lee H-G, Choi Y-J, Kim T-G, Cho C-S (2007) Proposed mechanisms of (−)-epigallocatechin-3-gallate for anti-obesity. Chem Biol Interact 167(2):85–98
30. Forester SC, Gu Y, Lambert JD (2012) Inhibition of starch digestion by the green tea polyphenol, (−)-epigallocatechin-3-gallate. Mol Nutr Food Res 56(11):1647–1654
31. Lambert JD et al (2003) Epigallocatechin-3-gallate is absorbed but extensively glucuronidated following oral administration to mice. J Nutr 133(12):4172–4177
32. Narumi K et al (2014) Simultaneous detection of green tea catechins and gallic acid in human serum after ingestion of green tea tablets using ion-pair high-performance liquid chromatography with electrochemical detection. J Chromatogr B Analyt Technol Biomed Life Sci 945–946:147–153
33. Cai Y, Anavy ND, Chow H-HS (2002) Contribution of presystemic hepatic extraction to the low oral bioavailability of green tea catechins in rats. Drug Metab Dispos 30(11):1246–1249
34. Zhang L, Zheng Y, Chow MSS, Zuo Z (2004) Investigation of intestinal absorption and disposition of green tea catechins by Caco-2 monolayer model. Int J Pharm 287(1–2):1–12
35. Krupkova O, Ferguson SJ, Wuertz-Kozak K (2016) Stability of (−)-epigallocatechin gallate and its activity in liquid formulations and delivery systems. J Nutr Biochem 37:1–12
36. Chow H-HS et al (2003) Pharmacokinetics and safety of green tea polyphenols after multiple-dose administration of epigallocatechin gallate and polyphenon E in healthy individuals. Clin Cancer Res 9(9):3312–3319
37. Yang CS, Wang X, Lu G, Picinich SC (2009) Cancer prevention by tea: animal studies, molecular mechanisms and human relevance. Nat Rev Cancer 9(6):429–439
38. Shtay R, Keppler JK, Schrader K, Schwarz K (2019) Encapsulation of (—)-epigallocatechin-3-gallate (EGCG) in solid lipid nanoparticles for food applications. J Food Eng 244:91–100
39. ud Din F et al (2017) Effective use of nanocarriers as drug delivery systems for the treatment of selected tumors. Int J Nanomedicine 12:7291–7309
40. Dag D, Oztop MH (2017) Formation and characterization of green tea extract loaded liposomes. J Food Sci 82(2):463–470
41. Shi M et al (2018) Food-Grade Encapsulation Systems for (−)-Epigallocatechin Gallate. Molecules 23(2):445
42. Dube A, Nicolazzo JA, Larson I (2011) Chitosan nanoparticles enhance the plasma exposure of (−)-epigallocatechin gallate in mice through an enhancement in intestinal stability. Eur J Pharm Sci 44(3):422–426

43. Ganesan P, Narayanasamy D (2017) Lipid nanoparticles: different preparation techniques, characterization, hurdles, and strategies for the production of solid lipid nanoparticles and nanostructured lipid carriers for oral drug delivery. Sustain Chem Pharm 6:37–56
44. Dzulhi S, Anwar E, Nurhayati T (2018) Formulation, characterization and in vitro skin penetration of green tea (Camellia sinensis L.) leaves extract-loaded solid lipid nanoparticles. J Appl Pharm Sci 8:57–62
45. Fangueiro JF et al (2016) Biopharmaceutical evaluation of epigallocatechin gallate-loaded cationic lipid nanoparticles (EGCG-LNs): in vivo, in vitro and ex vivo studies. Int J Pharm 502(1):161–169
46. Zhang J, Nie S, Wang S (2013) Nanoencapsulation enhances epigallocatechin-3-gallate stability and its anti-atherogenic bioactivities in macrophages. J Agric Food Chem 61(38):9200–9209
47. Zhang J, Nie S, Martinez-Zaguilan R, Sennoune SR, Wang S (2016) Formulation, characteristics and anti-atherogenic bioactivities of CD36-targeted epigallocatechin gallate (EGCG)-loaded nanoparticles. J Nutr Biochem 30:14–23
48. Smith A, Giunta B, Bickford PC, Fountain M, Tan J, Shytle RD (2010) Nanolipidic particles improve the bioavailability and alpha-secretase inducing ability of epigallocatechin-3-gallate (EGCG) for the treatment of Alzheimer's disease. Int J Pharm 389(1–2):207–212
49. Chen J et al (2017) Development and evaluation of resveratrol, vitamin E, and epigallocatechin gallate loaded lipid nanoparticles for skin care applications. Eur J Pharm Biopharm 117:286–291
50. Radhakrishnan R et al (2016) Encapsulation of biophenolic phytochemical EGCG within lipid nanoparticles enhances its stability and cytotoxicity against cancer. Chem Phys Lipids 198:51–60
51. Nafee N, Makled S, Boraie N (2018) Nanostructured lipid carriers versus solid lipid nanoparticles for the potential treatment of pulmonary hypertension via nebulization. Eur J Pharm Sci 125:151–162
52. Gerasimov OV, Boomer JA, Qualls MM, Thompson DH (1999) Cytosolic drug delivery using pH- and light-sensitive liposomes. Adv Drug Deliv Rev 38(3):317–338
53. Minnelli C et al (2018) A Poloxamer-407 modified liposome encapsulating epigallocatechin-3-gallate in the presence of magnesium: characterization and protective effect against oxidative damage. Int J Pharm 552(1):225–234
54. Chen W, Zou M, Ma X, Lv R, Ding T, Liu D (2019) Co-encapsulation of EGCG and quercetin in liposomes for optimum antioxidant activity. J Food Sci 84(1):111–120
55. Marwah M, Perrie Y, Badhan RKS, Lowry D (2019) Intracellular uptake of EGCG-loaded deformable controlled release liposomes for skin cancer. J Liposome Res 1–14. https://doi.org/10.1080/08982104.2019.1604746
56. Zou L, Peng S, Liu W, Chen X, Liu C (2015) A novel delivery system dextran sulfate coated amphiphilic chitosan derivatives-based nanoliposome: capacity to improve in vitro digestion stability of (−)-epigallocatechin gallate. Food Res Int 69:114–120
57. "Therapeutic efficacy of epigallocatechin gallate-loaded nanoliposomes against burn wound infection by methicillin-resistant Staphylococcus aureus - Abstract - Skin pharmacology and physiology 2013, Vol. 26, No. 2 - Karger Publishers." [Online]. Available: https://www.karger.com/Article/Abstract/345761. Accessed 23 June 2019
58. Luo X et al (2014) Optimization on condition of epigallocatechin-3-gallate (EGCG) nanoliposomes by response surface methodology and cellular uptake studies in Caco-2 cells | Nanoscale Research Letters | Full Text. [Online]. Available: https://nanoscalereslett.springeropen.com/articles/10.1186/1556-276X-9-291. Accessed 23 June 2019
59. Song Q et al (2014) Enhanced uptake and transport of (+)-catechin and (−)-epigallocatechin gallate in niosomal formulation by human intestinal Caco-2 cells. Int J Nanomed; 08-May-2014. [Online]. Available: https://www.dovepress.com/enhanced-uptake-and-transport-of%2D%2Dcatechin-and%2D%2D-epigallocatechin-gall-peer-reviewed-article-IJN. Accessed 23 June 2019
60. Ramdass SK et al (2014) Paclitaxel/epigallocatechin gallate coloaded liposome: a synergistic delivery to control the invasiveness of MDA-MB-231 breast cancer cells - ScienceDirect."

[Online]. Available: https://www.sciencedirect.com/science/article/pii/S0927776514006195?via%3Dihub. Accessed 23 June 2019
61. Iyisan B, Landfester K (2019) Polymeric Nanocarriers. In: Gehr P, Zellner R (eds) Biological responses to nanoscale particles. Springer, Cham, pp 53–84
62. Soppimath KS, Aminabhavi TM, Kulkarni AR, Rudzinski WE (2001) Biodegradable polymeric nanoparticles as drug delivery devices. J Control Release 70(1):1–20
63. Hong Z, Xu Y, Yin J-F, Jin J, Jiang Y, Du Q (2014) Improving the effectiveness of (−)-epigallocatechin gallate (EGCG) against rabbit atherosclerosis by EGCG-loaded nanoparticles prepared from chitosan and polyaspartic acid. J Agric Food Chem 62(52):12603–12609
64. Sanna V et al (2017) Targeted nanoparticles encapsulating (−)-epigallocatechin-3-gallate for prostate cancer prevention and therapy. Sci Rep 7:41573
65. Zeng L, Yan J, Luo L, Ma M, Zhu H (2017) Preparation and characterization of (−)-epigallocatechin-3-gallate (EGCG)-loaded nanoparticles and their inhibitory effects on Human breast cancer MCF-7 cells. Sci Rep 7:45521
66. Cano A et al (2019) Dual-drug loaded nanoparticles of Epigallocatechin-3-gallate (EGCG)/Ascorbic acid enhance therapeutic efficacy of EGCG in a APPswe/PS1dE9 Alzheimer's disease mice model. J Control Release 301:62–75
67. Zhang G, Zhang J (2018) Enhanced oral bioavailability of EGCG using pH-sensitive polymeric nanoparticles: characterization and in vivo investigation on nephrotic syndrome rats. Drug Des Devel Ther 12:2509–2518
68. Singh NA, Bhardwaj V, Ravi C, Ramesh N, Mandal AKA, Khan ZA (2018) EGCG nanoparticles attenuate aluminum chloride induced neurobehavioral deficits, beta amyloid and tau pathology in a rat model of Alzheimer's disease. Front Aging Neurosci 10:244
69. Lin Y-H, Feng C-L, Lai C-H, Lin J-H, Chen H-Y (2014) Preparation of epigallocatechin gallate-loaded nanoparticles and characterization of their inhibitory effects on *Helicobacter pylori* growth *in vitro* and *in vivo*. Sci Technol Adv Mater 15(4):045006
70. Cano A et al (2018) Epigallocatechin-3-gallate loaded PEGylated-PLGA nanoparticles: a new anti-seizure strategy for temporal lobe epilepsy. Nanomed Nanotechnol Biol Med 14(4):1073–1085
71. Huang H-Y et al (2018) Gelatin–epigallocatechin gallate nanoparticles with hyaluronic acid decoration as eye drops can treat rabbit dry-eye syndrome effectively via inflammatory relief. Int J Nanomed; 08-Nov-2018. [Online]. Available https://www.dovepress.com/gelatinndashepigallocatechin-gallate-nanoparticles-with-hyaluronic-aci-peer-reviewed-fulltext-article-IJN. Accessed 23 June 2019
72. Wu YR, Choi HJ, Kang YG, Kim JK, Shin J-W (2017) In vitro study on anti-inflammatory effects of epigallocatechin-3-gallate-loaded nano- and microscale particles. Int J Nanomedicine 12:7007–7013
73. Chavva S et al (2019) Epigallocatechin gallate-gold nanoparticles exhibit superior antitumor activity compared to conventional gold nanoparticles: potential synergistic interactions. Nano 9(3):396
74. Wu S et al (2018) Biosynthesis of flower-shaped Au nanoclusters with EGCG and their application for drug delivery. J Nanobiotechnol 16(1):90
75. Tsai L-C, Hsieh H-Y, Lu K-Y, Wang S-Y, Mi F-L (2016) EGCG/gelatin-doxorubicin gold nanoparticles enhance therapeutic efficacy of doxorubicin for prostate cancer treatment. Nanomedicine 11(1):9–30
76. Chen C-C, et al (2014) Improving anticancer efficacy of (–)-epigallocatechin-3-gallate gold nanoparticles in murine B16F10 melanoma cells. [Online]. Available: https://www.ncbi.nlm.nih.gov/pmc/articles/PMC4020885/. Accessed 24 June 2019
77. Zhang J et al (2014) Epigallocatechin-3-gallate (EGCG)-stabilized selenium nanoparticles coated with Tet-1 peptide to reduce amyloid-β aggregation and cytotoxicity. - PubMed - NCBI. [Online]. Available: https://www.ncbi.nlm.nih.gov/pubmed/24758520. Accessed 24 June 2019
78. Li Y et al (2018b) 'Cell-addictive' dual-target traceable nanodrug for Parkinson's disease treatment via flotillins pathway. Theranostics 8(19):5469–5481

79. Hsieh DS et al (2011) The treatment of bladder cancer in a mouse model by epigallocatechin-3-gallate-gold nanoparticles. - PubMed - NCBI." [Online]. Available: https://www.ncbi.nlm.nih.gov/pubmed/21782236. Accessed 25 June 2019
80. Avadhani KS et al (2017) Skin delivery of epigallocatechin-3-gallate (EGCG) and hyaluronic acid loaded nano-transfersomes for antioxidant and anti-aging effects in UV radiation induced skin damage. Drug Deliv 24(1):61–74
81. Shetty PK et al (2017) Skin delivery of EGCG and silibinin: potential of peptide dendrimers for enhanced skin permeation and deposition. AAPS PharmSciTech 18(6):2346–2357
82. Shafiei SS, Solati-Hashjin M, Samadikuchaksaraei A, Kalantarinejad R, Asadi-Eydivand M, Osman NAA (2015) Epigallocatechin gallate/layered double hydroxide nanohybrids: preparation, characterization, and in vitro anti-tumor study. PLoS One 10(8):e0136530
83. Lambert JD et al (2006) Dose-dependent levels of epigallocatechin-3-gallate in human colon cancer cells and mouse plasma and tissues. Drug Metab Dispos Biol 34(1):8–11

Clove Oil/Eugenol as the Nanotechnological Perspective for Healthcare Applications

Mohamad Taleuzzaman, Syed Sarim Imam, and Sadaf Jamal Gilani

1 Introduction

Eugenol (EUG) (Fig. 1) is a phenolic compound that comes in the chemical class of phenylpropanoids. It has the IUPAC name 4-allyl-2-methoxyphenol, chemical formula $C_{10}H_{12}O_2$, and molecular weight 164.2 g/mol. In the year 1929, eugenol was first isolated, and in the 1940s, their commercial production commenced in the United States [1].

A well-defined phytogenic bioactive component habitually occurs in large herbal plants. Plants like clove, cinnamon, tulsi, and pepper are the main sources of eugenol. There are a number of natural sources reported for the presence of eugenol such as *Myristica fragrans* Houtt. (nutmeg), *Cinnamomum verum* J. Presl (true cinnamon), *C. loureirii* Nees. (Saigon cinnamon), *Ocimum gratissimum* Forssk. (basil), and *Ocimum basilicum* L. (sweet basil). But the plant *Eugenia caryophyllata* (*S. Aromaticum*) have been reported for the principal source of eugenol as it contain about 45–90% of the total oil.

The daily intake of 2.5 mg/kg bodyweight for humans of eugenol is approved by the Food and Agriculture Organization (FAO) and the World Health Organization (WHO) [2]. Also, the US Food and Drug Administration (FDA) have confirmed eugenol as safe, and it is considered noncarcinogenic and non-mutagenic. Eugenol is present in various medicinal plants and hugely used for the prevention and treatment

M. Taleuzzaman (✉)
Department of Pharmaceutical Chemistry, Maulana Azad University, Jodhpur, Rajasthan, India

S. S. Imam
Department of Pharmaceutics, College of Pharmacy, King Saud University, Riyadh, Saudi Arabia

S. J. Gilani
College of Basic Health Sciences, Princess Nourah Bint Abdul Rahman University, Riyadh, Saudi Arabia

© Springer Nature Singapore Pte Ltd. 2020

M. Rahman et al. (eds.), *Nanomedicine for Bioactives*,
https://doi.org/10.1007/978-981-15-1664-1_16

Fig. 1 Chemical structure of eugenol

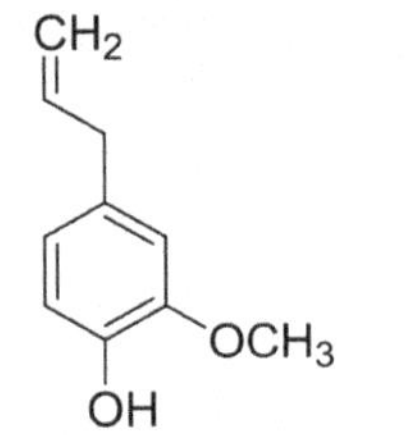

of different lifestyle-related risk factors. A number of biological properties like antiviral, antioxidant, anti-inflammatory, and anticancer have been reported. Antioxidant and anti-inflammatory actions are reported at low concentration, but at high concentration, it produces tissue-damaging free radicals. From ancient times, the extract of different parts of the plants having eugenol was endorsed for the treatment of various complications like bronchitis, diarrhea, skin diseases, cancer, hyperlipidemia, liver ailments, hyperglycemia, arthritis, cardiovascular diseases, and inflammation. Also, it is widely used in agricultural field to protect food from microbial growth during storage, considering this compound as pesticide and fumigant.

From more than 2000 years, clove has been used in ancient China as a spice and fragrance, and it is a well-known medicine for toothache [1]. It is also used as an industrial source to produce isoeugenol and methyl eugenol [3]. In the eastern Indonesia, a small island named Maluku is known as the "Spice Islands" from where the world trade for clove takes place [4]. There are a number of extraction methods available by which eugenol can be easily extracted from plants. Commercially, eugenol is derived from clove bud/leaf oil and cinnamon leaf oil or basil obtained through steam distillation, and it is further refined [5].

The plant parts leaves and buds of *Eugenia caryophyllata* were first used for eugenol extraction generally known as clove. To fulfill the demand in the market, this compound is synthesized in the laboratory through allylation of guaiacol; allyl chloride has a similar kind of functional property. This compound is partially soluble in water but highly soluble in an organic solvent.

The presence of significant amount of eugenol in the extract of several medicinal plants opens the gateway for the scientist interest in the research field. Due to the broad range of biological and functional properties, eugenol still stands priority for research. The World Health Organization declared eugenol as generally recognized as safe (GRAS), and it is considered as non-mutagenic.

2 Physical and Chemical Properties

Eugenol is generally well soluble in organic solvents and sparingly soluble in water (log P = 2.49). Chemically, it is a phenylpropanoid, weak acid and available in two isoforms: eugenol and isoeugenol. The color of eugenol is clear to pale yellow liquid with a characteristic and pleasant odor of cloves and a spicy pungent taste [1, 6]. It is also known as caryophyllic acid, allylguaiacol, 2-methoxy-4-(2-propenyl) phenol, and 4-allylcatechol-2-methyl ether. After oral administration, it is rapidly absorbed and metabolized and about completely excreted into urine as 4-hydroxy-3-methoxyphenyl-propane, isoeugenol, and other compounds [7]. The absorption takes place via the small intestine and is rapidly distributed in all organs. During the metabolism process, there is formation of conjugates taking place with sulfate, glucuronic acid (major), and glutathione studied in vitro with 1 mM concentration (lethal dose) [8].

3 Therapeutic Effects of Eugenol

There are a number of therapeutic actions that have been reported for eugenol and clove oil. They act on different diseases by different mechanisms. A brief summary of mechanism of action has been depicted in Table 1.

3.1 Anti-Infective Activity

3.1.1 Antibacterial Activity

From a century, the eugenol has been used to treat microbial infection. It has the ability to inhibit the growth of the microorganism and show a potent antibacterial potential against numerous strains of Gram-positive (*E. faecalis*, *S. epidermidis*, *S. pyogenes*, *S. aureus*, *S. pneumoniae*, *L. monocytogenes*, *B. cereus*, and *B. subtilis*) and Gram-negative (*E. coli*, *P. vulgaris*, *S. choleraesuis*, *S. typhi*, *P. aeruginosa*, *H. pylori*, and *Y. enterocolitica*) bacteria [25]. The agar well diffusion method has been used to conduct the microbial experiment and confirm the inhibition growth of the microorganism. It will act by damaging the cell wall and cell membrane of the microorganism, and cell lysis also takes place when the intracellular contents come out along with proteins and lipids [26, 27].

3.1.2 Antifungal Activity

A number of research study reported the antifungal activity of eugenol against various pathogens like *A. ochraceus*, *F. graminearum*, *F. moniliforme*, *P. citrinum*, *P. viridicatum*, *T. rubrum*, *T. mentagrophytes*, *C. tropicalis*, and *C. krusei*. Eugenol has a property to change the architecture of the envelope of fungal strain of *C. albicans* [28]. It inhibits the growth by interfering with the attachment and transition to the hyphal form, which decreases the colonization on the host tissues of *C. albicans*. The lipophilic nature of eugenol smoothly disperses between the fatty acyl chains of cell membranes, by which the fluidity and permeability of the cell membrane lead to changes in cell growth. Eugenol with methyl group has more antifungal activity, and the phenolic components, carvacrol, and eugenol are known to possess fungicidal characteristics [29].

Table 1 Eugenol/clove oil mechanism of action against different diseases

S. No.	Therapeutic properties	Mechanism of action	Important facts	References
1.	Antioxidant activity	Inhibition of DPPH radical and hydroxyl radical in concentration-dependent manner.	• Targeted iron-mediated lipid peroxidation and auto-oxidation of Fe^{2+}, also by isomer isoeugenol. • Nitric oxide scavenging activity. • Reversed neuronal excitotoxic.	[9, 10]
2.	Antibacterial activity	Harm the cell wall and membrane, leakage of cell content, i.e., proteins and lipids.	• Activity against various strains of Gram-positive and Gram-negative bacteria. • Greater effects observed when taken with vancomycin, gentamicin, and β-lactam. • Also, synergic interactions with cinnamate, cinnamaldehyde, thymol, and carvacrol.	[11, 12]
3.	Antifungal activity	Declined the adherence and metabolic activity of biofilms observed in *C. albicans*. Reduction of ergosterol biosynthesis followed by apoptosis in Candida cells leads to apoptosis.	• Combinations of eugenol and thymol led to a synergistic effect.	[13, 14]
4.	Antiviral activity	Inhibit viral replication and reduce viral infection specifically against herpes simplex-1 (HSV-1) and herpes simplex -2 (HSV-2).	• With acyclovir exhibits a promising antiviral property. • Acts against human cytomegalovirus (CMV), murine CMV (MCMV), and hepatitis C virus	[15, 16]
5.	Anti-parasitic activity	Cell shape modifications, presence of precipitates in the cytoplasm, autophagic vesicles, internalization of flagella and ventral disc, membrane blebs and intracellular/nuclear clearing.	• Methanolic extract of Piper betle containing eugenol exhibited antileishmanial potential against Leishmania donovani. • Exhibited antimalarial activity with an IC50 value of 753 μM against the chloroquine-resistant strain Plasmodium falciparum (FCR-3).	[17, 18]

(continued)

Table 1 (continued)

S. No.	Therapeutic properties	Mechanism of action	Important facts	References
6.	Anticancer activity	Eugenol-related biphenyl (S)-6, 6'-dibromo dehydrodieugenol elicits specific antiproliferative activity on neuroectodermal tumor cells by partially triggering apoptosis.	• Cytotoxic concentrations induced the reduction of ATP of oxidative stress and an increase in the polyamines and glycolytic metabolites. • Inhibited matrix metalloproteinase-9 activities in PMA-stimulated HT1080 cells via inactivation of ERK. • Combination with 5-fluorouracil exhibited more cytotoxicity against the cervical cancer cells (HeLa). • 5-Fluorouracil increased the number of cells in the S and G2/M phases when compared to treatment with the individual compounds alone.	[19–22]
7.	Anti-inflammatory	Inhibition of prostaglandin synthesis and neutrophil/ macrophage chemotaxis.	• Reduced inflammation by decreasing TNF-α and infiltration of neutrophils during pulmonary infection in animals.	[23, 24]

3.1.3 Antiplasmodial Activity and Antiviral Activity

The eugenol has been found effective against chloroquine-resistant strain *P. falciparum*. It has also a property to control the viral infection and replication and also been experimentally reported against the herpes simplex viruses, i.e., HSV-1 and HSV-2, finding the IC50 values between 16.2 mg/mL and 25.6 mg/mL [30].

3.2 Anti-inflammatory Activity

The anti-inflammatory potential of eugenol has been documented. The principal constituent's methoxyphenol of clove oil is found to inhibit cyclooxygenase (COX)-2 expression and tumor necrosis factor (TNF) signaling. COX-2 and 15-LOX enzyme expression increases as the production of inflammatory mediator macrophages and neutrophils. Both eugenol and acetyl eugenol are the major constituents of *Eugenia caryophyllata* and have the potential to inhibit the expression of COX-2 and 15-LOX enzymes. From clinical study data, it indicates that the immune-protective application of eugenol notably enhances [31].

3.3 Anticancer Activity

The chemo-protective potential of eugenol has been reported against the several types of cancer [32]. The anticancer activity of eugenol and their chemically synthesized derivatives is found against several types of cancer like gastric cancer, colon cancer, leukemia, breast cancer, prostate cancer, melanoma, and skin tumors proven by in vitro studies [8].

3.4 Neuroprotective and Anti-stress-Related Activity

The composition of most of the herbal medicine used for the Alzheimer disease (AD) has herb Acori Graminei Rhizoma which contains high amount of eugenol. It acts as a neuroprotective agent against amyloid-β-peptide, ischemia, and excitotoxicity by improving neuronal complications and inhibiting the transmission in sciatic nerves. It is also able to inhibit lipid peroxidation and prevent the reduction in the level of dopamine in experimental rats [33].

3.5 Analgesic Activity

The analgesic activity of eugenol alleviates tooth pain. It inhibits voltage-gated sodium channel (VGSC) in trigeminal ganglion (TG) neurons and activates transient receptor potential vanilloid subtype 1 (TRPV-1). The inhibitory effects of eugenol on the inhibition of TRPV-1 showed that eugenol inhibited voltage-activated Na^+ and Ca^{+2} channels [33].

4 Biological Activities of Eugenol in Animal Models

A number of well-known studies of biological activities like antimicrobial, antinociceptive, and antiviral have been reported for eugenol. It is effective for prophylaxis and treatment of vaginal and oral candidiasis in immunosuppressed rats [34]. Limited in vivo chemopreventive efficacy against free radical-mediated chronic diseases have been reported [35]. In skin cancer, it acts by inhibiting the activation of NF-jB [36]. No well-controlled clinical studies for various chronic diseases in human patients have yet been reported.

5 Biological Activities of Eugenol in Humans

The therapeutic application of eugenol was approved in monographs of the expert panel German Commission E published between 1983 and 1993. It is used to prevent infection and reduce pain and further widely used as zinc oxide-eugenol chelate cement in dentistry [37]. In periodontitis, it is used in intra-pocket delivery system as zinc oxide gel containing eugenol. In low concentration, it helps prevent bacterial infection by releasing the eugenol into the gingival pocket [38]. Isoeugenol, the dimer beta-O-4-dilignol, and another dimer DHDI have shown extreme, weak, and moderate allergenicity to the tested animal [39]. It is generally found as nonallergenic, but in sensitive individuals, it may cause low-grade tissue reactions. Many studies have been reported the use of eugenol for preventing chronic diseases in animal models, but still, their therapeutic use in humans remains unexplored.

6 Eugenol Nanoformulation and Their Biological Evaluation

In this chapter, the main emphasis is given on the latest developments in eugenol/clove oil formulation with improved therapeutic efficacy. In the past few years, there was a significant progress that took place in the development of eugenol formulations like liposome, nanoparticles, microemulsions, and micelles (Fig. 2). Eugenol exhibits poor physicochemical stability and limited aqueous solubility and leads to poor bioavailability. Similarly, a number of formulations have been proven to enhance the physicochemical stability of eugenol upon incorporating in different formulations (Table 2).

6.1 Antibacterial Action

The formulation of chitosan nanoparticles (ChNPs) of eugenol and carvacrol has been prepared and evaluated for their antioxidant and antibacterial properties [50]. The antioxidant activities were assayed using the diphenylpicrylhydrazyl (DPPH) method. The developed formulation showed good antibacterial assays against Gram-negative bacterium *Escherichia coli* (*E. coli*) and a Gram-positive bacterium *Staphylococcus aureus* (*S. aureus*). In another study, antimicrobial delivery of eugenol and trans-cinnamaldehyde was successfully developed by synthesizing spherical poly (DL-lactide-co-glycolide) (PLGA) nanoparticles. The nanoparticle was prepared with emulsion evaporation method in the presence of poly (vinyl alcohol) (PVA) as a surfactant. The formulation showed high eugenol and trans-cinnamaldehyde entrapment efficiency (EE) approximately 98% and 92%, respectively. Finally, the developed formulation proved to be efficient in inhibiting the growth of *Salmonella* spp. (Gram-negative bacterium) and *Listeria* spp. (Gram-positive bacterium) with concentrations ranging from 20 to 10 mg/mL [51]. In another formulation, eugenol-loaded solid lipid nanoparticles (SLN) was prepared and evaluated for antifungal activity for oral candidiasis in immunosuppressed rats. The formulation SLN (2) showed greater and faster drug release than SLN (1), due to higher encapsulation efficiency of SLN (2) (98.52%) as compared to SLN (1) (91.80%). The formulation also showed greater antifungal activity after administration into SLN [52]. The antioxidant and antimicrobial effects of eugenol nanoparticles in water-based microemulsion systems were evaluated through the developed

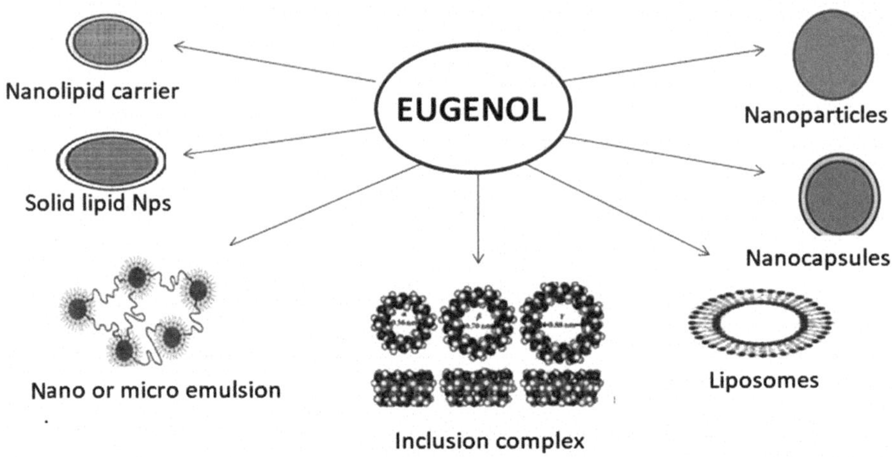

Fig. 2 Eugenol loaded different nano and lipidic formulations

formulation using the DPPH∗ free radical scavenging assay and the agar disc dilution method, respectively. The prepared eugenol nanoparticles in water-based microemulsion were found to be more effective as compared to clove essential oil (CEO) [49]. Further, the formulation polylactic acid-chitosan-magnetite-eugenol nanospheres was prepared and evaluated for antimicrobial effect. The formulation showed the size ranges between 20 and 80 nm. The formulation showed the effect by damaging the adherence and biofilm formation of *S. aureus* and *P. aeruginosa* bacteria strains. It also showed good biocompatibility and expedites the normal development of human endothelial cells [53].

Eugenol-loaded solid lipid nanoparticles (EG-SLNs) were prepared and converted into carbopol hydrogels to evaluate fungal infection. The formulation showed greater hydration on human cadaver skin compared with that of hydrogel. The formulation EG-SLNs were converted into the hydrogel but did not change in respect to the particle size and morphology. The formulation showed promising effect for epidermal targeting to treat fungal infections in the skin [54]. In another eugenol formulation, the ethanol injection method and dynamic high-pressure microfluidization method were used to encapsulate in nanoliposomes (EN). The developed formulation was evaluated for stability study. The results of the study revealed better storage stability at both poor temperature and 4 °C. The stability of formulation is evaluated by its nonsignificant variation in size diameter and encapsulation efficiency after 2 months of storage. The antibacterial activity was assessed against four common foodborne pathogenic bacteria (*S. aureus*, *E. coli*, *S. enterica serovar* Typhimurium, and *L. monocytogenes*) using the Luria broth and milk medium. The result of the study showed good activity against all the tested organism [55]. The study was performed with ketoprofen (KP) nanoemulsion (NE) with eugenol and NEG without KP, and their antibacterial activity concentrations were measured. The antibacterial study was evaluated by using the agar diffusion

Table 2 Eugenol/clove oil-encapsulated nanoformulations and their therapeutic efficacy

S. No	Formulation	Polymer	Activity	Findings	Reference
1.	Nanoparticles	Chitosan	Antifungal	The nano-encapsulated CEO demonstrated a superior performance against *Aspergillus niger*, isolated from spoiled pomegranate, compared with ChNPs and free oil.	[40]
2.	Microcapsules	Gelatin	Acaricidal	Mortality tests on *Dermatophagoides farinae* conducted on fabric with attached microcapsules showed that clove bud oil, containing a more phenolic monoterpenoid (eugenol), was more effective at reducing the live HDMs (94% mortality).	[41]
3.	Nanoemulsion	Triacetin, Tween-80, Labrasol, water.	Antibacterial	The antibacterial effects of F1 were highly comparable with standard amikacin against all bacterial strains.	[42]
4.	Micelle	poly (ethylene glycol), poly (ε-caprolactone)		A sustained release profile of CEO from CEO-loaded micelles. The results indicate the successful formulation of CEO-loaded PEG-b-PCL micelle nanoparticles.	[43]
5.	Emulsion	Tween 80	Visceral leishmaniasis	In vitro and ex vivo leishmanicidal effect of EE that occurred via an increase in production of NO without adverse effects on mammalian macrophages.	[44]

(continued)

Table 2 (continued)

S. No	Formulation	Polymer	Activity	Findings	Reference
6.	Nanoparticles	Mesoporous silica	Antimicrobial	The electrospun PHBV films with loadings above 10 wt.−% of mesoporous silica nanoparticles containing eugenol successfully inhibited the bacterial growth.	[45]
7.	Liposomes	Hyaluronic acid	Metastatic melanoma	Migration assay and proliferation study revealed significantly higher inhibition of cell migration and proliferation by Coated-Dacarbazine Eugenol Liposomes.	[46]
8.	Sealers	Zinc-oxide	Subcutaneous connective tissue reaction	Nano zinc-oxide eugenol sealer has histocompatibility properties comparable to conventional commercial sealers.	[47]
9.	Ethosomes	Soyaphosphotidyl choline, ethanol	Candidiasis	The ethosomal gel showed satisfactory antifungal activity against the fungus *C. albicans* compared to pure clove oil.	[48]
10.	Nanoparticles in alcohol-free microemulsion		Antioxidant and antimicrobial	Microemulsion improved the evaluated activities of CEO and eugenol compared with the crude counterparts.	[49]

method. The study was performed against *S. aureus* and *E. coli* bacteria. The cytokine levels of each treated group were evaluated and results showed much closed value. It suggests that developed KP-loaded NEG formulation has the potential antibacterial, analgesic, and anesthetic properties to combat periodontal disease and effectively treat periodontitis [56]. The research group [57] developed eugenol oil nanoemulsions and evaluated for the antimicrobial activity. They developed the formulation by using gum arabic and lecithin as food-grade natural emulsifiers. The optimized formulation showed particle size of 103.6 ± 7.5 nm by mixing aqueous phase (gum arabic, 0.5%; lecithin, 0.5% w/v) and eugenol oil (1.25% w/v). The prepared formulation did not adversely affect the antimicrobial activity of eugenol against *L. monocytogenes* and *S. enteritidis*. Recently, the evaluation of the antimicrobial activity of nanostructured lipid carriers (NLCs) encapsulated with natural molecules, such as eugenol, plumbagin, hydroquinone, alpha-asarone, and alpha-tocopherol, was conducted. The formulation was prepared by melt and ultrasonication method, and the result of the study revealed lower toxicity of loaded NLCs as compared to their free form on human cultured cells [58]. In the year 2018, Shao et al. prepared eugenol-loaded nanoemulsion by ultrasonication and emulsification techniques. The optimized eugenol formulation was prepared using the composition chitosan-eugenol ratio of 1:1, Tween-20 1% wt/v, ultrasonic power 450 W, and ultrasonic time 6 min. The developed nanoparticles were found in spherical shape and size distributed within 80–100 nm. The nanoparticles performed excellent antioxidant capacity and antimicrobial activity [59]. The prepared eugenol-entrapped ethosome nanoparticles (ELG-NPs) were evaluated for its antibacterial effects. The formulation showed particle size of 44.21 nm and entrapment efficiency of 82%. ELG-NPs with the optimized process (0.5% eugenol, 2% lecithin, and 30% ethanol) showed a satisfactory slow release and prolonged antibacterial action [60]. In another study, clove essential oil (CEO)-encapsulated chitosan nanoparticles (ChNPs) were prepared by an emulsion-ionic gelation technique to improve the antifungal efficacy. The optimized formulation on the basis of dynamic light scattering. The formulation showed the particle size distribution of around 40 and 100 nm for the most optimum unloaded and oil-loaded ChNPs. The CEO-nanoencapsulated chitosan nanoparticles showed a greater efficacy against *A. niger*, isolated from spoiled pomegranate, compared with ChNPs and free oil [61].

In another research report prepared eugenol nanoformulations and in vitro antimicrobial evaluated against *G. cingulata*. They prepared mancozeb (0.1 mg/mL) and eugenol (33 mg/mL)-loaded nanoemulsion using the spontaneous emulsification method. The developed formulations were evaluated for cytotoxicity, genotoxicity, and ecotoxicity tests in healthy human cells through MTT. The physicochemical evaluation result showed that the developed formulation revealed mean particle size between 200 and 300 nm, polydispersity index less than 0.3, negative zeta-potential, and acidic pH. The MTT assay results of the developed nanoencapsulation were able to avoid the reduction of the cell viability caused by mancozeb, while eugenol was shown to be safe for cell use in both free and nanostructured forms [62]. In another study, nanoencapsulation was performed by using whey protein or soy lecithin and maltodextrin in combination with oleic acid and chitosan. The eugenol encapsulation formulation was prepared by spray-drying method and evaluated for antioxidant and antimicrobial effects [63]. The formulations prepared with whey protein and lecithin showed greater eugenol encapsulation between 95 and 98% and better antibacterial effect against *E. coli* and *L. innocua* due to high loading.

6.2 Anticancer Activity

The evaluation of the antiperoxidative effect of eugenol lipid liposome nanoformulations was performed. The formulation inhibited the lipid peroxidation action in a dose-dependent manner. The effect was assessed by formation of thiobarbituric acid reactive substances studied in the liver, with preventive action on xanthine oxidase (XO) which mediated the xanthine(X)-Fe+3-ADP lipid peroxidation. The formulation eugenol liposomes inhibit the XO-X-Fe+3-ADP-mediated peroxidation in addition to quenching various radical species. The formulation enhanced the inhibition of enzyme to 85% at 10 micron concentration [64]. In another study, N-succinyl chitosan gel delivery system of microemulsified eugenol, honey, and sodium hyaluronate was developed and evaluated. The oral mucositis, the major side effects of cancer chemotherapy, was treated effectively with synergistic effects on various pathological factors. The formulation microemulsified eugenol (10% v/v), honey (10% v/v), and sodium hyaluronate (0.2% w/v) showed the optimum pH, spreadability, mucoadhesion strength, and viscosity [65]. The formulation nanoemulsions (NEs) loaded with eugenol (EU) were prepared and evaluated for apoptotic potential against liver and colon cancer cells. The MTT assay and flow cytometry were used to determine cell viability and apoptosis, respectively. The apoptosis of cells takes place via reactive oxygen species (ROS) generation, and its antiproliferative mechanisms are showed by EU-loaded NE [66]. In another research, poly (ε-caprolactone) lipid-core nanocapsule (LNC) or LNC loaded with acetyl eugenol (AcE-LNC) was evaluated for in vivo efficacy to treat B16F10-induced melanoma in mice. The study results revealed the pretreated melanoma cells with LNC and AcE-LNC; the neutrophils increase the production of reactive oxygen species (ROS), which causes cell death. The pretreatment with formulation did not cause melanoma cell death but leads to death in serum deprivation or hypoxia or the presence of neutrophils [67]. The formulation liposomes were prepared with hyaluronic acid (HA) using solvent injection method. The prepared liposomes were loaded with the combination of anti-melanoma agents (dacarbazine and eugenol). The prepared liposomes were evaluated for cell line study. The result of the study revealed coated dacarbazine eugenol liposomes showed higher inhibition of cell migration and proliferation. The cell migration assay and proliferation study were used to confirm the therapeutic potential, and the result showed that the developed formulation was found to be very effective against resistant and aggressive metastatic melanoma. The formulation showed very less unwanted cytotoxicity and also reduced doses of the chemotherapeutic agent [46].

6.3 Anti-inflammatory and Analgesic Activity

Anti-inflammatory activity has been reported for eugenol-loaded formulations. The research was conducted by preparation and evaluation of eugenol controlled-release mucoadhesive tablets for gingival application [68]. The mucoadhesive tablets were prepared using polymers like carbopol 934P and hydroxypropyl methylcellulose

(HPMC) K4M in the ratio of 1:2, 1:1, and 2:1 with eugenol (10 mg). The developed formulation showed controlled drug release mechanism for 8 hrs. The presence of carbopol in the formulation helps in prolonged drug release of eugenol, whereas the polymer HPMC retards the drug release. The in vivo results indicate the increased potential of eugenol as antibacterial, local analgesic, and anesthetic treatment.

The eugenol-loaded nanocapsules (NCs) were developed and evaluated for periodontal infections. The formulation eugenol-loaded polycaprolactone (PCL) NCs were prepared by solvent displacement method and evaluated for drug release study. The formulation showed biphasic in vitro drug release pattern and followed Michaelis-Menten-like model. The developed formulation was evaluated for percentage cell viability values, and the results were found near to 100 in the cell viability assay. This type of result confirmed that the prepared NCs are not cytotoxic. Further, the in vivo study results showed significant difference observed in the interdental papilla in comparison to the untreated, pure eugenol and placebo groups [69].

In this work, O/W eugenol nanoemulsion was designed for the evaluation of anti-inflammatory effects as a topical delivery system [70]. The topical eugenol nanoemulsion was prepared with different concentrations of eugenol (1%, 2%, and 4%) and further combination of eugenol (4%) and piroxicam (0.5%). The developed formulations were biologically assessed for carrageenan-induced paw edema in rats. The optimized formulation composition was found with eugenol as oil phase (2%), Tween-20 as surfactant (14%), and isopropyl alcohol as cosurfactant (14%) in water. The optimized nanoemulsion showed droplet size of 24.4 nm and polydispersity index of 0.3. The biological study result revealed that the prepared nanoemulsion significantly improved the anti-inflammatory activity after 1.5 h, compared with marketed piroxicam gel [71] have developed eugenol-loaded nanostructured systems and biologically assessed for the anti-inflammatory and antioxidant effect. The formulation nanostructured systems have been successfully used as topical drug delivery for skin diseases. Additionally, the formulation polymeric nanocarriers (NCEUG) were further prepared to improve the chemical and irritant characteristics of EUG. The irritant contact dermatitis (ICD) test in mice induced by TPA and treated group with NCEUG reduced significantly the ear edema in mice when compared to the EUG solution. These findings suggest that EUG is a promising bioactive molecule, and its nanoencapsulation seems to be an interesting approach for the treatment of ICD.

6.4 Other Diseases

The formulation microemulsions loaded with eugenol were formulated and assessed for fatty liver and dyslipidemia in high-fructose-fed rats. The comparative study was designed with the CO microemulsion (COM) and eugenol microemulsion (EM) and evaluated for physicochemical characterization. The developed formulation showed low size for both the formulation, i.e., 8.0 nm and 8.9 nm, respectively. The formulation was biologically evaluated after oral administration, and it revealed that CO conventional emulsion, COM, and EM showed a significant improvement in fatty liver and dyslipidemia with consequent help in the protection from cardiovascular diseases [72].

In another study, researchers have developed eugenol-encapsulated chitosan-coated PCL nanoparticles and assessed for brain uptake study by intranasal route. The formulation was developed through double emulsification-solvent evaporation technique, and the drug bioavailability in the rat brain was evaluated. The developed formulation showed enhanced bioavailability and also may help to treat cerebral ischemia [73].

The eugenol-encapsulated beta-cyclodextrin (beta-CD) was prepared by molecular inclusion method. The developed formulations were evaluated for the particle size, zeta-potential, thermal transition, and morphology after freeze-drying [74]. The developed formulation showed low particle size (340 nm), negative zeta-potential (−34.5 mV), and high inclusion efficiency (91.7%) for encapsulated eugenol. The relationship between retention rate of eugenol and time during release was described by a mathematical model of Avrami equation. The release mechanism and the release rate constant were rapidly elevated with increasing relative humidity and storage temperature.

In another study, the eugenol-loaded nanoemulsion using zein and sodium caseinate (NaCas) was prepared and evaluated by research group [75]. The prepared nanoemulsions showed a low particle diameter (109–139 nm) and a negative zeta-potential (from −28.5 to −35.8 mV) with spherical morphology. The formulation showed entrapment efficiency (EE) of 1% (v/v), and eugenol reached 84.24% by 2% (m/v) NaCas/zein at a mass ratio of 1:1. The result of the current study showed a promising clean and low-cost strategy to deliver lipophilic compounds containing the hydroxyl group.

Zhao et al. [76] prepared and evaluated ibuprofen eugenol ester phospholipid microemulsion system using oil phase, surfactants, and cosurfactants, respectively, The solubility of the optimized microemulsion formulation consisting of ibuprofen eugenol (6.4%), Miglyol 812 (9.6%), SbPC (6%), HS-15 (6%), PEG400 (8.4%), ethanol (3.6%), and distilled water (60% w/w) was about 21,000 times higher than that in water. The isolated rat intestinal membrane study was performed to check the effect of eugenol on colchicine transport by the nanoemulsion system. The nanoemulsion showed the average particle size of 41.2 ± 7.2 nm with the significant higher permeation of colchicine nanoemulsion across the intestinal membrane from that of the control group (0.2 mM colchicine). Finally, the coadministration of eugenol in colchicine nanoemulsion was done to enhance the colchicine bioavailability. The addition of eugenol showed the enhanced bioavailability by manyfold compared with the free colchicine solution [77].

7 Conclusion

Eugenol, a natural bioactive compound, has high potential as a therapeutic agent which can be incorporated in the nanodelivery systems for the treatment of several diseases. There are a number of eugenol/clove oil-loaded nanoformulations developed and evaluated for antibacterial, analgesic, anticancer, and cerebral ischemia. The developed nanoformulations have shown greater therapeutic effect after encapsulating into nanoformulations. The research on the properties of clove and its major constituent eugenol has been still continuing. A number of animal and human activities have also been reported for the eugenol. The difference in concentration may lead to a difference in effect such as having either an analgesic or an anesthetic effect.

References

1. Barceloux DG (2008) Medical toxicology of natural substances. In: Foods, fungi, medicinal herbs, plants and venomous animals. Wiley, Hoboken, NJ
2. World Health Organization (WHO) (1982) Evaluation of certain food additives and contaminants; twenty-sixth report of the joint FAO/WHO expert committee on food additives, WHO technical report series, No. 683; Geneva, Switzerland
3. Li YH, Sun ZH, Zheng P (2004) Determination of vanillin, eugenol and isoeugenol by RP-HPLC. Chromatographia 60:709–713
4. Chomchalow N (1996) Spice production in Asia–an overview. Presented at IBC Asian Spice Markets Conference, Singapore, 27–28
5. Bedoukian PZ (1986) Perfumery and flavouring synthetics, 3rd edn. Allured Publishing Corporation, Carol Stream, IL
6. Anees AK, Ubaid ur R, Moazzam RK, Amna S, Tariq M, Muneeb K (2017) Essential oil eugenol: sources, extraction techniques and nutraceutical perspectives. RSC Adv 7:32669–32681
7. Fischer IU, Von Unruh GE, Dengler HJ (1990) The metabolism of eugenol in man. Xenobiotica 20:209–222
8. Raja MRC, Srinivasan V, Selvaraj S, Mahapatra SK (2015) Versatile and Synergistic potential of eugenol: a review. Pharm Anal Acta 6:367
9. Mahapatra SK, Roy S (2014) Phytopharmacological approach of free radical scavenging and anti-oxidative potential of eugenol and Ocimum gratissimum Linn. Asian Pac J Trop Med 7S1:S391–S397
10. Tominaga H, Kobayashi Y, Goto T, Kasemura K, Nomura M (2005) DPPH radical scavenging effect of several phenylpropanoid compounds and their glycoside derivatives. Yakugaku Zasshi 125:371–375
11. Oyedemi SO, Okoh AI, Mabinya LV, Pirochenva G, Afolayan AJ (2009) The proposed mechanism of bactericidal action of eugenol, a a-terpineol and terpinene against Listeria monocytogenes, Streptococcus pyogenes, Proteus vulgaris and Escherichia coli. Afr J Biotechnol 8:1280–1286
12. Pei RS, Zhou F, Ji BP, Xu J (2009) Evaluation of combined antibacterial effects of eugenol, cinnamaldehyde, thymol, and carvacrol against E. coli with an improved method. J Food Sci 74:379–383
13. Doke SK, Raut JS, Dhawale S, Karuppayil SM (2014) Sensitization of Candida albicans biofilms to fluconazole by terpenoids of plant origin. J Gen Appl Microbiol 60:163–168
14. de Paula SB, Bartelli TF, Di Raimo V, Santos JP, Morey AT et al. (2014) Effect of eugenol on cell surface hydrophobicity, adhesion, and Biofilm of Candida tropicalis and Candida dubliniensis isolated from oral cavity of HIV-infected patients. Evid Based Complement Alternat Med 1–8. https://doi.org/10.1155/2014/505204
15. Kurokawa M, Nagasaka K, Hirabayashi T, Uyama S, Sato H et al (1995) Efficacy of traditional herbal medicines in combination with acyclovir against herpes simplex virus type 1 infection in vitro and in vivo. Antivir Res 27:19–37
16. Yukawa TA, Kurokawa M, Sato H, Yoshida Y, Kageyama S et al (1996) Prophylactic treatment of cytomegalovirus infection with traditional herbs. Antivir Res 32:63–70

17. Misra P, Kumar A, Khare P, Gupta S, Kumar N et al (2009) Pro-apoptotic effect of the landrace Bangla Mahoba of Piper betle on Leishmania donovani may be due to the high content of eugenol. J Med Microbiol 58:1058–1066
18. Van Zyl RL, Seatlholo ST, Van Vuuren SF, Viljoen AM (2006) The biological activities of 20 nature identical essential oil constituents. J Essent Oil Res 18:129–133
19. Pisano M, Pagnan G, Loi M, Mura ME, Tilocca MG et al (2007) Antiproliferative and pro-apoptotic activity of eugenol-related biphenyls on malignant melanoma cells. Mol Cancer 6:8
20. Behbahani M (2014) Evaluation of in vitro anticancer activity of Ocimum basilicum, Alhagi maurorum, Calendula officinalis and their parasite Cuscuta campestris. PLoS One 9:e116049
21. Koh T, Murakami Y, Tanaka S, Machino M, Onuma H et al (2013) Changes of metabolic profiles in an oral squamous cell carcinoma cell line induced by eugenol. In Vivo 27:233–243
22. Nam H, Kim MM (2013) Eugenol with antioxidant activity inhibits MMP-9 related to metastasis in human fibrosarcoma cells. Food Chem Toxicol 55:106–112
23. Kim SS, Oh OJ, Min HY, Park EJ, Kim Y et al (2003) Eugenol suppresses cyclooxygenase-2 expression in lipopolysaccharide-stimulated mouse macrophage RAW264.7 cells. Life Sci 73:337–348
24. Raghavenra H, Diwakr BT, Lokesh BR, Naidu KA (2006) Eugenol--the active principle from cloves inhibits 5-lipoxygenase activity and leukotriene-C4 in human PMNL cells. Prostaglandins Leukot Essent Fatty Acids 74:23–27
25. Hemaiswarya S, Doble M (2009) Synergistic interaction of eugenol with antibiotics against Gram-negative bacteria. Phytomedicine 16:997–1005
26. El-Abed S, Houari A, Latrache H, Remmal A, Koraichi SI (2011) In vitro activity of four common essential oil components against biofilm-producing Pseudomonas aeruginosa. Res J Microbiol 6:394–401
27. Moon SE, Kim HY, Cha JD (2011) Synergistic effect between clove oil and its major compounds and antibiotics against oral bacteria. Arch Oral Biol 56:907–916
28. Braga PC, Sasso MD, Culici M, Alfieri M (2007) Eugenol and thymol, alone or in combination, induce morphological alterations in the envelope of Candida albicans. Fitoterapia 78:396–400
29. Ahmad A, Khan A, Khan LA, Manzoor N (2010) In vitro synergy of eugenol methyleugenol with fluconazole against clinical Candida isolates. J Med Microbiol 59:1178–1184
30. Benencia F, Courrèges MC (2000) In vitro and in vivo activity of eugenol on human herpes virus. Phytother Res 14:495–500
31. Magalhaes CB, Riva DR, De Paula LJ, Brando-Lima A, Koatz VL, Leal-Cardoso JH, Zin WA, Faffe DS (2010) In vivo anti-inflammatory action of eugenol on lipopolysaccharide induced lung injury. J Appl Physiol 108:845–851
32. Noonan DM, Benelli R, Albini A (2007) Angiogenesis and cancer prevention: a vision. Recent Results Cancer Res 174:219–224
33. Guy PK, Vermaak I, Alvaro MV (2012) Eugenol—from the remote maluku islands to the international market place: a review of a remarkable and versatile molecule. Molecules 17:6953–6981
34. Fujisawa S, Murakami Y (2016) Eugenol and Its role in chronic diseases. In: Gupta SC et al (eds) Drug discovery from mother nature, advances in experimental medicine and biology. Springer, New York, pp 45–66
35. Jaganathan SK, Supriyanto E (2012) Antiproliferative and molecular mechanism of eugenol-induced apoptosis in cancer cells. Molecules 17:6290–6304
36. Kaur G, Athar M, Alam MS (2010) Eugenol precludes cutaneous chemical carcinogenesis in mouse by preventing oxidative stress and inflammation and by inducing apoptosis. Mol Carcinog 49:290–301
37. Kadoma Y, Murakami Y, Atsumi T, Ito S, Fujisawa S (2009) Cloves (Eugenol). In: Aggarwal BB, Kunnumakkara AB (eds) Molecular targets and therapeutic uses of species: modern uses for ancient medicine. World Scientific, Singapore, pp 117–148
38. Mahadlek J, Charoenteeraboon J, Phaechamud T (2010) Zinc oxide gels for periodontitis treatment. J Metal Mater Mineral 20(3):159–163

39. Takeyoshi M, Iida K, Suzuki K, Yamazaki S (2008) Skin sensitization potency of isoeugenol and its dimers evaluated by a non-radioisotopic modification of the local lymph node assay and guinea pig maximization test. J Appl Toxicol 28:530–534
40. Hasheminejad N, Khodaiyan F, Safar M (2019a) Improving the antifungal activity of clove essential oil encapsulated by chitosan nanoparticles. Food Chem 275:113–122
41. Kim JR, Sharma S (2011) Acaricidal activities of clove bud oil and red thyme oil using microencapsulation against HDMs. J Microencapsul 28:82–91
42. Anwer MK, Jamil S, Ibnouf EO, Shakeel F (2014) Enhanced antibacterial effects of clove essential oil by nanoemulsion. J Oleo Sci 63:347–354
43. Thonggoom O, Punrattanasin N, Srisawang N, Promawan N, Thonggoom R (2016) In vitro controlled release of clove essential oil in self-assembly of amphiphilic polyethylene glycol-block-polycaprolactone. J Microencapsul 33:239–248
44. Islamuddin M, Chouhan G, Want MY, Ozbak HA, Hemeg HA, Afrin F (2016) Immunotherapeutic potential of eugenol emulsion in experimental visceral leishmaniasis. PLoS Negl Trop Dis 10(10):1–23
45. Melendez-Rodriguez B, Figueroa-Lopez KJ, Bernardos A, Martínez-Máñez R, Cabedo L, Torres-Giner S, Lagaron JM (2019) Electrospun antimicrobial films of poly(3-hydroxybutyrate-*co*-3-hydroxyvalerate) containing eugenol essential oil encapsulated in mesoporous silica nanoparticles. Nanomaterials (Basel): 9(2). pii: E227. https://doi.org/10.3390/nano9020227
46. Mishra H, Mishra PK, Iqbal Z, Jaggi M, Madaan A, Bhuyan K, Gupta N, Gupta N, Vats K, Verma R, Talegaonkar S (2019) Co-delivery of eugenol and dacarbazine by hyaluronic acid-coated liposomes for targeted inhibition of survivin in treatment of resistant metastatic melanoma. Pharmaceutics. 11(4). pii: E163. https://doi.org/10.3390/pharmaceutics11040163
47. Omidi S, Javidi M, Zarei M, Mushakhian S, Jafarian A (2017) Subcutaneous connective tissue reaction to a new nano zinc-oxide eugenol sealer in rat model. Iran Endod J 12:64–69
48. Shetty S, Jose J, Kumar L, Charyulu RN (2019) Novel ethosomal gel of clove oil for the treatment of cutaneous candidiasis. J Cosmet Dermatol 18:862–869
49. Hamed SF, Sadek Z, Edris A (2012) Antioxidant and antimicrobial activities of clove bud essential oil and eugenol nanoparticles in alcohol-free microemulsion. J Oleo Sci 61:641–648
50. Chen F, Shi Z, Neoh KG, Kang ET (2009) Antioxidant and antibacterial activities of eugenol and carvacrol-grafted chitosan nanoparticles. Biotechnol Bioeng 104:30–39
51. Gomes C, Moreira RG, Castell-Perez E (2011) Poly (DL-lactide-co-glycolide) (PLGA) nanoparticles with entrapped *trans*-cinnamaldehyde and eugenol for antimicrobial delivery applications. J Food Sci 76:16–24
52. Garg A, Singh S (2011) Enhancement in antifungal activity of eugenol in immunosuppressed rats through lipid nanocarriers. Colloids Surf B Biointerfaces 87:280–288
53. Holban AM, Grumezescu V, Grumezescu AM, Vasile BŞ, Truşcă R, Cristescu R, Socol G, Iordache F (2014) Antimicrobial nanospheres thin coatings prepared by advanced pulsed laser technique. Beilstein J Nanotechnol 5:872–880
54. Garg A, Singh S (2014) Targeting of eugenol-loaded solid lipid nanoparticles to the epidermal layer of human skin. Nanomedicine 9:1223–1238
55. Peng S, Zou L, Liu W, Gan L, Liu W, Liang R, Liu C, Niu J, Cao Y, Liu Z, Chen X (2015) Storage stability and antibacterial activity of eugenol nanoliposomes prepared by an ethanol injection-dynamic high-pressure micro fluidization method. J Food Prot 78:22–30
56. Srivastava M, Neupane YR, Kumar P, Kohli K (2016) Nanoemulgel (NEG) of Ketoprofen with eugenol as oil phase for the treatment of ligature-induced experimental periodontitis in Wistar rats. Drug Deliv 23(7):2228–2234
57. Hu Q, Gerhard H, Upadhyaya I, Venkitanarayanan K, Luo Y (2016) Antimicrobial eugenol nanoemulsion prepared by gum arabic and lecithin and evaluation of drying technologies. Int J Biol Macromol 87:130–140
58. Cortesi R, Valacchi G, Muresan XM, Drechsler M, Contado C, Esposito E, Grandini A, Guerrini A, Forlani G, Sacchetti G (2017) Nanostructured lipid carriers (NLC) for the delivery of natural molecules with antimicrobial activity: production, characterisation and in vitro studies. J Microencapsul 34:63–72

59. Shao Y, Wu C, Wu T, Li Y, Chen S, Yuan C, Hu Y (2018) Eugenol-chitosan nanoemulsions by ultrasound-mediated emulsification: Formulation, characterization and antimicrobial activity. Carbohydr Polym 193:144–152
60. Jin P, Yao R, Qin D, Chen Q, Du Q (2019) Enhancement in antibacterial activities of eugenol-entrapped ethosome nanoparticles via strengthening its permeability and sustained release. J Agric Food Chem 67:1371–1380
61. Hasheminejad N, Khodaiyan F, Safari M (2019b) Improving the antifungal activity of clove essential oil encapsulated by chitosan nanoparticles. Food Chem 275:113–122
62. da Silva Gündel S, Dos Reis TR, Copetti PM, Favarin FR, Sagrillo MR, da Silva AS, Segat JC, Baretta D, Ourique AF (2019) Evaluation of cytotoxicity, genotoxicity and ecotoxicity of nanoemulsions containing Mancozeb and Eugenol. Ecotoxicol Environ Saf 169:207–215
63. Talón E, Lampi AM, Vargas M, Chiralt A, Jouppila K, González-Martínez C (2019) Encapsulation of eugenol by spray-drying using whey protein isolate or lecithin: release kinetics, antioxidant and antimicrobial properties. Food Chem 295:588–598
64. Nagababu E, Lakshmaiah N (1997) Inhibition of xanthine oxidase-xanthine-iron mediated lipid peroxidation by eugenol in liposomes. Mol Cell Biochem 166:65–71
65. Dhawan N, Kumar K, Kalia AN, Arora S (2014) N-succinyl chitosan as buccal penetration enhancer for delivery of herbal agents in treatment of oral mucositis. Curr Drug Deliv 11:415–425
66. Majeed H, Antoniou J, Fang Z (2014) Apoptotic effects of eugenol-loaded nanoemulsions in human colon and liver cancer cell lines. Asian Pac J Cancer Prev 15:9159–9164
67. Drewes CC, Alves AC, Hebeda CB, Copetti I, Sandri S, Uchiyama MK, Araki K, Guterres SS, Pohlmann AR, Farsky SH (2017) Role of poly(ε-caprolactone) lipid-core nanocapsules on melanoma-neutrophil crosstalk. Int J Nanomedicine 12:7153–7163
68. Jadhav BK, Khandelwal KR, Ketkar AR, Pisal SS (2004) Formulation and evaluation of mucoadhesive tablets containing eugenol for the treatment of periodontal diseases. Drug Dev Ind Pharm 30:195–203
69. Pramod K, Aji Alex MR, Singh M, Dang S, Ansari SH, Ali J (2016) Eugenol nanocapsule for enhanced therapeutic activity against periodontal infections. J Drug Target 24:24–33
70. Esmaeili F, Rajabnejhad S, Partoazar AR, Mehr SE, Faridi-Majidi R, Sahebgharani M, Syedmoradi L, Rajabnejhad MR, Amani A (2016) Anti-inflammatory effects of eugenol nanoemulsion as a topical delivery system. Pharm Dev Technol 21:887–893
71. de Araújo Lopes A, da Fonseca FN, Rocha TM, de Freitas LB, Araújo EVO, DVT W, RCP LJ, LKAM L (2018) Eugenol as a promising molecule for the treatment of dermatitis: antioxidant and anti-inflammatory activities and its nanoformulation. Oxidative Med Cell Longev 2018:8194849
72. Al-Okbi SY, Mohamed DA, Hamed TE, Edris AE (2014) Protective effect of clove oil and eugenol microemulsions on fatty liver and dyslipidemia as components of metabolic syndrome. J Med Food 17:764–771
73. Ahmad N, Ahmad R, Alam MA, Ahmad FJ (2018) Quantification and brain targeting of eugenol-loaded surface modified nanoparticles through intranasal route in the treatment of cerebral ischemia. Drug Res 68:584–595
74. Seo EJ, Min SG, Choi MJ (2010) Release characteristics of freeze-dried eugenol encapsulated with beta-cyclodextrin by molecular inclusion method. J Microencapsul 27:496–505
75. Wang L, Zhang Y (2017) Eugenol nanoemulsion stabilized with zein and sodium caseinate by self-assembly. J Agric Food Chem 65:2990–2998
76. Zhao X, Chen D, Gao P, Ding P, Li K (2005) Synthesis of ibuprofen eugenol ester and its microemulsion formulation for parenteral delivery. Chem Pharm Bull (Tokyo) 53(10):1246–1250
77. Shen Q, Wang Y, Zhang Y (2011) Improvement of colchicine oral bioavailability by incorporating eugenol in the nanoemulsion as an oil excipient and enhancer. Int J Nanomedicine 6:1237–1243

Regulatory Considerations for Safety of Nanomaterials

Jasreen Kaur, Harpreet Singh, and Madhu Khatri

1 Introduction

In the scientific community, the definition of nanomaterials (NMs) and their regulatory evaluations have always been an area of active scientific and political debate. The nanomaterials tend to exhibit properties that are quite different from the ones exhibited by their non-nanoforms. This has raised various questions about the risks they possess to human health and environment. Various definitions of nanomaterials have been proposed by various committees around the world, but it is a matter of fact that these definitions, which focus primarily on the size factor, may be insufficient from a risk evaluation standpoint because other important elements are not included. These elements such as dose, surface area, and chemical composition should be considered while determining whether a nanomaterial needs additional caution for their use and disposal. There is a dire need to set up basic regulations for the controlled release and disposal of nanomaterials. Various factors need to be considered to set up these guidelines (Fig. 1). The first factor that needs to be considered is the definition of nanoparticles (NPs). As discussed above, the definition of nanoparticles plays an important role in setting up proper guidelines for their use

J. Kaur
University Institute of Engineering and Technology (UIET), Panjab University, Chandigarh, India

Centre for Nanoscience and Nanotechnology, Panjab University, Chandigarh, India

H. Singh
University Institute of Engineering and Technology (UIET), Panjab University, Chandigarh, India

M. Khatri (✉)
University Institute of Engineering and Technology (UIET), Panjab University, Chandigarh, India

Wellcome Trust/DBT India Alliance, Hyderabad, India
e-mail: madhuk@pu.ac.in

© Springer Nature Singapore Pte Ltd. 2020
M. Rahman et al. (eds.), *Nanomedicine for Bioactives*,
https://doi.org/10.1007/978-981-15-1664-1_17

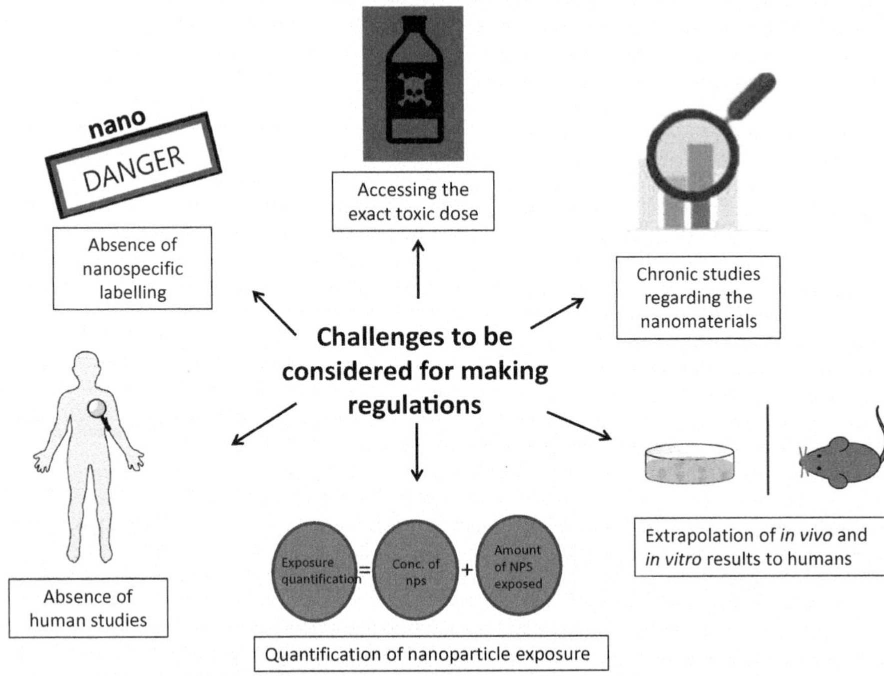

Fig. 1 Challenges to be considered by regulatory authorities

and disposal as the toxicity of nanomaterials tends to change with various factors. Secondly, the physiochemical characterization of nanomaterials also plays an important role in setting up regulatory policies for nanomaterials as toxicities of nanomaterials tend to change with slight modification in various factors such as size, surface area, and porosity. Some nanomaterials are toxic at one size while their toxicity subsides at a greater rate at a different size. Several studies have been performed on this basis and convincing results have been observed. Other important factor that needs to be considered is the process by which these nanomaterials enter into the environment and human body as the difference in the process of entry of these nanomaterials may lead to different outcomes. These nanoparticles cause different physiological disorders when they enter through different routes of entry; therefore, it is mandatory to consider different routes for setting up certain guidelines. The most important factor that should be considered for setting up guidelines is the toxicity of these nanomaterials. Different nanomaterials tend to have toxicities, such as DNA damage and protein misfolding, which have been tested using various in vitro and in vivo models. Therefore, it is mandatory to give these nanomaterials a hazard ranking which will be useful for forming regulations for their use. Lastly, many guidelines have been set up in last few years that need refinement and renewal after the consideration of abovementioned factors. These factors have been discussed in detail in the following sections.

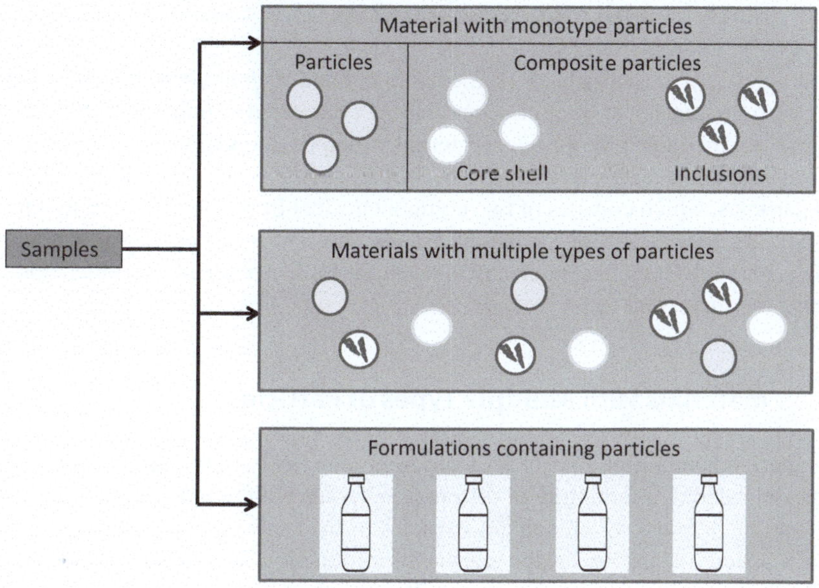

Fig. 2 Types of nanomaterials in suspension

2 Identification and Information of Nanoparticles

The first step in testing the safety of nanomaterials is their adequate and unequivocal identification. In case of materials that are not present in nanorange, identification is possible on the basis of chemical identity and molecular structure, but in case of nanomaterial, physicochemical characterization is mandatory. There are various definitions and criteria for identification of nanomaterials, but size distribution and surface area are considered as essential parameters for their identification. In previous studies, nanomaterials have already been categorized in various contexts such as risk assessment, detection in foods, description of nanomaterial data resources, or toxicology. Further grouping of nanomaterials can be done depending on their chemical nature, toxicology in human beings, importance in commercial aspects, or various modes of action in a physiological context. The generated measurement results, which are obtained by various characterizations, should then be sufficiently robust to serve as a basis for whether a material meets a regulatory size criterion for nanomaterial.

The material categorization scheme (MCS) has been introduced which is technique driven and proves as a pragmatic approach to facilitate the identification of nanomaterials for regulatory purposes (Fig. 2). Materials are divided into groups, according to the following categories.

2.1 Material with Monotype Particles

All particles have essentially the same chemical and structural composition. These particles may consist of a single chemical or compound element or different elements or compounds with the same internal structure also known as composite particles. These composite particles can be further divided into many forms like core shell particles, which consists of two components that lie within the other that forms the shell [1]. Particles with inclusions are also present in which one component is phase-separated from the other and one phase is dispersed in the other, thereby forming the inclusions [2].

2.2 Materials with Multiple Types of Particles

All materials containing particles of different chemical or structural compositions can be considered as a mixture of different materials with monotype particles [3]. It may contain articles, which contain particles of the same or different types and are given a special shape or design during its production. This shape and size will decide its function to a greater degree than its chemical composition. They may consist of nanomaterials of one or several types depending upon the chemical substances in different physical phases (liquid/solid/ gaseous) [4]. Similarly, a formulation is a combination of different chemicals that do not react chemically with each other and are chosen to form a product with desirable characteristics. For example, a sunscreen containing titanium dioxide (TiO_2) nanoparticles is considered as a formulation, and a tennis racquet with incorporated carbon nanotubes (CNTs) is considered as an article.

Labeling of nanomaterials is also an integral part of identification and information of nanomaterials. It is the action of putting the label on a product, which gives important information about the content and composition of the product, from where the product originated, the dangerous substances present in the product, and also information on the proper use of nanomaterials. The main objective of labeling of product is to guarantee access is provided to customers regarding certain information, so that they can make informed decisions to make a choice and/or to take precautions regarding their health or the environment. A product containing nanomaterials should be clearly mentioned so that it gives the consumer the liberty to select or avoid such products. This can be achieved by ingredients labeling or by positive or negative nanoclaims.

3 Characterization and Physicochemical Properties of Nanomaterials

Nanoparticles are different from their bulk material counterparts as they possess distinct physicochemical properties such as size, shape, surface properties, molecular weight, composition, purity, identity, solubility, and stability. These factors are critically relevant to particular physiological interactions that these nanomaterials undergo, and therefore, these may provide benefits in medical applications as well as other areas of industries such as electronics and semiconductors. In order to set up regulations for the safety of these nanomaterials, it is important to understand how the different physicochemical characteristics of nanomaterials affect their distribution in the environment and human beings and their behavior. This demands various robust techniques for studying the different physicochemical characteristics of nanomaterials. The different techniques used for characterization of nanomaterials, based on their different features, are described in the following sections.

3.1 Chemical Characterization

3.1.1 Chemical Composition of Nanomaterials

The chemical composition is a fundamental descriptor for nanomaterials, and it can be reported as a regulatory requirement in all nanosafety studies. The methods to be used for determining the chemical composition should be carefully selected for safety measurements. The properties to be taken into consideration are the composition of the nanomaterials and impurities, as some techniques seem to be relevant only for one or the other type of materials [2]. The techniques used for checking the chemical composition are listed in Table 1.

3.1.2 Surface Chemistry

It refers to the composition of nanomaterials in terms of elements and functional groups. A specific surface chemistry may be calculated and checked through one or more treatment steps under well-defined conditions [5]. It is one of the relevant steps in regulatory considerations of nanomaterials. It is well known that nanomaterials and nanostructured materials have a much larger specific surface area (SSA) as compared to bulk materials. Therefore, whenever any interaction occurs between the surface and its surroundings including the ones arising from specific chemical composition, it is proved to be more relevant for nanomaterials than for bulk materials [6]. "Surface chemistry" part generally focuses on the chemical composition of the particle surfaces and not functional groups. The techniques used for checking the chemical composition are listed in Table 1.

3.2 Physical Characterization (Table 1)

3.2.1 Particle Size and Distribution

The particle size and particle size distribution are highly relevant in consideration for regulatory purpose because their values define whether the particle lies in nanorange. This is considered as the first step for physicochemical characterization of nanomaterials [7]. Size of the nanomaterials cannot be elucidated by a single technique, but several techniques are used based on various physical phenomena (e.g., based on particle imaging, sedimentation phenomena, or light scattering) [8]. All

Table 1 Various methods for physicochemical characterization of nanomaterials

Methods	Advantages	Disadvantages
Chemical composition		
EDX(EDS) (energy dispersive X-ray)	• Suitable for detecting elements with an atomic number more than 5 • Useful to retrieve information on spatial atomic distribution	• Not suitable for nanoparticles of complex composition, in complex matrices • Not suitable for large aggregates
ICP OES (inductively coupled plasma optical emission spectrophotometry)	• Provides quantitative information about the concentration of an element within the sample • 70 elements can be analyzed simultaneously	• Emission lines of different elements interfere which makes the analysis of complex samples difficult
ICP MS (inductively coupled plasma mass spectrometry)	• ICP-MS is more sensitive than ICP-OES	• Performance for the lighter elements poor due to background atmosphere in the flame-based measurement • No information on the crystal structure of the material is provided
XPS (X-ray photoelectron spectroscopy)	• Suitable for characterizing the chemical composition of uncoated materials such as metal oxides and CNTs	• Works only if surface layer (0–10 nm) is representative for the core material
XRD (X-ray diffraction)	• Reliable method for identification of their chemical composition	• Not useful for quantifying stoichiometric variations. • Best used in combination with other direct methods
Surface chemistry		
XPS (X-ray photoelectron spectroscopy)	• Determination of the elemental composition of nanomaterial surfaces • Suitable for measuring the composition of a 0–10 nm surface layer	• Suitable for materials which are stable under ultrahigh vacuum conditions
ToF SIMS (time-of-flight secondary ion mass spectrometry)	• Suitable for analysis of the composition of solid surfaces and thin films • Very sensitive qualitative technique (in the range of ppm to ppb) • Elemental, isotopic, or molecular composition of the surface analysis can be done to a depth of 1–2 nm	• Difficulty in distinguishing between the properties of the core and surface of the particle • Sample mounting (on a carbon tape) limits the analysis of C and Si in the sample
AES (auger electron spectroscopy)	• Information gathered on the elemental composition of a sample in the 2–20 nm depth range	• Ultrahigh vacuum is required

(continued)

Table 1 (continued)

Methods	Advantages	Disadvantages
XRF (X-ray fluorescence spectroscopy)	• Applied for chemical composition analysis • Identifies a wide range of elements with detection limits (LOD) in the ppm range • No particular treatment required for samples before analysis, except removal of surface contamination	• Vacuum conditions required for detection of light elements
Particle size distribution		
DLS (dynamic light scattering)	• Nondestructive, ensemble technique to measure particle size distribution • Rapid and more reproducible measurement • Measurement in any liquid media is possible	• Quantity of smaller particles underestimated in mixtures and polydisperse samples • Analysis depends on the dispersibility and state of dispersion of the sample • Data evaluated based on the assumption that the particles are spherical Cannot distinguish between individual particles and aggregates/agglomerates
CLS (centrifugal liquid sedimentation)	• Particle density-based ensemble technique • Depends on the dispersibility and state of dispersion of the sample • Information on the Stokes diameter of the particles also provided	• Cannot distinguish between individual particles and aggregates/agglomerates
TEM (transmission electron microscopy)	• Give proper evaluation of size and shape of the materials including agglomeration and aggregation	• Suitable for nanoparticles that are not affected by the vacuum or electron beam used and do not agglomerate on the grid Ultrathin sample required • Expensive equipment
SEM	• Sample preparation is easier than for TEM	• Suitable for nanoparticles that are not affected by the vacuum or electron beam used and do not agglomerate on the grid. • Expensive equipment. • Provides information on conductive materials only
UV-vis (ultra-violet–visible analysis)	• Suitable for measurement of agglomeration/aggregation	• Suitable only if particles are monodispersed and perfectly spherical

(continued)

Table 1 (continued)

Methods	Advantages	Disadvantages
SMPS (scanning mobility particle sizer)	• Suitable for measurement of aerosols • Provides information on the size of particles, agglomerates, and aggregates	• With the assumption that the particles are spherical
AFM (atomic force microscopy)	• Gives 3D information on the particles • Direct measurement of samples in dry or aqueous environments	• Time consuming and poor sampling • Analysis limited to the outer surface only
(U)SAXS ((ultra)small angle X-ray scattering)	• Suitable for investigating biological samples and complex matrix materials • Simple sample preparation	• Cannot be used as an only technique for size determination • Confirmation of results from other techniques required
Specific surface area		
BET (surface area calculated from gas adsorption, based on the Brunauer-Emmett-Teller method)	• Suitable for nanomaterials which do not absorb the gas used	• Applicable only for dry solid samples
SAXS (small angle X-ray scattering)	• Suitable for investigating biological samples and complex matrix • Simple sample preparation	• Not recommended as a primary method to determine specific surface area
Porosity		
Gas adsorption method, modeledbyBET(Brunauer-Emmett-Teller theory)	• For determining porosity is standardized for non-microporous materials	• Applicable only for dry solid samples
Mercury intrusion porosimetry	• Give proper information on porosity	• High pressure required for measurement • Deformation of the material tested, so not suitable for metal-containing nanomaterials
Crystalline phase and size		
XRD (X-ray diffraction)	• Best suitable method to determine crystallinity of nanomaterials • Well-established method	• No size determination possible
SAXS	• Delivers structural information of nanoparticles • Confirmation that materials are amorphous or nonhomogeneous	• Relatively low resolution
TEM	• Crystalline structures can be seen at very high magnifications	• Considered as a "pushing the method" • Require significant beam time to measure a number of particles

(continued)

Table 1 (continued)

Methods	Advantages	Disadvantages
DSC (differential scanning calorimetry)	• Suitable method for investigating phase transitions	• No information on the type of crystallinity or size of crystals provided
Shape		
TEM/SEM	• Sufficient resolution is provided for reliable image and thereby information on the shape of the nanoparticles	• Gives a 2D projection of 3D particles, so avoiding bias due to preferential orientation effects should be considered
AFM/STM	• Provide 2.5D images of particles	• Shape analysis prone to errors. Time consuming and expensive

the methods used for determining particle size and distribution have various advantages and limitations, and therefore, multiple techniques need to be used in order to get a complete overview of the material [9].

The different techniques used for size determination require that the nanoparticles remain in a particular state so that they are accessible to the different techniques of measurement [10]. Nanomaterials are generally available in two main states: as dry powders or in suspension. Therefore, sample preparation is an important step for characterization of, for example, the dispersion of these nanomaterials in the medium to be used for studies [11]. This may lead to agglomeration and increase in their size that is where size distribution analysis plays an indispensable role. Most nanoparticles tend to be not in a perfect shape and may have many imperfections that may completely change their toxic potential [12]. Many characterization techniques till date are not nanospecific; thus, it is mandatory to look for techniques that are relevant to the particular nanomaterials.

3.2.2 Specific Surface Area

The (volume) specific surface area [(V)SSA] is considered to be relevant in a regulatory framework as it can be used as a parameter to decide whether a given material is a nanomaterial [7]. Nanoparticles tend to have more surface area than their bulk counterparts due to their small size. This increased surface area can lead to a drastic change these nanomaterials have with different environments and human body as compared to its bulk counterparts [10]. Various techniques are discussed in Table 1.

3.2.3 Porosity

Porosity is also relevant in a regulatory context as this parameter is relevant to the inner surface of a nanomaterial. For porosity, several pore sizes are described as macropores, mesopores, and micropores, and therefore, different methods need to be used to measure the different types [13]. Standardized measurement methods for porosity have been proposed by ISO (e.g., [14–16]). Various techniques are discussed in Table 1.

3.2.4 Crystallite Size and Crystalline Phase

Crystalline phase, and for crystalline materials also crystallite size, also acts as a regulatory end point (e.g., [17, 18]), and this plays an important role in characterization of the material. It is well-established that materials with the same composition but having different crystalline structure tend to exhibit different toxicological pathways or reactivity to various materials.

3.2.5 Shape

Shape is an important characterization technique in regulatory considerations. It has been observed that nanomaterials have various optical properties related to its shape. Some particles tend to have well-defined shapes, while the others do not. Moreover, sometimes, the particle's shape within the nanomaterials varies significantly. The measurement of the shape of nanomaterials is challenging because the concept of shape is not easily quantifiable. Furthermore, various methods of particle size determination give results that are based on the assumption that the particles are having well-defined particles. Thus, in order to interpret the measurements correctly, it is mandatory to elucidate the real particle [10, 18–20]. In order to quantify shape, various descriptors such as sphericity, convexity, aspect ratio, and fractal have been devised.

4 Entry of Nanomaterials in Environment

In order to set up regulatory guidelines, knowledge about the hazard potential of nanomaterials is not sufficient. Proper knowledge about the release of nanomaterials and the resulting exposure is mandatory. New studies in recent years have provided new insights about the behavior and effects of nanomaterials in the environment. Identification of processes and mechanisms has been done for the description of their behavior and their impact in the environment. These findings are summarized in the following subsection.

4.1 Entry in the Environment

In addition to the chemical composition of a nanomaterial, the potential hazards of nanomaterials can be determined by properties such as its size, geometry, crystal structure, and surface properties (e.g., charge, surface chemistry) [21]. The properties of nanomaterials can be easily influenced by various environmental parameters such as salinity, pH value, and natural organic substances. It can affect the mobility, bioavailability, and the toxic effects of nanomaterials in the environment [50]. Manufactured nanomaterials tend to enter the environment through various ways such as intentional as well as unintentional release. In order to assess the environmental exposure of environment to nanomaterials, it is important to gain an insight into various products and applications containing nanomaterials and their release over their life cycle, i.e., production, use, transport, recycling, and waste disposal [49]. In addition, it is important to know about the role of nanomaterials in the various products and applications [22].

Nanomaterials are being used in certain industries such as printing industry, paints, fabrics, and healthcare products, including sunscreens and cosmetics, and therefore find a portal for entry into the environment [23] (Fig. 3). Emitted particles gradually deposit on land and surface water bodies, and therefore, they have the potential to contaminate soil, thereby migrating into surface and groundwater, which finally leads to their interaction with biota [24]. Aquatic systems receive their share from particles in solid wastes, effluents released from wastewater, direct discharges, or accidental spillages by wind or rainwater runoff. The biggest risk for environmental release is associated with spillages occurring due to transportation of manufactured nanoparticles from production facilities to other manufacturing sites [25]. Several investigations have already been conducted to study the uptake, accumulation, and persistence of these nanomaterials in environmental organisms. Most of these studies were conducted using invertebrates such as water fleas, fishes, and earthworms [26]. It has been found that these nanomaterials have potential to accumulate in low levels in these organisms, and the excretion of these nanomaterials plays an important role for their entry into the environment.

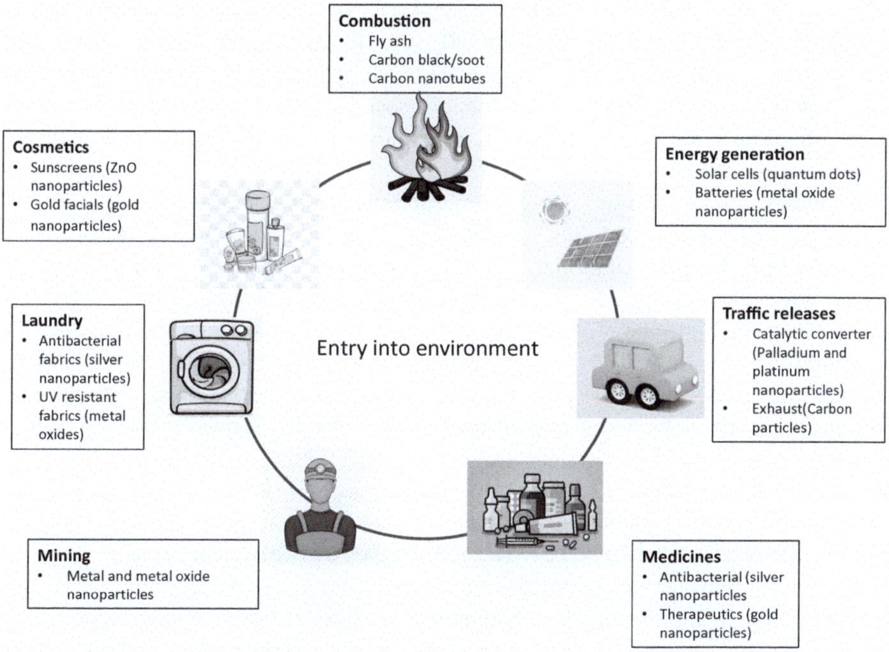

Fig. 3 Ports of entry of nanomaterials into environment

4.2 Entry to the Human Body

In order to set up certain guidelines for safety regarding nanomaterials, it is imperative to consider the routes of entry of nanomaterials into the human body so that proper safety protocols can be developed and implemented. The primary route of entry of these nanomaterials into human body is through inhalation exposure, which can potentially occur to airborne nanomaterials. These nanomaterials tend to agglomerate into larger particles, which leads to the alteration in their properties, and therefore alter their behavior in lieu of their potential exposure and entry into the human body [27, 28]. They can deposit in respiratory systems causing severe toxicities due to their high surface area, morphology, etc. Several particles are formed after the degradation of nanomaterials, which can also possess risk if they exhibit biological activity similar to nanoparticles [29]. Nanoparticles tend to have high deposition efficiencies in lungs of healthy individuals as well as the ones suffering from diseases such as asthma or chronic obstructive pulmonary diseases (Fig. 4).

The skin can also serve as one of the portals for entry of nanomaterials into the human body. This exposure can be intentional or unintentional depending on the surroundings of the individuals to which they are being exposed. Intentional exposure to these nanomaterials includes the application of lotions, creams, and detergents containing silver nanomaterials as one of the common ingredients on a daily basis. Even the sunscreens contain nano-TiO_2 and nano-ZnO for their water-repellant and UV protection properties [30, 31]. The latest wound dressings contain nanomaterials such as silver and copper oxide nanoparticles due to their antimicrobial properties [32]. Several unintentional exposures include the dermal contact of skin with anthropomorphic substances, which are released during the manufacture, or combustion of these nanomaterials [33].

Another route of exposure is ingestion; humans can ingest nanoparticles directly in the form of food ingredients, additives, and supplements [34]. Additionally, a large number of inhaled nanoparticles have a tendency to move with the help of mucociliary cells, through the trachea through which it enters the stomach. NPs can also get absorbed into the bloodstream and consequently can reach each organ after crossing the epithelium. TiO_2 or silicates/aluminosilicates are used on a large basis in food industries and in the pharmaceutical industry as an additive. TiO_2 is also present in sweets as an additive with the quantity ranging between 1 and 5 μg/mg [35, 36].

5 Toxicity and Environmental Risks of Nanomaterials

In order to set up certain rules and regulations for safety of these nanomaterials, it is mandatory to have a prior knowledge about the toxicity of various nanomaterials available in the environment. Toxicity of nanomaterials will be specific to the type of base material, shape, coating, and size. In order to study the toxic effects of nanomaterials, various cell lines, conditions for their culture, and incubation times are being used.

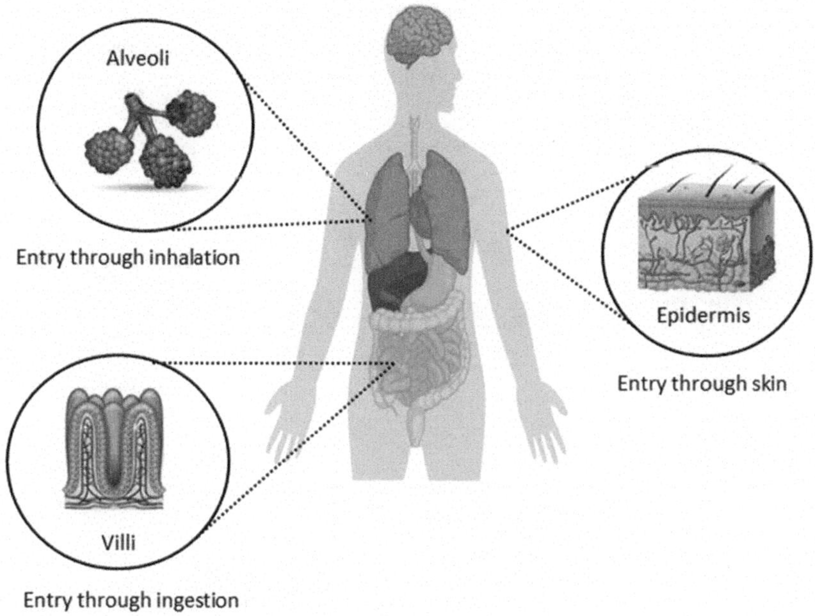

Fig. 4 Portals of entry of nanomaterials into human body

Various biological models including cell lines, *Saccharomyces cerevisiae*, *Daphnia magna*, *Drosophila*, zebra fishes, and mice models are being used for the toxicity testing of nanomaterials. Extensive range of nanotechnology products is present, and therefore, these are broken down into different classes mainly metal, metal oxides, and carbon nanomaterials (Fig. 5).

5.1 Metal Nanoparticles

Metallic nanoparticles are the most extensively used class of engineered nanomaterials. According to previous studies, gold has been known to be safe after performing various assays for studying cellular uptake and cellular toxicity of nanomaterials. Previously scientists investigated the intracellular uptake of different types of gold nanoparticles and found that spherical nanoparticles were not inherently toxic to the human cells, in spite of the fact that they were taken up by the cells [47]. Similar studies were done with immune cells, and it was found that these nanoparticles were not toxic; in fact, they lead to a reduction in the amount of potentially harmful reactive oxygen species in the cells [49]. Exposure to silver leads to the deposition of silver in the skin and other parts of the body leading to breathing problems, allergic reactions such as rashes, swelling, and inflammation, etc. Size-dependent interaction of silver nanoparticles was shown in previous studies where viability metrics tend to decrease with increasing dose of Ag-15 nm and Ag-30 nm nanoparticles [37].

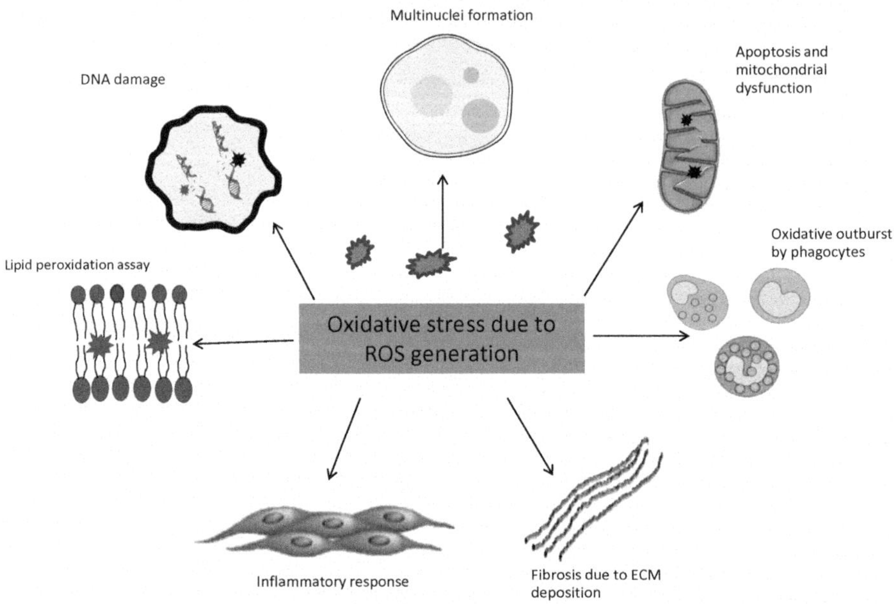

Fig. 5 Mechanism of toxicity of nanomaterials in human body

Similarly, it was reported that particle composition also has a significant effect on toxicity of metallic nanoparticles in aquatic organisms mainly zebra fish and daphnia species. These were exposed to different nanomaterials, namely, silver, copper, aluminum, nickel, and cobalt, and it was observed that nanosilver and nanocopper caused toxicity in all organisms at lethal concentrations of about 40 μg/L [38, 39]. Many manufacturers across the world are working on the clothing articles that have used nanosilver as one of the antimicrobial agent, which may have some environmental effects upon release of silver released from socks into water. The results have shown both ionic and colloidal silvers released from the socks [40].

5.2 Metal Oxide Nanoparticles

Metal oxide nanoparticles are used wisely as additives in pharmaceuticals, food colorants, and cosmetics. Various groups have worked to decode the toxicity of these nanomaterials, like Shah et al. [41] who studied the inhalation exposure of titanium dioxide nanoparticles. They showed that these nanoparticles tend to form aggregates and they showed inflammatory response in animals. Similarly, studies were done to evaluate the toxicity of TiO_2 nanoparticles using NRK52E cells [42]. A significant increase in oxidative stress in these cells was observed at highest concentrations with a dose-dependent increase. They also showed that nanoparticles penetrate in the plasma membrane and got localized in periregions of nuclear membrane, which emphasized on the fact that these nanomaterials have direct interaction with cellular molecule, which causes adverse biological responses in cells [43].

Phytotoxicity studies were also done which showed both positive and negative effect of these nanomaterials on higher plants. It was observed that alumina nanoparticles inhibited the root elongation of corn, cucumber, soya bean, cabbage, and carrot [44]. Lin and Xing [45] confirmed the toxicity of ZnO nanoparticles by examining cell internalization and upward translocation by *Lolium perenne* (ryegrass). Their result showed that ZnO nanoparticles reduced the ryegrass biomass significantly, shrank the root tips, and vacuolated the root epidermal and cortical cells. Freshwater microalgae, *Pseudokirchneriella subcapitata*, was used to evaluate the comparative toxicity of nanoparticulate ZnO, bulk ZnO, and $ZnCl_2$, which found that toxicities caused by these nanomaterials were comparable. In one of the studies, seven industrially important nanoparticles were used for toxicity testing in a human mesothelioma and a rodent fibroblast cell lines. Their toxicities were compared to the effect of nontoxic amorphous silica and toxic crocidolite asbestos [39]. Iron oxide nanoparticles were found to be the most toxic ones. Studies were done to test the nanotoxicity of different metal oxide nanoparticles by assessing their cytotoxicity and ability to cause DNA damage and oxidative stress. Copper oxide nanomaterials were found to be the most toxic with high toxicity and DNA damage, followed by zinc oxide. Titanium oxide was found to just cause DNA damage [51].

5.3 Carbon Nanomaterials

Carbon nanotubes (CNTs) are emerging as a new class of multifunctional building blocks, which are accounting for the development of nanotechnology. The number of multiwalled carbon nanotubes (MWCNTs) being used in industries is increasing due to the presence of industrial-scale facilities for the relatively low-cost production, and therefore, exposures to these tend to increase significantly in the coming years. Several studies have been done to examine the uptake and potential hazards of CNTs to humans and other biological systems. It has been observed in several studies that CNTs induce inflammatory responses in human T-cells. Liu et al. [46] also studied cytotoxicity of carbon-based nanomaterials and demonstrated that water-soluble CNTs did not have any toxic effects when tested in a wide variety of immune system cells. Several studies have demonstrated that ultrafine carbon particles show lung penetration and have the ability to cross the blood-brain barrier, therefore causing toxicity in the central nervous system.

Dose plays an important role in toxicity testing of nanomaterials. By definition, dose is the amount of a substance administered to the individual at one time. However, other parameters are also necessary to characterize the exposure these nanomaterials give according to the dose. A dose can either be considered as a measurement of environmental exposure or the quantity of substance administered to an individual over a period of time. Dose can be divided into two types, namely, adsorbed dose, i.e., the amount of dose that primarily enters the body through various routes of exposure or is taken up by organs, and the administered dose, i.e., the quantity that is administered orally or by injection. In case of nanotoxicity studies, the adsorbed dose plays an important role, as it is the exact dose that is toxic to the individuals. These differences in definition lie because all substances that enter the body are not necessarily adsorbed by it. An organism's age and body size affects the toxic effects of a given dose of nanomaterials. Thus, an organism's body weight plays an important role in dose evaluation.

6 Current Regulations for Nanomaterial Usage and Disposal

Several steps have already been taken to combat the risks associated with unavoidable exposure of humans to the nanomaterials. Various regulations via legislation, laws, and rules have been implemented all around the world by several government organizations to minimize or avoid the risks associated with nanomaterial exposure. However, no specific international regulation on protocols or legal rules for production, handling, or labeling of these nanomaterials, toxicity testing, and evaluation of the environmental exposure exists. In case of biomedical fields, several medical standards do exist in relevance to their ethics and environmental safety. It has been observed that the USA and the European Union (EU) have set up strong regulatory bodies and guideline legislation to cope up the risks caused by these nanomaterials. Several developments have been made by the European Commission in reference to the nanomaterials use (D'Silva and Van 2016). The legislation and guidelines for nanomaterial use have been employed inside European countries so that conformity exists across legislative areas, and also, treatment of nanomaterials is the same among all the sectors they are being used. The definition of nanomaterials given by European Commission means, " a materials which may be natural, incidental or manufactured containing particles, which are either present in an unbound state or as an aggregate, and also in which 50 percent or more of the particles have one or more external dimension in size ranging from 1 nm to 100 nm." The EU has also formed a committee named the Scientific Committee on Emerging and Newly Identified Health Risks (SCENIHR) to estimate the risks associated with nanoparticle exposure. EU cosmetic Regulation 1223/2009 defines the term nanomaterial as "an intentionally manufactured, insoluble and bioresistant materials that has one or more dimension in the range of 1nm to 100 nm." It is to be noted that cosmetics face regulation from various committees, namely, from the USFDA's Federal Food, Drug, and Cosmetic Act (FFDCA), Voluntary Cosmetic Registration Program (VCRP), Personal Care Products Council (PCPC), REACH, Scientific Committee on Consumer Safety (SCCS), EU Cosmetics Product Notification Portal (CPNP), and International Cooperation on Cosmetic Regulation (ICCR). These regulations from the USA, EU, and Japan state that exposure via cosmetics is of great concern for industries producing these as scientific policymakers, primarily because of their daily use worldwide. In the USA, several agencies, namely, the US Environmental Protection Agency (USEPA), the Institute for Food and Agricultural Standards (IFAS), and the Food and Drug Administration (FDA), have set up various protocols to combat various risks associated with nanomaterials and their related products. The FDA has been working on identifying various sources of nanomaterials, their environmental exposures, and several risks associated with these nanomaterials followed by the strategies to mitigate these risks. The medical usage of hazardous nanomaterials is taken care by the European Medicines Agency (EMEA) and US Food and Drug Administration (USFDA). Special edition called the Nanomaterials Policy Recommendations report covers several ideas to reduce the risks related to NMs in food-related industries. This report also emphasizes on the adoption of detailed public policy of nanomaterial usage, nanomaterial safety analysis, supplier standards evaluation, and other methods by several companies using nanomaterials. Various organic suppliers, which constitute the UK Soil Association, the Canada General Standards Board, and the Biological Farmers of Australia, have already put a ban on use of nanomaterials in food.

7 Conclusion

Nanoparticles are being used almost every day worldwide, therefore leading to a dire need to set up certain rules and guidelines, so that these nanomaterials do not pose a threat to human and environmental life. It can be concluded from abovementioned studies that various factors need to be considered in order to set up various regulations for the use and disposal of nanoparticles.

References

1. Hodoroaba V, Ghanem A (2019) Nanoscale advances a technique-driven materials categorisation scheme to support regulatory identification. Nanoscale Adv 1:781–791. https://doi.org/10.1039/c8na00175h
2. Wohlleben W, Mielke J, Bianchin A, Ghanem A (2017) Reliable nanomaterial classification of powders using the volume specific surface area method. J Nano Res 19:61. https://doi.org/10.1007/s11051-017-3741-x
3. Szakal C, Tsytsikova L, Carlander D, Duncan TV (2014) Measurement methods for the oral uptake of engineered nanomaterials from human dietary sources: summary and outlook. Compr Rev Food Sci Food Saf 13(4):669–678
4. Arts JHE, Hadi M, Irfan M, Keene AM, Kreiling R, Lyon D (2015) A decision-making framework for the grouping and testing of nanomaterials (DF4nanoGrouping). *Reg Tox and Pharma 71*(2):S1–S27. https://doi.org/10.1016/j.yrtph.2015.03.007
5. Patel P, Shah J (2017) Safety and Toxicological Considerations of Nanomedicines: The Future Directions. Curr Clin Pharmacol 12(2):73–82
6. OECD (2016) Testing Programme of Manufactured Nanomaterials—Dossiers and Endpoints. Available at http://www.oecd.org/chemicalsafety/nanosafety/dossiersand-endpoints-testing-programme-manufactured-nanomaterials.htm
7. EC (2011) Commission Recommendation of 18 October 2011 on the Definition of Nanomaterial (2011/696/EU). pp. 38–40; OJ L 275
8. ISO 17867:2015 (2015) Preview Particle Size Analysis—Small-angle X-ray Scattering
9. Linsinger T, Roebben G, Gilliland D, Calzolai L, Rossi F, Gibson N, Klein C (2012) Requirements on measurements for the implementation of the European commission. Regul Toxicol Pharmacol 92:8–28

10. US-EPA (2017) Chemical substances when manufactured or processed as nanoscale materials: TSCA reporting and recordkeeping requirements. https://www.regulations.gov/document?D=EPA-HQ-OPPT-2010-0572-0137
11. Gilliland D, Gibson P, Hempelmann U, Vicent Pena JB, Kund K, Wohlleben W, Koch T, Burke A, Mcnulty G, Hartl-Gunselmann A, Knobl S, Reisinger M, Sokull-Klüttgen B, Stamm H, Liewald H (2014) Basic comparison of particle size distribution measurements of pigments and fillers using commonly available industrial methods. EUR. https://doi.org/10.2788/21024.26916
12. EC (2012) Regulation (EU) 528/2012 of the European Parliament and of the Council of 22 May 2012 Concerning the making available on the market and use of biocidal products. OJ L 167/1
13. OECD (2014) OECD Series on the Safety of Manufactured Nanomaterials, No. 41. Report of the OECD Expert Meeting on the Physico-chemical Properties of Manufactured Nanomaterials and Test Guidelines. OECD, Paris, France, pp. 56. ENV/JM/MONO (2014) 15. Available at http://www.oecd.org/officialdocuments/publicdisplaydocumentpdf/?cote=ENV/JM/MONO(2014)15&docLanguage=En
14. ISO 15901–1:2016 (2016) Evaluation of Pore Size Distribution and Porosity of Solid Materials by Mercury Porosimetry and Gas Adsorption—Part 1: Mercury Porosimetry
15. ISO 15901–15902:2006 (2006) Pore Size Distribution and Porosity of Solid Materials by Mercury Porosimetry and Gas Adsorption—Part 2: Analysis of Mesopores and Macropores by Gas Adsorption
16. ISO 15901–15903:2007 (2007) Pore Size Distribution and Porosity of Solid Materials by Mercury Porosimetry and Gas Adsorption—Part 3: Analysis of Micropores by Gas Adsorption
17. ECHA (2017a) Guidance on Information Requirements and Chemical Safety Assessment. Appendix R7-1 Recommendations for Nanomaterials Applicable to Chapter R7a Endpoint Specific Guidance, Version 2.0, May 2017. European Chemicals Agency. https://echa.europa.eu/documents/10162/13632/appendix_r7a_nanomaterials_en.pdf/
18. EFSA (2011) Guidance on the risk assessment of the application of nanoscience and nanotechnologies in the food and feed chain. EFSA J 9(5):2140. https://doi.org/10.2903/j.efsa.2011.2140. 36 pp
19. EC (2006) Regulation (EC) No 1907/2006 of the European Parliament and of the Council of 18 December 2006 Concerning the Registration, Evaluation, Authorisation and Restriction of Chemicals (REACH), Establishing a European Chemicals Agency, Amending Directive 1999/45/EC and Repealing Council Regulation (EEC) No 793/ 93 and Commission Regulation (EC) No 1488/94 as Well as Council Directive 76/ 769/EEC and Commission Directives 91/155/EEC, 93/67/EEC, 93/105/EC and 2000/21/EC. OJ L 396:1–849
20. ECHA (2017b) How to Prepare Registration Dossiers that Cover Nanoforms: Best Practices, Version 1.0, May 2017. https://echa.europa.eu/documents/10162/13655/how_to_register_nano_en.pdf
21. Hansen SF, Baun A (2012) European Regulation Affecting Nanomaterials - Review of Limitations and Future Recommendations. Dose Response 10(3):364–383
22. Shandilya N et al (2015) Emission of titanium dioxide nanoparticles from building materials to the environment by wear and weather. Environ Sci Technol 49(4):2163–2170
23. Gondikas AP et al (2014) Release of TiO_2 nanoparticles from sunscreens into surface waters: a one-year survey at the old Danube recreational Lake. Environ Sci Technol 48(10):5415–5422
24. Hartmann NB et al (2014) Environmental fate and behavior of nanomaterials—new knowledge on important transformation processes. Environmental Project No. 1594 of the Danish Environmental Protection Agency
25. Hou WC et al (2013) Biological accumulation of engineered nanomaterials: a review of current knowledge. Environ Sci Process Impacts 15(1):103–122
26. Walser T et al (2012) Persistence of engineered nanoparticles in a municipal solid waste incineration plant. Nat Nanotech 7:520–524

27. Miller R, Raftis JB, Langrish JP, McLean SG, Samutrtai P, Connell SP, Wilson S, Vesey AT, Fokkens PHB, Boere AJF et al (2017) Inhaled nanoparticles accumulate at sites of vascular disease. ACS Nano 11:4542–4552. https://doi.org/10.1021/acsnano.6b08551
28. Van der Zande M, Vandebriel RJ, Van Doren E, Kramer E, Herrera Rivera Z, Serrano-Rojero CS, Gremmer ER, Mast J, Peters RJ, Hollman PC et al (2012) Distribution, elimination, and toxicity of silver nanoparticles and silver ions in rats after 28-day oral exposure. ACS Nano 6:7427–7442. https://doi.org/10.1021/nn302649p
29. Pujalté I, Dieme D, Haddad S, Serventi AM, Bouchard M (2017) Toxicokinetics of titanium dioxide (TiO_2) nanoparticles after inhalation in rats. Toxicol Lett 265:77–85. https://doi.org/10.1016/j.toxlet.2016.11.014
30. Bennett SW, Zhou D, Mielke R, Keller AA (2012) Photoinduced disaggregation of TiO_2 nanoparticles enables transdermal penetration. PLoS One 7:e48719. https://doi.org/10.1371/journal.pone.0048719
31. Crosera M, Prodi A, Mauro M, Pelin M, Florio C, Bellomo F, Adami G, Apostoli P, De Palma G, Bovenzi M et al (2015) Titanium dioxide nanoparticle penetration into the skin and effects on HaCaT cells. Int J Environ Res Public Health 12:9282–9297. https://doi.org/10.3390/ijerph120809282
32. Tak YK, Pal S, Naoghare PK, Rangasamy S (2015) Shape-dependent skin penetration of silver nanoparticles: does it really matter? Sci Rep 5:16908. https://doi.org/10.1038/srep16908
33. Park YH, Bae HC, Jang Y, Jeong SH, Lee HN, Ryu WI, Yoo MG, Kim YR, Kim MK, Lee JK et al (2013) Effect of the size and surface charge of silica nanoparticles on cutaneous toxicity. Mol Cell Toxicol 209:67–74. https://doi.org/10.1007/s13273-013-0010-7
34. Yoshida T, Yoshioka Y, Takahashi H, Misato K, Mori T, Hirai T, Nagano K, Abe Y, Mukai Y, Kamada H et al (2014) Intestinal absorption and biological effects of orally administered amorphous silica particles. Nanoscale Res Lett 9:532. https://doi.org/10.1186/1556-276X-9-532
35. De Matteis V, Cascione M, Brunetti V, Toma CC, Rinaldi R (2016) Toxicity assessment of anatase and rutile titanium dioxide nanoparticles: the role of degradation in different pH conditions and light exposure. Toxicol In Vitro 37:201–210. https://doi.org/10.1016/j.tiv.2016.09.010
36. Hadrup N, Lam HR (2014) Oral toxicity of silver ions, silver nanoparticles and colloidal silver—A review. Regul Toxicol Pharmacol 68:1–7. https://doi.org/10.1016/j.yrtph.2013.11.002
37. Asharani PV, Wu YL, Gong Z, Valiyaveettil S (2008) Toxicity of silver nanoparticles in zebrafish models. Nanotechnology 19. https://doi.org/10.1088/0957-4484/19/25/255102
38. Haque E, Ward A (2018) Zebrafish as a model to evaluate nanoparticle toxicity. Nanomaterials 8(7):561
39. Liu J, Fan D, Wang L, Shi LI, Ding J, Chen Y, Shen S (2014) Effects of ZnO, CuO, Au, and TiO_2 nanoparticles on Daphnia magna and early life stages of Zebrafish Danio rerio. Environ Prot Eng 40(1):431–439
40. Böhme S, Baccaro M, Schmidt M, Potthoff A, Stärk HJ, Reemtsma T, Kühnel D (2017) Metal uptake and distribution in the zebrafish (Danio rerio) embryo: differences between nanoparticles and metal ions. Environ Sci Nano 4(5):1005–1015
41. Shah SNA, Shah Z, Hussain M, Khan M (2017) Hazardous effects of titanium dioxide nanoparticles in ecosystem. Bioinorg Chem Appl 2017:4101735
42. Valentini X, Absil L, Laurent G, Robbe A, Laurent S, Muller R, Nonclercq D (2017) Toxicity of TiO 2 nanoparticles on the NRK52E renal cell line. Mol Cell Toxicol 13(4):419–431
43. Sajid M, Ilyas M, Basheer C, Tariq M, Daud M, Baig N, Shehzad F (2015) Impact of nanoparticles on human and environment: review of toxicity factors, exposures, control strategies, and future prospects. Environ Sci Pollut Res 22(6):4122–4143
44. Arruda SCC, Silva ALD, Galazzi RM, Azevedo RA, Arruda MAZ (2015) Nanoparticles applied to plant science: a review. *Talanta 131*:693–705
45. Lin DH, Xing BS (2007) Phytotoxicity of nanoparticles: inhibition of seed germination and root elongation. Environ Pollut 150:243–250
46. Liu Y, Zhao Y, Sun B, Chen C (2012) Understanding the toxicity of carbon nanotubes. Acc Chem Res 46(3):702–713

47. Chithrani BD, Ghazani AA, Chan WCW (2006) Determining the size and shape dependence of gold nanoparticle uptake into mammalian cells. Nano Lett 6:662–668
48. D'Silva J, van Calster G (2009) Taking temperature—A review of European Union regulation in nanomedicine. Eur J Health Law 16:249–269. https://doi.org/10.1163/157180909X453071
49. Kaegi R et al (2008) Synthetic TiO_2 nanoparticle emission from exterior facades into the aquatic environment. Environ Pollut 156(2):233–239
50. SCENIHR, Risk Assessment of Products of Nanotechnologies. 2009.; http://ec.europa.eu/health/ph_risk/committees/04_scenihr/docs/scenihr_o_023.pdf. Accessed 17 July 2017
51. Kaur J, Khatri M, Puri S (2019) Toxicological evaluation of metal oxide nanoparticles and mixed exposures at low doses using zebra fish and THP1 cell line. Environ Toxicol 34(4):375–387

Vitamin-Loaded Nanocarriers as Nutraceuticals in Healthcare Applications

Madhu Gupta, Radha Aggarwal, Neha Raina, and Azmi Khan

1 Introduction

Vitamins are nutrients that come under the class of organic compounds and are considered as essential micronutrients that are generally deprived of providing energy but are highly required for proper functioning of body and are not produced in the human body but play a vital role for conducting the normal functions. Vitamins are indispensable for metabolism, growth and development as well as for the proper functioning of the body. Vitamin D and B2 are vitamins that are generated in the body; rest all vitamins are attained from the food [1]. Vitamins are cardinal part of the food consumed by humans as they have the capacity of regulating various biochemical reactions taking place in human body. Vitamins that are existent and established are 13 in number; in addition their roles in the physiological reactions cannot be denied and are categorised on the basis of their solubility into two classes. The vitamins that are lipophilic in nature comprise vitamins, namely A, E, K, and D in conjugation with the carotenoids displaying the functional properties of vitamin A [2]. Furthermore, the vitamins having hydrophilic nature comprise vitamin C, the group of B vitamins which includes thiamin (B1), riboflavin (B2), nicotinic acid (B3), pyridoxine (B6), pantothenic acid (B5), folate, and cyanocobalamin (B12), which act jointly with vital functions of body. Vitamins having distinct structure demonstrate distinctive biological actions that are entitled as vitamers, based on their structure. The provitamins do not exhibit characteristics of vitamins themselves and are organic compounds that are transformed into vitamins by metabolism process [3]. Sunlight is responsible for carrying out vitamin D synthesis in the body; additionally, vitamin B_3 formation can only be possible by an adequate amount of tryptophan in proteins that elicits its synthesis. All classes of vitamins are produced by plants other than cyanocobalamin and hence it ought to be included in the diet of humans [4]. Despite the fact that vitamins are needed in small proportions, for

M. Gupta (✉) · R. Aggarwal · N. Raina · A. Khan
Delhi Pharmaceutical Sciences and Research University, New Delhi, India

© Springer Nature Singapore Pte Ltd. 2020

M. Rahman et al. (eds.), *Nanomedicine for Bioactives*,
https://doi.org/10.1007/978-981-15-1664-1_18

ensuring that an adequate number of vitamins is obtained from the diet, foods that are enriched with vitamins can be ingested by individuals, for instance, in the type of functional foods in conjugation with vitamins as nutraceuticals. Furthermore, a few vitamins that are extracted from marine resources for instance seaweeds possess antioxidant action in addition to other benefits related to health for example reducing blood pressure, helping in preventing cardiovascular disorders, or moderating the risk of deadly disease like cancer [5]. The dominant role of vitamins in the physiological system comprises of improving the immunity of body in addition to eyesight, strengthening dermal health, and helps preventing cancer (vitamin A); responsible for developing the immunity counter to various microorganisms, mitigate disorders like depression in addition to anxiety, chances of stroke are reduced, and unwinding premenstrual syndrome; keeping common cold or other viral infections at bay, maintaining healthy skin, accelerate wound healing, regulating the cholesterol and blood glucose level, minimising nervous system disorders (vitamin C) [6] averting the deadly cancer and diseases of cardiovascular system along with escalating strong bones and oral health (vitamin D); keep in check the neural disorders, for instance, Alzheimer and further abnormalities related to neurons, stamina develops, and averting disorders related to skin (vitamin E) [7]. The use of nanoscience in appropriate delivery of vitamins has attained considerable attention nowadays, and nanoencapsulation is projected as groundbreaking technique for site-specific delivery of vitamins and hence shielding these bioactive components from degradation in damaging environment [8]. Nanotechnology is a briskly developing zone having unique properties and broader applications. U.S. National Nanotechnology Initiative describes nanoparticles (NPs) in the size range of 1–100 nm. Various techniques of nanotechnology used for the liposoluble vitamins include inclusion complexation, emulsion systems, and liposome solid lipid nanoparticles. For nutraceutical encapsulation, for vitamins that are liposoluble various systems are reported for delivery including liposomes, nanoemulsions, lipid nanoparticles, polymeric micelles, and nanoparticles [9]. Nanoformulations that are substantially recommended by numerous experimental results are lipid-based nanoformulations because of their absorption to a greater extent when ingested and enhanced stability in addition to lesser degradation in the gastrointestinal tract. The technique of nanoencapsulation is extensively employed for protecting and releasing numerous bioactive ingredients in controlled manner; predominantly, pharmaceuticals and food bioactive components (nutraceuticals) can be successfully incorporated into formulations, hence resulting in enhanced bioavailability. The process of entrapment of nutrients at the nanoscale is termed as nanoencapsulation [10]. Thus, nanoencapsulation is an effective technique for providing absorbance at relatively elevated rate in addition to delivery of nutrients in comparison to microscale delivery [11]. Nanocarriers (<1000 nm in general, <100 nm in pharmaceutical sector) are capable of providing better bioavailability because of improved surface-to-volume ratios. As well as with reference to technological aspect, nanocarriers are efficiently used for enhancing the solubility of nutraceuticals that are hydrophobic in nature. Food-grade nanocarriers are being frequently employed for vitamin encapsulation. In nanodelivery systems polymeric compounds that are natural in origin are generally employed for instance cyclodextrins, albumin, alginate, gelatin, collagen, chitosan, and pectin [12]. Numerous formulation approaches employed for effective delivery of nutraceuticals that are encapsulated comprise nanoparticles, nanocrystals, nanocapsules, micelles, etc. Nutraceutical is a term derived from both nutrition and pharmaceutical, given by Stephen DeFelice. Stephen DeFelice further described nutraceutical as "a food or portion of food that is capable of providing medical or health benefits together with preclusion of disease and treatment" [13]. Nutraceuticals are considered as nutritional ingredients that are able to deliver medicinal or physiological benefits apart from nutritional requirements and contain a broad variety of components for instance phenolic components, vitamins, and lipids [14]. Still the medicinal or physiological benefits and efficacy of nutraceuticals are maintained only by preserving their bioavailability, described as a portion of an ingested compound that is well absorbed and reaches effectively in systemic circulation. Many factors are responsible for reduced bioavailability of a compound, consisting of reduced permeability and/or solubility inside the gastrointestinal (GI) tract, inadequate residence time in stomach, and lack of stability at the time of food processing [15].

Nanotechnology widely employed in food and nutraceuticals aids in providing taste enhancement, flavour, smell, colour, texture, and thickness of food materials along with nutraceutical compounds, and benefits of applying nanotechnology in the domain of food and nutrition area have effectively increased the rate of absorption in addition to bioavailability of nutraceuticals and supplements used for health benefits and development of new packaging materials for packing food materials with enhanced properties like mechanical strength. In the current situation, several nutraceutical products are facing problem in delivery of nutrients aimed at better usage in vivo [16]. Principally, nanotechnology is designed for eradication of the problem of ineffective medications and nutraceutical delivery. By applying techniques of nanotechnology, it is easier to solve the basic obstructions of nutraceutical delivery. Nanodelivery systems aid in targeting the specific site. Nanodelivery systems helps to achieve the desired size and surface activity; in addition charge of nanostructures of molecules can also be derived [17]. The difficulties associated with absorption and solubility of the nutrients could be easily reduced as a result of advanced nanotechnology. In future, it is considered as a way intended for ameliorating advancement of nutraceutical products used presently [18]. Nanotechnology helps in attaining favourable properties. The properties thus derived can be advantageous for altering the challenging biological process into the desirable process [19]. Initially, when there is reduction in size, the surface area of molecule expands as a consequence of increased desired biological process. Secondly, the alteration of nutraceutical components into nanomolecule can eradicate undesired biological effects; for example, the technique of nanoencapsulation used for vitamins and antioxidants aids in reducing gastrointestinal damage following ingestion [20]. For this reason, it can conclude that for surmounting in vivo obstructions the nanodelivery system can be successfully employed. The vital aspect of nanotechnology utilised for food and nutraceuticals has been the encapsulation in addition to release of bioactive materials in a controlled way. The primary difficulties of incorporating nutraceuticals for instance β-carotene, curcumin, and vitamins A, D, E, and K in food stuffs are their chemical unstableness at the stage of food processing (e.g. photosensitivity, oxygen, thermal sensitivity) or inside GI tract and reduced solubility in aqueous medium along with decreased bioavailability. Altogether the factors mentioned above have effect on nutraceuticals' functionality, and subsequently, their therapeutic benefits cannot be determined even after ingesting them in high quantities. Additionally, when incorporated into food products directly, nutraceuticals possessing obnoxious sensory properties (e.g. polyunsaturated fatty acids) can undesirably disturb food properties besides shelf life. On the other hand, few compounds (e.g. vitamin C) can produce interactions with other food components that are undesirable, affecting food appearance, texture, mouthfeel, stability, and bioavailability of vital components negatively. Thus, nutraceutical encapsulation is compulsory for (1) protecting them in contradiction of adverse external factors; (2) permitting their incorporation into foodstuffs; (3) masking unpleasant taste; (4) permitting their release in controlled way; and (5) protecting and making best use of their functional properties, i.e. for ensuring that they are preserved in indigenous bioactive form till they are delivered to the desired site of action [21].

2 Vitamins Reported as Nutraceuticals

Vitamins play a dominant role in the development of human body. Various vitamins are used as nutraceuticals in the healthcare sector. Numerous investigational studies were conducted on vitamins used as nutraceuticals.

2.1 Vitamin C

Vitamin C comes under the category of water-soluble vitamins and cannot be synthesised in the human body. The concentration of vitamin C in the body can be maintained by consumption of dietary or other sources of vitamin C. Oxidation of vitamin C results in the production of dehydroxy-ascorbic acid and glucose transporters shifted dehydroxy-acid into the cells; here it goes under reduction process and gets converted into its original form, i.e. ascorbic acid for the use of different mechanisms by the cells [22]. From the various studies it has been found that the consumption of supplements of vitamin C increases its concentration in the skin and blood plasma [23]. Vitamin C has a protective action against the oxidative damage in cell membranes, DNA, and tissues of human body because of its powerful free radical scavenger and antioxidant activity. Vitamin C is also a donor of electron and an important cofactor in the process of collagen hydroxylation and promotes the extracellular and intracellular collagen maturation [24]. Vitamin C exhibits reduced damage caused by oxidation (induced by UVB radiations) and skin neoplasms induced by UV radiations in mice and provides protection of keratinocytes in humans against lipid peroxidation induced by UVA radiations. Marker of oxidative stress known as malondialdehyde and its contents in the human skin also get decreased due to the presence of vitamin C in the body [23, 25]. Green peppers, raw red, kiwifruit, broccoli, oranges, Brussels sprouts, grapefruits, and strawberries are the vegetable sources having high concentration of vitamin C [26].

2.2 Vitamin E

Vitamin E comes under the category of fat-soluble vitamins. α-Tocopherol (αT) is the form of vitamin E which is the biologically active, most abounding, and leading form utilised in various metabolisms of human body [27, 28]. αT provides the protection of skin from the damage by UVB radiations with the process of altering the ROS formation and free radical scavenging, surface and cell membrane stabilisation, reduction of apoptotic cell count, and minimisation of nuclear factor (kappa B) activation [24, 29]. UV radiation exposure leads to reduction of αT concentration within the skin that causes the promotion of skin ageing process. However, by using the oral and topical route of administration, the concentration of αT in skin could be increased [27]. High-concentration oral sources of vitamin E occurring naturally are seeds of plants such as almonds, sunflower seeds, walnuts, peanuts, sesame seeds, pistachios, pecans, and vitamin E that are available in small amount in vegetables and fruits [30, 31]. According to numerous American national surveys it was discovered that the consumption of vitamin E by Americans is less as compared to the amount mentioned by their RDA. Consumption of vitamin E in dietary sources is not associated with any type of adverse effects but it is reported that it can cause haemorrhage and alteration of process of coagulation of blood in animals due to the consumption of αT in higher doses [32]. However, some more studies are required for the determination of vitamin E's ideal allowance on a daily basis in order to achieve maximum benefits of vitamin E for the skin [26].

2.3 Seaweed Vitamins

Seaweeds are known as an excellent source of various fat-soluble vitamins like β-carotene with activity of vitamin A and vitamin E and water-soluble vitamins such as vitamin B1, B2, B12, and vitamin C [33]. Seaweeds differ in its composition and the profile of vitamins depending upon the salinity and geographical area, presence of light, algal species, seawater temperature, time period of year, etc. [34, 35]. Therapeutic efficacy of vitamins in seaweed principally depends on the individual solubility of vitamins that affects the absorption of that vitamin in the intestinal region of body as well as the tissue uptake. Fat-soluble vitamins show different absorption and bioavailability depending upon whether these vitamins are consumed along with the lipid containing foodstuffs or not. Because of the solubility in fat, these fat-soluble vitamins absorb in the intestine with the same path as that of fat present in diet. In addition, the form of vitamin present in the food is also equally important for the absorption and bioavailability [36]. It is stated that seaweeds in a quantity of 100 g is an enough source that provides vitamins A, B2, and B12 more than required on a daily basis but completes only two-third daily requirement of vitamin C [37]. Red seaweeds (Porphyra, Palmaria) are rich source of provitamin A with considerable amounts of vitamins B1, B2, and B12 (also found in green seaweeds). However, the overall content of vitamins in brown seaweeds (Laminaria, Undaria) is very less; still the vitamin C content of brown seaweeds is found to be high [34]. Seaweed compounds having activity against oxidation are widely employed in the cosmetics, pharmacological, and food industries in comparison with other compounds of seaweeds. Seaweeds provide protection against oxidation due to some compounds including carotenoids (β-carotene), vitamin C (ascorbic acid), and vitamin E (a-tocopherol) and partly due to niacin and vitamin B1 [33].

2.4 Vitamin D

True meaning of vitamins does not include vitamin D in a technical aspect. It is more of a precursor of calcitriol (active form of vitamin D) and less of a micronutrient. The active form of vitamin D is a hydrophobic seco-steroid hormone [38]. Deficiency of vitamin D occurs in the whole world associated with numerous health-related consequences. Around 30–60% population of North American and Europe grieves from the problem of deficiency of vitamin D because of the inadequate exposure of sunlight and extremely less source of vitamin D in their diet [39]. To get rid of this problem, the peoples of Western countries started adding vitamin D in products of beverages and foods like bread, soft drinks, and milk, to enhance the nutritional value of these items. So, deficiency of vitamin D is associated with various health problems which commence the new development of functional diets and therapies utilising nanotechnologies for the amalgamation of vitamin D into products of pharmaceuticals and food industry with no harm of activity and therapeutic efficacy of their products [40].

2.5 Carotenoids

In present time, carotenoids gain enormous importance as an excellent functional component of food. Carotenoids perform different functions like key role in photosynthetic plants and animals and as pigments of natural origin but they are considered as components offering numerous health benefits to promote health [41]. Carotenoids provide most official health benefits for humans by function of provitamin A, maternal and infant nutrition contribution [42], improvements of cognitive functions of brain, pivotal activity to promote heart and eye health [43], UV light protection to prevent skin cancer [44], stimulation and modulation of immune system [45], and effects on the process of gene translation and transcription [46]. The principal benefit of carotenoids in a systematic way is the potent activity against oxidation [41].

3 Vitamin-Loaded Nanocarrier Systems Employed in Nutraceuticals

Nanocarriers are carrier systems of drugs and nutraceuticals that are colloidal in nature with a particle size less than 500 nm. The ratio of surface area to volume within nanocarriers is high that enables them to modify some characteristics of drugs like basic nature and activity of drug in biological system, reduction of toxicity, controlled release, improved profile of pharmacokinetics, and delivery of drug at a targeted site in a system for delivery of drugs. Furthermore, the physical and chemical properties of nanocarriers may be improved or adjusted by changing the particle size, shapes, composition (organic/inorganic or hybrid), and properties related to surface phenomenon (functional group, attachment of targeting moieties, surface charge, coating of PEGylation or other materials) of nanocarriers. Nanocarriers are designed with a target of best therapeutic efficacy in low doses with minimal side effects as possible [47]. From the numerous investigations carried out worldwide, vitamin-loaded nutraceuticals were investigated for the treatment for different kinds of cancer. But these nutraceuticals also show some limitations like poor solubility in aqueous medium and poor permeation through the skin responsible for the poor availability of drugs in the human biological system [48, 49]. Various kinds of nanocarriers which have been explored in recent years in order to modulate the pharmacokinetics of vitamin-loaded nutraceuticals are as follows (Fig. 1).

3.1 Liposomes

Liposomes are lipid vesicles that are spherical in shape and comprise a centre core which is aqueous in nature and enclosed within the bilayers of lipids. Liposomes differ in the number of lipid bilayers; it can be one or more than one membrane of bilayer made from lipids of synthetic or natural origin. The synthetic polymer used in lipid bilayers is generally polyethylene glycol (PEGylated or stealth liposomes)

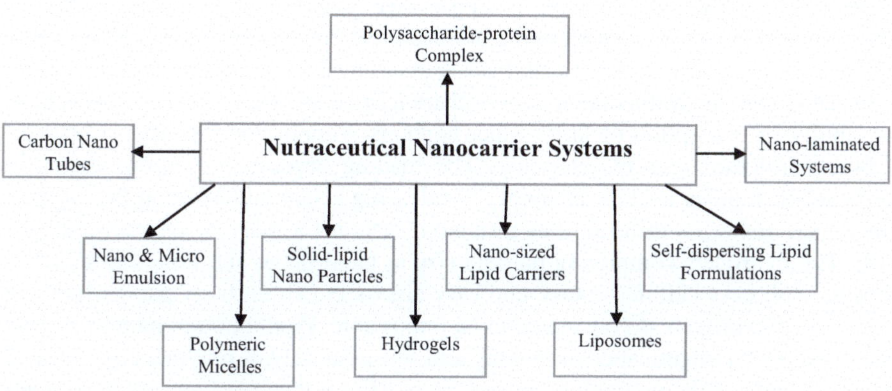

Fig. 1 Various nanocarrier systems used for delivery of vitamin-loaded nutraceuticals

which extends the half-life of liposomes in systemic circulation [6]. As a modern nanocarrier system, liposomes offer several advantages as compared to conventional systems of drug delivery. The advantages of liposomes include improved rate and amount of drug delivery, protection of the active form of drug from the atmospheric conditions, performance features of drug product at improved rate, effective treatment with minimal toxicity of systemic fluid, supply of expensive drug formulation at low cost [7], and drugs enclosed in liposomes exhibiting modified pharmacokinetic profile in comparison to the drug freely available in solution [8]. Nanoliposomes are those liposomes having size in nanoscale which again enhance all the advantages of liposomes listed above due to the fact that nanoliposomes are very widely used as carriers for delivery of different agents active biologically such as vaccines, cosmetics, drugs, and nutraceuticals [7, 20].

Liu and Park prepared chitosan-coated nanoliposomes loaded with vitamin E using sonication method. From the investigation it was found that these nanosize liposomes are promising vitamin E carriers because it exhibits higher loading efficiency of vitamin E over 99% and increase the stability by preventing the chemical degradation of vitamin E [10].

Hamishehkar et al. formulated vitamin D3-loaded nanoliposomes by encapsulating the bioactives in the lipid-based carriers using thin-film hydration-sonication method offering advantages like protection against oxidation and provision of better stability of vitamin D3 [11].

Nanoliposomes provide numerous advantages; still there are some disadvantages related to nanoliposomes. The main disadvantage of nanoliposomes is consideration of these as foreign matters by systemic circulation of body; that is why they cleared by RES at a rapid rate sometimes without showing required therapeutic efficacy. Moreover, disintegration of nanoliposomes occurs due to electrostatic, van der Walls, and hydrophobic forces [20].

3.2 Hydrogels

In present time, hydrogels derived from polymers with bio origin gained enormous attention particularly as a challenging nutraceutical delivery system [12]. Networks of water-loving polymers bind with hydrogen or covalent bonds, physical entanglements, and van der Walls interactions and are termed as hydrogels. Hydrogels have the ability of absorption of large water contents without any disturbance in their network structure present internally. The loading capacity of nutraceuticals in hydrogels is high as well as the ability of releasing active ingredient at the targeted site which makes its utilisation as a suitable system for delivery of nutraceuticals [13]. The hydrogels comprising particle size in nano-range are termed as nanohydrogels with easy cellular uptake and easy access in certain body parts where the presence of hydrogels is considered to be impossible. Production of nanohydrogels is carried out by chemical cross-linking and self-assembly of polymers by physical means of hydrophilic or amphiphilic polymers. High capacity of nutraceutical loading, high stability, high aqueous content, and ability to respond for the atmospheric stimuli are some chief advantages of nanohydrogels [14, 19].

Eslami et al. prepared hydrogels by entrapment of vitamin D3 in two carbohydrate bioploymers and optimised with calcium alginate with three independent variables by utilising response surface methodology. After investigations it was revealed that microcapsules of alginate after coating with chitosan exhibited the slow rate of release of vitamin D3 as compared to non-coated microcapsules of alginate in the buffer of natural pH and the release of vitamin D3 is achieved at a controlled rate in both the polymers [50].

Spizzirri et al. prepared flavonoid-based hydrogels of vitamin B complex to make them more stable against oxidation. Loading of vitamin B complex was done in catechin-based cross-linked hydrogels. Results of this study confirm an excellent improvement in the stability of vitamin B complex based on catechin as compared to hydrogels based on polyacrylic acid [51].

3.3 Polymeric Micelles

Polymeric micelles (PMs) prepared through the self-assembly of di- or tri-block copolymers are amphiphilic in nature without natural origin in a milieu of aqueous materials and size ranges from 10 to 100 nm [15]. Both hydrophilic and hydrophobic sections are present in the block copolymers used in the preparation of PMs that are responsible for their amphiphilic nature. When these amphiphilic block copolymers are mixed with an aqueous medium, formation of micelles takes place above certain polymer concentrations referred to as critical micelle concentration (CMC). The central core of PMs consists of hydrophobic segment and shell portion comprises hydrophilic segment. For the loading and delivery of vitamins used in the cancer treatment, PMs are considered as a promising nanocarrier system [16]. PMs offer various advantages like high capacity of drug loading and enhanced solubility of drug in water resulting in low dose and low toxicity [14]. Casein micelles are the most commonly used PMs for the encapsulation of vitamins and other nutraceuticals [19].

Moeller et al. prepared polymeric micelles of casein loaded with vitamin using spray and freeze-drying method at pilot scale. The casein micelles show a better stability of vitamin D2 after the storage of 4 months and a high loading efficiency as 90% of vitamin D was encapsulated in dried casein micelles [20].

3.4 Nano- and Microemulsion

Among all the colloidal nanocarrier systems used for the delivery of vitamins, interest in micro- and nanoemulsions has been increased day by day because production of these systems from the dietary-grade ingredients is very easy and simple. Although there are several similarities between nano- and microemulsions still some noteworthy difference between these systems also exists.

Microemulsions are defined as the colloidal dispersion system with better thermal stability comprising small size and spherical shape particles (oils, emulsifying agents, and co-surfactants) dispersed in a medium aqueous in nature [52].

Appearance of microemulsion is clear and transparent and may be slightly turbid with higher capacity to dissolve lipophilic compounds, easiness to formulate, and easy-to-scale up application for commercialisation of product [53].

Nanoemulsion is also a colloidal dispersion but unstable thermodynamically prepared from dispersion of one liquid as small-size spherical droplets of radius less than 100 nm into another liquid and both liquids are immiscible to each other [52]. Techniques used to produce nanoemulsions include homogenisation done at high pressure or method of self-emulsification of both liquids [21, 54]. This system used for delivery of vitamins offers various advantages such as kinetically and thermally stable emulsion, optical transparency, comparatively easy formation, absorption at a fast rate after administration by oral route, and probability of controlling the size of globules [19].

McClements et al. synthesised nanoemulsion loaded with vitamin E by the method of spontaneous emulsification and after investigation it was revealed that the particle size of vitamin E was decreased up to less than 50 nm [55].

Ali et al. formulated naringenin nanoemulsion loaded with vitamin E and evaluated for its efficacy against Parkinson's disease. From the results of this study, it was revealed that the prepared nanoemulsion was physically stable with maximal in vitro release of drug via intranasal route [56].

3.5 Solid Lipid Nanoparticles (SLNs)

Solid lipid nanoparticles (SLNs) are nanocarriers with size ranging from 50 to 1000 nm [57]. SLNs are composed of fully crystallised lipid droplets with an internal crystalline structure in an organised pattern and accommodation of biologically active components in the matrix made from lipids [58]. SLNs offer numerous advantages like release of active component at the targeted site and at a controlled rate, higher capacity of drug loading, good stability, low homogenisation rate, and good biocompatibility and biodegradability [19]. Nutraceuticals can be incorporated in the SLNs with the help of any of the three different types of models named as (1) homogenous matrix model (solid solution model), (2) bioactive-enriched shell model, and (3) bioactive-enriched core model [59, 60].

San Martin-Gonzalez and Patel prepared the SLNs of vitamin D2 (ergocalciferol) and characterised for the drug loading. From the results it was revealed that loading of vitamin D2 was increased with reduction of turbidity in the preparation [61].

Gohla et al. formulated SLNs loaded with vitamin A for topical use and investigated for the release properties of vitamin A. From the investigation it was found that release of vitamin A from the delivery is increased and also a slow rate of release was successfully achieved [62].

3.6 Self-Dispersing Lipid Formulations (SDLF)

Oil solutions and mixtures of oils, co-solvents, co-surfactants, and surfactants which are isotropic in nature are termed as SDLFs. SDLFs do not exist as emulsions but get emulsified and change into emulsions when they come in contact with biological fluids or aqueous medium [19, 54]. Dispersion of SDLFs can be carried out in liposomes and other vesicles to form lamellar phases or in hexosomes and cubosomes for hexagonal or cubic phases, respectively [53]. SDLFs provide several advantages such as enhancement of solubility of nutraceuticals in the gastrointestinal tract by enhancing their adsorption at sites of GIT. Availability of SDLFs in biological system depends on lipophilic behaviour of nutraceuticals, lipid digestion, and mean diameter of droplets and lipid [63].

Zaghloul et al. carried out a comparative study between three different oral dosage forms (self-nanoemulsified nanocarriers (SNEED)) loaded with vitamin A in rats and investigated the bioavailability of vitamin A. From the results it was evaluated that the extent and the rate of absorption of vitamin A were enhanced in tablets and capsules filled with SNEED in comparison to capsules enclosing the vitamin A oily solution. Also, vitamin A absorption was improved in clinical studies carried out on human volunteers [64].

3.7 Polysaccharide-Protein Complex

Combination of more than two biopolymers (polysaccharides) and proteins is called as polysaccharide-protein complex also termed as precipitates and coacervates. Water-rich open structures formed by biopolymers charged oppositely. Formation of precipitates is also done by biopolymers charged oppositely but their formation is promoted when interactions between biopolymers are strong and enough for the release of counter-ions and water [14].

Leng et al. carried out the encapsulation of vitamin B12 (water-soluble) and vitamin E (fat-soluble) together with single W/O/W multiple emulsion. After that the ability of polysaccharide-whey protein isolate (WPI) complexes was investigated to control the release rate of both vitamins in a synergistic way. The two polysaccharides used in combination with WPI were *k*-carrageenan and low methoxyl pectin. From the results it was found that complexes of polysaccharide-WPI in comparison to pure WPI and polysaccharides showed control release behaviour of vitamin B12 from the W/O/W emulsion system confirming the synergism of oil phase and polysaccharide-WPI complexes (Table 1) [65].

4 Pharmaceutical Technological Methods for Preparation of Vitamin-Loaded Nanocarriers Used in Nutraceuticals

Technologies used for encapsulation of vitamins in nanocarriers are novel and unique field in the pharmaceutical, nutraceutical, and food industries providing numerous benefits, e.g. release of enclosed active component at a controlled rate, enhanced shelf stability, and higher bioavailability. Some biopolymers are approved at the commercial scale for encapsulation process of vitamins. Biopolymers based on carbohydrates like cyclodextrins and starch also provide protection of material

Table 1 Pharmaceutical technological methods used for loading of vitamins in nanocarriers as nutraceuticals

Nanocarrier technology	Dosage form	Vitamin type	Advantages	Reference
Freeze-drying method	PMs	Vitamin A	Higher binding of vitamin A and casein	[66]
Hot homogenisation	SLNs	Vitamin B12	Enhanced activity of B12-loaded SLNs against cancer cells as compared to free B12	[67]
Simple ionic gelation method	Polysaccharide-protein complexes	Vitamin D3	High loading and encapsulating efficiency	[68]
High-pressure homogenisation	Nanoemulsion	Vitamin E	Stability of nanoemulsions depends highly on the type of surfactant	[69]
Spontaneous emulsification method	Nanoemulsion	Vitamin E	Physically stability of emulsion and targeting efficiency through intranasal route	[70]
Self-assembly	Hydrogels	Vitamin E	Potential local action against ageing	[71]

from the atmospheric conditions which are sensitive for it [72]. The different well-known techniques used for encapsulation of vitamins in nanocarriers which are used conventionally and in modern age are as follows:

4.1 Spray Drying

Conventional methods used for encapsulation of vitamins having the capacity to produce powder in encapsulated form with varying size of particles are most commonly used for encapsulation of vitamins soluble in lipids at a commercial scale [73]. The dosage form used for microencapsulated powder of vitamins includes tablets having the ability to provide protection oxidation [74].

4.2 Spray Chilling and Spray Cooling

Both spray chilling and spray cooling techniques comprise vitamin diffusion in a wax or fat in melted form. These techniques are considered satisfactory for the encapsulation of vitamins only soluble in lipids or fats [72].

4.3 Emulsion Technique

In this process of encapsulation, vitamins are dispersed into a phase of immiscible liquid (shell material). Secondly, formation of shell occurs by making some adjustments around the vitamins scattered in solution. Oil in water is the two-phase system prevailing widely in the encapsulation process of vitamins [75].

4.4 Fluidised Bed Coating

It is also known as air suspension coating. Solid particles are suspended in an upward-moving flow of air which can be either cool or hot. Afterwards, the solid particles are sprayed through the top of the atomised particles of coating wall material, which can be molten or dissolved in an evaporable solvent. The coating's material can be cellulose derivatives, dextrin, emulsifiers, lipids, protein derivatives, and starch derivatives. This method is prevalent in nutritional supplements that contain encapsulated versions of vitamin C, vitamin B complex, and a variety of vitamin/mineral premixes [76].

4.5 Liposome Entrapment

Vesicular structure made up of one or more layers including 100% entrapment of water phase in a membrane prepared from phospholipids is called liposomes [72]. Kirby et al. prepared liposomes of vitamin C (encapsulated) using the technique of liposome entrapment with higher efficiency [77].

4.6 Coacervation

In this technique, separation of material of coating (liquid phase) is done from the solution of polymeric materials and a layer of coating material is wrapped uniformly around the particle present in the core. De-solvation, cross-linking, and thermal methods are used for the solidification of the above structure. At the end, microcapsules in the form of discrete particles are obtained by the process of filtration [78]. Comunian et al. formulated coacervates of vitamin A, microencapsulated in acacia and gelatin, and the effect of different process variables on the process of microencapsulation was investigated [79].

4.7 Ionotropic Gelation

Method of ionic gelation depends on cross-linking of polyelectrolytes in ionic environment which forms beads of hydrogels known as gelispheres that are defined as cross-linked spherical shaped entity of hydrophilic polymers exhibiting huge gelation and index of swelling in in vivo condition of biological fluids. Controlled release rate of vitamins is achieved by the relaxation of hydrophilic polymers used in this process. Vitamins and some compounds active biologically are encapsulated in these gelispheres which retain the natural structure of encapsulated materials [72].

5 Characterisation Methods for Vitamin-Loaded Nanocarriers Used in Nutraceuticals

Morphology of nanosize particles can be determined with the help of some commonly used characterisation techniques such as atomic force microscopy (AFM), cryogenic transmission electron microscopy (cryo-TEM), and transmission electron microscopy (TEM). Researchers and scientists extensively used the dynamic light scattering (DLS) technique to predict the distribution of particle size of nanosized vitamins. Doppler anemometry is employed for measuring the zeta potential providing information about the size of particles in nanoscale and their stability by in vivo studies. Fourier transform infrared spectroscopy (FTIR) is used for analysis of modification that occurs on the surface of nanosized particles. Chromatography or spectroscopy under a range of pressure at predefined wavelengths may be utilised for determining the characteristics (quantitatively) of vitamins entrapped in the nanosize capsules [80].

Vilanova and Solans characterised complexes of β-cyclodextrin (β-CDs) molecules and vitamin A palmitate (VAP) through UV-Vis and FTIR spectroscopy.

According to the spectroscopy results, as the concentration of β-CDs increased, the solubility of VAP declined continuously, and thus a less water-soluble complex formed (ultimately two β-CD molecules encapsulate a unit VAP molecule). For the investigation of functional groups participating in inclusion complexation, FTIR assay was carried out. As a result, at the bands of 3050 and 950 cm^{-1} (belonging to =CH bond) the double bond was not present showing that the VAP is incorporated within the moiety of =CH section [81].

6 Regulatory Issues

Therapeutic benefits proclaimed by nutraceutical products must comply with regulatory requirements for products having medicinal products in terms of efficacy, marketing authorisation procedures, safety, and quality testing [82]. Utilisation of nanodelivery systems for preparing the food should not be limited only to fresh food; instead healthier food is also in trend despite that there is possibility of causing serious health issues [83]. Numerous regulatory bodies comprising Food and Drug Administration (FDA), European Food and Safety Authority (EFSA), Occupational Safety and Health Administration (OSHA), Environmental Protection Agency (EPA), National Institute for Occupational Safety and Health (NIOSH), US Patent and Trademark Office (USPTO), US Department of Agriculture (USDA), and Consumer Product Safety Commission (CPSC) supervise nanosystem application in food products [84]. In the year 2012, two draft guidance documents were released by FDA with reference to nanotechnology. In the documents categories of food and cosmetics are covered but dietary supplements were not included in it. As stated by the guidance document of FDA, if the chemical and physical properties are changed then it also affects the bioavailability of food substances and these changes also influence toxicity level of food substances. As per draft guidance, its new dietary ingredient (NDI), nanotechnology was regarded as an example of a procedure that is able to create a new dietary ingredient with addition of notification required by FDA [85]. According to the EC Food Law Regulation, numerous points need to be kept in consideration while devising nanomaterials in food applications; for instance nanomaterials must be free from toxicity and heavy metals in addition to mycotoxins [86]. Directive 89/107/EEC states that nanomaterials designed for utilising in packaging of food products must be evaluated initially as a direct food additive [87].

7 Role of Vitamin as Nutraceuticals in Healthcare System

Nutraceuticals are also considered as substances that are biologically active or components of either substances defined chemically or activity of known therapeutic index commonly acceptable for the contribution of therapeutic effect of drug. Because of pre-assumed profile of safety, therapeutic efficacy and possible nutritional effects of nutraceuticals attract the interest of researchers and scientists day by day. Recently, interest in the nutraceuticals increases because of numerous benefits related to good health and these have become favourable alternatives for utilisation in the medicinal science of modern era [88]. Vitamin as nutraceuticals plays an important role in the management of the following disorders:

7.1 Alzheimer's Disease (AD)

Most commonly AD occurs in the form of dementia. Neurodegeneration diseases for example Huntington's disease, AD, and Parkinson's disease occur chiefly due to the oxidative stress. From the numerous researches and studies, it has been established that low risk of AD is associated with taking a diet highly enriched with antioxidants like vitamin C and vitamin E because it is easy to prevent the disease as compared to treat it [89].

7.2 Obesity

Obesity occurs in the condition when the amount of fat in body increases up to unhealthy amount. Obesity is a risk factor responsible for large number of disorders like hypertension, congestive heart failure, angina pectoris, hyperlipidemia, osteoarthritis, reduced fertility, renal vein thrombosis, respiratory problems etc. [90]. Dietary supplement has the significant ability to promote loss of fat in the body and lowering of body weight include the mix of glucomannan, fenugreek, chitosan, vitamin C and *Gsylvestre* [91].

7.3 Cancer

Cancer becomes a major cause of worry among all health-related issues in the countries which are on developing stage. Cancer can be preventing with the help of healthy diet and lifestyle [92, 93]. Increased consumption of fruits and vegetables enhanced the level of components having antioxidative effects such as vitamin C, vitamin E, selenium, lycopene and several phytochemicals [94]. Smoking is very harmful in patients suffering from cancer because antioxidative components of food like vitamin E, beta-carotene, retinyl palmitate, retinol, isotretoin or acetylcysteine becomes ineffective due to smoking. From the studies carried out for the treatment of cancer indicate the role of nutraceuticals in the prevention of cancer [95].

References

1. ud Din F, Aman W, Ullah I, Qureshi OS, Mustapha O, Shafique S, Zeb A (2017) Effective use of nanocarriers as drug delivery systems for the treatment of selected tumors. Int J Nanomedicine 12:7291
2. Bethune SJ, Schultheiss N, Henck JO (2011) Improving the poor aqueous solubility of nutraceutical compound pterostilbene through cocrystal formation. Cryst Growth Des 11(7):2817–2823
3. McClements DJ, Li F, Xiao H (2015) The nutraceutical bioavailability classification scheme: classifying nutraceuticals according to factors limiting their oral bioavailability. Annu Rev Food Sci Technol 6:299–327
4. Arora D, Jaglan S (2016) Nanocarriers based delivery of nutraceuticals for cancer prevention and treatment: a review of recent research developments. Trends Food Sci Technol 54:114–126
5. Liu G, Huang W, Babii O, Gong X, Tian Z, Yang J, Wang Y, Jacobs RL, Donna V, Lavasanifar A, Chen L (2018) Novel protein–lipid composite nanoparticles with an inner aqueous compartment as delivery systems of hydrophilic nutraceutical compounds. Nanoscale 10(22):10629–10640
6. Torchilin VP (2005) Recent advances with liposomes as pharmaceutical carriers. Nat Rev Drug Discov 4(2):145
7. Deshpande PP, Biswas S, Torchilin VP (2013) Current trends in the use of liposomes for tumor targeting. Nanomedicine 8(9):1509–1528
8. Malam Y, Loizidou M, Seifalian AM (2009) Liposomes and nanoparticles: nanosized vehicles for drug delivery in cancer. Trends Pharmacol Sci 30(11):592–599
9. Bernela M, Kaur P, Ahuja M, Thakur R (2018) Nano-based delivery system for nutraceuticals: the potential future. In: Advances in animal biotechnology and its applications. Springer, Singapore, pp 103–117
10. Liu N, Park HJ (2009) Chitosan-coated nanoliposome as vitamin E carrier. J Microencapsul 26(3):235–242
11. Mohammadi M, Ghanbarzadeh B, Hamishehkar H (2014) Formulation of nanoliposomal vitamin D3 for potential application in beverage fortification. Adv Pharm Bull 4(Suppl 2):569
12. Abaee A, Mohammadian M, Jafari SM (2017) Whey and soy protein-based hydrogels and nano-hydrogels as bioactive delivery systems. Trends Food Sci Technol 70:69–81
13. Mokhtari S, Jafari SM, Assadpour E (2017) Development of a nutraceutical nano-delivery system through emulsification/internal gelation of alginate. Food Chem 229:286–295
14. McClements DJ, Decker EA, Park Y, Weiss J (2009) Structural design principles for delivery of bioactive components in nutraceuticals and functional foods. Crit Rev Food Sci Nutr 49(6):577–606

15. Zhu Y, Liao L (2015) Applications of nanoparticles for anticancer drug delivery: a review. J Nanosci Nanotechnol 15(7):4753–4773
16. Biswas S, Kumari P, Lakhani PM, Ghosh B (2016) Recent advances in polymeric micelles for anti-cancer drug delivery. Eur J Pharm Sci 83:184–202
17. Gothwal A, Khan I, Gupta U (2016) Polymeric micelles: recent advancements in the delivery of anticancer drugs. Pharm Res 33(1):18–39
18. Gonçalves RF, Martins JT, Duarte CM, Vicente AA, Pinheiro AC (2018) Advances in nutraceutical delivery systems: from formulation design for bioavailability enhancement to efficacy and safety evaluation. Trends Food Sci Technol 78:270–291
19. Shin GH, Kim JT, Park HJ (2015) Recent developments in nanoformulations of lipophilic functional foods. Trends Food Sci Technol 46(1):144–157
20. Moeller H, Martin D, Schrader K, Hoffmann W, Lorenzen PC (2018) Spray-or freeze-drying of casein micelles loaded with vitamin D2: studies on storage stability and in vitro digestibility. LWT 97:87–93
21. Livney YD (2015) Nanostructured delivery systems in food: latest developments and potential future directions. Curr Opin Food Sci 3:125–135
22. Subramani T, Yeap SK, Ho WY, Ho CL, Omar AR, Aziz SA, Rahman NM, Alitheen NB (2014) Vitamin C suppresses cell death in MCF-7 human breast cancer cells induced by tamoxifen. J Cell Mol Med 18(2):305–313
23. McArdle F, Rhodes LE, Parslew R, Jack CI, Friedmann PS, Jackson MJ (2002) UVR-induced oxidative stress in human skin in vivo: effects of oral vitamin C supplementation. Free Radic Biol Med 33(10):1355–1362
24. Wu X, Cheng J, Wang X (2017) Dietary antioxidants: potential anticancer agents. Nutr Cancer 69(4):521–533
25. Tebbe B, Wu S, Geilen CC, Eberle J, Kodelja V, Orfanos CE (1997) L-ascorbic acid inhibits UVA-induced lipid peroxidation and secretion of IL-1α and IL-6 in cultured human keratinocytes in vitro. J Investig Dermatol 108(3):302–306
26. Souyoul SA, Saussy KP, Lupo MP (2018) Nutraceuticals: a review. Dermatol Ther 8(1):5–16
27. Evans JA, Johnson EJ (2010) The role of phytonutrients in skin health. Nutrients 2(8):903–928
28. Jiang Q (2014) Natural forms of vitamin E: metabolism, antioxidant, and anti-inflammatory activities and their role in disease prevention and therapy. Free Radic Biol Med 72:76–90
29. Maalouf S, El-Sabban M, Darwiche N, Gali-Muhtasib H (2002) Protective effect of vitamin E on ultraviolet B light–induced damage in keratinocytes. Mol Carcinog 34(3):121–130
30. McLaughlin PJ, Weihrauch JL (1979) Vitamin E content of foods. J Am Diet Assoc 75(6):647–665
31. Dreher ML (2012) Pistachio nuts: composition and potential health benefits. Nutr Rev 70(4):234–240
32. Vitamin E (3 Nov 2016) Fact sheet for health professionals. https://ods.od.nih.gov/factsheets/VitaminE-HealthProfessional/. Accessed 17 May 2017
33. Škrovánková S (2011) Seaweed vitamins as nutraceuticals. In: Advances in food and nutrition research, vol 64. Academic Press, London, pp 357–369
34. Mabeau S, Fleurence J (1993) Seaweed in food products: biochemical and nutritional aspects. Trends Food Sci Technol 4(4):103–107
35. Norziah MH, Ching CY (2000) Nutritional composition of edible seaweed Gracilaria changgi. Food Chem 68(1):69–76
36. Weber P, Bendich A, Machlin LJ (1997) Vitamin E and human health: rationale for determining recommended intake levels. Nutrition 13(5):450–460
37. Ortiz J, Romero N, Robert P, Araya J, Lopez-Hernández J, Bozzo C, Navarrete E, Osorio A, Rios A (2006) Dietary fiber, amino acid, fatty acid and tocopherol contents of the edible seaweeds Ulva lactuca and Durvillaea Antarctica. Food Chem 99(1):98–104
38. Krishnan AV, Feldman D (2011) Mechanisms of the anti-cancer and anti-inflammatory actions of vitamin D. Annu Rev Pharmacol Toxicol 51:311–336. https://doi.org/10.1146/annurev-pharmtox-010510-100611

39. Ginter E, Simko V (2009) Vitamin D deficiency, atherosclerosis and cancer. Bratisl Lek Listy 110(12):751–756
40. Ramalho MJ, Coelho MA, Pereira MC (2017) Nanoparticles for delivery of vitamin D: challenges and opportunities. In: Gowder S (ed) A critical evaluation of vitamin D—clinical overview. IntechOpen, Croatia, p 11
41. Rostamabadi H, Falsafi SR, Jafari SM (2019) Nanoencapsulation of carotenoids within lipid-based nanocarriers. J Control Release 298:38–67
42. Monsen ER (2000) Dietary reference intakes for the antioxidant nutrients: vitamin C, vitamin E, selenium, and carotenoids. J Acad Nutr Diet 100(6):637
43. Chew EY, Clemons TE, SanGiovanni JP, Danis R, Ferris FL, Elman M, Antoszyk A, Ruby A, Orth D, Bressler S, Fish G (2013) Lutein+ zeaxanthin and omega-3 fatty acids for age-related macular degeneration: the Age-Related Eye Disease Study 2 (AREDS2) randomized clinical trial. JAMA 309(19):2005–2015
44. Stahl W, Sies H (2012) β-Carotene and other carotenoids in protection from sunlight. Am J Clin Nutr 96(5):1179S–1184S
45. Jonasson L, Wikby A, Olsson AG (2003) Low serum β-carotene reflects immune activation in patients with coronary artery disease. Nutr Metab Cardiovasc Dis 13(3):120–125
46. Sharoni Y, Linnewiel-Hermoni K, Zango G, Khanin M, Salman H, Veprik A, Danilenko M, Levy J (2012) The role of lycopene and its derivatives in the regulation of transcription systems: implications for cancer prevention. Am J Clin Nutr 96(5):1173S–1178S
47. Neubert RH (2011) Potentials of new nanocarriers for dermal and transdermal drug delivery. Eur J Pharm Biopharm 77(1):1–2
48. Mishra BB, Patel BB, Tiwari S (2010) Colloidal nanocarriers: a review on formulation technology, types and applications toward targeted drug delivery. Nanomedicine 6(1):9–24
49. How CW, Rasedee A, Manickam S, Rosli R (2013) Tamoxifen-loaded nanostructured lipid carrier as a drug delivery system: characterization, stability assessment and cytotoxicity. Colloids Surf B: Biointerfaces 112:393–399
50. Eslami M, Shahedi M, Fathi M (2018) Development of hydrogels for entrapment of vitamin D3: physicochemical characterization and release study. Food Biophys 13(3):284–291
51. Spizzirri UG, Cirillo G, Curcio M, Altimari I, Picci N, lemma F (2013) Stabilization of oxidable vitamins by flavonoid-based hydrogels. React Funct Polym 73(8):1030–1037
52. McClements DJ (2012) Nanoemulsions versus microemulsions: terminology, differences, and similarities. Soft Matter 8(6):1719–1729
53. Augustin MA, Hemar Y (2009) Nano-and micro-structured assemblies for encapsulation of food ingredients. Chem Soc Rev 38(4):902–912
54. Dima Ş, Dima C, Iordăchescu G (2015) Encapsulation of functional lipophilic food and drug biocomponents. Food Eng Rev 7(4):417–438
55. Saberi AH, Fang Y, McClements DJ (2013) Fabrication of vitamin E-enriched nanoemulsions: factors affecting particle size using spontaneous emulsification. J Colloid Interface Sci 391:95–102
56. Gaba B, Khan T, Haider MF, Alam T, Baboota S, Parvez S, Ali J (2019) Vitamin E loaded naringenin nanoemulsion via intranasal delivery for the management of oxidative stress in a 6-OHDA Parkinson's disease model. Biomed Res Int 2019:2382563
57. Müller RH, Mäder K, Gohla S (2000) Solid lipid nanoparticles (SLN) for controlled drug delivery—a review of the state of the art. Eur J Pharm Biopharm 50(1):161–177
58. Katouzian I, Esfanjani AF, Jafari SM, Akhavan S (2017) Formulation and application of a new generation of lipid nano-carriers for the food bioactive ingredients. Trends Food Sci Technol 68:14–25
59. Fathi M, Mozafari MR, Mohebbi M (2012) Nanoencapsulation of food ingredients using lipid based delivery systems. Trends Food Sci Technol 23(1):13–27
60. Ramteke KH, Joshi SA, Dhole SN (2012) Solid lipid nanoparticle: a review. IOSR J Pharm 2(6):34–44
61. Patel MR, San Martin-Gonzalez MF (2012) Characterization of ergocalciferol loaded solid lipid nanoparticles. J Food Sci 77(1):N8–N13

62. Jenning V, Schäfer-Korting M, Gohla S (2000) Vitamin A-loaded solid lipid nanoparticles for topical use: drug release properties. J Control Release 66(2–3):115–126
63. Pathak K, Raghuvanshi S (2015) Oral bioavailability: issues and solutions via nanoformulations. Clin Pharmacokinet 54(4):325–357
64. Taha E, Ghorab D, Zaghloul AA (2007) Bioavailability assessment of vitamin A self-nanoemulsified drug delivery systems in rats: a comparative study. Med Princ Pract 16(5):355–359
65. Li B, Jiang Y, Liu F, Chai Z, Li Y, Li Y, Leng X (2012) Synergistic effects of whey protein–polysaccharide complexes on the controlled release of lipid-soluble and water-soluble vitamins in W1/O/W2 double emulsion systems. Int J Food Sci Technol 47(2):248–254
66. Mohan MS, Jurat-Fuentes JL, Harte F (2013) Binding of vitamin A by casein micelles in commercial skim milk. J Dairy Sci 96(2):790–798
67. Genç L, Kutlu HM, Güney G (2015) Vitamin B12-loaded solid lipid nanoparticles as a drug carrier in cancer therapy. Pharm Dev Technol 20(3):337–344
68. Teng Z, Luo Y, Wang Q (2013) Carboxymethyl chitosan–soy protein complex nanoparticles for the encapsulation and controlled release of vitamin D3. Food Chem 141(1):524–532
69. Ozturk B, Argin S, Ozilgen M, McClements DJ (2014) Formation and stabilization of nanoemulsion-based vitamin E delivery systems using natural surfactants: Quillaja saponin and lecithin. J Food Eng 142:57–63
70. Pangeni R, Sharma S, Mustafa G, Ali J, Baboota S (2014) Vitamin E loaded resveratrol nanoemulsion for brain targeting for the treatment of Parkinson's disease by reducing oxidative stress. Nanotechnology 25(48):485102
71. Gallardo V, Muñoz M, Ruíz MA (2005) Formulations of hydrogels and lipogels with vitamin E. J Cosmet Dermatol 4(3):187–192
72. Katouzian I, Jafari SM (2016) Nano-encapsulation as a promising approach for targeted delivery and controlled release of vitamins. Trends Food Sci Technol 53:34–48
73. Jafari SM, He Y, Bhandari B (2007) Encapsulation of nanoparticles of d-limonene by spray drying: role of emulsifiers and emulsifying techniques. Dry Technol 25(6):1069–1079
74. Shi XY, Tan TW (2002) Preparation of chitosan/ethylcellulose complex microcapsule and its application in controlled release of vitamin D2. Biomaterials 23(23):4469–4473
75. Wang R, MacGillivray BC, Macartney DH (2009) Stabilization of the base-off forms of vitamin B12 and coenzyme B12 by encapsulation of the α-axial 5, 6-dimethylbenzimidazole ligand with cucurbit [7] uril. Dalton Trans 18:3584–3589
76. Risch SJ, Reineccius G (1995) Encapsulation and controlled release of food ingredients. In: ACS symposium series. American Chemical Society, USA
77. Kirby CJ, Whittle CJ, Rigby N, Coxon DT, Law BA (1991) Stabilization of ascorbic acid by microencapsulation in liposomes. Int J Food Sci Technol 26(5):437–449
78. Gibbs F, Kermasha S, Alli I, Mulligan CN (1999) Encapsulation in the food industry: a review. Int J Food Sci Nutr 50(3):213–224
79. Comunian TA, Abbaspourrad A, Favaro-Trindade CS, Weitz DA (2014) Fabrication of solid lipid microcapsules containing ascorbic acid using a microfluidic technique. Food Chem 152:271–275
80. Garti N (ed) (2008) Delivery and controlled release of bioactives in foods and nutraceuticals. Elsevier, Amsterdam
81. Vilanova N, Solans C (2015) Vitamin A palmitate–β-cyclodextrin inclusion complexes: characterization, protection and emulsification properties. Food Chem 175:529–535
82. Pandey M, Verma RK, Saraf SA (2010) Nutraceuticals: new era of medicine and health. Asian J Pharm Clin Res 3(1):11–15
83. Pradhan N, Singh S, Ojha N, Shrivastava A, Barla A, Rai V, Bose S (2015) Facets of nanotechnology as seen in food processing, packaging, and preservation industry. Biomed Res Int 2015:365672
84. Qi L, Xu Z, Jiang X, Hu C, Zou X (2004) Preparation and antibacterial activity of chitosan nanoparticles. Carbohydr Res 339(16):2693–2700

85. Javeri I (2016) Application of "nano" nutraceuticals in medicine. In: Nutraceuticals. Academic Press, London, pp 189–192
86. Scampicchio M, Ballabio D, Arecchi A, Cosio SM, Mannino S (2008) Amperometric electronic tongue for food analysis. Microchim Acta 163(1–2):11–21
87. Sondi I, Salopek-Sondi B (2004) Silver nanoparticles as antimicrobial agent: a case study on E. coli as a model for Gram-negative bacteria. J Colloid Interface Sci 275(1):177–182
88. Khan MU, Elhassan GO, Khalilullah H, Eid EE (2016) Nutraceuticals–an opportunity of healthcare: A review. Trop J Pharm Res 5(11):229–239
89. Caterson ID, Gill TP (2002) Obesity: epidemiology and possible prevention. Best Pract Res Clin Endocrinol Metab 16(4):595–610
90. Woodgate DE, Conquer JA (2003) Prevalence of self-treatment with complementary products and therapies for weight loss: a randomized, cross-sectional study in overweight and obese patients in Colombia. Curr Ther Res 64:248–262
91. Kasbia GS (2005) Functional foods and nutraceuticals in the management of obesity. Nutr Food Sci 35(5):344–352
92. Willis MS, Wians FH Jr (2003) The role of nutrition in preventing prostate cancer: a review of the proposed mechanism of action of various dietary substances. Clin Chim Acta 330(1–2):57–83
93. Nkondjock A, Ghadirian P (2004) Dietary carotenoids and risk of colon cancer: case-control study. Int J Cancer 110(1):110–116
94. Weese JS, Arroyo L (2003) Bacteriological evaluation of dog and cat diets that claim to contain probiotics. Can Vet J 44(3):212
95. Oksanen PJ, Salminen S, Saxelin M, Hämäläinen P, Ihantola-Vormisto A, Muurasniemi-Isoviita L, Nikkari S, Oksanen T, Pörsti I, Salminen E, Siitonen S (1990) Prevention of travellers diarrhoea by Lactobacillus GG. Ann Med 22(1):53–56

Evidence-Based Review on Clinical Potential of Thymoquinone in Breast Cancer

Md Abul Barkat, Harshita, Faheem Hyder Pottoo, Sarwar Beg, Mahfoozur Rahman, and Farhan J. Ahmad

Abbreviations

AGP	α_1-Acid glycoprotein
ALL	Acute lymphoblastic leukemia
ALT	Alanine transaminase
ALT	Alanine aminotransferase
ASCO	American Society of Clinical Oncology
AST	Aspartate aminotransferase
BC	Breast cancer
CEA	Carcinoembryonic antigen
ER	Estrogen receptor
GSH	Glutathione

M. A. Barkat
Department of Pharmaceutics, College of Pharmacy, University of Hafr Al-Batin, Al Jamiah, Hafr Al-Batin 39524, Saudi Arabia

Nanomedicine Research Lab, School of Pharmaceutical Education & Research, Jamia Hamdard, New Delhi, India

Harshita (✉)
Department of Pharmaceutics, College of Pharmacy, University of Hafr Al-Batin, Al Jamiah, Hafr Al-Batin 39524, Saudi Arabia

F. H. Pottoo
Department of Pharmacology, College of Clinical Pharmacy, Imam Abdulrahman Bin Faisal University (Formerly University of Dammam), Dammam, Saudi Arabia

S. Beg · F. J. Ahmad
Nanomedicine Research Lab, School of Pharmaceutical Education & Research, Jamia Hamdard, New Delhi, India

M. Rahman
Department of Pharmaceutical Sciences, Shalom Institute of Health & Allied Sciences, SHUATS, Allahabad, Uttar Pradesh, India

© Springer Nature Singapore Pte Ltd. 2020

M. Rahman et al. (eds.), *Nanomedicine for Bioactives*,
https://doi.org/10.1007/978-981-15-1664-1_19

HER2	Human epidermal growth factor receptor 2
LDH	Lactate dehydrogenase
MDR	Multidrug resistant
NPs	Nanoparticles
PAI-1	Plasminogen activator inhibitor 1
PBD	Polo-box domain
Plk1	Polo-like kinase 1
PR	Progesterone receptor
ROS	Reactive oxygen species
Se	Selenium
SOD	Superoxide dismutase
TQ	Thymoquinone
UPA	Urokinase plasminogen activator
Vas	Volume of distribution at steady state

1 Introduction

Cancer is a cluster of diseases involving uncharacteristic growth of cell and its ability to spread easily throughout the body. Despite the great accomplishments in the fight against cancer over the past decades, unfortunately it remains as the leading cause of death worldwide and is regarded as a deadly disease second only to cardiovascular diseases [1]. As per the reports of American Cancer Society, there are about 74 different kinds of malignancy. Among the various types of cancer, breast cancer (BC) is the second most prevalent type of cancer, with approximately 252,710 new cases of invasive BC in women and 2470 in men in the year of 2017. In 2017, due to invasive BC approximately 40,610 deaths of women in the USA have been recorded. BC is a complex heterogeneous disease, characterized by high degree of mortality and morbidity, making it a major health problem, and it is caused due to either genetic and environmental factors or a combination of both [2–5]. A number of risk factors have been identified which are directly or indirectly associated with BC, but the clinical representation of BC patients differs individually. However few factors are responsible for the high risk of BC in women than others [6]. In spite of significant improvement in the treatment choices for BC during the last few decades, the death rates of women due to BC in the USA and worldwide are still a major concern [3].

For the treatment, screening, and vigilance of BC, the American Society of Clinical Oncology (ASCO) has suggested 13 different categories of tumor markers [6]. The tumor markers like CA 15-3, CA 27.29, carcinoembryonic antigen (CEA), estrogen receptor (ER), progesterone receptor (PR), human epidermal growth factor receptor-2 (HER2), urokinase plasminogen activator (UPA), plasminogen activator inhibitor 1 (PAI-1), and multiparameter assays for gene expression exhibited evidence-based clinical usefulness and were used in clinics for BC diagnosis and treatment. But, other categories of tumor markers such as P53, cathepsin D, cyclin E, and nestin are also employed in the screening and management of BC but they showed no sufficient evidence-based clinical usefulness [6].

Besides the surgery, there are various types of interventions available for the management of BC such as chemotherapy and radiation therapy [7]. These traditional treatment options are generally costly and exhibit poor selectivity with prolonged profile of side effects and other safety concerns; also, sometimes they show limited therapeutic potential in the management of BC [7]. The safety concerns and toxicity associated with the traditional therapies warrant research to develop better drugs and delivery systems to improve the overall efficacy and minimize the chemotherapy-related toxicity. As elaborated in a comprehensive review, the natural products have a significant role in the management of cancer owing to minimum or no toxicity and multifarious mechanisms of action [8].

New drugs and drug formulations are incessantly being formulated for the management of ER-positive BC. TQ is one of the therapeutic candidates that have shown anticancer properties as per the reports on in vitro and in vivo models [9, 10]. TQ (2-isopropyl-5-methylbenzo-1,4-quinone) is a plant-derived bioactive compound which was isolated and documented in the 1960s and 1970s from the volatile oil of *Nigella sativa* (*N. sativa*; black seed) [9, 11]. It is also found together with its few derivatives such as diTQ, thymohydroquinone, and thymol [9–11]. It has shown a variety of properties including antioxidant, anti-inflammatory, anticarcinogenic, and chemo-sensitizing [2, 12]. Due to the anticancer potential, it has been reported for use in the management of several types of cancer like ovarian, breast, and colon cancers [13]. TQ promotes or causes apoptosis of tumor cells through the induction of higher Bax/Bcl-2 ratio in MCF7, HCT-116, and HL-60 cancer cells. Such research reports were based on flow cytometry, western blot analysis, and multicolor fluorescence in situ hybridization (mFISH) [13]. In this chapter, we aim to describe the TQ's clinical importance and its potential in the treatment of BC.

2 Clinical Presence of TQ in BC

Ghosheh et al. isolated TQ from *N. sativa* seed volatile oils and apart from TQ, other similar compounds, namely diTQ, thymohydroquinone, and thymol, were also quantified from black cumin [14]. Over a decade ago, several researchers showed great interest to explore the anticancer potentials of TQ. Since then, numerous studies have been performed to assess and explore the anticancer or chemopreventive role and exact molecular mechanism of action of TQ in a variety of tumor cell lines and different types of cancer in animal models (Fig. 1). In BC cell lines (MDA-MB-468 and T47D), TQ inhibits the PI3K/Akt signaling, facilitates the G1 arrest, and promotes apoptosis [15]. TQ also impeded p53-mutated acute lymphoblastic leukemia (ALL) cells through the activation of a p73-dependent signaling pathway (mitochondrial and cell cycle checkpoint); this pathway afterward targets the anti-apoptotic and epigenetic integrator UHRF1, which is crucial for cell cycle progression [16]. Moreover, it was noted that the downregulation of cyclic nucleotide phosphodiesterase PDE1A is the prime function of p73 and UHRF1 deregulation in TQ-persuaded ALL cell apoptosis [17]. In ALL cell line CEM-ss, TQ also promotes the generation of reactive oxygen species (ROS) and HSP70,

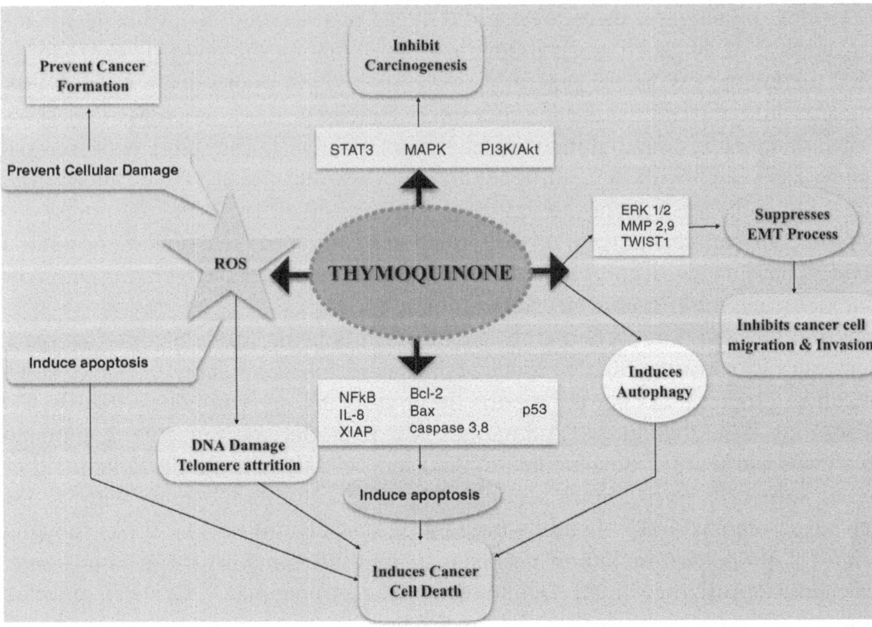

Fig. 1 Molecular mechanism of thymoquinone action in breast cancer

downregulates Bcl-2, upregulates Bax, and triggers caspase 3 and caspase 8 for instigating apoptosis [18]. A recent research report reveals that TQ promotes induction of apoptosis in MCF-7BC cells through upregulation of p53 expression [19]. Moreover, TQ extensively enhanced the expression of miR-34a through p53 as well as downregulation of Rac1 expression, and subsequently depolymerization of actin along with actin cytoskeleton disruption [20]. The disruption of actin cytoskeleton causes a significant cutback in the formation of lamellipodia and filopodia on the surfaces of cell, as a result of inhibiting the cell migration [21].

In addition, TQ also shows its therapeutic efficacy by regulating the physiological and biochemical phenomena implicated in ROS generation. In normal cells, TQ exerts its action as an efficient antioxidant and impedes the generation of superoxide radicals and lipid peroxidation, or enhances the functions of various antioxidant enzymes such as catalase, glutathione (GSH), superoxide dismutase (SOD), glutathione transferase, and quinone reductases [12, 22–24]. In cancer cells, however, TQ promotes the ROS production and demotes the GSH generation in a dose-dependent manner [25–27]. Therefore TQ shows a dual role in tumors which depends on the cellular feature; it may perform as an antioxidant or a prooxidant. TQ is considered as a rare and unique bioactive compound which targets different proteins and restrains the protein–protein interactions [28]. It interacts with the several antioxidant enzymes like GSH, NADH, and NADPH and produces two efficient antioxidants, namely glutathionyl-dihydro-thymoquinone and dihydroTQ [29]. It also interacts with human α_1-acid glycoprotein (AGP), bovine serum albumin (BSA), and phosphoserine/phosphothreonine recognition site of the polo-box domains (PBDs) [30]. TQ interferes with the function of polo-like kinase 1 (Plk1) by jamming its PBD-dependent contact, consequently interfering with its localization [28]. Proteins with PBDs and GSH are found to show potential targets for chemotherapy [27, 31].

Unluckily, there are only few research reports available on the pharmacokinetic and pharmacodynamic characteristics of TQ. In a recent research finding, TQ was available for action in the rat plasma for up to 12 h after administered orally [32]. In a study, TQ was accumulated in the nuclei of potoroo kidney cells employing a streptavidin/biotin detection system [27]. Research-based evidence has revealed the localization of TQ analog in the nuclei and Golgi apparatus compartments employing a very specific visualization technique based on alkyne-azide cycloaddition [27]. The recognition of all TQ-targeting sites and its plasma distribution profile in vivo can significantly improve the understanding of its molecular mechanism, which warrants further studies. TQ is available in tautomeric forms like keto and enol form separately, and its mixtures. The keto form is considered as a major fraction (~90%) and it is accountable for the desired pharmacological action [33]. TQ is a hydrophobic compound, and hence its solubility is a big challenge for formulation scientist because it directly affects the bioavailability. Moreover, solubility of TQ is time dependent, its solubility in aqueous medium at 24 h ranges from 549 to 669 μg/ml, and enhancement in the solubility has been seen at 72 h as 665–740 μg/ml [34]. So, recently researchers have been continuously striving to develop novel analogs of TQ with improved bioavailability and pharmacological action. A few synthetic analogs of TQ and its pharmacological action have been summarized in Table 1 [10]. Variety of routes recommended for the administration of TQ include intravenous (i.v.) [10], intraperitoneal (i.p.) [44–46], oral, as well as subacute and subchronic administration [10, 44, 47, 48].

However, oral administration may possibly cause biotransformation due to first-pass metabolism by the metabolizing enzymes (phase 2 enzymes) in the liver for example a quinine reductase (DT-diaphorase) that accelerates two-electron reduction of TQ into a hydroquinone [49]. In a study, El-Dakhakhny found and reported an LD_{50} of 10 mg/kg for TQ upon i.p. administration in rats [50]. In another study, researcher reported that no alteration has been seen in the biochemical parameters such as serum alanine transaminase (ALT) and lactate dehydrogenase (LDH) upon i.p. administration of TQ at different dose concentrations of 4, 8, 12.5, 25, and 50 mg/kg in mice [51]. In the same study, higher dose of TQ (>50 mg/kg body weight) administered via i.p. route was found lethal in mice with the LD_{50} of 90.3 mg/kg [51].

In several research reports based on the therapeutic outcome of TQ for the determination of anti-inflammatory, anticancer, antioxidant, as well as cytoprotective functions, researchers have generally employed the doses between 5 and 12.5 mg/kg via i.p. route in mice and rats with no consideration about patient safety or toxicity profile [10, 51, 52]. In several studies, researchers have suggested 10–100 mg/kg body weight dose of TQ via i.p. route with no report of lethal toxicities or deaths [10, 53–55]. As per the research reports, the maximum tolerated dose for TQ via i.p. route was 22.5 mg/kg in male and 15 mg/kg in female rats, while the dose of 250 mg/kg

Table 1 Thymoquinone potential on different animal models with breast cancer

Dose	Test compound/ formulation	Disease model	Activity	References
10–50 μM	TQ	In human MCF-7 breast cancer cells	TQ reactivated p21 and Maspin, induced pro-apoptotic gene Bax, downregulated the anti-apoptotic gene Bcl-2 and arrested the cell cycle at G2/M phase	[11, 35]
0–100 μl/ml	TQ-encapsulated multilamellar Nio-Au	In tamoxifen-resistant and Akt-overexpressing MCF7 breast cancer cells	Enhanced apoptosis by inducing p53 and inhibiting MDM-2 expression	[11, 36]
6.25–100 μM with paclitaxel	TQ and paclitaxel combination	In breast cancerous female Balb/c mice	Upregulation of tumor-suppressor genes, p21, Brca1, and Hic1. Upregulation of the growth factors, Vegf and Egf, while downregulation of caspase-3, caspase-7, and caspase-12 and PARP and a reduction in phosphorylated p65 and Akt1 expression	[11, 37]
Polyethylene glycol TQ nanoparticles (<50 nm)	PEG4000-TQ-Nps	In MCF-7, HBL-100 breast cancer cell	miR-34A upregulation-mediated downregulation of Rac1 expression followed by actin depolymerization, thereby disrupting the actin cytoskeleton leading to the reduction in the lamellipodia and filopodia formation on cell surfaces, thus retarding cell migration	[11, 20]
15 mg TQ	Molecular micelle-modified poly (D,L-lactide-co-glycolide) (PLGA) nanoparticles	In MDA-MB-231 breast cancer cell	Interferes with PI3K/Akt signaling and promotes G(1) arrest	[15, 38, 39]
7 mg TQ	TQ-loaded liposomes (TQ-LP)	MCF-7 and T47D breast cancer cell lines	Suppressing the proliferation of breast cancer cell	[40]

(continued)

Table 1 (continued)

Dose	Test compound/ formulation	Disease model	Activity	References
–	Thymoquinone-fatty acid conjugates	HL-60 human leukemia, multidrug-resistant KB-V1/ Vbl cervix, 518A2 melanoma [16], and MCF-7/ Topo breast cancer cells	Induced apoptosis associated with DNA laddering, a decrease in mitochondrial membrane potential, and a slight increase in reactive oxygen species	[41]
10 mM TQ	TQ	In MCF-7, MDA-MB-231, MDA-MB-468, and T-47D breast cancer cell	G1-phase arrest till 24-h incubation with TQ while extended exposure showed phase shift to subG1 indicating apoptosis, supported by suppression of cyclin D1, cyclin E, and cyclin-dependent kinase inhibitor p27 expression. Immunoblot and membrane potential studies revealed mitochondrial impairment behind apoptotic process with upregulation of Bax, cytoplasmic cytochrome c and procaspase-3, PARP cleavage along with Bcl-2, Bcl-xL, and surviving downregulation	[15]
3.125–100 mM TQ	TQ	In MCF-7 and MDA-MB-231 breast cancer cell	Induces apoptosis TQ shows higher cytotoxicity against MDA compared to MCF-7. TQ modulates TNF receptor superfamily genes in MDA differentially and in a stronger manner. TNF receptor superfamily genes activate NF-kB signaling, levels of P65 and its phosphorylation were evaluated in MCF-7 and MDA. P65 phosphorylation was not detected in MCF-7; however TQ reduced both full length and phosphorylated P65 in MDA	[42]

(continued)

Table 1 (continued)

Dose	Test compound/ formulation	Disease model	Activity	References
	TQ	In MCF-7, MDA-MB-231, and BT-474 breast cell	TQ inhibited breast cancer cell proliferation efficiently, revealing it to be a potentially effective suppressor of tumor cell growth, invasion, and migration. It also highlighted for the first time that the anticancer effects of TQ involve the activation of the PPAR-γ pathway via binding to the receptor, and affecting its regulated gene products TQ able to increase PPAR-γ activity and downregulate the expression of the genes for Bcl-2, Bcl-xL, and survivin in breast cancer cells	[43]

(in both male and female rats) was reported for oral route [56]. The clearance observed after administration of TQ via i.v. route was 7.19 ml/kg/min whereas the calculated volume of distribution at steady state (Vss) was 700.90 ml/kg. At variance, upon administration of TQ via oral route, the apparent clearance value and Vss were 12.30 ml/min/kg and 5109.46 ml/kg. The estimated absorption half-life ($T_{1/2}$) of TQ was about 217 min, and it was quickly eliminated from plasma [57]. TQ shows a sharp peak (λ_{max}) at 254–257 nm during UV-vis spectroscopic analysis and thus it is considered as a highly photosensitive compound even after short-term exposure which ultimately leads to the chemical degradation that is independent of pH and types of solvent. Moreover, it shows stability problem at alkaline pH in aqueous medium, and therefore, such stability issues increase upon increment in pH of the solution, while minimum degradation has been seen at lower pH (in acidic solution) [33]. In the pharmacokinetics and pharmacodynamics dealing of TQ, drug-protein interactions play a crucial role which greatly affects its pharmacological action and the calculated TQ plasma protein bindings in both human and rabbit were 98.99% and 99.19%, respectively [57]. Thus, TQ is considered as a less stable compound and shows fast elimination from plasma as well as delayed or slower absorption via oral route of administration [58]. In a study, Lupidi et al. assessed and reported the TQ-human serum albumin (HSA) interactions without any chemical changes which can affect the secondary structure of HAS. The thermodynamic study showed that the TQ-HSA association (hydrophobic interactions) was a spontaneous process, in which intermolecular forces work to make the complex stable [58, 59]. Moreover,

El-Najjar et al. reported the effect of TQ association with BSA and α1-AGP on its anticancer activity [60]. The research reports suggested that the covalent interaction of TQ-BSA causes the loss of TQ anticancer efficacy against tested tumor cells, while upon TQ-AGP interaction TQ retains its anticancer activity. The lack of evidence-based bioavailability and pharmacokinetic parameters and drug delivery issues slow down the clinical utilization of TQ in clinics. Therefore, more exploration is warranted to enhance the scientific clarity and better insight into TQ therapeutic efficacy meant for future clinical establishment.

In a study, Alobaedi et al. reported that the combination therapy of TQ with resveratrol synergizes the anticancer potential against BC cell in mice. The therapeutic action of such combination against BC cell is exerted through promotion of apoptosis, inhibition of angiogenesis, and immunomodulation. The combination therapy may also lead to the significant reduction in the size of tumor with a percentage healing of 60% [2]. In another study, Ghahramanloo et al. found that the TQ potently exerted its action by inhibiting the viability of the human invasive BC cell lines (MDA-MB-231) via apoptosis induction as exhibited through acridine orange/propidium iodide-stained cells at dissimilar time points following TQ treatment. TQ action on MDA-MB-231 cells led to the significant improvement in intracellular ROS generation, which plays a crucial role in tumor cell death [3].

Ravindran et al. used nanotechnology-based drug delivery approach and prepared TQ-loaded poly(DL-lactide-co-glycolide) nanoparticles (NPs) and tried against a number of cancer cell lines including BC cell line (MCF7). The obtained results concluded that the nanoencapsulation of TQ improved the anticancer potential than free TQ because it elevated the bioavailability and cellular uptake of TQ [61, 62]. In addition, Ganea et al. also evaluated the anticancer efficacy of TQ-loaded poly(DL-lactide-co-glycolide) NPs on other BC cell line, namely MDAMB231 which showed a better efficacy than free TQ [38]. In another research report, Odeh et al. evaluated the anticancer activity of TQ-loaded liposomal particles on BC cell line (MCF7) which caused the inhibition of cell proliferation in MCF7 [40, 61]. In another study, Bhattacharya et al. formulated TQ-encapsulated polymeric NPs employing biodegradable, hydrophilic polymers such as polyvinylpyrrolidone and polyethylene glycol to improve the TQ solubility, heat and light resistance, and systemic bioavailability, which can greatly enhance the anticancer effect [20]. This nanoparticle can persuade BC cell killing and reduce migration. Myristic acid-chitosan-based nanogels of TQ were developed by Dehghani et al. for the management of human breast adenocarcinoma cell MCF-7. Captivatingly, this nanogel formulation showed improved action against cancer cell than TQ alone [39, 63].

A further crisis of TQ in clinical practice has been observed due to its safety concern in human. Upon i.p. and oral administration of TQ in mice the observed LD_{50} were 104.7 and 870.9 mg/kg, while in rats the LD_{50} was observed to be 57.5 and 794.3 mg/kg i.p. and oral administration, respectively [64]. However, TQ shows its action against cancer cell even at very minute concentrations, just about <10 mg/kg [39, 64, 65]. Therefore, safety concern might not pose a big crisis. The above literature reports support the TQ utility in the management of BC [3].

3 Current Status of TQ in Clinical Trials

The preclinical reports reveal the therapeutic significance of TQ in terms of enhancement in the pharmacological action of anticancer agents and also show protective action for normal tissues/cells against chemotherapy-induced toxicity. For instance, TQ provides protection and ameliorates against nephro- and cardiotoxicity due to the use of cisplatin, ifosfamide, and doxorubicin [66–68]. Still no clinical trial has been conducted to assess the therapeutic significance and potential of TQ in humans. Of interest, Gali-Muhtasib et al. observed that TQ persuades great improvement in the levels of p16 protein within 2 h of administration; this study is of interest, since as reported by Gali-Muhtasib et al. and Hochhauser, regulation of p16 protein expression enhances the tumor sensitivity to cancer therapy [69–71].

In a study, Barron et al. examined and reported proliferation of osteoblast cells (MG 63) upon treatment of TQ and selenium (Se) in combination. The obtained results exhibit the decrease in cell proliferation, improved cellular damage, and reduced alkaline phosphatase and GST levels, signifying that the combination-dose administration of TQ and Se possibly will be an efficient treatment choice to treat human osteosarcoma cells [72]. Worthen et al. reported that TQ and its derivatives show cytotoxic effect in a number of parental and multidrug-resistant (MDR) human cancer cell lines [73]. Recently, in a study researcher reported about the chemosensitizing function of TQ to conventional anticancer drugs in vitro/in vivo in an orthotopic model of pancreatic malignancy. The in vitro evaluations revealed that the pre-exposure of cancerous cells with TQ (25 μM) for 48 h and subsequently gemcitabine or oxaliplatin treatment caused 60–80% inhibition in comparison to 15–25% inhibition while gemcitabine or oxaliplatin taken alone. TQ synergized pancreatic cancer cell death through downregulation of NF-κB, Bcl-2 family, and NF-κB-dependent anti-apoptotic protein members such as XIAPs and surviving, fascinatingly, TQ was capable of downregulating NF-κB in vitro, causing chemosensitization [74, 75].

Till date, there is no clinical trial reported on TQ by the US Government. But in an Arabian Phase I trial report, TQ (10 mg/kg/day) was recommended safe and shows better patient compliance, but at this dose, TQ shows no significant anticancer potential [76]. TQ application in humans is restricted because of its chemical properties and reduced membrane penetration capacity. Poor aqueous solubility of TQ significantly affects its bioavailability and potential. Moreover, due to its high lipophilicity TQ possesses tough challenges for formulation scientist [40]. Several researches have been conducted to conquer the pharmacokinetic issues of TQ and its toxicity. A number of studies revealed that the TQ shows synergistic effects in combination with other conventional anticancer agents, which possibly will be an interesting parameter to analyze in future clinical trials of TQ [45]. Clinical trial-based investigation is required to corroborate the anticancer efficacy of TQ through better exploration of different molecular targets and its molecular mechanism.

4 Safety Concern of TQ

Medicinal plants and bioactive phytoconstituents have vital importance in the health management due to its low or minimal toxicity. Several researches have been conducted to explore the detailed toxicological profile of TQ in vitro and in vivo [77–79]. In an animal model (rats and mice), the oral and i.p. lethal dose (LD_{50}) has been reported by earlier scientists [51, 64, 79]. The safety, tolerability, and toxicity profile of *N. sativa* at maximum doses have already been demonstrated in human clinical trials. In a study, researcher reported that TQ (0–10 μM) shows no cytotoxicity to fibroblast-like synoviocytes [80]. In a study, research reports revealed that *N. sativa* did not show any hepatotoxicity, and supplementations of *N. sativa* decreased the concentration of alanine amino transferase (ALT) and aspartate amino transferase (AST) in treated rats in comparison to control group [81]. In another study researcher suggested that the suitable oral dose of TQ for humans is 0.6 mg/kg/day [80], and the recommended dose of extract of *N. sativa* for postmenopausal women is 0.05 mg/kg/day [82]. In an experimental study, the LD_{50} of TQ was 2.4 g/kg (1.52–3.77, 95% CL) upon oral administration [42, 79, 83]. The conclusion drawn from several evidence-based research reports is that the *N. sativa fixed* oil shows no fatal toxicities at the dose range up to 10 ml/kg/day [79, 83]. Powder extract of *N. sativa* seed did not show any toxicity and was considered as safe and tolerated at maximum doses (28 g/kg orally) and its oil was also safe at a certain dose (28.8 ml/kg) in rabbits and rats, respectively [79, 84].

Poor scientific base and preclinical data with TQ show the maximum tolerated dose (MTD) that is explained as the maximum dose that is safe in animals without any adverse effects; for TQ toxicity is considered as a limitation in applying TQ in clinical practice to cure cancer. However, preliminary data concerning MTD for TQ in rats indicate lower MTD values in female rats than in male rats [56]. The probable hazard of gender sensitivity to TQ toxicity with female BC patients can be overcome by choosing combinatorial therapy of TQ with doxorubicin that is frequently applied to treat BC [61, 85]. In an animal model, there is a number of studies which reported an antioxidant protective effect of TQ in doxorubicin-triggered cardiotoxicity without any interference with the action of doxorubicin in cancer management [61, 86]. Furthermore, TQ in conjunction with ionizing radiation (γ-radiation) showed a synergistic effect on BC cells in vitro [61, 87]. It is to be noted that evidence shows that TQ causes inhibition of phosphorylation of ERK and MAPK pathways in many cancer cells, which is also an essential fact for bone and teeth formation. Thus, during developmental phases, chemotherapy with TQ is a major concern. However, the LD_{50}, pharmacokinetic behavior, and a few other reports suggest that TQ toxicity is not only dependent on its concentration, but also on its route administration [42].

5 Conclusion

Based on the preceding account, the therapeutic outcome of TQ should not be overthrown despite poor clinical settings considering its bioavailability that needs to be pursued. Once a concord on its bioavailability comes out, planned Phase I clinical trials to corroborate its expediency in chemosensitization and recognize it as chemopreventive or chemoprotective agent should be considered for various site-specific malignancies. In the meantime, attempts should go on with focusing on lab research to investigate further in-depth consideration on its mechanism of action at molecular level as well as to plan strategy to develop clinically effective TQ analogs and TQ in combination with currently used chemotherapeutic agents where TQ or its analogs can potentiate the anticancer potential of various currently approved anticancer agents.

References

1. Torre LA, Siegel RL, Ward EM, Jemal A (2016) Global cancer incidence and mortality rates and trends—an update. Cancer Epidemiol Biomarkers Prev 25(1):16–27
2. Alobaedi OH, Talib WH, Basheti IA (2017) Antitumor effect of TQ combined with resveratrol on mice transplanted with breast cancer. Asian Pac J Trop Med 10(4):400–408
3. Ghahramanloo KH, Noordin MI, Kamalidehghan B et al (2017) Involvement of NF-κB and HSP70 signaling pathways in the apoptosis of MDA-MB-231 cells induced by TQ: an in vitro study. Life Sci J 14(5):61–70
4. DeSantis CE, Fedewa SA, Sauer AG, Kramer JL, Smith RA, Jemal A (2016) Breast cancer statistics, 2015: convergence of incidence rates between black and white women. CA Cancer J Clin 66(1):31–42
5. American Cancer Society (2017) Cancer facts and figures 2017. American Cancer Society, Atlanta, GA
6. Kabel AM (2017) Tumor markers of breast cancer: new prospective. J Oncol Sci 3(1):5–11
7. Shapiro CL, Recht A (2001) Side effects of adjuvant treatment of breast cancer. N Engl J Med 344(26):1997–2008
8. Talib WH (2011) Anticancer and antimicrobial potential of plant-derived natural products. In: Rasooli I (ed) Phytochemicals: bioactivities and impact on health. IntechOpen, Croatia, pp 141–158
9. Razavi BM, Hosseinzadeh H (2014) A review of the effects of *Nigella sativa* L. and its constituent, TQ, in metabolic syndrome. J Endocrinol Invest 37:1031–1040
10. Darakhshan S, Pour AB, Colagar AH, Sisakhtnezhad S (2015) TQ and its therapeutic potentials. Pharmacol Res 95-96:138–158
11. Islam MT, Sultana N, Riaz TA et al (2016) TQ is knocking at the door of clinical trial. Int Arch Med 9(122):1–25
12. Banerjee S, Padhye S, Azmi A et al (2010) Review on molecular and therapeutic potential of TQ in cancer. Nutr Cancer 62(7):938–946
13. Motaghed M, Al-Hassan FM, Hamid SS (2014) TQ regulates gene expression levels in the estrogen metabolic and interferon pathways in MCF7 breast cancer cells. Int J Mol Med 33:8–16
14. Ghosheh OA, Houdi AA, Crooks PA (1999) High performance liquid chromatographic analysis of the pharmacologically active quinones and related compounds in the oil of the black seed (Nigella sativa L.). J Pharm Biomed Anal 19:757–762

15. Rajput S, Kumar BN, Dey KK, Pal I, Parekh A, Mandal M (2013) Molecular targeting of Akt by TQ promotes G(1) arrest through translation inhibition of cyclin D1 and induces apoptosis in breast cancer cells. Life Sci 93:783–790
16. Alhosin M, Abusnina A, Achour M et al (2010) Induction of apoptosis by TQ in lymphoblastic leukemia Jurkat cells is mediated by a p73-dependent pathway which targets the epigenetic integrator UHRF1. Biochem Pharmacol 79:1251–1260
17. Abusnina A, Alhosin M, Keravis T et al (2011) Down-regulation of cyclic nucleotide phosphodiesterase PDE1A is the key event of p73 and UHRF1 deregulation in TQ-induced acute lymphoblastic leukemia cell apoptosis. Cell Signal 23:152–160
18. Salim LZ, Mohan S, Othman R et al (2013) TQ induces mitochondria-mediated apoptosis in acute lymphoblastic leukaemia *in vitro*. Molecules 18:11219–11240
19. Dastjerdi MN, Mehdiabady EM, Iranpour FG, Bahramian H (2016) Effect of TQ on P53 gene expression and consequence apoptosis in breast cancer cell line. Int J Prev Med 7:66
20. Bhattacharya S, Ahir M, Patra P et al (2015) PEGylated-TQ-nanoparticle mediated retardation of breast cancer cell migration by deregulation of cytoskeletal actin polymerization through miR-34a. Biomaterials 51:91–107
21. Ke X, Zhao Y, Lu X et al (2015) TQ inhibits hepatocellular carcinoma growth *in vitro* and *in vivo* via repression of Notch signaling. Oncotarget 6:32610–32621
22. Woo CC, Kumar AP, Sethi G, Tan KH (2012) TQ: potential cure for inflammatory disorders and cancer. Biochem Pharmacol 83:443–451
23. Mansour MA (2000) Protective effects of TQ and desferrioxamine against hepatotoxicity of carbon tetrachloride in mice. Life Sci 66:2583–2591
24. Badary OA, Gamal El-Din AM (2001) Inhibitory effects of TQ against 20-methylcholanthrene-induced fibrosarcoma tumorigenesis. Cancer Detect Prev 25:362–368
25. Gali-Muhtasib H, Kuester D, Mawrin C et al (2008) TQ triggers inactivation of the stress response pathway sensor CHEK1 and contributes to apoptosis in colorectal cancer cells. Cancer Res 68(14):5609–5618
26. Dergarabetian EM, Ghattass KI, El-Sitt SB et al (2013) TQ induces apoptosis in malignant T-cells via generation of ROS. Front Biosci (Elite Ed) 15(5):706–719
27. Schneider-Stock R, Fakhoury IH, Zaki AM, El-Baba CO, Gali-Muhtasib HU (2014) TQ: fifty years of success in the battle against cancer models. Drug Discov Today 19(1):18–30
28. Reindl W, Yuan J, Krämer A, Strebhardt K, Berg T (2008) Inhibition of polo-like kinase 1 by blocking polo-box domain-dependent protein–protein interactions. Chem Biol 15:459–466
29. Khalife KH, Lupidi G (2007) Nonenzymatic reduction of TQ in physiological conditions. Free Radic Res 41(2):153–161
30. Yin Z, Song Y, Rehse PH (2013) TQ blocks pSer/pThr recognition by Plk1 Polo-box domain as a phosphate mimic. ACS Chem Biol 8(2):303–308
31. Yuan J, Sanhaji M, Krämer A et al (2011) Polo-box domain inhibitor poloxin activates the spindle assembly checkpoint and inhibits tumor growth in vivo. Am J Pathol 179(4):2091–2099
32. Pathan SA, Jain GK, Zaidi SM et al (2011) Stability-indicating ultra-performance liquid chromatography method for the estimation of TQ and its application in biopharmaceutical studies. Biomed Chromatogr 25(5):613–620
33. Salem ML (2005) Immunomodulatory and therapeutic properties of the Nigella sativa L. seed. Int Immunopharmacol 5(13–14):1749–1770
34. Salmani JM, Asghar S, Lv H, Zhou J (2014) Aqueous solubility and degradation kinetics of the phytochemical anticancer TQ; probing the effects of solvents, pH and light. Molecules 19:5925–5939
35. Parbin S, Shilpi A, Kar S et al (2015) Insights into the molecular interactions of TQ with histone deacetylase: evaluation of the therapeutic intervention potential against breast cancer. Mol Biosyst 12(1):48–58
36. Rajput S, Puvvada N, Kumar BN et al (2015) Overcoming Akt induced therapeutic resistance in breast cancer through siRNA and TQ encapsulated multilamellar gold niosomes. Mol Pharm 12:4214–4225

37. Şakalar C, İzgi K, İskender B et al (2016) The combination of TQ and paclitaxel shows antitumor activity through the interplay with apoptosis network in triple-negative breast cancer. Tumour Biol 37(4):4467–4477
38. Ganea GM, Fakayode SO, Losso JN, van Nostrum CF, Sabliov CM, Warner IM (2010) Delivery of phytochemical TQ using molecular micelle modified poly(D,L lactide-co-glycolide) (PLGA) nanoparticles. Nanotechnology 21(28):285104
39. Asaduzzaman Khan M, Tania M, Fu S, Fu J (2017) Thymoquinone, as an anticancer molecule: from basic research to clinical investigation. Oncotarget 8(31):51907–51919
40. Odeh F, Ismail SI, Abu-Dahab R, Mahmoud IS, Al BA (2012) TQ in liposomes: a study of loading efficiency and biological activity towards breast cancer. Drug Deliv 19(8):371–377
41. Effenberger K, Breyer S, Schobert R (2010) Terpene conjugates of the Nigella sativa seed-oil constituent TQ with enhanced efficacy in cancer cells. Chem Biodivers 7(1):129–139
42. Barkat MA, Abul H, Ahmad J, Khan MA, Beg S, Ahmad FJ (2017) Insights into the targeting potential of thymoquinone for therapeutic intervention against triple-negative breast cancer. Curr Drug Targets 19(1):70–80
43. Woo CC, Loo SY, Gee V et al (2011) Anticancer activity of TQ in breast cancer cells: possible involvement of PPAR-γ pathway. Biochem Pharmacol 82(5):464–475
44. Abdel-Fattah AM, Matsumoto K, Watanabe H (2000) Antinociceptive effects of Nigella sativa oil and its major component, TQ, in mice. Eur J Pharmacol 400(1):89–97
45. El Gazzar M, El Mezayen R, Marecki JC, Nicolls MR, Canastar A, Dreskin SC (2006) Anti-inflammatory effect of thymoquinone in a mouse model of allergic lung inflammation. Int Immunopharmacol 6(7):1135–1142
46. El-Mahmoudy A, Shimizu Y, Shiina T, Matsuyama H, El-Sayed M, Takewaki T (2005) Successful abrogation by TQ against induction of diabetes mellitus with streptozotocin via nitric oxide inhibitory mechanism. Int Immunopharmacol 5(1):195–207
47. Hosseinzadeh H, Parvardeh S (2004) Anticonvulsant effects of TQ, the major constituent of Nigella sativa seeds, in mice. Phytomedicine 11:56–64
48. Pari L, Sankaranarayanan C (2009) Beneficial effects of TQ on hepatic key enzymes in streptozotocin-nicotinamide induced diabetic rats. Life Sci 85:830–834
49. Nagi MN, Almakki HA (2009) TQ supplementation induces quinone reductase and glutathione transferase in mice liver: possible role in protection against chemical carcinogenesis and toxicity. Phytother Res 23:1295–1298
50. El-Dakhakhny M (1965) Studies on the Egyptian Nigella sativa L. Some pharmacological properties of the seeds' active principle in comparison to its dihydro compound and its polymer. Arzneimittelforschung 15(10):1227–1229
51. Mansour MA, Ginawi OT, El-Hadiyah T, El-Khatib AS, Al-Shabanah OA, Al-Sawaf HA (2001) Effects of volatile oil constituents of Nigella sativa on carbon tetrachloride-induced hepatotoxicity in mice: evidence for antioxidant effects of TQ. Res Commun Mol Pathol Pharmacol 110(3–4):239–251
52. Nagi MN, Alam K, Badary OA, al-Shabanah OA, al-Sawaf HA, al-Bekairi AM (1999) TQ protects against carbon tetrachloride hepatotoxicity in mice via an antioxidant mechanism. Biochem Mol Biol Int 47(1):153–159
53. Kanter M (2008) Nigella sativa and derived TQ prevents hippocampal neurodegeneration after chronic toluene exposure in rats. Neurochem Res 33:579–588
54. Al-Majed AA, Al-Omar FA, Nagi MN (2006) Neuroprotective effects of TQ against transient forebrain ischemia in the rat hippocampus. Eur J Pharmacol 543(1–3):40–47
55. Kanter M (2011) TQ attenuates lung injury induced by chronic toluene exposure in rats. Toxicol Ind Health 27:387–395
56. Abukhader MM (2012) The effect of route of administration in TQ toxicity in male and female rats. Indian J Pharm Sci 74:195–200
57. Alkharfy KM, Ahmad A, Khan RM, Al-Shagha WM (2015) Pharmacokinetic plasma behaviors of intravenous and oral bioavailability of TQ in a rabbit model. Eur J Drug Metab Pharmacokinet 40(3):319–323

58. Lupidi G, Scire A, Camaioni E et al (2010) TQ, a potential therapeutic agent of Nigella sativa, binds to site I of human serum albumin. Phytomedicine 17(10):714–720
59. Lupidi G, Camaioni E, Khalifé H et al (2012) Characterization of TQ binding to human α acid glycoprotein. J Pharm Sci 101:2564–2573
60. El-Najjar N, Ketola RA, Nissilä T et al (2011) Impact of protein binding on the analytical detectability and anticancer activity of TQ. J Chem Biol 4(3):97–107
61. AbuKhader MM (2013) TQ in the clinical treatment of cancer: fact or fiction? Pharmacogn Rev 7(14):117–120
62. Ravindran J, Nair HB, Sung B, Prasad S, Tekmal RR, Aggarwal BB (2010) TQ poly (lactide-co-glycolide) nanoparticles exhibit enhanced antiproliferative, anti-inflammatory, and chemo-sensitization potential. Biochem Pharmacol 79:1640–1647
63. Dehghani H, Hashemi M, Entezari M, Mohsenifar A (2015) The comparison of anticancer activity of TQ and nanoTQ on human breast adenocarcinoma. Iran J Pharm Res 14:539–546
64. Al-Ali A, Alkhawajah AA, Randhawa MA, Shaikh NA (2008) Oral and intraperitoneal LD50 of TQ, an active principle of Nigella sativa, in mice and rats. J Ayub Med Coll Abbottabad 20:25–27
65. Khan MA, Tania M, Wei C et al (2015) TQ inhibits cancer metastasis by downregulating TWIST1 expression to reduce epithelial to mesenchymal transition. Oncotarget 6:19580–19591
66. Badary OA, Nagi MN, al-Shabanah OA et al (1997) TQ ameliorates the nephrotoxicity induced by cisplatin in rodents and potentiates its antitumor activity. Can J Physiol Pharmacol 75:1356–1361
67. al-Shabanah OA, Badary OA, Nagi MN et al (1998) TQ protects against doxorubicin-induced cardiotoxicity without compromising its antitumor activity. J Exp Clin Cancer Res. 17:193–198
68. Badary OA, Abdel-Naim AB, Abdel-Wahab MH, Hamada FM (2000) The influence of TQ on doxorubicin-induced hyperlipidemic nephropathy in rats. Toxicology 143:219–226
69. Gali-Muhtasib HU, AbouKheir WG, Kheir LA, Darwiche N, Crooks PA (2004) Molecular pathway for TQ-induced cell-cycle arrest and apoptosis in neoplastic keratinocytes. Anticancer Drugs 15:389–399
70. Gali-Muhtasib H, Roessner A, Schneider-Stock R (2006) TQ: a promising anti-cancer drug from natural sources. Int J Biochem Cell Biol 38:1249–1253
71. Hochhauser D (1997) Modulation of chemosensitivity through altered expression of cell cycle regulatory genes in cancer. Anticancer Drugs 8:903–910
72. Barron J, Benghuzzi H, Tucci M (2008) Effects of TQ and selenium on the proliferation of mg 63 cells in tissue culture. Biomed Sci Instrum 44:434–440
73. Worthen DR, Ghosheh OA, Crooks PA (1998) The in vitro anti-tumor activity of some crude and purified components of black seed, Nigella sativa L. Anticancer Res 18:1527–1532
74. Banerjee S, Zhang Y, Ali S et al (2005) Molecular evidence for increased antitumor activity of gemcitabine by genistein in vitro and in vivo using an orthotopic model of pancreatic cancer. Cancer Res 65:9064–9072
75. Banerjee S, Zhang Y, Wang Z et al (2007) In vitro and in vivo molecular evidence of genistein action in augmenting the efficacy of cisplatin in pancreatic cancer. Int J Cancer 120:906–917
76. Al-Amri AA, Bamoasa AO (2009) Phase I safety and clinical activity of TQ in patients with advanced refractory malignant disease. Shiraz E-Med J 10:107–111
77. Qadri SM, Mahmud H, Foller M, Lang F (2009) TQ-induced suicidal erythrocyte death. Food Chem Toxicol 47(7):1545–1549
78. Khader M, Bresgen N, Eckl PM (2009) *In vitro* toxicological properties of TQ. Food Chem Toxicol 47(1):129–133
79. Rahmani AH, Alzohairy MA, Khan MA, Aly SM (2014) Therapeutic implications of black seed and its constituent TQ in the prevention of cancer through inactivation and activation of molecular pathways. Evid Based Complement Alternat Med 2014:1–13
80. Vaillancourt F, Silva P, Shi Q, Fahmi H, Fernandes JC, Benderdour M (2011) Elucidation of molecular mechanisms underlying the protective effects of TQ against rheumatoid arthritis. J Cell Biochem 112(1):107–117

81. Dollah HMA, Parhizkar S, Latiff LA, Hassan MHB (2013) Toxicity effect of Nigella sativa on the liver function of rats. Adv Pharm Bull 3(1):97–102
82. Valizadeh N, Zakeri HR, Shafiee A (2009) The effect of *Nigella sativa* extract on biochemical bon markers in osteopenic postmenopausal women. Iran J Endocrinol Metab 10:570–580
83. Khanna T, Zaidi FA, Dandiya PC (1993) CNS and analgesic studies on *Nigella sativa*. Fitoterapia 64(5):407–410
84. Zaoui A, Cherrah Y, Mahassini N, Alaoui K, Amarouch H, Hassar M (2002) Acute and chronic toxicity of *Nigella sativa* fixed oil. Phytomedicine 9(1):69–74
85. Effenberger-Neidnicht K, Schobert R (2011) Combinatorial effects of TQ on the anticancer activity of doxorubicin. Cancer Chemother Pharmacol 67:867–874
86. Nagi MN, Mansour MA (2000) Protective effect of TQ against doxorubicin induced cardiotoxicity in rats: a possible mechanism of protection. Pharmacol Res 41:283–289
87. Velho-Pereira R, Kumar A, Pandey BN, Jagtap AG, Mishra KP (2011) Radiosensitization in human breast carcinoma cells by TQ: role of cell cycle and apoptosis. Cell Biol Int 35:1025–1029

The manufacturer's authorised representative in the EU is Springer Nature Customer Service Centre GmbH, Europaplatz 3, 69115 Heidelberg, Germany. If you have any concerns regarding our products, please contact ProductSafety@springernature.com

Printed and bound by CPI Group (UK) Ltd, Croydon, CR0 4YY

30/09/2026

02225249-0004